AF411307

Selected Papers from the 19th International Conference on Calorimetry in Particle Physics (CALOR 2022)

Selected Papers from the 19th International Conference on Calorimetry in Particle Physics (CALOR 2022)

Editors

Fabrizio Salvatore
Antonella De Santo
Iacopo Vivarelli

MDPI • Basel • Beijing • Wuhan • Barcelona • Belgrade • Manchester • Tokyo • Cluj • Tianjin

Editors

Fabrizio Salvatore	Antonella De Santo	Iacopo Vivarelli
University of Sussex	University of Sussex	University of Sussex
Sussex	Sussex	Sussex
UK	UK	UK

Editorial Office
MDPI
St. Alban-Anlage 66
4052 Basel, Switzerland

This is a reprint of articles from the Special Issue published online in the open access journal *Instruments* (ISSN 2410-390X) (available at: https://www.mdpi.com/journal/instruments/special_issues/calor22).

For citation purposes, cite each article independently as indicated on the article page online and as indicated below:

LastName, A.A.; LastName, B.B.; LastName, C.C. Article Title. *Journal Name* **Year**, *Volume Number*, Page Range.

ISBN 978-3-0365-8394-5 (Hbk)
ISBN 978-3-0365-8395-2 (PDF)

Contents

About the Editors

Fabrizio Salvatore

Fabrizio Salvatore is a Professor of Physics at the University of Sussex and former Head of the Experimental Particle Physics group. He has been involved in many different projects at the forefront of physics, leading work in various areas of research—from physics analysis to more technical projects. He is the Principal Investigator on the 2022-25 EPP Consolidated Grant (multi-million pounds grant), which supports the research of the entire EPP group. As a member of the ATLAS Experiment at the Large Hadron Collider (LHC) he held various coordination roles over the years in physics analysis groups, as well as in the core software of the ATLAS Trigger system. His main research interests are looking for physics beyond the standard model at the Large Hadron Collider (Supersymmetry, Exotics models), trigger and data acquisition for current and future experiments, and calorimeter detectors at future colliders.

Antonella De Santo

Antonella De Santo became the first female Full Professor of Physics at the University of Sussex in 2013, having joined the university in 2009 to set up a new research group working on the ATLAS experiment at CERN's Large Hadron Collider, which she continues to lead. Her research is currently focused on the search for new physics beyond the Standard Model (including electroweak SUSY searches and searches for vector-like leptons) using multileptonic signatures at ATLAS. In the UK, prior to joining Sussex, Antonella held teaching and research positions at Royal Holloway, University of London (also working on the ATLAS experiment, with a focus on early SUSY searches and on triggering) and the University of Oxford (focusing on the characterisation of the multi-anode photomultipliers used in the Near Detector of the MINOS neutrino oscillation experiment at Fermilab, and the study of the performance of the RPC detectors used in the HARP hadro-production experiment at CERN). Before that, she was a CERN Research Fellow (with leadership in data analysis and data preparation for the NOMAD neutrino oscillations experiment), and, for a brief period, a research assistant at the University of Pisa, where she also obtained her PhD in 1997 with a thesis presenting the first analysis exploring short-baseline muon-neutrino to electron–neutrino oscillations using NOMAD data, placing early constraints on the so-called LSND anomaly. In recognition of her achievements, in 2017 Antonella was awarded a Royal Society Wolfson Research Merit Award.

Iacopo Vivarelli

Iacopo Vivarelli, Professor, is a member of the ATLAS Collaboration, one of the two general-purpose experiments using data from the Large Hadron Collider (LHC). For many years he has been searching for evidence of new physical phenomena in the LHC collisions. In particular, he has been looking for the production of the so-called supersymmetric partners of the known elementary particles. The lack of evidence for new phenomena led him to investigate the elephant in the room, that is, the production of the heaviest elementary particle, the top quark. He is particularly interested in rare production processes, where pairs of tops are produced together with a vector boson. The study of these processes could bring a new understanding to the interaction of the top quark with the rest of the Standard Model particles. One of his passions is calorimetry, that is, the process of stopping and destroying particles to measure their energy. He is working together with colleagues in Italy, the US and Korea to build a new type of calorimeter that should dramatically improve the energy resolution of these devices.

Preface to "Selected Papers from the 19th International Conference on Calorimetry in Particle Physics (CALOR 2022)"

In these proceedings, we summarise the state of the art of research and development (R&D) in calorimetry for current and future high-energy physics experiments.

During the five days of talks, the many presentations given at the conference have sparked very interesting conversations about the future of detector development for particle physics, nuclear physics, and space-based experiments.

A lot of emphasis was put on presentations from early career researchers, who had the opportunity for discussion during the conference and social events with the main experts in calorimetry R&D, simulation tools for calorimetry, and particle/nuclear physics.

The conference was sponsored by CAEN (https://www.caen.it/), the University of Sussex School of Mathematical & Physical Sciences (http://www.sussex.ac.uk/mps/) and Instruments/MDPI, who presented three prizes for best presentation from an early career researcher, best poster, and best presentation overall.

We trust that the selection of papers we present in this Special Issue will give a very comprehensive summary of the status of R&D for future detectors, future detection techniques, and tools for simulation of high energy particle interactions.

Fabrizio Salvatore, Antonella De Santo, and Iacopo Vivarelli
Editors

 *instruments*

Article

Performance and Calibration of the ATLAS Tile Calorimeter

Tomas Davidek on behalf of the ATLAS Collaboration

IPNP, Faculty of Mathematics and Physics, Charles University, 180 00 Prague 8, Czech Republic; tomas.davidek@mff.cuni.cz

Abstract: The Tile Calorimeter (TileCal) is the central hadronic calorimeter of the ATLAS experiment at the LHC. This sampling device is made of steel plates acting as absorber and scintillating tiles as active medium. The wavelength-shifting fibers collect the light from scintillators and carry it to the photomultiplier tubes (PMTs). The analogue signals from the PMTs are amplified, shaped and digitized by sampling the signal every 25 ns and stored on detector until a trigger decision is received. The TileCal front-end electronics read out the signals produced by 9852 channels, whose dynamic range covers the interval from 30 MeV to 2 TeV. Each stage of the signal propagation from scintillation light to the signal reconstruction is monitored and calibrated. During LHC Run-2, high-momentum isolated muons and isolated hadrons have been used to study and validate the electromagnetic scale and the hadronic response, respectively. The time resolution was studied with multi-jet events. Results of performance studies that address calibration, stability, energy scale, uniformity and time resolution are presented.

Keywords: ATLAS; Tile Calorimeter; calorimeter calibration; calorimeter performance

Citation: Davidek, T., on behalf of the ATLAS Collaboration. Performance and Calibration of the ATLAS Tile Calorimeter. *Instruments* **2022**, *6*, 25. https://doi.org/10.3390/instruments6030025

Academic Editors: Fabrizio Salvatore, Alessandro Cerri, Antonella De Santo and Iacopo Vivarelli

Received: 26 July 2022
Accepted: 15 August 2022
Published: 20 August 2022

Publisher's Note: MDPI stays neutral with regard to jurisdictional claims in published maps and institutional affiliations.

1. Introduction

The TileCal [1] is a hadronic calorimeter of the ATLAS experiment [2] that explores proton–proton (pp) and heavy ion collisions at the LHC [3] performed at the highest energies ever achieved in a laboratory. It provides essential input to the measurements of energies and directions of jets, isolated hadrons, hadronically decaying τ-leptons and of the missing transverse momentum. The TileCal also contributes to the muon identification and provides information to the first level trigger.

This sampling device is made of alternating layers of steel and scintillating tiles and covers the ATLAS region $|\eta| < 1.7$. Mechanically, the calorimeter is divided into a central part (Long Barrel) and two Extended Barrels. Each part consists of 64 modules shown in Figure 1. The readout cells are organized into three radial layers, whose depths are 1.5, 4.1, 1.8 and 1.5, 2.6, 3.3 nuclear interaction lengths in the Long Barrel and Extended Barrels, respectively. The pseudo-projective cells are segmented in pseudorapidity ($\Delta\eta = 0.1$ and $\Delta\eta = 0.2$ in the outermost radial layer) and azimuth ($\Delta\phi = 2\pi/64 \approx 0.1$ as defined by the module geometry). The light from scintillating tiles is collected by wavelength-shifting (WLS) fibers on both sides of each module. Each cell is readout by two photo-multipliers (PMTs), whose signals are further processed by the bi-gain (high/low-gain for smaller/larger signals) readout electronics.

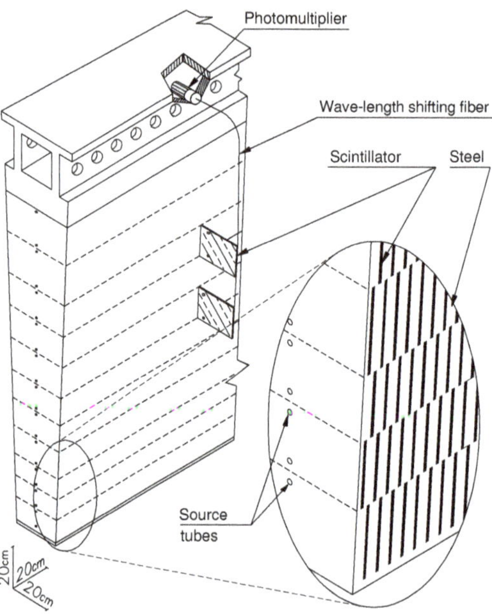

Figure 1. A schema of the mechanical assembly and optical readout of one Tile Calorimeter module. Each module consists of 11 radial rows of tiles of different sizes that are grouped into 3 radial readout layers (3-6-2 rows of tiles in the Long Barrel, 3-4-4 rows of tiles in the Extended Barrels). The pseudo-projective cell geometry in pseudorapidity is achieved by grouping the corresponding WLS fibers onto one PMT.

2. Calibration and Signal Reconstruction

The TileCal exploits three dedicated calibration systems that are briefly described below. Each system monitors different stage of the signal processing chain.

2.1. Cesium

The cesium system measures the signal induced by a ^{137}Cs radioactive source that passes through all tiles. Dedicated calibration runs allow us to calibrate the optical system (tiles and WLS fibers) and PMTs. The signal is read out through dedicated slow electronics that integrate the PMT currents over 10 to 20 ms.

The channel response is equalized by adjusting the gain of the PMTs via their high voltage settings. The equalization is performed once at the beginning of the Run-2 data-taking period. The changes in the Cs response are tracked with calibration constants (C_{Cs}) with a precision of about 0.3%. The Cs response evolution during Run-2 reflects the optics component degradation due to radiation dose especially pronounced in the innermost radial layer A (Figure 2) as well as the PMT gain variations discussed in Section 2.2.

2.2. Laser

Short laser pulses (The laser pulses have very similar shape to those from collision data, with a FWHM of 50 ns.) are simultaneously sent to all PMTs in order to monitor their gain and to measure possible non-linearities of their response (As more than 99.6% of all PMTs show non-linearity below 1%, no such correction is applied during the LHC Run-2.). Dedicated standalone runs are performed daily for this purpose. Laser events are also exploited during collision runs during the LHC abort gaps to track possible fast PMT gain changes as well as for monitoring the time calibration (Section 2.6).

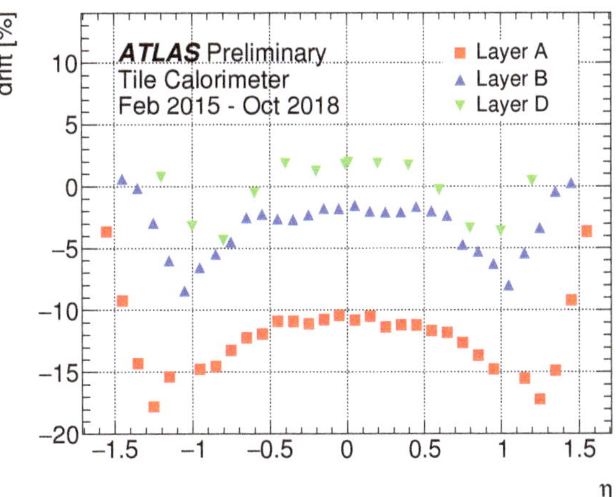

Figure 2. The drift of the Cs response between the beginning and the end of Run-2, shown for cells in different radial layers and as a function of the cells' pseudorapidity [4]. Cells close to $|\eta| \approx 1.1$ are exposed to higher radiation doses with respect to more central cells in each radial layer.

The laser calibration provides per-channel constants (C_{Las}) that determine the PMT response relative to the last Cs calibration. The precision of the laser calibration system is at the level of 0.5%. Down-drifts are observed during collision periods, while the PMT gain recovers during the off-beam periods (Figure 3). The largest gain drifts are observed in PMTs reading cells exposed to the highest radiation dose in the innermost layer (A).

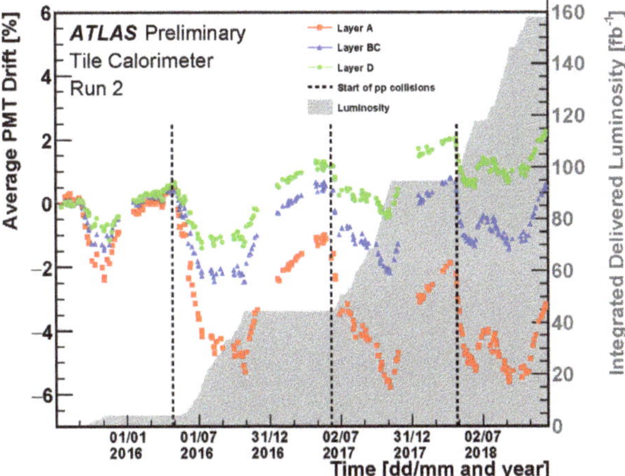

Figure 3. The evolution of the mean relative laser response of three radial layers as a function of time. The dashed lines indicate the beginning of the *pp* collision period in each year [4].

2.3. Charge Injection

A charge injection system (CIS) injects pulses of specified charge into the fast readout electronics, spanning the whole dynamic range of both gains. It provides an amplitude-to-charge conversion factor (C_{CIS}) for every channel and gain and is also used to map non-linearities in the readout electronics.

The overall CIS precision is approximately 0.7%, and it shows very good time stability (0.05% in individual channels) over the whole Run-2.

2.4. Minimum Bias System

This system integrates the PMT response to soft inelastic interactions over many bunch-crossings. It shares the same readout path with the cesium system. As these interactions are symmetric in azimuth, this system allows for the calibration of the special cells (so-called E-cells) that are not accessible by the Cs source.

Since the minimum bias signal is proportional to the instantaneous luminosity, it is also used for the luminosity measurements [5].

2.5. Combined Energy Calibration

Figure 4 displays the response evolution of individual calibration systems. The Cs and minimum bias results are in a very good agreement as expected. The difference between Cs and laser response is due to the optical system degradation, which amounts to approximately 10% in the most irradiated cell A13 during the LHC Run-2.

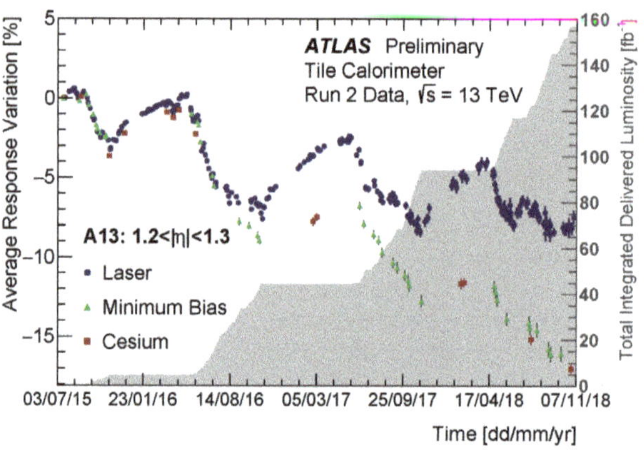

Figure 4. The variation of the average response to cesium, laser and minimum bias for the cell A13, as a function of time during the entire Run-2 period [4].

The TileCal energy in each channel is reconstructed at the electromagnetic (EM) scale using the formula

$$E\,[\text{GeV}] = \frac{A\,[\text{ADC counts}]}{C_{\text{Cs}} \cdot C_{\text{Las}} \cdot C_{\text{CIS}}[\text{ADC counts/pC}] \cdot C_{\text{TB}}[\text{pC/GeV}]}, \tag{1}$$

where A stands for the pulse amplitude reconstructed from seven consecutive samples using the Optimal Filtering (OF) algorithm [6], and C_{Cs}, C_{Las} and C_{CIS} are the constants corresponding to the individual calibration systems. The last factor C_{TB} defines the EM scale and was determined in dedicated beam tests, linking the total measured charge in the calorimeter with the electron beam energy.

2.6. Time Calibration

The goal of the time calibration is to ensure that particles traveling from the ATLAS interaction point at the speed of light produce signals with a phase $t_0 \approx 0$ in every channel. This feature is important for the time-of-flight measurement as well as for the proper energy reconstruction with the OF algorithm, since the OF weights depend on the phase.

The time calibration is performed on a per-channel basis with splash events (Special events are when a single LHC beam hits the collimator about 140 m upstream from the ATLAS detector. Lots of particles are produced and pass through the detector approximately parallel to the beam axis.) and initial pp collisions. Only cells/channels associated to reconstructed jets are used in order to avoid bias from non-collision background. Since the mean cell time slightly depends on the deposited energy (Section 3.3), only events with channel energies between 2 and 4 GeV are selected for the calibration.

The stability of the calibration is monitored with two complementary methods, the laser events shot during the abort gaps of collision runs (Section 2.2) as well as with the physics collision data. Two examples of identified problems are shown in Figure 5. These problems are then corrected in the data used for physics analyses. The time stability is better than 1 ns.

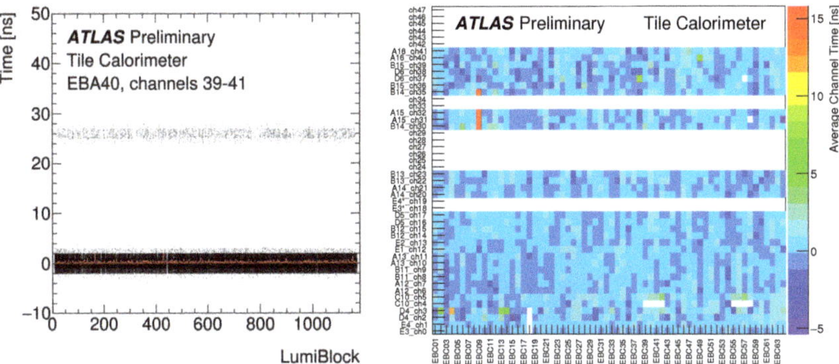

Figure 5. A problem in electronics causes a time offset corresponding to 1 bunch-crossing (25 ns) in a group of three channels in about 1% of events, as determined with the laser events (**left** panel) [4]. Another problem causes the reconstructed time in a group of six channels being off by a constant value in all events. The example plot shows affected channels 30–35 (channels 33, 34 are not used) in the module EBC09 and comes from the collision data monitoring (**right** panel) [7].

3. Performance

The performance of the TileCal during Run-2 was checked with isolated muons and hadrons, the time resolution was addressed with jets. The improved muon identification at the first level trigger was checked with muons from Z decays.

3.1. Response to Isolated Muons

The TileCal EM scale and response uniformity was checked with isolated muons originating from the W decays. Only muon with momenta between 20 and 80 GeV are considered, since they lose their energy predominantly by ionization, hence their response scales almost linearly with the path length through the cell. The muon tracks are measured in the Inner Detector [2] and extrapolated through TileCal taking into account the detector material and magnetic field [8].

The muon response in each cell is evaluated using the ratio of deposited energy ΔE over the muon path length Δx through that cell. As the distribution of $\Delta E / \Delta x$ is non-Gaussian, a truncated mean (Mean value of the distribution where 1% of events with the highest values are removed.) is taken as the measure of the cell response. In order to reduce the residual non-linearity of the truncated mean, the double ratio

$$R \equiv \frac{(\Delta E / \Delta x)_{\text{data}}}{(\Delta E / \Delta x)_{\text{MC}}} \tag{2}$$

is considered in the analysis. The cell response uniformity for one cell across the azimuth is shown in Figure 6. It amounts to 2.4%, and it is consistent between different cell types. Furthermore, all radial layers show a response consistent with $R = 1$ within 2%.

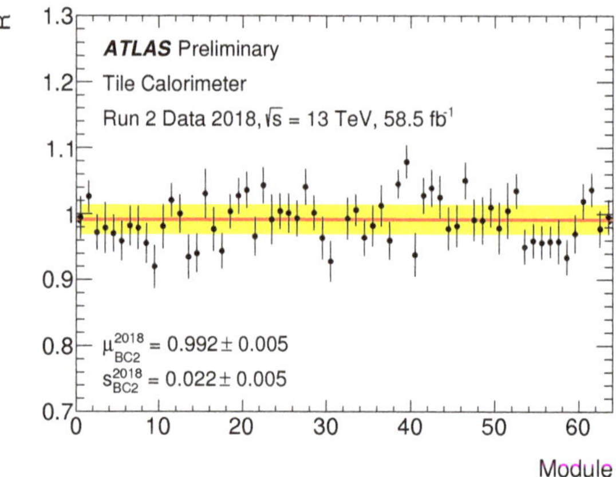

Figure 6. The double ratio R as a function of the TileCal module for the cell type BC2 as obtained with muons from W decay. The error bars represent the statistical uncertainty. The red line indicates the mean relative response (μ) and s (yellow band) displays additional uncertainty as determined from Gaussian likelihood fit performed on the R values for this cell type [7].

The time stability of the response to isolated muons is addressed by comparing the double ratio R between three periods (2015 + 2016, 2017 and 2018) of Run-2. The results indicate very good stability at the level of few percents.

3.2. Response to Isolated Hadrons

The response to isolated charged hadrons was investigated with the ratio of the calorimeter energy E and associated track momentum p. This ratio was compared between data and MC simulations.

Isolated tracks are selected in the Inner Detector, and their momenta p are measured there. Only tracks with $p > 2$ GeV are used. The calorimeter energy E is determined at the EM scale from calorimeter clusters associated to that isolated track (Clusters are built from calorimeter cells whose distance ΔR measured from the cluster center satisfies $\Delta R \equiv \sqrt{(\eta_{\text{cell}} - \eta_{\text{cluster}})^2 + (\phi_{\text{cell}} - \phi_{\text{cluster}})^2} < 0.2.$). Since this study focuses on the TileCal response, the EM calorimeter signal corresponding to that track is required to be compatible with that of a minimum ionizing particle (MIP). Muons and neutral particles are removed with further calorimeter selection criteria.

The obtained E/p ratio is shown in Figure 7 as a function of pseudorapidity and track momentum for low pile-up data, i.e., where the average number of interactions per bunch-crossing is about two. A ratio $E/p < 1$ is observed, reflecting the non-compensated nature of the TileCal. A good agreement between data and MC is observed for these low pile-up data. In the central region ($|\eta| < 1$), the E/p data/MC ratio is approximately 0.98. Similar results were reported by another ATLAS analysis [9]. A larger difference is observed in the region $|\eta| \approx 1.5$, where the data/MC ratio is affected by the so-called crack scintillators. These scintillators are designed to correct for the energy lost in the dead material between the barrel and end-cap calorimeters. Imperfections in the dead material description and a less efficient MIP-like selection in EM calorimeters in this region lead to larger discrepancies between data and MC simulations.

3.3. Time Resolution

The time performance was studied with jets. As mentioned in Section 2.6, only cells associated to the reconstructed jets are used for this purpose. Jets are required to point to the TileCal and to have a minimum transverse momentum of 20 GeV.

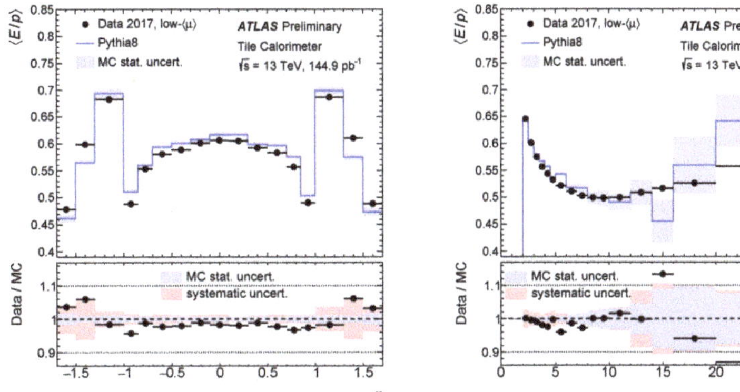

Figure 7. The TileCal response to isolated charged hadrons characterized by the energy over momentum ratio E/p, as a function of pseudorapidity (**left**) and momentum (**right**). The lower panels show the ratio of data to MC simulation, systematic uncertainties cover the effects of residual contamination from neutral particles and energy mis-measurements due to energy losses in front of the calorimeter [7].

The mean cell time is stable across the four data-taking years as shown in Figure 8, left panel. It slightly decreases with the deposited energy due to neutrons and the slow hadronic component of the developed shower. The cell time resolution, determined as the Gaussian width of the corresponding time distribution, improves with energy as expected and approaches 0.4 ns at high energies (Figure 8, right). The RMS of the cell time distribution reflects the non-Gaussian tails and depends on the pile-up conditions, while the Gaussian widths are rather stable.

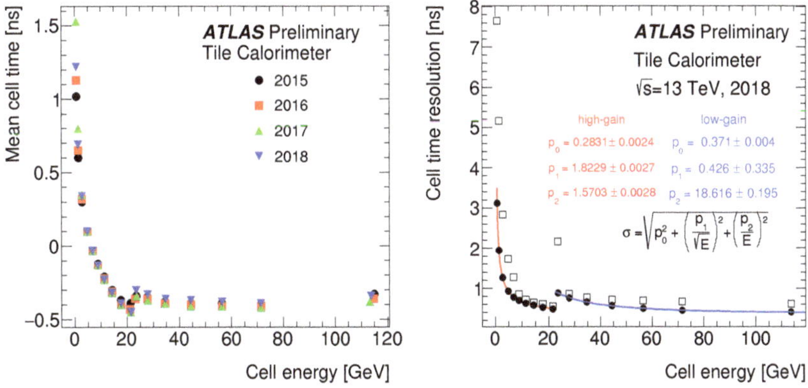

Figure 8. The mean cell time (**left**) and cell time resolution (**right**) as a function of the cell energy, as measured in cells associated to reconstructed jets. The statistical uncertainties are not visible since they are smaller than the symbols. The small discontinuity close to 20 GeV corresponds to the high/low-gain transition [7].

3.4. Tile-Muon Trigger

Special Tile-muon Digitizer Boards (TMDBs) have been in operation since the beginning of 2018. They provide the coincidence of signals from the outermost TileCal layer cells and the Thin Gap Chambers of the ATLAS muon system [2] in order to improve the background rejection in the first level muon trigger in the region $1.05 < |\eta| < 1.30$.

The pseudorapidity distribution of the reconstructed particles obtained with the Tile-muon trigger is compared to that of the first level muon trigger in Figure 9. The total trigger rate reduces by about 6%, but significant reduction is observed in the region $1.05 < |\eta| < 1.30$ where the TMDBs are installed. Studies with $Z \to \mu\mu$ events show that this improvement is achieved at a cost of at most 2.5% inefficiency, compatible with the expected geometrical inefficiency due to thin gaps in azimuth between the TileCal modules [10].

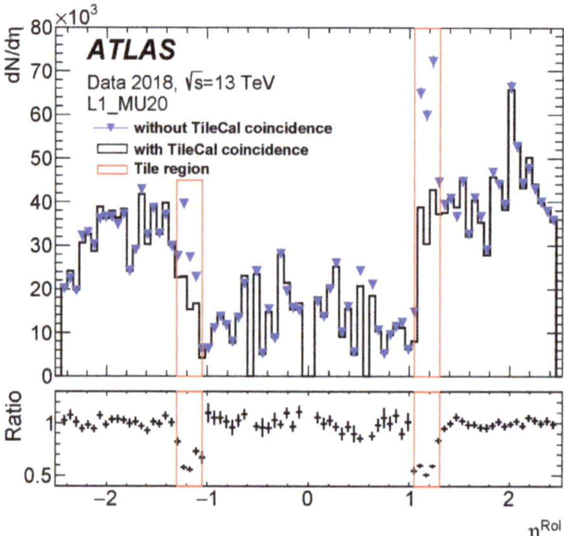

Figure 9. The pseudorapidity distribution, reconstructed using online information, of particles with transverse momentum above 20 GeV in events selected with the first level muon trigger (blue triangles) and the Tile-muon trigger (black line). Their ratio is shown in the lower pad. The coincidence region is highlighted with redrectangles [10].

4. Conclusions

The ATLAS Tile Calorimeter calibration and performance during LHC Run-2 has been presented. The individual calibration systems achieve precisions better than 1%, and the combined calibration guarantees very good response stability. The time calibration exhibits a stability better than 1 ns due to the extensive monitoring.

The TileCal performance has been assessed with isolated particles and jets. The EM scale settings and the response uniformity is verified with isolated muons and single hadrons. The cell time resolution is measured with jets. It approaches 0.4 ns for energies above 100 GeV. The performance of the Tile-muon trigger, operational since 2018, has improved the background rejection in the first level muon trigger.

Funding: This research was partly funded by the grant LTT17018 of the Ministry of Education, Youth and Sports of the Czech Republic and the grant UNCE/SCI/013 of the Charles University.

Conflicts of Interest: The author declares no conflict of interest.

References

1. ATLAS Collaboration. Operation and performance of the ATLAS Tile Calorimeter in Run 1. *Eur. Phys. J. C* **2018**, *78*, 987. [CrossRef] [PubMed]
2. ATLAS Collaboration. The ATLAS Experiment at the CERN Large Hadron Collider. *J. Instrum.* **2008**, *3*, S08003. [CrossRef]
3. Evans, L.; Bryant, P. LHC Machine. *J. Instrum.* **2008**, *3*, S08001. [CrossRef]
4. ATLAS Tile Calorimeter Calibration Public Plots. Available online: https://twiki.cern.ch/twiki/bin/view/AtlasPublic/ApprovedPlotsTile (accessed on 14 August 2022).

5. ATLAS Collaboration. Luminosity determination in pp collisions at $\sqrt{s} = 7$ TeV using the ATLAS detector at the LHC. *Eur. Phys. J. C* **2011**, *71*, 1630. [CrossRef]
6. Fullana, E.; Castelo, J.; Castillo, V.; Cuenca, C.; Ferrer, A.; Higón, E.; Iglesias, C.; Munar, A.; Poveda, J.; Ruiz-Martinez, A.; et al. *Optimal Filtering in the ATLAS Hadronic Tile Calorimeter*; Technical report; CERN: Geneva, Switzerland, 2005.
7. ATLAS Tile Calorimeter Collision Data Public Plots. Available online: https://twiki.cern.ch/twiki/bin/view/AtlasPublic/TileCaloPublicResults (accessed on 14 August 2022).
8. Lund, E.; Bugge, L.; Gavrilenko, I.; Strandlie, A. Track parameter propagation through the application of a new adaptive Runge-Kutta-Nystroem method in the ATLAS experiment. *J. Instrum.* **2009**, *4*, P04001. [CrossRef]
9. ATLAS Collaboration. Measurement of the energy response of the ATLAS calorimeter to charged pions from $W^{\pm} \to \tau^{\pm}(\to \pi^{\pm}\nu_{\tau})\nu_{\tau}$ events in Run 2 data. *Eur. Phys. J. C* **2022**, *82*, 223 [CrossRef]
10. ATLAS Collaboration. Performance of the ATLAS muon triggers in Run 2. *J. Instrum.* **2020**, *15*, P09015. [CrossRef]

 *instruments*

Project Report

Calorimetry in a Neutrino Observatory: The JUNO Experiment

Beatrice Jelmini on behalf of the JUNO Collaboration

Dipartimento di Fisica e Astronomia "Galileo Galilei", Università degli Studi di Padova & INFN Sezione di Padova, 35131 Padova, Italy; beatrice.jelmini@pd.infn.it

Abstract: The Jiangmen Underground Neutrino Observatory (JUNO) is a multipurpose experiment under construction in southern China; detector completion is expected in 2023. JUNO is a homogeneous calorimeter consisting of a target mass of 20 kt of an organic liquid scintillator, aiming to detect antineutrinos from reactors to investigate the neutrino oscillation mechanism. The scintillation and Cerenkov light emitted after the interaction of antineutrinos with the liquid scintillator is seen by a compound system of 20 inch large PMTs and 3 inch small PMTs, with a total photo-coverage of 78%. A dual-calorimetry technique is developed based on the presence of the two independent photosensor systems which are characterized by different average light level regimes, resulting in different dynamic ranges. Thanks to this novel technique, an unprecedented high light yield, and in combination with a comprehensive multiple-source and multi-position calibration campaign, JUNO is expected to reach energy-related systematic uncertainties below 1% and an effective energy resolution of 3% at 1%, required for the neutrino oscillation analysis.

Keywords: liquid scintillator; PMTs; calibration; energy nonlinearity; energy resolution; dual calorimetry

Citation: Jelmini, B., on behalf of the JUNO Collaboration. Calorimetry in a Neutrino Observatory: The JUNO Experiment. *Instruments* **2022**, *6*, 26. https://doi.org/10.3390/instruments6030026

Academic Editors: Fabrizio Salvatore, Alessandro Cerri, Antonella De Santo and Iacopo Vivarelli

Received: 31 July 2022
Accepted: 22 August 2022
Published: 24 August 2022

Publisher's Note: MDPI stays neutral with regard to jurisdictional claims in published maps and institutional affiliations.

1. Introduction: The JUNO Experiment

The Jiangmen Underground Neutrino Observatory (JUNO) experiment [1,2] is a neutrino multipurpose experiment currently under construction in southern China; detector completion is expected in 2023.

The JUNO experiment is located about 700 m underground (1800 m.w.e) and consists of a spherical central detector (CD), surrounded by a Cerenkov water pool and surmounted by a top tracker and the calibration house, as shown in Figure 1a. The Cerenkov water pool shields the CD from environmental radioactivity and, together with the top tracker, works as an active muon veto. The CD is an acrylic 35.4 m diameter wide spherical vessel filled with 20 kt of liquid scintillator, supported by a stainless-steel latticed shell. A system of photomultiplier tubes (PMTs) is installed on the supporting latticed shell to detect the scintillation and Cerenkov light produced by (anti)neutrino interactions in the liquid scintillator. The PMT system consists of 17,612 20 inch large PMTs (LPMTs) [3] and 25,600 3 inch small PMTs (SPMTs), ensuring a total photo-coverage of 78%.

The main goal of the JUNO experiment is to probe the neutrino oscillation mechanism [4] by detecting electron antineutrinos produced in nuclear reactors, located at an average distance of 52.5 km. Figure 1b shows the non-oscillated spectrum (black line) and the oscillated spectra for the two possible neutrino mass orderings (red and blue lines), as expected to be seen by the JUNO experiment. In the first year of data taking, JUNO aims to measure three of the oscillation parameters, Δm_{31}^2, Δm_{21}^2, and $\sin^2 \theta_{12}$, with unprecedented sub-percent precision [5]; furthermore, it aims to determine the neutrino mass ordering with a 3 σ significance in 6 years of data taking.

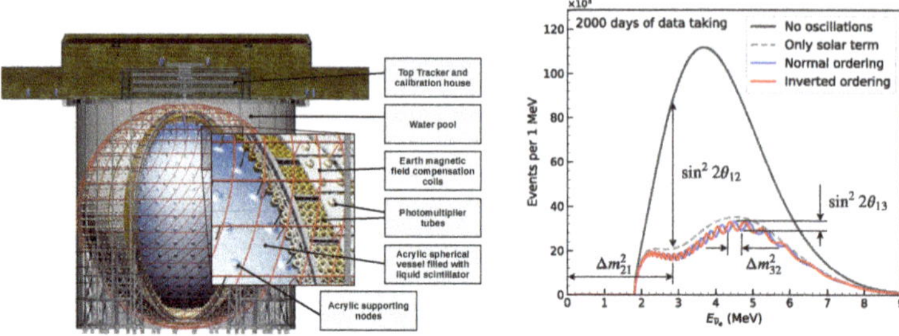

(**a**) *JUNO detector scheme.*   (**b**) *Reactor spectrum at JUNO.*

Figure 1. (**a**) Scheme of the JUNO detector [5]. (**b**) Oscillated spectra expected to be seen in JUNO for the two neutrino mass orderings (red and blue lines). The expected spectrum without neutrino oscillations is also shown (black line). An average baseline of 52.5 km and a data-taking time of 6 years were considered in the evaluation of the spectra [5].

As can be seen from Figure 1b, a precise and accurate measurement of the antineutrino energy is essential in order to distinguish the two mass orderings. However, both the detection mechanism and the readout system introduce nonlinear biases and smearing effects that could affect the detected spectrum, thus spoiling JUNO physics potential. Consequently, an effective energy resolution of 3% at 1 MeV and energy-related systematics controlled within 1% are required. To study the detector response and meet the requirements, an extensive multiple-source and multi-position calibration strategy [6] was developed and is described in Section 2. A dual-calorimetry calibration [7,8] was also developed in order to take into account nonlinearities introduced by the readout system and is described in Section 3.

All results presented in this proceedings were obtained using JUNO official simulation software based on SNiPER [9].

2. JUNO Calibration Strategy

2.1. Detector Response: Liquid Scintillator Nonlinearity and Positional Nonuniformity

A calibration strategy [6] was developed to calibrate the liquid scintillator nonlinearity (LSNL) of the target material and the positional nonuniformity (NU), caused by the optical attenuation of light given the very large size of the JUNO detector.

There are two contributions to the energy LSNL: quenching effect and the Cerenkov light. The quenching effect is characteristic of the scintillation mechanism and is dominant in the low-energy part of the spectrum; it is usually parametrized with the semi-empirical Birks' law. The fraction of detected primary Cereknov light is sub-dominant in JUNO, since most of the Cerenkov light is absorbed by the liquid scintillator and possibly re-emitted as secondary scintillation light; nonetheless, it needs to be considered and introduces nonlinearities at high energies.

Due to the very large size of the JUNO detector, there is a nonlinear response as a function of the position of the neutrino interaction within the liquid scintillator volume. Figure 2a shows the average number of PEs obtained by simulating 2.22 MeV gammas as a function of the distance from the detector center, R, for different polar angles θ; the plot is representative of the positional NU of the detector response. From the figure, the value of the detected light yield at the center of the detector is $Y_0 = 1345$ PE/MeV [6]. The value of the detected light yield reported in this proceedings is taken from Ref. [6] and is obtained by simulating 2.22 MeV gammas at the center of the detector. The latest simulation results suggest that the detected light yield for the JUNO experiment may be about 20 % higher. The reasons for the increase in the detected light yield are to be found in a higher photon detection efficiency of the PMTs [3], and updates of the PMT optical model [10] and the central detector geometry [11] in the JUNO official simulation software. For increasing radius R, an increase in the detected light yield is observed, reaching a maximum towards the edge of the detector and eventually dropping due to leakage effect and internal total reflection effect at the detector boundary. The increase in the region between 6 and 15 m is due to a combination of the optical attenuation of photons propagating through the detector, which follows an exponential law, and the variation in the active photon coverage [6].

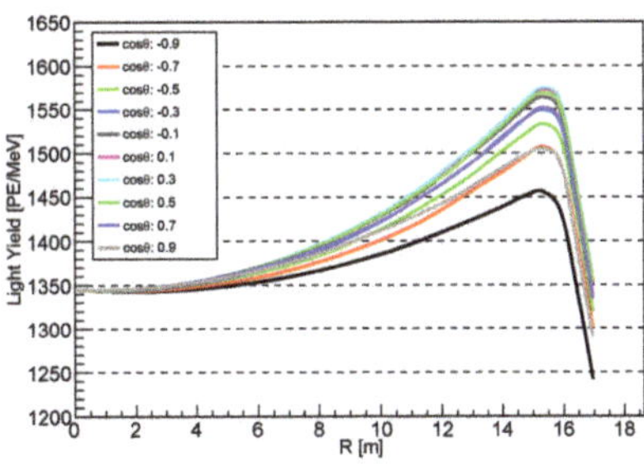

(a) *Positional Nonuniformity.*

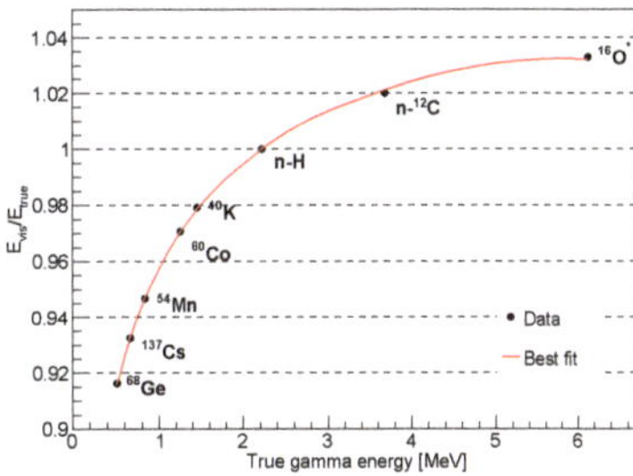

(b) *Liquid Scintillator Nonlinearity.*

Figure 2. (**a**) Expected light yield in JUNO as a function of the distance from the detector center, showing the positional nonuniformity of the detector response; different lines refer to different polar angles, highlighting a difference between the upper and lower hemispheres [6]. (**b**) Liquid scintillator nonlinearity expected for the JUNO experiment due to the quenching effect and Cereknov light contribution. Black points represent simulated data, while the red line is the best fit line; see text for more details [6].

2.2. Multiple-Source and Multi-Position Calibration Campaign

To calibrate the LSNL, several radioactive sources are needed to cover the energy range of interest for the neutrino oscillation analysis.

We plan to deploy the following gamma sources: ^{137}Cs (0.662 MeV γ), ^{54}Mn (0.835 MeV γ), ^{60}Co (1.173+1.333 MeV γ), and ^{40}K (1.461 MeV γ). A positron source, ^{68}Ge, will also be used to have two low-energy 0.511 MeV gamma-rays. We also plan to deploy two neutron sources, AmBe and AmC, which will provide higher energy gamma-rays (4.43 MeV and 6.13 MeV, respectively) and neutrons, which will then be captured by hydrogen nuclei, thus producing a 2.22 MeV gamma, or by ^{12}C nuclei, producing one 4.94 MeV gamma-ray or a 3.68 MeV and a 1.26 MeV gamma-ray. The continuous spectrum of a cosmogenic background, ^{12}B, will be used to cover the high-energy part of the reactor spectrum, thus allowing full coverage of the energy range of interest.

Results based on the full JUNO simulation are shown in Figure 2b. The black points represent simulated data, showing the peaks corresponding to the various radioactive sources listed above. The red line is the best fit curve obtained by using a Daya Bay model based on a four-parameter function [12]:

$$\frac{E_{\text{vis}}}{E_{\text{true}}} = \frac{p_0 + p_3/E_{\text{true}}}{1 + p_1 e^{-p_2 E_{\text{true}}}}. \tag{1}$$

The Daya Bay model used in the fit can properly describe the LSNL, with a residual bias between simulated data and the best fit curve less than 0.2%.

To calibrate the positional NU, radioactive sources will be deployed in several positions in the whole liquid scintillator volume, thanks to a robust and flexible calibration hardware. The calibration hardware [6] is located in the calibration house above the central detector, as shown in Figure 1a, and consists of several independent subsystems, as shown in Figure 3a.

The automatic calibration unit (ACU) was developed to calibrate the detector along the central vertical axis z; the positioning of the source along the z axis is expected with a precision better than 1 cm. Due to its simplicity and robustness, we plan to use the ACU to frequently deploy several sources and monitor the stability of the energy scale during data-taking.

The guide tube system (GT) consists of a tube looped outside of the acrylic sphere containing the liquid scintillator along a longitudinal circle; a positioning precision of 3 cm is expected for this system. The GT system was developed to calibrate the CD nonuniformity at the boundary.

We also plan to use two cable loop systems (CLSs) to deploy sources to off-axis positions; the two systems will be installed in two opposite half-planes. An independent ultrasonic system is employed to position the sources with a precision of 3 cm.

Finally, a remotely operated vehicle (ROV) will deploy a radioactive source almost anywhere in the CD volume with a precision of 3 cm. The ROV serves as a supplemental system to the ACU, GT, and CLSs, and will be used to study local effects or azimuthal dependence where the CLSs cannot provide sufficient information.

An extensive study [6] was performed to find the optimal set of calibration points which minimizes the effective energy resolution, Equation (3), introduced in the next section. Figure 3b shows an optimal set of 250 random calibration points in a half vertical plane of the JUNO CD that can be used to correct the detector nonuniformity.

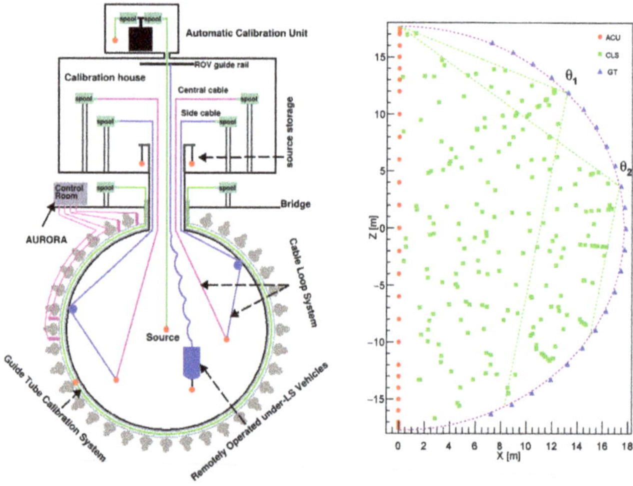

(**a**) *Calibration hardware.*   (**b**) *Nonuniformity calibration points*

Figure 3. (**a**) Overview of the calibration hardware for the JUNO experiment. See text for details on the individual subsystems [6]. (**b**) An optimal set of 250 random calibration points in a half vertical plane of the JUNO CD needed to properly correct the position nonuniformity of the response [6].

2.3. Energy Resolution

The energy resolution is parametrized with the following standard equation:

$$\frac{\sigma_E}{E} = \sqrt{\left(\frac{a}{\sqrt{E}}\right)^2 + b^2 + \left(\frac{c}{E}\right)^2}. \tag{2}$$

The term a is the stochastic term, dominated by the statistical fluctuations in the emission of optical photons and subsequent detection on the PMT photocathode, thus being related to the total light yield value. b is the constant term which is dominated by the positional NU and can be controlled thanks to the multi-position calibration campaign presented in Section 2.2. The last term, c, is dominated by the contribution of the dark noise from the PMTs. Results from the full MC simulation are shown in Figure 4.

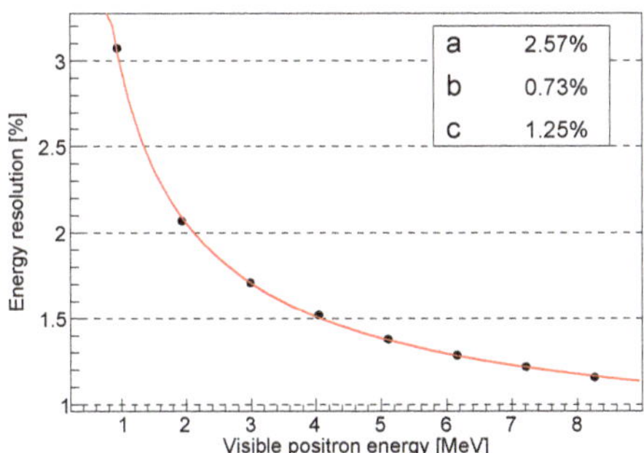

Figure 4. Energy resolution expected for the JUNO experiment obtained by full MC simulation of positrons at different energies. The red line is the best fit curve obtained by using Equation (2). Best fit values for the three parameters are shown in the box [6].

In JUNO, we usually express the energy resolution in terms of an effective resolution term:

$$\tilde{a} \equiv \sqrt{(a)^2 + (b \times 1.6)^2 + \left(\frac{c}{1.6}\right)^2}. \tag{3}$$

The factors 1.6 and 1/1.6 reflect the fact that the second term is not improving with the visible energy, and that the third term declines quickly with the energy [6]. From the most recent simulation, $\tilde{a}$ is found to be less than 3%, thus meeting the requirements for the reactor antineutrino spectrum analysis.

3. Dual-Calorimetry Calibration

Large PMTs are expected to cover a wide dynamic range, from 0 to 100 photoelectrons, and perform charge measurement through charge integration. The charge integration relies on the sampling of the waveform, which is performed by the front-end and read-out large PMT electronics [2] located underwater near the PMTs and equipped with 14 bit 1 GHz flash analog-to-digital converters (ADCs). Both the PMTs and the readout electronics could introduce additional instrumental nonlinearities, which could in principle be different from channel to channel.

A dual-calorimetry calibration (DCC) [7,8] was developed to correct the large-PMT channel-level nonlinearity and relies on the additional small-PMT system. Small PMTs are placed between large PMTs and, given their small size, they mainly work in the single-photoelectron regime, allowing charge measurement through photoelectron counting. Thus, with good approximation, they constitute a linear system which can be used as an in-detector reference for the large-PMT system.

The DCC can be performed by using a tunable ultraviolet light source covering the full range of the reactor spectrum and by comparing the channel-level large-PMT response to the response of the small-PMT system. The response can be factorized into three terms, as follows [8]:

$$R^{\mathrm{L-channel}} = R_{\mathrm{LSNL}} \cdot R_{\mathrm{NU}}^{\mathrm{L}} \cdot R_{\mathrm{QNL}}^{\mathrm{L}}; \tag{4}$$

$$R^{\mathrm{S-system}} = R_{\mathrm{LSNL}} \cdot R_{\mathrm{NU}}^{\mathrm{S}} \cdot R_{\mathrm{QNL}}^{\mathrm{S}}, \tag{5}$$

where the superscripts L and S stand for large PMT and small PMT, respectively.

R_{LSNL} is the term related to the LS nonlinearity, which is the same for both systems since they are exposed to the same event, hence the same energy deposition. $R_{\mathrm{NU}}^{\mathrm{L/S}}$ is the term describing the positional nonuniformity, which in principle can be different for the two PMT systems. The UV source will be located at the center of the detector; given the fixed location of the source, the nonuniformity terms are constants, thus can be factorized out. Finally, $R_{\mathrm{QNL}}^{\mathrm{L/S}}$ is the term related to the charge nonlinearity.

In the end, if we compare the response of the large-PMT channel and small-PMT system we are left with the terms related to the charge response:

$$\frac{R^{\text{L-channel}}}{R^{\text{S-system}}} = \frac{R^{\text{L}}_{\text{QNL}}}{R^{\text{S}}_{\text{QNL}}}.$$

(6)

Figure 5 shows an example of the application of the DCC technique at event level, where large-PMT channel-level nonlinearities are corrected. The dotted black line represents the ideal case with zero instrumental nonlinearity. The solid red line represents a 2% event-level instrumental nonlinearity originated by assuming an extreme scenario, with 50% channel-level nonlinearity in the 0 to 100 photoelecton range. The dashed blue line shows the corrected instrumental nonlinearity obtained after the application of the channel-wise DCC technique: the instrumental nonlinearity is reduced to <0.3%.

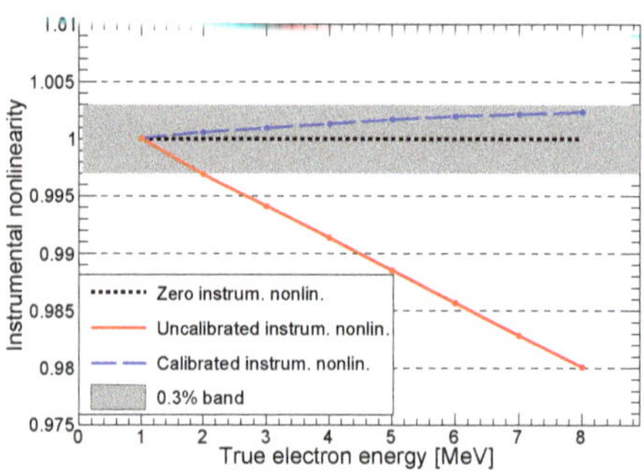

Figure 5. Example of the application of the dual-calorimetry calibration technique [6,8]. The solid red line shows an extreme scenario of a 50% channel-level nonlinearity resulting in a 2% event-level nonlinearity; the dashed blue line represents the same scenario after applying the channel-level correction, resulting in a residual <3% event-level nonlinearity.

4. Conclusions

The JUNO experiment is expected to reach a precise and accurate measurement of the antineutrino energy for the neutrino oscillation analysis. This result will be achieved thanks to a multiple-source and multi-position calibration system to correct the liquid scintillator nonlinearity and the positional nonuniformity. A dual-calorimetry calibration technique will also be used to correct large-PMT channel-level nonlinearity by using the small PMT system as an in-detector linear reference.

Funding: This research received no external funding.

Data Availability Statement: Not applicable.

Conflicts of Interest: The author declares no conflict of interest.

Abbreviations

The following abbreviations are used in this manuscript:

ACU	Automatic calibration unit
CD	Central detector
CLS	Cable loop system
DDC	Dual-calorimetry calibration
GT	Guide Tube
JUNO	Jiangmen Underground Neutrino Observatory
LSNL	Liquid scintillator nonlinearity
NU	Nonuniformity
PMT	Photomultiplier tube

QNL   Charge nonlinearity
ROV   Remotely operated vehicle

References

1. An, F.; An, G.; An, Q.; Antonelli, V.; Baussan, E.; Beacom, J.; Bezrukov, L.; Blyth, S.; Brugnera, R.; Buizza Avanzini, M.; et al. Neutrino Physics with JUNO. *J. Phys. G Nucl. Part. Phys.* **2016**, *43*, 30401. [CrossRef]
2. The JUNO Collaboration. JUNO physics and detector. *Prog. Part. Nucl. Phys.* **2022**, *123*, 103927. j.ppnp.2021.103927. [CrossRef]
3. The JUNO Collaboration; Abusleme, A.; Adam, T.; Ahmad, S.; Ahmed, R.; Aiello, S.; Akram, M.; An, F.; An, G.; An, Q.; et al. Mass Testing and Characterization of 20-inch PMTs for JUNO. *Eur. Phys. J. C* 2022, *submitted*, https://arxiv.org/abs/2205.08629.
4. Workman, R.L.; Burkert, V.D.; Crede, V.; Klempt, E.; Thoma, U.; Tiator, L.; Agashe, K.; Aielli, G.; Allanach, B.C.; Amsler, C.; et al. Review of Particle Physics. *Prog. Theor. Exp. Phys.* **2022**, 83C01.
5. The JUNO Collaboration; Abusleme, A.; Adam, T.; Ahmad, S.; Ahmed, R.; Aiello, S.; Akram, M.; An, F.; An, G.; An, Q.; et al. Sub-percent Precision Measurement of Neutrino Oscillation Parameters with JUNO. *Chin. Phys. C* 2022, *submitted*, https://arxiv.org/abs/2204.13249.
6. The JUNO Collaboration; Abusleme, A.; Adam, T.; Ahmad, S.; Ahmed, R.; Aiello, S.; Akram, M.; An, F.; An, G.; An, Q.; et al. Calibration strategy of the JUNO experiment. *J. High Energ. Phys.* **2021**, *4*, 1–33. [CrossRef]
7. He, M. Double Calorimetry System in JUNO. In Proceedings of the TIPP 2017, Beijing, China, 22 May 2022. [CrossRef]
8. Han, Y. Dual Calorimetry for High Precision Neutrino Oscillation Measurement at JUNO Experiment. Ph.D. Thesis, Université de Paris, Paris, France, 2020.
9. Lin, T; Zou, J.; Li, W.; Deng, Z.; Fang, X.; Cao, G.; Huang, X.; You, Z.; On Behalf of the JUNO Collaboration. The Application of SNiPER to the JUNO Simulation. *J. Phys. Conf. Ser.* **2017**, *898*, 42029. [CrossRef]
10. Wang, Y.G.; Cao, G.; Wen, L.; Wang, Y.F. A new optical model for photomultiplier tubes. *Eur. Phys. J. C* **2022**, *82*, 1–14. [CrossRef]
11. Zhi, W. The Central Detector of JUNO. In Proceedings of the Neutrino 2022 Virtual Conference, Seoul, Korea, 30 May 2022. [CrossRef]
12. An, F.P.; Balantekin, A.B.; Band, H.R.; Beriguete, W.; Bishai, M.; Blyth, S.; Rosero, R. Spectral measurement of electron antineutrino oscillation amplitude and frequency at Daya Bay. *Phys. Rev. Lett.* **2014**, *112*, 61801. [CrossRef] [PubMed]

 *instruments*

Article

RADiCAL—Precision Timing, Ultracompact, Radiation-Hard Electromagnetic Calorimetry

Thomas Anderson [1], Thomas Barbera [2], Bradley Cox [1], Paul Debbins [3], Maxwell Dubnowski [1], Kiva Ford [2], Maxwell Herrmann [3], Chen Hu [4], Colin Jessop [2], Ohannes Kamer-Koseyan [3], Alexander Ledovskoy [1], Yasar Onel [3], Carlos Perez-Lara [1], Randal Ruchti [2,*], Daniel Ruggiero [2], Daniel Smith [2], Mark Vigneault [2], Yuyi Wan [2], Mitchell Wayne [2], James Wetzel [3], Liyuan Zhang [4] and Ren-Yuan Zhu [4]

[1] Department of Physics, University of Virginia, Charlottesville, VA 22904, USA
[2] Department of Physics and Astronomy, University of Notre Dame, Notre Dame, IN 46556, USA
[3] Department of Physics, University of Iowa, Iowa City, IA 52242, USA
[4] Department of Physics, California Institute of Technology, Pasadena, CA 91125, USA
* Correspondence: rruchti@nd.edu; Tel.: +1-574-302-7181

Abstract: To address the challenges of providing high-performance calorimetry in future hadron collider experiments under conditions of high luminosity and high radiation (FCC-hh environments), we conducted R&D on advanced calorimetry techniques suitable for such operation, based on scintillation and wavelength-shifting technologies and photosensor (SiPM and SiPM-like) technology. In particular, we focused our attention on ultra-compact radiation-hard EM calorimeters based on modular structures (RADiCAL modules) consisting of alternating layers of the very dense absorber and scintillating plates, read out via radiation hard wavelength shifting (WLS) solid fiber or capillary elements to photosensors positioned either proximately or remotely, depending upon their radiation tolerance. RADiCAL modules provide the capability to measure simultaneously and with high precision the position, energy and timing of EM showers. This paper provides an overview of the instrumentation and photosensor R&D associated with the RADiCAL program.

Keywords: fast-timing; electromagnetic calorimetry; radiation-hard detectors

Citation: Anderson, T.; Barbera, T.; Cox, B.; Debbins, P.; Dubnowski, M.; Ford, K.; Herrmann, M.; Hu, C.; Jessop, C.; Kamer-Koseyan, O.; et al. RADiCAL—Precision Timing, Ultracompact, Radiation-Hard Electromagnetic Calorimetry. *Instruments* **2022**, *6*, 27. https://doi.org/10.3390/instruments6030027

Academic Editors: Fabrizio Salvatore, Alessandro Cerri, Antonella De Santo and Iacopo Vivarelli

Received: 10 August 2022
Accepted: 23 August 2022
Published: 25 August 2022

1. Introduction

The R&D objective and goals of RADiCAL are focused on the development of precision EM calorimetry for future hadron colliding beam experiments and are directed toward addressing the Priority Research Directions (PRD) for calorimetry listed in the DOE Basic Research Needs (BRN) workshop for HEP instrumentation [1].

Approach: To construct an array of ultracompact RADiCAL modules to explore the potential of ultracompact EM calorimetry capable of precision timing, energy and position measurements and specialized particle identification in high radiation fields [2]. Reaching the objective requires R&D on radiation-hard and fast-response scintillators, wavelength shifters and fiberoptic elements and photosensors.

Objective: To establish a performance baseline for RADiCAL modules through instrumentation capable of delivering an EM energy resolution approaching $\sigma/E = 10\%/\sqrt{E} \oplus 0.3/E \oplus 0.7\%$ [3], a timing resolution $\sigma_t < 50$ ps [4] and position resolution for the shower centroid within a few mm [5]. This initial effort uses radiation-hard optical elements but is instrumented with currently available SiPM photosensors, which are adequate for beam tests to establish and characterize a performance baseline for the RADiCAL technique but will not be performant if placed in high radiation areas. Once the time/energy/position baseline is established, R&D will be focused on the further development and refinement of radiation-hard optical components and new photosensors to replace those with vulnerabilities.

Our goals are:

1. To develop candidate instrumentation capable of operation in the FCC-hh endcap region up to $|\eta| \leq 4$ with an EM energy resolution indicated above, noting that for $|\eta| \leq 2.5$, the environmental conditions are expected to be 100 Mrad ionization dose and 3×10^{16} 1 MeV n_{eq}/cm^2 [3]. This effort will include further optical material development and R&D on radiation-hard photosensors to replace conventional SiPM;
2. To identify future directions and candidate instrumentation with the potential for operation at the FCC-hh in endcap/forward regions, where the operating conditions are foreseen to be a sobering 500 Grad ionization dose and 5×10^{18} 1 MeV n_{eq}/cm^2 [3]. To reach and function in this domain will require further creative innovations in optical materials and photosensor development.

2. Materials and Methods

Schematics of RADiCAL modules are shown in Figure 1, consisting of interleaved layers of LYSO:Ce tiles of 1.5 mm thickness and tungsten plates of 2.5 mm thickness, each of cross-sectional area 14×14 mm^2, stacked to a total depth of 114 mm corresponding to 25 X_o or 0.9 λ. The Molière radius of the structure is 13.7 mm, resulting in an ultracompact structure both transversely and longitudinally. The light produced in the LYSO:Ce tiles is then wavelength-shifted (WLS) and collected using specialized rad-hard (high OH$^-$ content) quartz capillaries containing liquid wave shifter or organic plastic or ceramic filaments in the capillary cores. The use of radiation-hard materials is essential, and extensive measurements of the radiation hardness of the scintillators LYSO:Ce, wave-shifting liquids and ceramics and capillaries have been studied extensively [2,6–9].

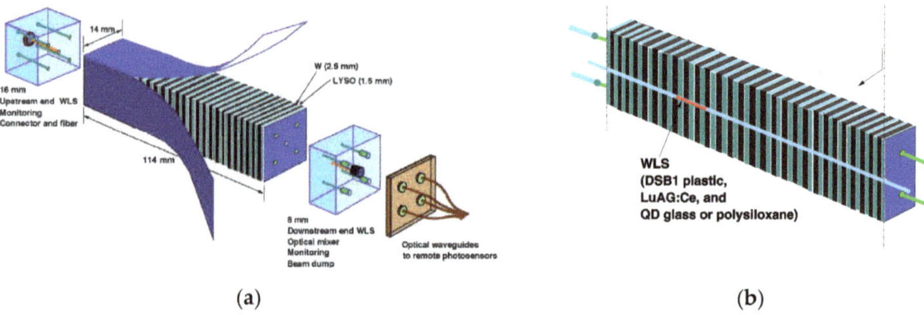

Figure 1. (a) Schematic of a RADiCAL module which is scaled to the Moliére radius of EM showers in a W/LYSO:Ce sampling calorimeter. The beam enters from the upper left and exits the lower right in this view. Light from the scintillation tiles is wave-shifted in capillaries or filaments that penetrate through the full length of the module to photosensors positioned at the ends of the module or remotely via fiberoptic waveguides. (b) A half-section of a RADiCAL module indicating two energy (E-type) capillaries penetrating the full length of the module and a timing (T-type) capillary superimposed to show its WLS filament (highlighted in red) positioned in the region of shower max for an EM shower.

Two types of WLS capillaries are utilized: E-type for energy measurement in which the WLS runs the full length of the module, and T-Type for precision timing measurement in which WLS is positioned locally in the region of EM shower maximum.

E-type capillaries are of 1.0 mm outer diameter and 0.4 mm inner diameter, with their cores filled with EJ309/DSB1 liquid wavelength shifter over their full length, and are read out with SiPM and low-gain amplifiers at the downstream ends only. For improved radiation hardness [6], LuAG:Ce WLS filaments of 114 mm in length and 1.15 mm in diameter will replace the liquid-filled capillaries with two benefits: photosensors can be placed at both upstream and downstream ends to measure the signal timing; the use of liquid WLS is avoided.

T-type capillaries are of 1.15 mm outer diameter and 0.95 mm inner diameter with 15 mm long and 0.9 mm diameter DSB1 organic plastic WLS filament positioned in the core at shower max. The remainder of the capillary core upstream and downstream is filled with quartz rods and fused to the capillary wall forming a solid quartz waveguide. For improved radiation hardness, rad-hard LuAG:Ce WLS filaments of 1.0 mm diameter and 15 mm length are positioned near the shower max. These filaments are optically connected via quartz rods of 1.0 mm diameter to photosensors positioned at upstream and downstream ends, should rad-hard devices be available. In both cases, the signals from the SiPMs are amplified at high gain for timing measurement and at low gain for local energy measurement at shower max.

Figure 2 show a schematic of the upstream face of a RADiCAL module, as a particle beam would see it as it enters the module. Indicated are the transverse placement locations of four WLS capillaries/filaments, which can be configured in different ways to measure energy, time and position. For shower position determination, the spatial localization of an EM shower is provided by the signal amplitudes of the energy measurements from the capillaries. Note that at shower max, the EM shower radius is significantly smaller than the Molière radius (R_M = 13.7 mm) and is given approximately by the radiation length of the structure (X_0~4.5 mm) [5]. Capitalizing on this, the shower position can be localized within a module to within a few mm, beneficial for event reconstruction under high pileup conditions in endcap and forward regions of experiments and for distinguishing nearby showers from decays of highly boosted objects.

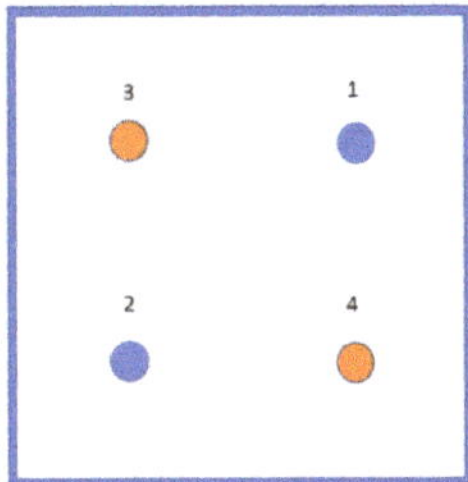

Figure 2. Schematic of capillary placement in a RADiCAL module as seen from the beam entry (upstream) face. In the arrangement shown, positions 1 and 2 are for T-type capillaries; positions 3 and 4 are for E-type capillaries; however, all four capillaries could be used for energy measurement (E-type) or for timing measurement (T-type), depending upon experimental preference.

3. Discussion of Experimental Results

For the CMS Endcap Upgrade Down Select Process in preparation for HL-LHC (held in 2015), a 4 × 4 array of such RADiCAL modules was tested in the CERN H4 beamline with electrons of energies 20 GeV < E < 200 GeV. For that specific test, all 16 of the modules were instrumented with E-type capillaries only. More recently, at the Fermilab Test Beam Facility (FTBF), using beam electrons of energy 12 GeV < E < 28 GeV, in December 2021 and June 2022, tests were carried out with a single RADiCAL module instrumented with two E-type and two T-type capillaries and a single RADiCAL module instrumented with four T-type capillaries. Data analysis from these recent tests is currently in progress.

3.1. Experimental Measurement of Energy Resolution

For the beam tests of a 4 × 4 array carried out at CERN in the H4 beamline, the active volume of the array was 56 mm × 56 mm × 114 mm, and there were 64 independent readout channels for this structure, with 4 E-Type DSB1 WLS capillaries per module, each capillary connected by a clear fiber waveguide to its own individual HPK SiPM photosensor having 15 μm square pixels. Figure 3 show the structure of the array, and Figure 4 display the measured energy resolution. In these studies, the (rad-hard) DSB1 WLS capillaries were compared with (non-rad-hard) 0.94 mm diameter Y11 WLS fibers indicating comparable results.

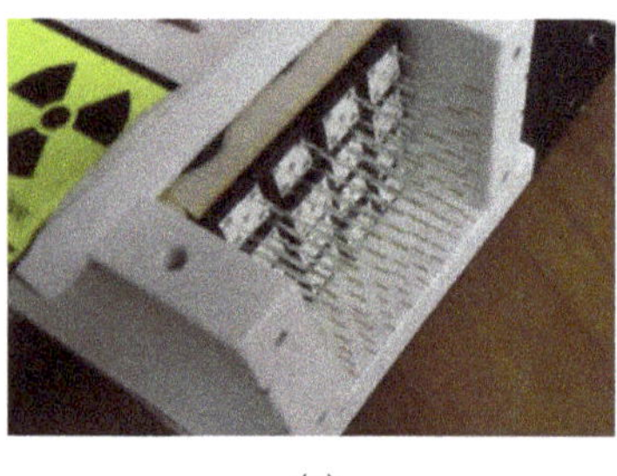

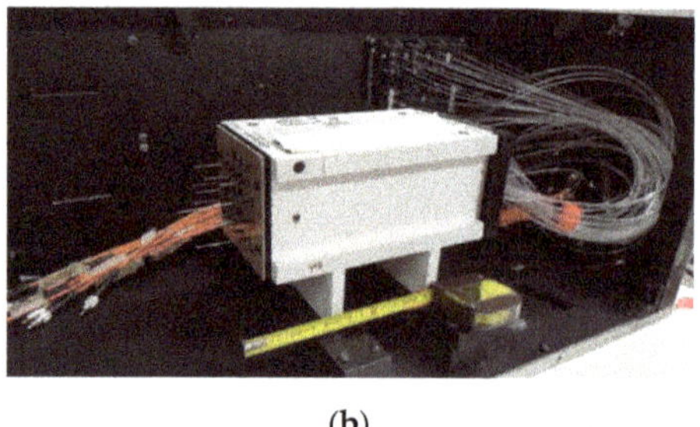

(**a**) (**b**)

Figure 3. (**a**) The 4 × 4 modular array during assembly. (**b**) Fiber optic waveguides used to transmit the WLS light to SiPM photosensors positioned outside of the beam region for this specific test. In the picture, the incoming electron beam is incident from the left on the 4 × 4 modular array enclosed within the white mechanical housing.

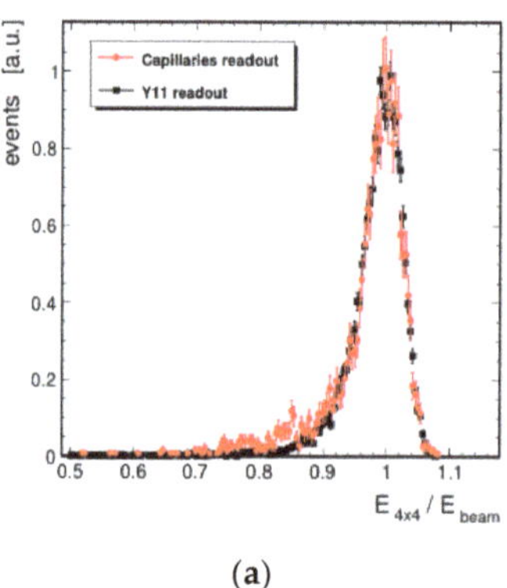

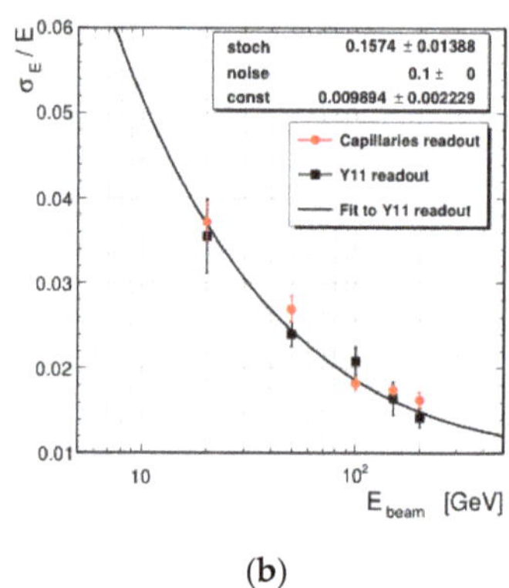

(**a**) (**b**)

Figure 4. (**a**) Energy resolution for the 4 × 4 W/LYSO array using capillary WLS readout out (red) compared to Y11 double clad WLS fiber readout (black), as measured using an electron beam in the CERN H4 beam line. Electron beam energy is 100 GeV. (**b**) Energy resolution as a function of beam energy for the 4 × 4 W/LYSO array. These results indicate performance leading to a 1% constant term.

The energy resolution achieved was: $\sigma_E/E = 15.7\%/\sqrt{E} \oplus 0.1/E \oplus 1\%$. The desired constant term and stochastic term indicated in Section 1 above can be improved by increasing the sampling fraction (the thickness of the LYSO:Ce scintillation tiles), by improved coupling between the scintillation tiles and the wave shifting capillaries and by slightly extending the length of the module (number of W and LYSO:Ce layers in the module).

3.2. Expectations for Timing and Spatial Resolution

A GEANT4 simulation has been carried out for a RADiCAL module of the type shown in Figure 1b [4,5] and guided and qualified by the measurements described in Section 3.2 above. A WLS capillary (T-type) is assumed to be inserted into the center of the module with the DSB1 WLS filament located in the region of EM shower maximum for 50 GeV electron showers. In this study, the simulation assumed only a single SiPM for readout, which is positioned at the downstream end of the WLS timing capillary, while in reality, timing measurement is possible with SiPM placed at both upstream and downstream ends of such capillaries.

The shower max timing signal is derived from a region of very small transverse size $r \sim X_0$ (see Figure 5a), a region whose radius is significantly smaller than the Molière radius.

In this small region, there are ~100 charged (shower) particles, clearly distinguishing EM signals from mip signals due to charged hadrons and providing a large and time-localized optical pulse. The timing resolution will be dominated by the rise time of this signal and the detected light yield within the first nanosecond of the optical pulse. Figure 5b indicate that, in simulation, timing resolutions of 30 ps $< \sigma_\tau <$ 50 ps could be achievable with RADiCAL modules. By reading out the light from both upstream and downstream ends of timing capillaries and using several timing capillaries per module, it is the objective of the ongoing beam tests at the Fermilab FTBF to verify the actual achievable timing performance.

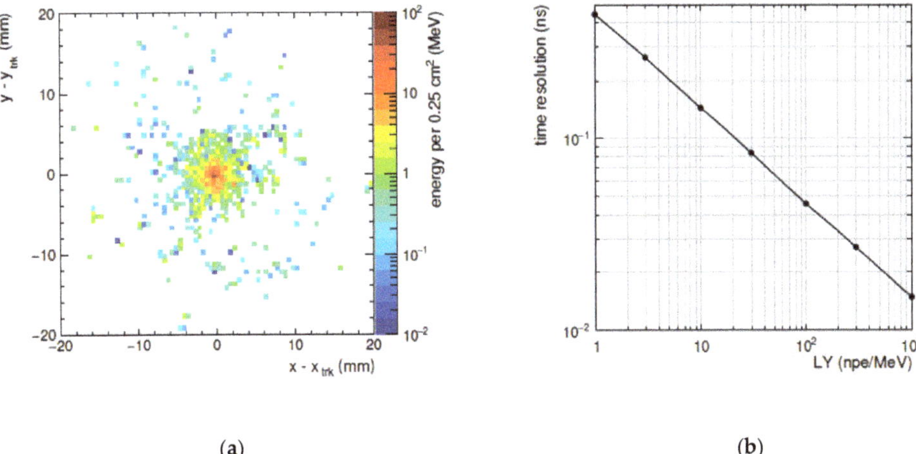

(a)   (b)

Figure 5. (**a**) Location of the energy in a RADiCAL module at shower max for 50GeV electrons in GEANT4 simulation. The module itself occupies a square region of 14×14 mm^2 at the center of the plot. (**b**) Timing resolution vs. detected light yield in photoelectrons per MeV, simulated for a 50 GeV electron shower. Downstream readout only in this GEANT4 simulation study.

4. Conclusions

The pattern recognition power and the potential for high-resolution measurement of both timing and energy of EM objects (electrons, positrons and gammas) in arrays of RADiCAL modules is a potentially promising technique for EM Calorimetry in future high-luminosity hadron collider experiments such as the FCC-hh. The energy resolution of a 4×4 array of RADiCAL modules was measured in a high-energy electron beam at CERN and found to be $\sigma_E/E = 15.7\%/\sqrt{E} \oplus 0.1/E \oplus 1\%$. This performance can be improved toward the desired resolution for FCC-hh by several methods: for the stochastic term, by improving the sampling fraction by using thicker scintillation tiles and further optimization of the optical coupling between the scintillating tiles and the wave shifters in the capillaries by increasing the thickness of the shifter; for the constant term, by increasing the overall module length toward 29 X_0 by increasing the number of tungsten and LYSO:Ce layers. When assessing the timing resolution of RADiCAL modules, measuring the timing at the shower max represents new territory under study. Currently underway are beam measurements of the timing resolution and spatial precision for EM showers at the Fermilab FTBF to verify how closely the RADiCAL module conforms to expectations from the GEANT4 simulation.

Author Contributions: Conceptualization and Methodology—B.C., A.L., Y.O., C.P.-L., R.R. and R.-Y.Z.; Validation: T.A., T.B., B.C., P.D., K.F., C.H., R.R., D.R., D.S., M.V., Y.W., L.Z. and R.-Y.Z.; Software: C.P.-L. and A.L.; Formal Analysis: M.D., M.H., O.K.-K., Y.W. and J.W.; Data Curation—C.J.; Project Administration—M.W.; Funding Acquisition—B.C., C.J., Y.O., R.R., M.W. and R.-Y.Z. All authors have read and agreed to the published version of the manuscript.

Funding: Research has been supported in part by the US Department of Energy under grant DE-SC0017810, the US National Science Foundation under grant NSF-PHY-1914059, the University of Notre Dame Resilience and Recovery Grant Program and by QuarkNet for High School Teacher and Student support.

Data Availability Statement: Data are not yet available as beam testing and data collection and analysis are currently in progress.

Acknowledgments: We thank the University of Notre Dame Radiation Laboratory Glass shop for capillary fabrication and the staff of the Fermilab Test Beam Facility for their support during the recent beam tests.

Conflicts of Interest: The authors declare no conflict of interest.

References

1. Fleming, B. Basic Research Needs for High Energy Physics Detector Research & Development. Available online: https://science.osti.gov/-/media/hep/pdf/Reports/2020/DOE_Basic_Research_Needs_Study_on_High_Energy_Physics.pdf (accessed on 24 December 2019).
2. Anderson, T.; Barbera, T.; Blend, D.; Chigurupati, N.; Cox, B.; Debbins, P.; Dubnowski, M.; Herrmann, M.; Hu, C.; Ford, K.; et al. RADiCAL—Precision-timing, Ultracompact, Radiation-hard Electromagnetic Calorimetry. *arXiv* **2022**, arXiv:2203.12806.
3. Aleksa, M.; Allport, P.; Bosley, R.; Faltova, J.; Gentil, J.; Goncalo, R.; Helsens, C.; Henriques, A.; Karyukhin, A.; Kiesseler, J.; et al. Calorimeters for the FCC-hh. *arXiv* **2019**, arXiv:1912.09962.
4. Ledovskoy, A. *Shashlik Timing*; Report to the RADiCAL Group: Geneva, Switerland, 2021. Available online: https://notredame.box.com/s/vjpbvpxn2sff72y6y4wvue3nnhcucgtx (accessed on 17 January 2021).
5. Ledovskoy, A. *RADiCAL Studies*; Report to the RADiCAL Group: Geneva, Switzerland, 2021. Available online: https://notredame.box.com/s/p9afzt9hxp2yg22vmoycfir8fnnt7tzk (accessed on 12 September 2021).
6. Hu, C.; Li, J.; Yang, F.; Jiang, B.; Zhang, L.; Zhu, R.-Y. LuAG ceramic scintillators for future HEP experiments. *Nucl. Instrum. Methods Phys. Res. Sect. A Accel. Spectrometers Detect. Assoc. Equip.* **2020**, *954*, 161723. [CrossRef]
7. Hu, C.; Lu, N.; Zhang, L.; Zhu, R.-Y.; Bornheim, A.; Narvaez, L.; Trevor, J.; Spiropulu, M. Gamma-ray and Neutrno-Induced Photocurrent and Readout Noise in LYSO+SiPM Packages. *IEEE Trans. Nucl. Sci.* **2021**, *68*, 1244–1250. [CrossRef]
8. Hu, C.; Yang, F.; Zhang, L.; Shu, R.-Y.; Kapustinsky, J.; Nelson, R.; Wang, Z. Proton-Induced Radiation Damage in LYSO and BaF$_2$ Crystal Scintillators. *IEEE Trans. Nucl. Sci.* **2018**, *65*, 1018–1024. [CrossRef]
9. Hu, C.; Yang, F.; Zhang, L.; Zhu, R.-Y.; Kapustinsky, J.; Nelson, R.; Wang, Z. Neutron-Induced Radiation Damage in LYSO, BaF$_2$ and PWO Crystals. *IEEE Trans. Nucl. Sci.* **2020**, *67*, 1086–1092. [CrossRef]

Article

Development of the ATLAS Liquid Argon Calorimeter Readout Electronics and Machine Learning for the HL-LHC

Julia Gonski and on behalf of the ATLAS Liquid Argon Calorimeter Group

Nevis Laboratories, Columbia University, Irvington, NY 10533, USA; julia.gonski@cern.ch

Abstract: The High Luminosity era of the Large Hadron Collider (LHC) starting in 2029 promises exciting discovery potential, giving unprecedented sensitivity to key new physics models and precise characterization of the Higgs boson. In order to maintain current performance in this challenging environment, the ATLAS liquid argon electromagnetic calorimeter will get entirely new electronics that reads out the entire detector with full precision at the LHC frequency of 40 MHz, and provides high granularity trigger information, while withstanding high operational radiation doses. New results will be presented from both front-end and off-detector component development, along with highlights from machine learning applications. The future steps and outlook of the project will be discussed, with an eye towards installation in the ATLAS cavern beginning in 2026.

Keywords: ATLAS; LAr; upgrade; calorimeter; readout; electronics

Citation: Gonski, J.; on behalf of the ATLAS Liquid Argon Calorimeter Group Development of the ATLAS Liquid Argon Calorimeter Readout Electronics and Machine Learning for the HL-LHC. *Instruments* **2022**, *6*, 28. https://doi.org/10.3390/instruments6030028

Academic Editors: Fabrizio Salvatore, Alessandro Cerri, Antonella De Santo and Iacopo Vivarelli

Received: 28 July 2022
Accepted: 23 August 2022
Published: 26 August 2022

Publisher's Note: MDPI stays neutral with regard to jurisdictional claims in published maps and institutional affiliations.

1. Introduction

The ATLAS liquid argon (LAr) calorimeter [1,2] measures the energy and timing of photons, electrons, and hadrons that are produced by proton-proton collisions in the Large Hadron Collider (LHC). It is a sampling calorimeter with 182,468 cells in an accordion geometry of active (LAr) and absorber (lead) material, segmented into three longitudinal layers in the barrel with dedicated endcap detectors to cover the high $|\eta|$ range. Figure 1 shows a schematic view of the LAr calorimeter in the ATLAS detector, along with a diagram of the geometry and dimensions of a calorimeter slice.

The LAr calorimeter readout electronics system samples the cells at the LHC bunch crossing (BC) frequency of 40 MHz, and sends a digitized pulse off the detector for signal analysis and triggering. This readout system is separated into on- and off-detector components. The front-end board (FEB) is located directly on the cryostat that provides cooling for the LAr cells, in order to optimize the analog performance of the electronics. It is therefore subject to an environment with high radiation doses, a high magnetic field, and limited access during run periods, presenting a variety of challenges in the electronics design. Signals from the FEB are sent out of the ATLAS cavern to the off-detector electronics, which apply digital filtering to extract energy and time for each cell and pass salient information to the trigger and data acquisition systems.

The LHC is scheduled to undergo an upgrade beginning in 2029 to deliver a higher instantaneous luminosity up to 7.5×10^{34} cm^{-2} s^{-1}, with approximately 200 simultaneous collisions expected in each BC. This leads to an increase in pileup, which refers to energy deposits from other simultaneous collisions or collisions in the adjacent bunch crossings. Two kinds of pileup can affect the measurement of LAr signals. In-time pileup comes from overlaid proton collisions within the same BC. Out-of-time pileup consists of energy that is leftover in the calorimeter from previous BCs, which accumulates because the LAr pulse signal takes approximately 25 BCs to read out.

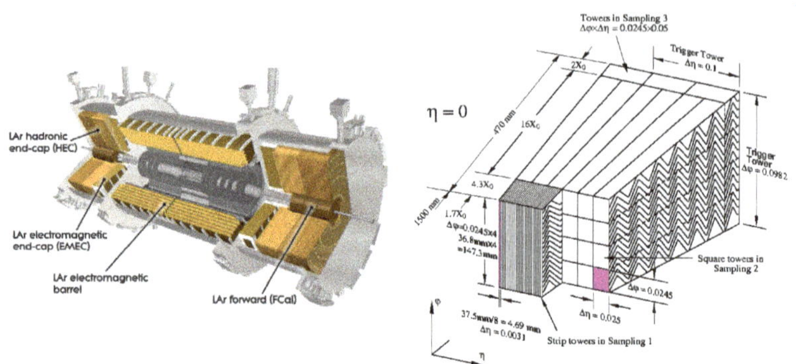

Figure 1. Schematic view of the different components of the ATLAS LAr calorimeter system (**left**), and a cut-out view of the calorimeter in the barrel including dimensions and coordinates (**right**) [1,2].

The density of detector signals in this High Luminosity LHC (HL-LHC) will present new challenges for the subsystems, which will need to accommodate higher trigger rates and radiation doses. While the LAr cells will continue to perform within required specifications throughout the nearly ten year lifetime of the HL-LHC, the readout must be fully re-designed and replaced (the only exception is the cold pre-amplification and summing circuit system of the hadronic endcap, which will remain unchanged from its current state.) to continue delivering high-quality LAr data. This amounts to 1524 FEBs, 122 calibration boards, and all off-detector electronics. The new LAr readout will provide information from the entire calorimeter at full precision for more powerful trigger decisions. It has been organized into two stages: Phase-I, which has already been installed in the ATLAS cavern and is being commissioned for Run 3 of the LHC, and Phase-II, which is scheduled for installation in 2026 [3].

Design choices for the new LAr readout are motivated by key physics drivers of the HL-LHC physics program. Probing the TeV mass scale for new particles means that the calorimeter must be capable of providing precise measurements for very high energy electromagnetic decay byproducts. Further, the goal of better characterizing the Higgs boson requires excellent reconstructed mass resolution of the $H \to \gamma\gamma$ process. Specifically, the new readout must ensure that photons from $H \to \gamma\gamma$ are mostly digitized on HIGH gain. This modifies the existing readout scheme where photons coming from a Higgs have a typical energy range at the value where the gain scale switches from MEDIUM to HIGH, thereby minimizing the gain inter-calibration systematic on the Higgs mass measurement. This leads to a readout design with only two gain scales (rather than the current three gain scales implemented in Run 2) [3].

The upgraded ATLAS trigger scheme must also be taken into consideration for the LAr readout design, as the off-detector electronics are responsible for providing LAr information to the trigger. The increase in trigger rate and latency for the HL-LHC motivates the adoption of a free-running all digital design for the LAr readout with no on-detector pipeline. The full data rate for the LAr calorimeter corresponds to approximately 350 Tbps. This new scheme allows for a sharper trigger turn-on to the efficiency plateau, while maintaining the ability to trigger on low momentum objects.

2. HL-LHC Readout Components

Figure 2 shows a block diagram of the full LAr calorimeter readout scheme for ATLAS in the HL-LHC. The physics goals of the experiment dictate specifications for the system. The readout must be able to handle a wide range of energies that may be deposited in a single cell during collisions, bounded by approximately 50 MeV at the lower end from electronic noise and reaching a maximum of approximately 3 TeV from electrons or photons produced in the decay of a new particle with mass $\mathcal{O}(10)$ TeV. Therefore, the readout must have a high dynamic range, in this case 16-bits with 11-bit precision. This is implemented in two overlapping 14-bit gain scales. The electronic noise must be less than the typical energy deposited by a minimum ionizing particle passing through the LAr calorimeter. Stringent nonlinearity requirements are imposed to ensure accurate measurements, specifically <0.1% for cell energies up to approximately 300 GeV. Finally, as the electronics will not be replaced throughout the HL-LHC operating period, they must remain performant over the full expected radiation dose, corresponding to a total ionizing dose of 1400 Gy (safety factor 1.5), and a non-ionizing energy loss of $<4.1 \times 10^{13}$ neutron equivalent per cm^2 (safety factor 2).

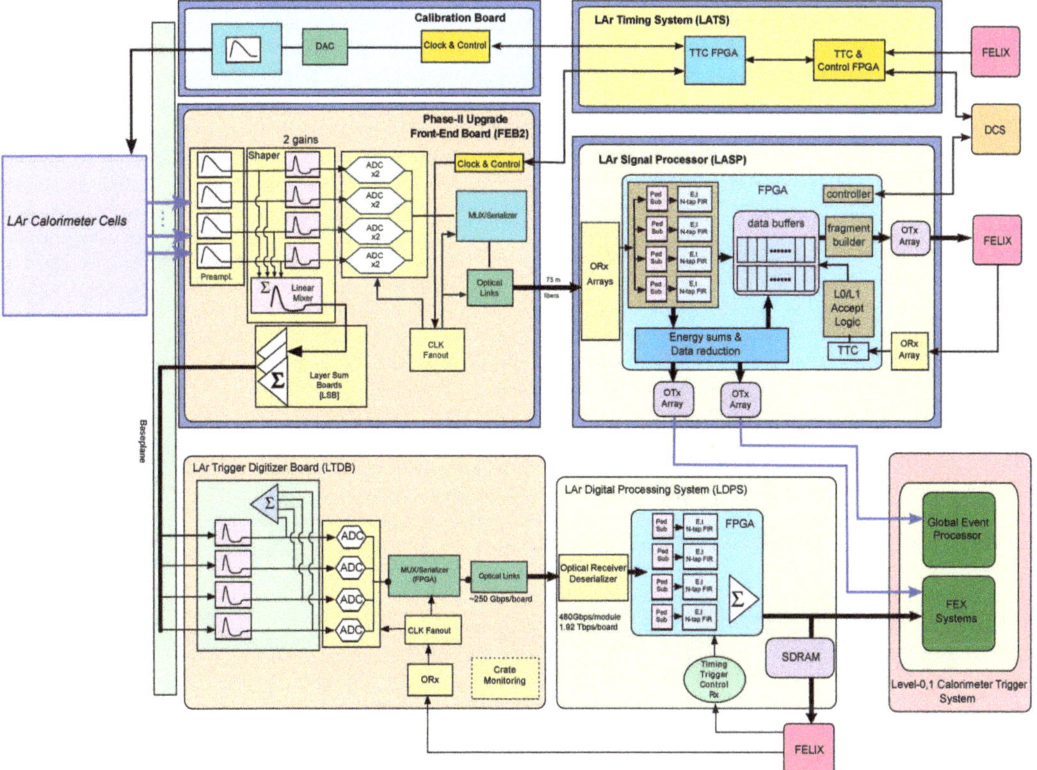

Figure 2. Block diagram of the LAr readout in the HL-LHC, including all on- and off-detector electronics.

2.1. Front End

The front end on-detector LAr electronics comprises the 2nd generation FEBs (FEB2) and the calibration boards. Each FEB2 and calibration board will have 128 data channels, and utilize the I2C configuration protocol. The unique requirements for the readout lead to the development of four full-custom application specific integrated circuits (ASICs) to be used on these boards. These are the preamplifier/shaper (PA/S), analog–digital converter (ADC), and calibration chips CLAROC and LADOC, each of which is described in detail below.

2.1.1. Preamplifier/Shaper

The PA/S chip is the first step in the readout chain after data leaves the LAr calorimeter cell. It performs analog processing on signals, namely amplification, splitting into two gain scales, and the application of a bipolar CR-(RC)^2 shaping function to create the desired LAr pulse shape. This generates differential outputs that are passed to both the next element in the front-end readout chain, as well as to the L0 trigger.

The candidate chosen for the PA/S on the FEB2 is the *ALFE*, which is the prototype custom ASIC built in 130 nm CMOS TSMC technology with 4 channels per ASIC. It has a tuneable input impedance to match the varying cell size across the calorimeter, as well as tuneable time constants for the shaping function, and can perform four channel summing for the hardware trigger. Figure 3 shows an image of the ALFE2 chip die, with specific circuit components highlighted, along with a picture of the testboard uses to make performance measurements. ALFE testing has resulted in an integral non-linearity <0.1%, as well as very low equivalent noise input of <150 nA and low crosstalk (<20 mV for the 50 Ω input impedance configuration). Radiation testing also revealed that the ALFE maintains good performance after a 12 kGy dose, when only a 1.4 kGy dose is anticipated. As this prototype version of the PA/S chip is well within the necessary specifications, it will be re-packaged in a ball gate array (BGA) for the FEB2 prototype. Pre-production of ASICs can also begin, in preparation for an ultimate yield of approximately 80 thousand chips total (including spares).

Figure 3. Die image of the PA/S ALFE2 pre-prototype ASIC with key circuit elements outlined (**left**), and the corresponding ALFE2 testboard with soldered chip (**right**).

2.1.2. Analog-Digital Converter

The PA/S passes differential signals to the ADC chip, which digitizes the incoming data at the LHC clock frequency of 40 MHz. The ASIC used for the HL-LHC is the eight channel *COLUTA* chip in 65 nm CMOS. The required 14-bit dynamic range is achieved by a 3-bit multiplying digital-analog converter (DAC) followed by 12-bit Successive Approximation Register (SAR), along with a Digital Data Processing Unit (DDPU) that applies calibration bit weights and serially transmits the digitized data. Figure 4 shows an image of the die for the prototype version, CV4, and the corresponding testboard for performance measurements.

CV4 delivers 1.2 ADC counts of noise on the pedestal, and an effective number of bits (ENOB) of 11.8 (11.5) for sine waves of 5 (8) MHz carrier frequencies. This exceeds the requirement of 11 ENOBs across the full dynamic range, and the CV4 also meets other specifications on nonlinearity and radiation tolerance. As with the ALFE, the CV4 will also be packaged in BGA for placement on the FEB2 prototype, and pre-production is scheduled to begin shortly.

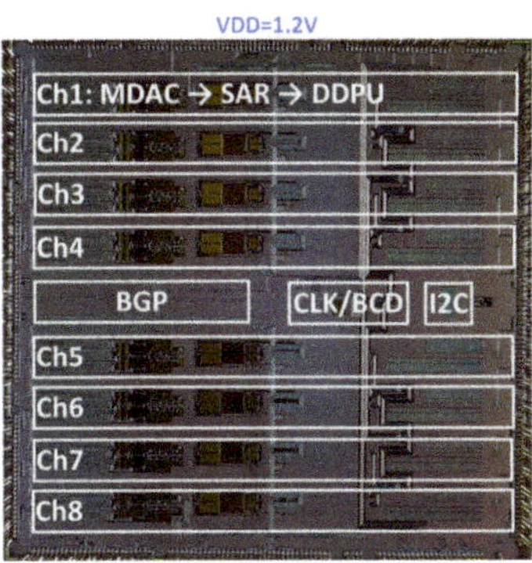

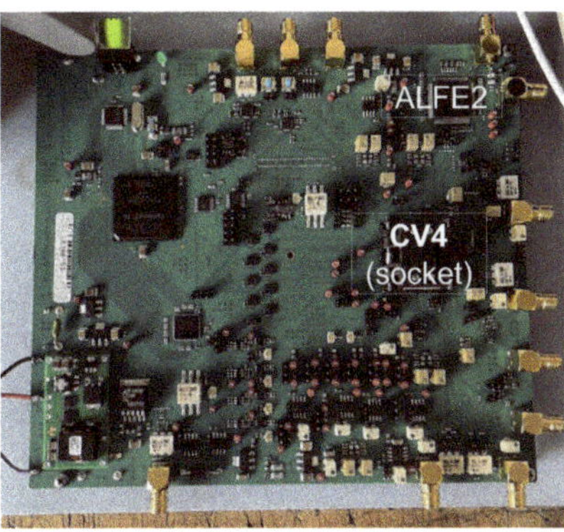

Figure 4. Die image of the ADC CV4 pre-prototype ASIC with key circuit elements outlined (**left**), and the corresponding CV4 testboard with socketed chip (**right**).

2.1.3. FEB2 Pre-Prototype

The integration of the FEB2 custom electronics into the full readout chain is tested with the pre-prototype of the FEB2. This so-called "slice" testboard has only 32 of 128 channels instrumented, enabling performance measurements of LAr pulses propagated through the full readout chain, as well as coherent noise and clock/configuration testing. Figure 5 shows a diagram of the data flow on the slice testboard, and an image of the board in its test setup. The slice testboard allowed for full validation of the slow control, monitoring, and redundancy of the bidirectional clock and control links, ensuring a robust configuration protocol and clock distribution.

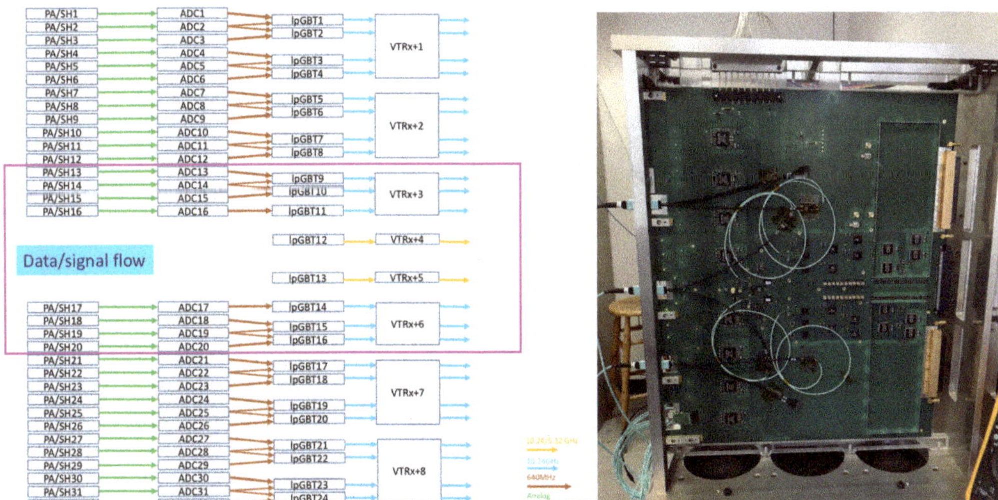

Figure 5. Data flow diagram of LAr signals on the FEB2, with the 32 channel pre-prototype slice testboard highlighted in magenta (**left**), and an image of the slice testboard (**right**).

To assess the reconstruction capability of the slice testboard, energy and timing measurements of each LAr pulse are computed using optimal filtering coefficients (OFCs) [4]. These are computed using precision knowledge of the LAr pulse shape, allowing for robustness against distortion due to pileup. The OFCs are applied to samples from four points on the signal waveform, separated by 25 ns as dictated by the ADC frequency, and the energy and timing of the pulse can be calculated from linear combinations of the OFCs and sample values. Figure 6 shows pulses read out by the slice testboard at a variety of amplitudes, along with the obtained resolution on the energy measurement over the full dynamic range of the system. For the highest energy pulses, where the energy resolution is expected to be the best, the resolution defined as σ_E/E reaches approximately 0.02%, well below the specification of 0.25%. The timing resolution for these large pulses is approximately 50 ps, which is dominated by the clock jitter on the board. Multi-channel performance can also be studied, and the measured coherent noise was found to be less than a few percent of the total pulse size, with a low cross talk well below the percent level.

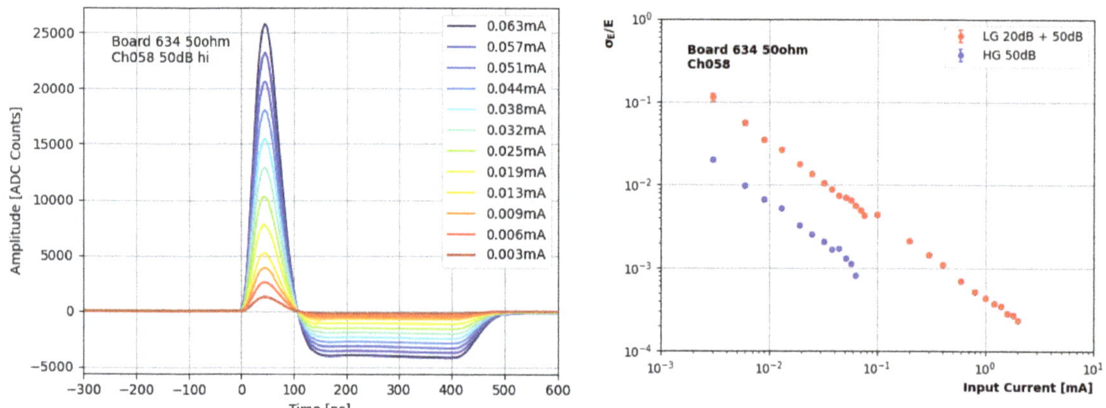

Figure 6. Plot of reconstructed LAr pulses from the slice testboard across the full dynamic range (**left**), and the corresponding energy resolution σ_E/E as a function of input current (**right**).

The next steps in FEB2 development include the design of the 128-channel prototype board, at which point tests can be performed with the proposed power distribution and on-detector crate. Production will consist of 1627 boards including spares.

2.1.4. Calibration System

For maximal performance, the readout system must be calibrated with the injection of a precise calibration pulse. Key specifications for this system include an integral non-linearity <0.1% in the FEB2 high gain, <0.2% in the intermediate range (end of high gain up to 250 mA), and <1% in the high current range (250–300 mA), along with a uniformity <0.25%. The circuit that generates the calibration pulse contains two custom ASICs. The CLAROC creates the pulse by opening a high-frequency switch; as the requirements on pulse size mean that this chip needs a 7.5 V power source, this chip is designed in 180 nm XFAB technology. The LADOC is a custom 16-bit DAC that is used to command the switch with built-in 130 nm TSMC technology. The test setup with these ASICs indicates that they meet nearly all required specifications, with the exception of linearity and radiation hardness, which are anticipated to be resolved in prototype versions. A 32 channel pre-prototype board CABANON, analogous to the slice testboard, measured cross-talk <0.1% of the signal, in line with specifications, and allowed for a power distribution test comparing two different DC/DC convertor candidates. An image of this pre-prototype is given in Figure 7. Next generation versions of the CLAROC and LADOC have been designed to overcome non-linearity and radiation hardness issues, and are currently being fabricated. These chips will be packaged in BGA as with the FEB2 ASICs, and a 128-channel prototype design will be forthcoming, harmonizing the development of both front-end system boards.

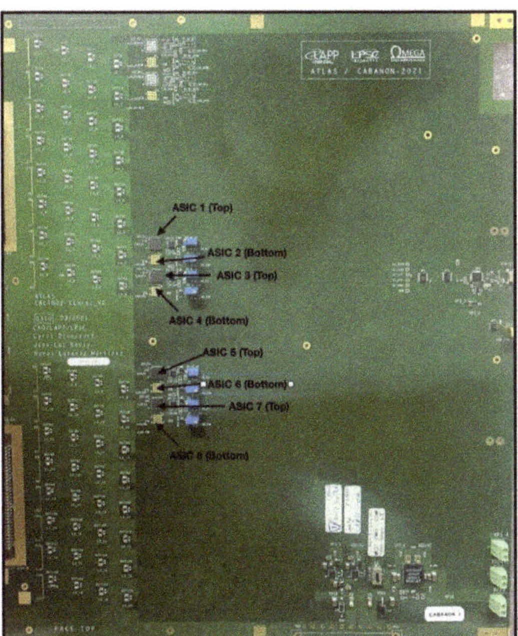

Figure 7. Image of the CABANON pre-prototype test calibration board, with ASIC locations highlighted.

2.2. Off Detector

The back-end electronics are located off the detector in the USA15 counting room, with no radiation from collisions. They consist of two systems: the LAr Timing System (LATS), and the LAr Signal Processor (LASP). The FEB2 boards connect to the off-detector electronics via optical links driven by lpGBT serializers.

2.2.1. Timing System

The LATS is responsible for the trigger, timing, and control (TTC) distribution to the front-end, configuration based on the lpGBT protocol, and monitoring of all 1524 FEB2 and 122 calibration boards, requiring a total of 3192 links. This is performed by the LATOURNETT board, which consists of 13 field programmable gate arrays (FPGAs). 12 of these are matrix FPGAs with both transmitters and receivers, each of which receives data from 1 FEB2, and the last FPGA has a centralized control role. The firmware for both of the FPGAs has been validated in simulation, and the power-up sequence has been verified with a dedicated testboard. Further testing will continue in parallel to the submission of a prototype board design, followed by integration tests with the on-detector electronics.

2.2.2. Signal Processor

The LASP applies digital filtering to digitized waveforms received from FEB2, in order to calculate the energy and time of the pulse. It then transmits this high-level information to both the trigger and DAQ at 25 Gbps. This operation must be done online in two FPGAs on the LASP Main Blade combined with a Smart Rear Transition Module (SRTM). Between 200 and 300 of each board will be produced, with 6-8 FEB2 boards per blade.

A testboard has been produced with full capability, allowing for the validation of power sequencing, I2C configuration sensors, clock distribution, and FPGA configuration. A photograph of the test setup, with the LASP Main Blade and SRTM, is given in Figure 8. A complex firmware design has been produced that is highly modular, allowing for flexible integration of the number of FEB2 boards read by each FPGA. Measurements with these first testboards will inform the prototype design, particularly on the choice of FPGA, due to limitations on resource usage and a tight power budget.

Figure 8. Image of the LASP Main Blade and the STRM, with the SRTM outlined in yellow.

3. Machine Learning Highlights on FPGAs

Signal processing on the LASP can benefit from machine learning (ML) techniques applied online at the FPGA level. Such advances are motivated by the increased pileup, which can degrade the energy and timing resolution as well as the performance of the trigger. Studies into the application of ML to mitigate pileup degradation utilize simulated digital pulses in a bunch train, which are used as input to three different architectures. The implementation described here achieves 37 neural nets (NNs) on one FPGA for a LASP running at 400 MHz, processing 10 channels each, with further optimizations ongoing [5].

The first architecture investigated is a convolution neural net (CNN), which has two separate phases for the tasks of pulse tagging and energy reconstruction. The first tagging layer is trained to detect energy deposits 3σ above the electronic noise (240 MeV) using pulse samples for eight bunch crossings. The output of this tagging layer is a detection probability, which is passed along with the original sample sequence to the energy reconstruction layer, trained to reconstruct the amount of energy deposited in each cell. Parameters and architecture of the CNN are optimized based on high efficiency for detecting large energy deposits, high rejection of background signals, and good energy resolution.

Two sequence modeling architectures are also considered, a vanilla recurrent neural net (RNN) and long short-term memory (LSTM). These allow the modeling of data as a sequence of BCs, providing a better characterization particularly of out-of-time pileup. The LSTM demonstrates superior management of information through long sequences but its complexity means that strict limits on network size must be imposed for an FPGA implementation. The LSTM is used with both single BC inputs and a sliding window technique which incorporates up to four inputs from the current pulse and one from the previous pulse. The vanilla RNN can only be used with sliding window inputs as it does not have enough complexity to converge in a single cell.

Figure 9 shows a comparison of the performance of the various ML architectures, where the difference between the true and reconstructed transverse cell energy is compared for all methods. For reference, the analytical OFC method is included. All ML-based reconstructions outperform the legacy method in both accuracy and precision. Shown also is the performance comparison between NNs implemented via software-based calculations and via FPGA firmware, given as the relative difference in energy reconstruction between VHDL and Keras simulations. Very good agreement is observed between FPGA and software algorithms is observed, indicating that an NN-based reconstruction is capable of processing LAr signals in the HL-LHC readout and clearing the way for further development of these techniques in the LASP.

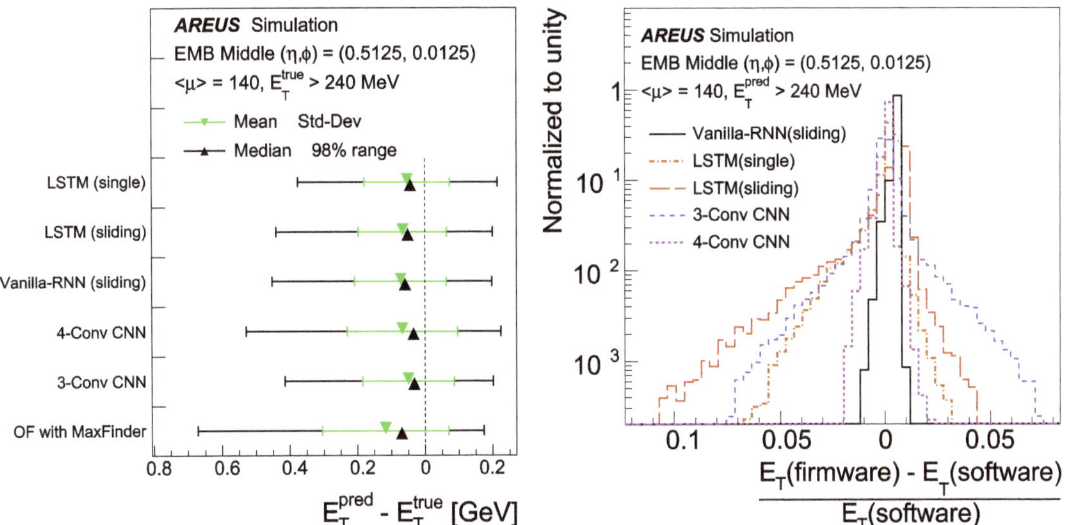

Figure 9. Performance of ML signal processing methods implemented in FPGAs on the LASP, specifically the true vs. reconstruction transverse energy for all ML and the legacy methods (**left**), and the relative energy deviation of firmware ML implementations from the software results (**right**) [5].

4. Conclusions

A status report is presented on the upgrade of the ATLAS LAr calorimeter for the HL-LHC era. The specifications of the upgrade are motivated by physics drivers for future LHC runs, leading to a free-running architecture that reads out the entire LAr calorimeter with full precision at the LHC clock of 40 MHz. Updates are given for all on- and off-detector components, namely the PA/S and ADC front-end ASICs, the pre-prototype FEB2 and calibration boards, the LATS, and the LASP. Comprehensive and performant results are presented utilizing pre-prototypes of key custom chips and boards. In the coming years, the final prototype design of all components must be designed, produced, and shown to continue meeting specifications in performance testing. Production and integration tests will also ramp up in preparation for the installation of all upgraded electronics into the ATLAS cavern starting in 2026.

Funding: The author is supported by the National Science Foundation under Grant No. PHY-2013070.

Data Availability Statement: Not applicable.

Conflicts of Interest: The authors declare no conflict of interest.

References

1. ATLAS Collaboration. The ATLAS Experiment at the CERN Large Hadron Collider, JINST 3 S08003. 2008. Available online: https://iopscience.iop.org/article/10.1088/1748-0221/3/08/S08003/meta (accessed on 27 July 2022).
2. ATLAS Collaboration. ATLAS Liquid Argon Calorimeter: Technical Design Report, CERN-LHCC-96-41. 1996. Available online: https://cds.cern.ch/record/331061 (accessed on 27 July 2022).
3. ATLAS Collaboration. ATLAS Liquid Argon Calorimeter Phase-II Upgrade: Technical Design Report, CERN-LHCC-2017-018, ATLAS-TDR-027. 2017. Available online: https://cds.cern.ch/record/2285582 (accessed on 27 July 2022).

4. Cleland, W.E.; Stern, E.G. Signal Processing Considerations for Liquid Ionization Calorimeters in a High Rate Environment. *Nucl. Instrum. Methods Phys. Res. A: Accel. Spectrom. Detect. Assoc. Equip.* **1994**, *338*, 467–497. [CrossRef]
5. Aad, G.; Berthold, A.-S.; Calvet, T.; Chiedde, N.; Fortin, E.M.; Fritzsche, N.; Hentges, R.; Laatu, L.A.O.; Monnier, E.; Straessner, A.; et al. Artificial Neural Networks on FPGAs for Real-Time Energy Reconstruction of the ATLAS LAr Calorimeters, ATL-LARG-PROC-2021-001. 2021. Available online: https://cds.cern.ch/record/2775033 (accessed on 27 July 2022).

34

Article

Upgrade of the CMS Barrel Electromagnetic Calorimeter for the High Luminosity LHC

Charlotte Cooke [1,2] on behalf of the CMS Collaboration

1 School of Physics, University of Bristol, Bristol BS8 1TH, UK; charlotte.cooke@bristol.ac.uk
2 Rutherford Appleton Laboratory, Didcot OX11 0QX, UK

Abstract: The high luminosity upgrade of the LHC (HL-LHC) at CERN will provide unprecedented instantaneous and integrated luminosities of up to 7.5×10^{34} cm^{-2}s^{-1} and 4500 fb^{-1}, respectively, from 2029 onwards. To cope with the extreme conditions of up to 200 collisions per bunch crossing, and increased data rates, the on- and off-detector electronics of the CMS electromagnetic calorimeter (ECAL) will be replaced. A dual gain trans-impedance amplifier and an ASIC providing two 160 MHz ADC channels, gain selection, and data compression will be used. The lead tungstate crystals and avalanche photodiodes (APDs) in the current ECAL will keep performing well and will therefore be maintained. The noise increase in the APDs, due to radiation-induced dark currents, will be minimised by reducing the ECAL operating temperature from 18 °C to around 9 °C. Prototype HL-LHC electronics have been tested and have shown promising results. In two test beam periods using the CERN SPS H4 beamline and an electron beam, the new electronics achieved the target energy resolution and a timing resolution consistent that is consistent with our requirements of 30 ps timing for energies greater than 50 GeV.

Keywords: electromagnetic calorimeter; compact muon solenoid; high luminosity large hadron collider

Citation: Cooke, C., on behalf of the CMS Collaboration. Upgrade of the CMS Barrel Electromagnetic Calorimeter for the High Luminosity LHC. *Instruments* **2022**, *6*, 29. https://doi.org/10.3390/instruments6030029

Academic Editors: Fabrizio Salvatore, Alessandro Cerri, Antonella De Santo and Iacopo Vivarelli

Received: 29 July 2022
Accepted: 22 August 2022
Published: 27 August 2022

Publisher's Note: MDPI stays neutral with regard to jurisdictional claims in published maps and institutional affiliations.

1. Introduction

The CMS electromagnetic calorimeter (ECAL) is a lead tungstate (PbWO$_4$) crystal calorimeter which provides excellent energy resolution in the harsh radiation environment of the LHC [1]. Specifically, it has achieved a 1% mass resolution for the Higgs Boson in the $\gamma\gamma$ decay channel [2]. It is a hermetic and compact detector with coverage up to $|\eta| = 3$, split into barrel and endcap regions as shown in Figure 1. The barrel region contains 61,200 crystals across 36 supermodules and uses avalanche photodiodes (APDs) as photodetectors. The endcap region contains 14,648 crystals in four half disk dees and uses vacuum phototriodes. The barrel covers the region $|\eta| < 1.48$ while the endcap covers $1.48 < |\eta| < 3$.

The main objective of the high luminosity LHC (HL-LHC) is to deliver a much larger dataset for physics analysis to the LHC experiments. The CMS ECAL will have to maintain physics performance for an integrated luminosity of 4500 fb^{-1}, a peak luminosity of 7.5×10^{34} cm^{-2}s^{-1} and for 200 pileup interactions, which is the number of simultaneous proton–proton collisions per LHC bunch crossing. In Run 3 (the current data taking period), CMS is expected to reach an integrated luminosity of 30 fb^{-1}, a peak luminosity of 2.2×10^{34} cm^{-2}s^{-1} and up around 60 pileup. This has several implications. Firstly, the trigger latency will be increased from 4 μs to 12.5 μs, which will allow the use of tracks in the Level 1 (L1) trigger [3]. The L1 trigger rate will also be increased by a factor of 7.5, going from 100 kHz to 750 kHz, with a high level trigger (HLT) rate of up to 10 kHz. Event reconstruction will be more challenging with much higher in-time (due to collisions in the same bunch crossing) and out-of-time pileup (due to collisions in different bunch crossings). The detector will also be subject to more radiation, leading to more noise in the form of increased APD leakage current and crystal transparency loss. There will also be an increased rate of anomalous signals ("spikes"), which are caused by hadrons impacting directly on the APDs.

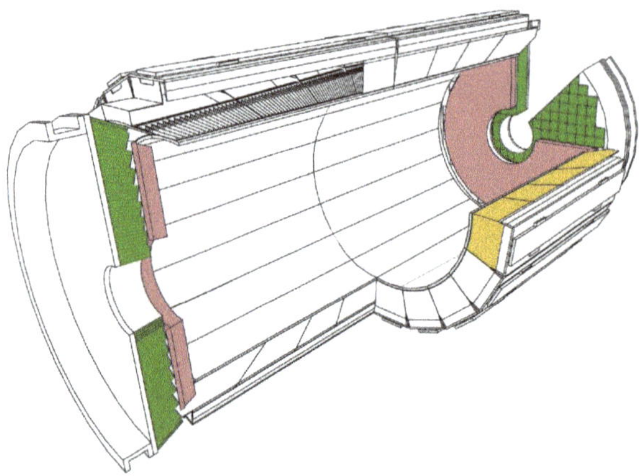

Figure 1. Layout of the CMS ECAL. One of the 36 barrel supermodules is highlighted in yellow, and the endcaps are highlighted in green.

2. Materials and Methods

The current ECAL on-detector electronics is comprised of a very front end (VFE) and front end (FE). The VFE card, which serves five readout channels, contains multi-gain pre-amplifiers (MPGAs), charge sensitive amplifiers with 3 output gain values: $\times 1$, $\times 6$ and $\times 12$, and multi-channel ADCs which have a resolution of 12 bit/gain value and a sampling frequency of 40 M samples per second. The FE card, which receives data from up to five VFE cards, controls the data pipeline and trigger primitive generation, which is information used by the L1 trigger processors. In this system the trigger data granularity is an array of 5×5 crystals.

The plan for the HL-LHC is to replace the endcaps with a high granularity calorimeter because of the radiation damage that will be sustained, particularly at high $|\eta|$ [4]. All ECAL barrel supermodules will be refurbished during long shutdown 3 (LS3), between 2026 and 2028. The lead tungstate crystals and APDs in the barrel will be retained, though the operating temperature will be reduced from 18 °C to 9 °C to keep noise levels below 250 MeV. This new operating temperature requires new coolant distribution pipes carrying chilled water at 6 °C, though the current cooling distribution inside the supermodule will not require any modifications. Further cooling would require an extra chiller with significantly larger capacity. The on-detector electronics will be replaced with new radiation hard ASICs with faster pulse shaping and a factor of 4 increase in the sampling rate. A schematic detailing the HL-LHC electronics can be seen in Figure 2. This will reduce impact of out-of-time pileup and limit the increase in APD noise, as well as improve spike rejection at L1 via pulse shape discrimination. For energies greater than 50 GeV, 30 ps timing resolution will be achieved.

A new streaming front-end board will provide single crystal information to the L1 trigger via high speed radiation hard optical links (lpGBT). The new off-detector board will allow the use of more advanced algorithms in high performance FPGAs.

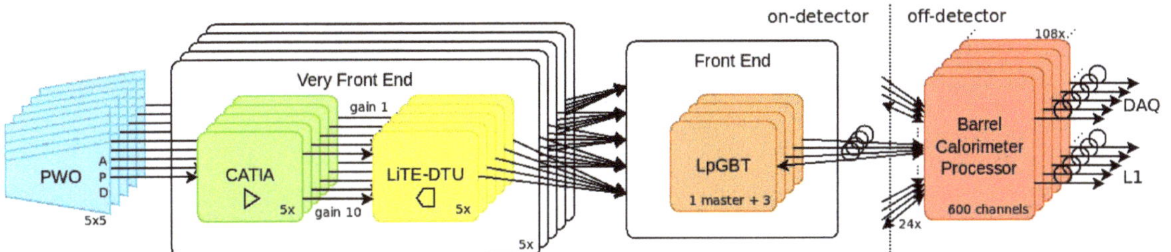

Figure 2. Schematic showing the layout for the CMS ECAL on- and off-detector readout for HL-LHC for a 5×5 crystal matrix.

3. Results

3.1. Lead Tungstate Crystal Longevity

The main concern for the ECAL crystals in the HL-LHC is ageing due to radiation damage. The scintillation mechanism is not affected by radiation, however radiation reduces the crystal transparency and therefore light output. This effect is monitored and corrected for using a dedicated light injection system. The change in light output between 2011 and 2018 can be seen in Figure 3. The relative response to laser light decreases with time due to radiation damage. This reduction is most pronounced at high η, while the ECAL barrel still retains $\sim$90% of its response. Monte Carlo simulations have been used to predict the light output in the HL-LHC era and these have been validated using test beam data studying the effect of hadron irradiation on crystal response. These results can be seen in Figure 4. The target integrated luminosity of 4500 fb^{-1} at the end of the HL-LHC would lead to a relative light yield between 25% and 45% depending on η. This is comparable to the current operating conditions in the ECAL endcap, giving confidence that we will be able to operate the barrel successfully with this level of light output.

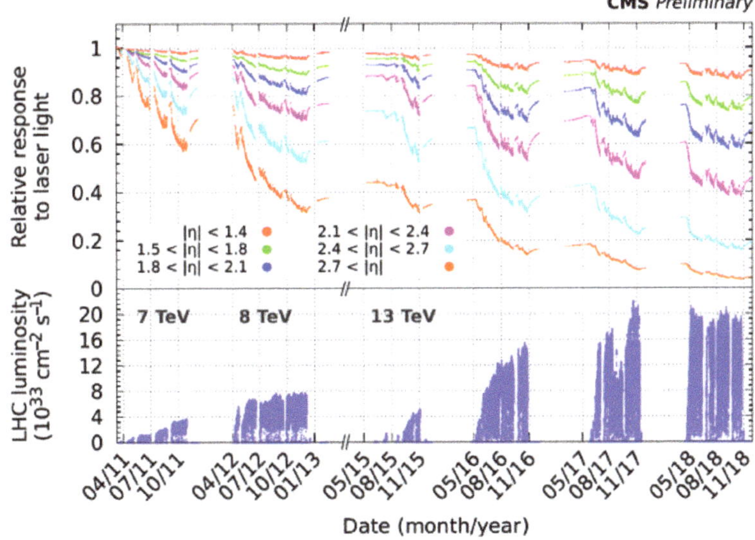

Figure 3. The relative response of the CMS ECAL to laser light and LHC instantaneous luminosity as a function of time plotted in different η ranges [5].

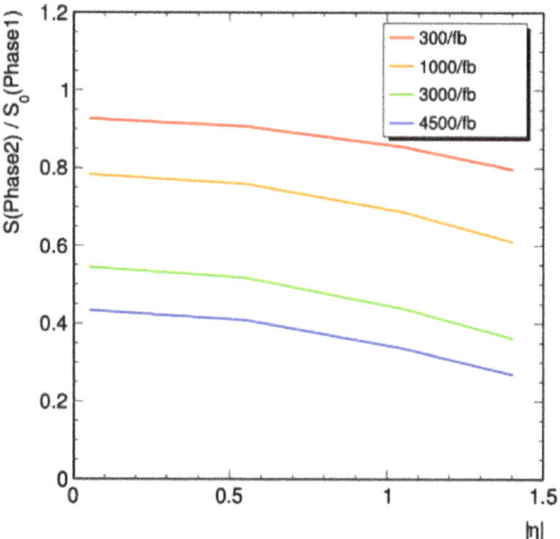

Figure 4. Predicted crystal light output relative to conditions at the start of CMS running, plotted as a function of of η for various integrated luminosities for 50 GeV photon showers with HL-LHC electronics at 9 °C [6].

3.2. Avalanche Photodiode Longevity

There are two causes of radiation damage to APDs, gamma rays and hadrons. Gamma rays create surface defects that increase the surface current and reduce the quantum efficiency. Hadrons create bulk damage that causes an increase in the bulk current. The main concern for the HL-LHC is the increased dark current, as the electronic noise depends on the square root of bulk current. It can be mitigated by reducing the operating temperature, as shown in Figure 5.

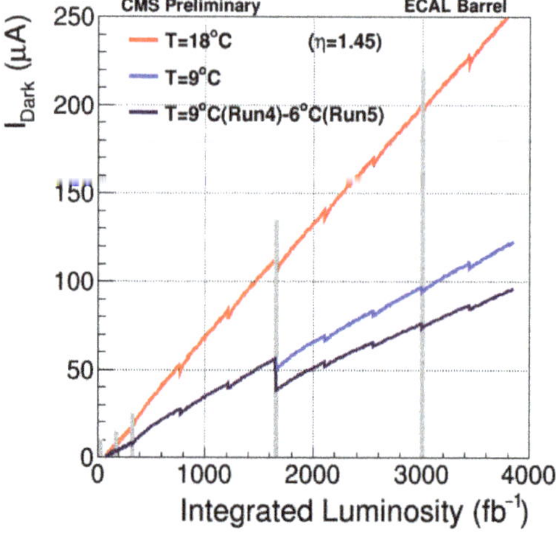

Figure 5. Dark current for different operating conditions. The current for APDs operated at 18 °C is shown in red, while that for APDs at 9 °C is shown in blue. A third possibility is shown in purple, where the APDs are operated at 9 °C in Run 4 (the first data taking period of HL-LHC), with the temperature reduced to 6 °C in Run 5 (the second data taking period of HL-LHC, after a shutdown which allows for experimental upgrades). The vertical shaded lines indicate long shutdowns [6].

3.3. Spike Rejection

Spikes are large isolated signals due to hadron interactions within the APD volume. Figure 6 shows a spike recorded in CMS, which shows a large signal in one readout channel. The pulse shapes caused by spikes and electromagnetic showers can be seen in Figure 7. In the current system, the spikes and EM showers are similar in both the start time and duration of the pulse. However in the HL-LHC the upgraded electronics will provide more discrimination between the two, as the spikes will have a faster rise and a shorter pulse shape, which will allow spike signals to be rejected. Spikes will dominate the L1 trigger rate at the HL-LHC if not suppressed. The expected event rates for the HL-LHC with 200 pileup events per bunch crossing can be seen in Figure 8. The ECAL will implement a spike killing algorithm in an off-detector FPGA and send this information as part of the trigger primitives to achieve >99.9% spike rejection in the L1 trigger. This level of spike rejection will leave less than 1 kHz residual spikes to trigger L1 for signals with $E_T > 5$ GeV.

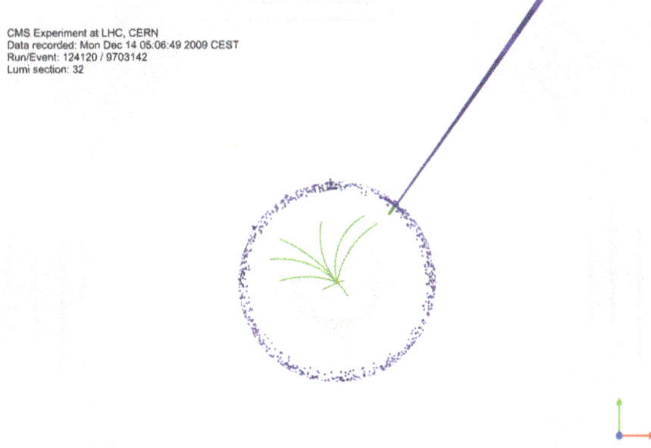

Figure 6. Event display showing a spike signal where a large amount of energy is deposited in a single ECAL readout channel. Tracks are shown in green and the energy deposited in the ECAL is shown in blue, with the spike event at approximately 1 o'clock.

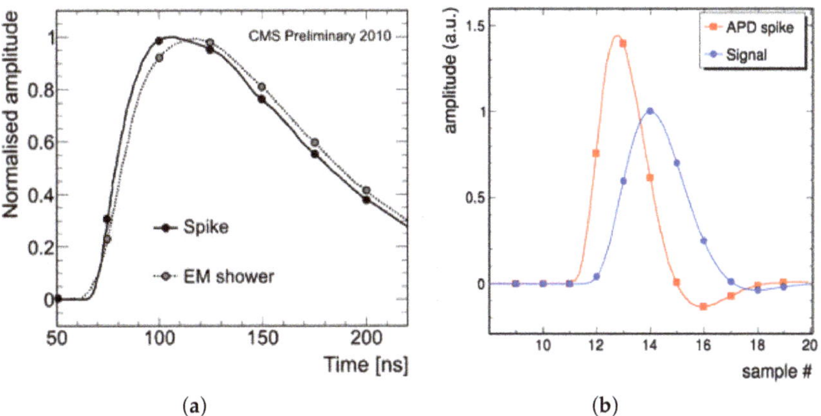

Figure 7. (**a**) Representative spike and EM shower pulse shapes in the current CMS ECAL, with the spike shown in black and the EM shower shown in grey. (**b**) Spike and EM shower pulse shapes in the CMS ECAL during the HL-LHC, with the spike pulse shown in red and the EM pulse shown in blue. HL-LHC pulse shapes are from the measured response of early prototype electronics convoluted with the known APD and scintillation pulse shapes.

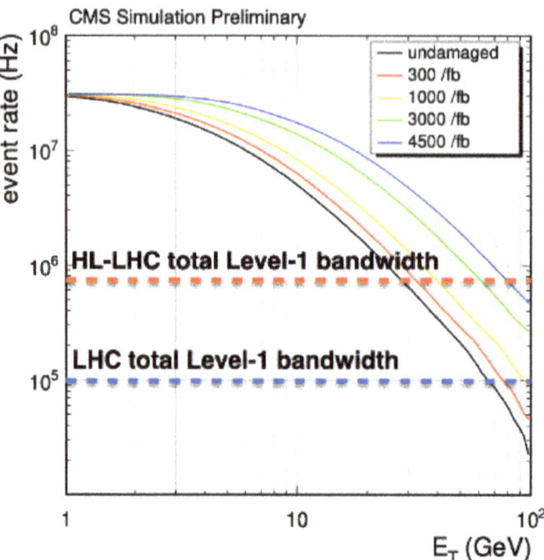

Figure 8. Expected rate of events for the HL-LHC in the CMS ECAL with 200 pileup, showing the energy deposits above specific thresholds after an integrated luminosity of 300, 1000, 3000 and 4500 fb^{-1} without any spike suppression applied. The L1 bandwidth for the current LHC and the HL-LHC are also shown on this plot [6].

3.4. Impact of Precision Timing

It will be challenging to maintain reconstruction performance with 140–200 pileup events per bunch crossing. For the decay channel $H \rightarrow \gamma\gamma$, the primary vertex efficiency will be reduced from 75% to 30% in the absence of detector upgrades. However, improved vertex localisation is possible with precise (30 ps) timing capabilities. This precise timing gives a sensitivity gain of about 10% on $H \rightarrow \gamma\gamma$ resolution and fiducial cross-section relative to no precise timing case with 140 pileup events. With precision timing (ECAL + MIP timing) we can obtain similar results to Run 2 conditions, as shown in Figure 9. MIP timing will be provided by a new MIP timing detector, details of which can be found in [7].

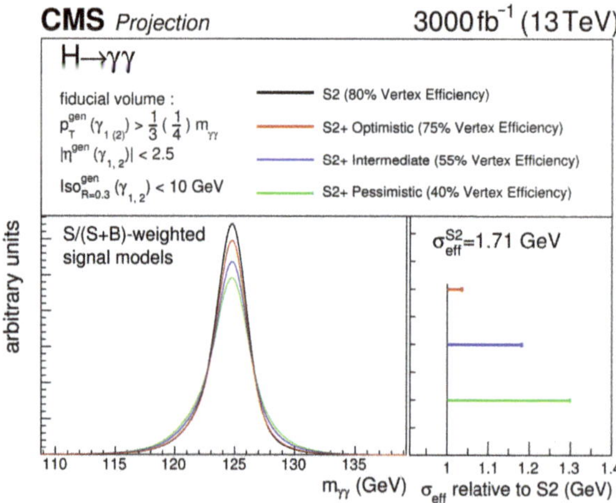

Figure 9. Lineshape for the H→ $\gamma\gamma$ signal in the four scenarios: no precise timing (green), precise timing in calorimeter (blue), precise timing in calorimeter and MIP timing (red) and Run 2 conditions (black) [8].

3.5. HL-LHC Electronics

As shown in Figure 2, the VFE in the HL-LHC readout system comprises of two ASICS namely the CATIA and LiTE-DTU. The CATIA is a pre-amplifier ASIC, using a trans-impedance amplifier (TIA) architecture with minimal pulse shaping. Faster pulse shaping is important for precise timing and improved spike rejection capabilities. It has two output gain values: $\times 1$ and $\times 10$. The CATIA V1 has been used in test beams with good results, and initial tests of CATIA V2.1 (the latest version) also show very promising results.

The LiTE-DTU is a data conversion, compression and transmission ASIC. It uses two 12-bit ADCs with 160 MS/s data conversion. It also has lossless data compression, a look-ahead algorithm where a sample saturation check prevents samples from different gains in the same APD signal timeframe being mixed. An effective number of bits (ENOB) scan has been performed as function of frequency with the results shown in Figure 10. These results match the expected ADC performance. In February, we received approximately 600 packaged LiTE-DTU v2. All 600 of these have been tested, with 98% passing. We have also performed a single event upset (SEU) test at CRC Louvain with no I2C errors and no PLL loss of lock.

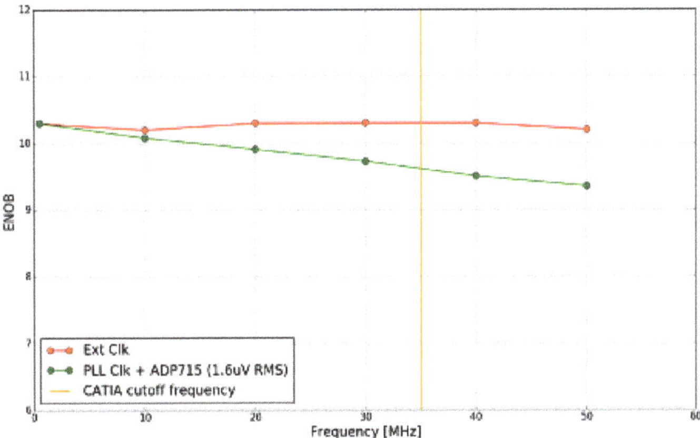

Figure 10. ADC ENOB as a function of the input frequency. The red data points show the ENOB using an external clock, while the green data points show the ENOB using a PLL clock and ADP715 (1.6 V RMS). The yellow line shows the CATIA cut off frequency.

We have also completed combined tests using both the CATIA v2.0 and the LiTE-DTU v2.0, using both single channel and multi channel test boards. Preliminary results show that the system is working well, with initial noise and timing measurements compatible with 30 ps timing for $E > 50$ GeV. The new LiTE-DTU features were also shown to be working. These features include test-pulse generation for CATIA, a secondary calibration scheme and ADC calibration with CATIA baseline-like levels. We are close to starting production for both ASICs.

The very front end (VFE) contains $5 \times$ CATIA and $5 \times$ LiTE DTU chips. It is responsible for the calibration of ADC values using CATIA's reference voltage and embedded multiplexer, as well as readout of temperature sensors (via FE) for components including APDs, CATIAs and PCBs. A pilot run of 8 VFE v3 boards has been tested, with a larger production launched at the beginning of May. The low voltage regulator (LVR) supplies voltages to the ASICS and FE board. It is a radiation tolerant ASIC that will readout APD leakage current, using DCDC converters (specifically bPOL12s [9] and linPOL12). The design is magnetic field tolerant and has low noise. The current versions of the VFE and LVR are tentatively the final versions.

The front end (FE) allows streaming of full granularity data off-detector at 40 MHz, which is not possible in the current detector. It sends a clock to the VFE directly from the lpGBT controller and uses I2C via a controller-responder chain. The FE also monitors the APD dark current. The FE v3 is close to the final version.

The backend electronics is comprised of a barrel calorimeter processor (BCP) board, which combines trigger and DAQ functionality and provides clock and control signals to the FE electronics. Each board handles signals from 600 crystals and uses commercially available FPGAs. Algorithms are being developed using high level synthesis to produce trigger primitives. The BCP v1 uses one KU115 FPGA and is currently being used for integration tests with VFE/FE boards and DAQ. However, the BCPv2 will use one VU13P instead of two KU115's. This will provide the BCP with nearly three times the memory, 30% more logic cells and 11% more digital signal processing. Schematics for the BCP v2 are currently under development. Testing results for the BCP v1 (including timing performance) are good.

3.6. Test Beam Results

In 2021, we tested a single ECAL tower (5 × 5 crystal matrix) with prototype HL-LHC electronics, specifically the CATIA v1.2 and LiTE DTU v1.2, at the CERN SPS H4 beamline using an electron beam with energies from 25 to 250 GeV. The test beam setup is shown in Figure 11. The time resolution measurement is performed by comparing the time measured by a single channel to that of an external timing reference detector placed along the beamline, with the results shown in Figure 12. The constant term of about 12 ps meets the requirements for the HL-LHC design. A timing resolution of 30 ps for E > 50 GeV is shown to be achievable. Figure 13 shows the energy resolution using test beam data from 2018, where we used CATIA v1 and a commercial 160 MHz ADC. The setup was otherwise identical to that used in 2021. The energy resolution meets the HL-LHC requirements, with a constant term of <0.5%.

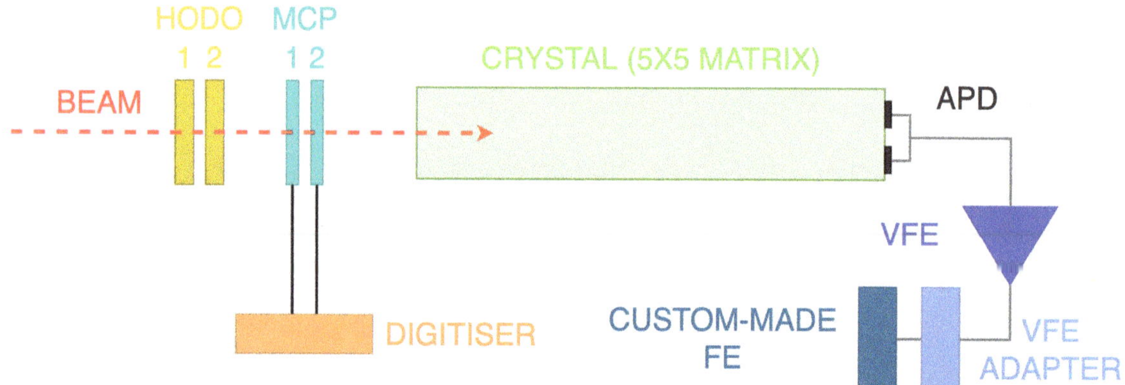

Figure 11. Test beam setup. Hodoscopes (hodo) were used to measure the beam position, while microchannel plate detectors (MCP) were used for external timing measurements. In 2018 the VFE was a commercial ADC, while in 2021 the LiTe-DTU was used.

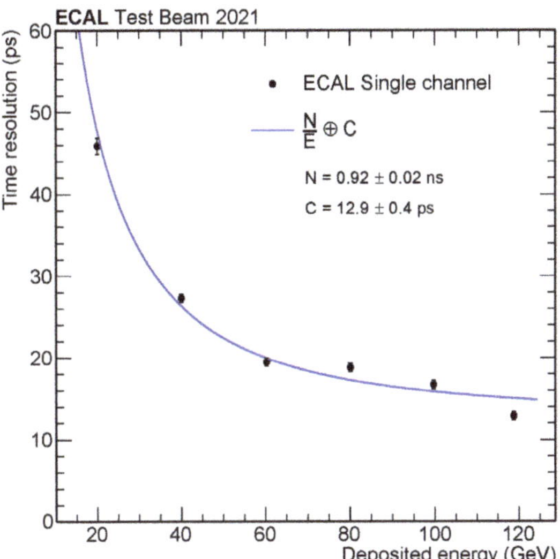

Figure 12. Time resolution as function of the deposited energy obtained in 2021 test beam data. The solid blue line represents the fit with the timing resolution function $N/E \oplus C$, where N denotes the noise, C the constant term, and E the deposited energy.

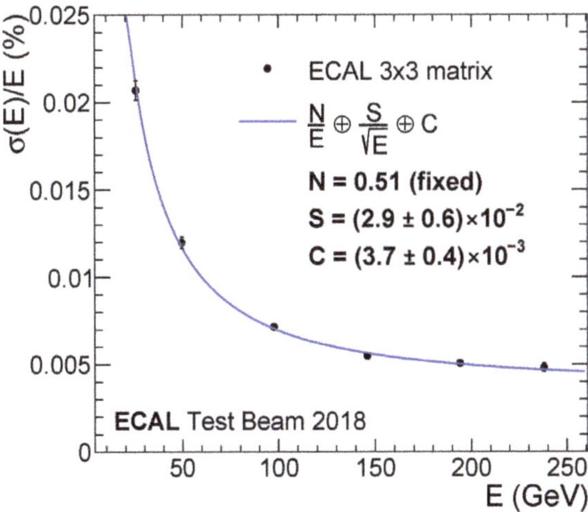

Figure 13. Energy resolution of a 3×3 crystal matrix as a function of the deposited energy obtained in 2018 test beam data. The solid blue line represents the fit with the energy resolution function $N/E \oplus S/\sqrt{E} \oplus C$, where N denotes the noise, S the stochastic term, C the constant term, and E the deposited energy.

4. Conclusions

Both the on- and off-detector electronics will be replaced in the barrel region of the CMS ECAL for the HL-LHC to maintain the current performance. Full featured ASICs have been received and initial tests show good results. The plan for this year includes a full system test of around 400 channels using a spare barrel supermodule. Towards the end of the year we will use this supermodule with the prototype HL-LHC electronics in a test beam with electrons and pions. We will also have an engineering design review to provide a green-light for the production of the front-end ASICs and electronic boards. Development of the BCPv2 and associated firmware will continue. We will have production ready (and tested) versions of the ASICs and on-detector boards by the end of the year.

All 36 ECAL barrel supermodules will be extracted at the start of LS3. The current electronics will be exchanged for their HL-LHC counterparts and associated services by dedicated teams on the surface, with every supermodule being fully tested before reinstallation. This whole procedure will take approximately one year.

All barrel calorimeter components are on track for installation during LS3.

Funding: The author is supported by the Science & Technology Facilities Council (STFC). This work was done within the context of the CMS Collaboration.

Institutional Review Board Statement: Not applicable.

Informed Consent Statement: Not applicable.

Data Availability Statement: Not applicable.

Conflicts of Interest: The author declares no conflict of interest.

References

1. CMS Collaboration. The CMS experiment at CERN LHC. *JINST* **2008**, *3*, S08004.
2. Cipriani, M.; on behalf of the CMS Collaboration. Photon Detection with the CMS ECAL in the Present and at the HL-LHC and Its Impact on Higgs-Boson Measurement. Available online: https://cds.cern.ch/record/2708020/ (accessed on 6 July 2022).
3. Yates, B.R.; for the CMS Collaboration. A CMS Level-1 Track Finder for the HL-LHC. Available online: https://arxiv.org/abs/2110.02826 (accessed on 6 July 2022).
4. CMS Collaboration. The Phase-2 Upgrade of the CMS Endcap Calorimeter, CERN-LHCC-2017-023. Available online: https://cds.cern.ch/record/2293646 (accessed on 6 July 2022).
5. CMS Collaboration. CMS ECAL Response to Laser Light, CMS-DP-2019-005. Available online: https://cds.cern.ch/record/2668200 (accessed on 7 July 2022).
6. CMS Collaboration. The Phase-2 Upgrade of the CMS Barrel Calorimeters, CERN-LHCC-2017-011. Available online: https://cds.cern.ch/record/2283187 (accessed on 6 July 2022).
7. CMS Collaboration. A MIP Timing Detector for the CMS Phase-2 Upgrade, CERN-LHCC-2019-003. Available online: https://cds.cern.ch/record/2667167?ln=en (accessed on 17 August 2022).
8. CMS Collaboration. Projected Performance of Higgs Analyses at the HL-LHC for ECFA 2016. Available online: https://cds.cern.ch/record/2266165 (accessed on 6 July 2022).
9. Faccio, F.; Michelis, S.; Blanchot, G.; Ripamonti, G.; Cristiano, A. The bPOL12V DCDC converter for HL-LHC trackers: Towards production readiness. In Proceedings of the TWEPP 2019 Topical Workshop on Electronics for Particle Physics, Santiago De Compostela, Spain, 2–6 September 2019; p. 070.

Article

Energy Reconstruction and Calibration of the MicroBooNE LArTPC

Richard Diurba on behalf of the MicroBooNE Collaboration

Laboratory for High Energy Physics, University of Bern, 3012 Bern, Switzerland; microboone_info@fnal.gov

Abstract: MicroBooNE uses a liquid argon time projection chamber (LArTPC) for simultaneous tracking and calorimetry. Neutrino oscillation experiments plan to use LArTPCs over the next several decades. A challenge for these current and future experiments lies in characterizing detector performance and reconstruction capabilities with thorough associated systematic uncertainties. This work includes updates related to LArTPC detector physics challenges by reviewing MicroBooNE's recent publications on calorimetry and its applications. Highlights include discussions on signal processing, calorimetric calibration, and particle identification.

Keywords: liquid argon detectors; liquid argon calibration; signal processing; particle identification

Citation: Diurba, R. Energy Reconstruction and Calibration of the MicroBooNE LArTPC. *Instruments* **2022**, *6*, 30. https://doi.org/10.3390/instruments6030030

Academic Editors: Fabrizio Salvatore, Alessandro Cerri, Antonella De Santo and Iacopo Vivarelli

Received: 29 July 2022
Accepted: 22 August 2022
Published: 29 August 2022

Publisher's Note: MDPI stays neutral with regard to jurisdictional claims in published maps and institutional affiliations.

1. Introduction

MicroBooNE is an 85 metric tonne liquid argon time projection chamber (LArTPC) that operated from 2015 to 2021 at the Fermi National Accelerator Laboratory (Fermilab) [1]. The detector sits roughly 500 m on-axis from the Fermilab Booster Neutrino Beam and expects an on-axis flux of neutrinos with energies of approximately 0.2 to 2 GeV [2]. LArTPCs operate by drifting ionized electrons from particles traversing the detector in an electric field to readout panels. The drifting of ionized electrons onto, in the case of MicroBooNE, arrays of wire planes, allows for simultaneous tracking and calorimetry of charged particles traversing the argon. For simulation and data, it uses the LArSoft software kit for decoding data packets, simulating the detector, and processing the reconstruction [3,4]. The energy deposition of particles in the detector is simulated with GEANT4 [5–7]. The simulation event generator for neutrino interactions is GENIE [8–11]. For cosmic ray muons, it is CORSIKA [12].

These proceedings intend to review the work of MicroBooNE over the years to develop a robust and thorough signal processing and calibration scheme for precision hit-by-hit calorimetry. The article will finish with examples of applications of how MicroBooNE exploits its precision calorimetry data for higher-level reconstruction, such as particle identification and shower clustering.

2. Signal Processing

The LArTPC of MicroBooNE has three wire planes. Two wires planes, Plane 0 and Plane 1, have angles ±60 degrees from the third plane, Plane 2. Plane 0 and Plane 1 measure ionized electrons in the argon using the induced signal of electrons traveling towards and away from these induction wire planes. The drifting electrons end on Plane 2, which reads a signal from the ionized electrons by collecting them on the wire. The combination of the two induction planes and one collection wire plane allows for three-dimensional tracking of the ionized electrons that remain from the passage of a charged particle in the argon. The electric field at MicroBooNE operates at 273 V/cm [1].

The wire planes undergo noise filtering and then two-dimensional deconvolution to eliminate effects from the electronics and sharpen signals to isolated wires. Figure 1 shows a data neutrino event candidate going through each stage of signal processing from raw waveforms, noise filtering, and deconvolution [13,14]. Concurrently with noise filtering, the electronic response calibration is applied. With the implementation of electronics response filtering and calibration, the blue bands in Figure 1 disappear, eliminating extraneous reconstructed hits surrounding the candidate neutrino interaction vertex. Without removing these extraneous signals, the reconstruction may incorrectly identify extra tracks and incorrect vertexes. The 2D deconvolution intends to sharpen signals and extract the ionization charge from the smearing of the electronics response. The process includes a Fourier transformation and a low-pass filter [13,14]. Downstream reconstruction can then form hits from these waveforms. An example would be to use Gaussian functions to extract the full charge [15].

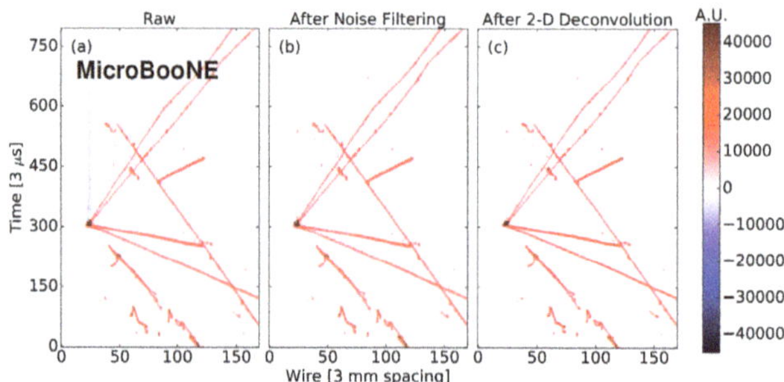

Figure 1. Candidate neutrino interaction event display from MicroBooNE data through each stage of signal processing from raw data (**a**), noise filtering (**b**), and finally the event with both noise filtering and 2D deconovolution (**c**). Figure was taken from [14].

The wire response simulated depends on the liquid argon property values used. For example, the diffusion of drift electrons alters how the simulation generates simulated waveforms. If the diffusion constants between data and simulation differ, then the waveforms between simulation and data may vary enough to propagate discrepancies to higher-level reconstruction variables. A method developed by MicroBooNE to factor in differences in waveforms in simulation and data is to modify the waveforms as a function of track variables [16]. The process is twofold. First, ratios of the hit charge and hit width of reconstructed waveforms are made from hits on cosmic muon tracks as a function of drift distance (X), height (Y), distance across the length of the detector (Z), and angles (θ_{XZ}, θ_{YZ}). These hits come from a Gaussian fit to regions of interest on the waveform [16].

Figure 2 shows the ratio for all three wire planes as a function of position across the drift distance (x). The trajectories of the cosmic ray muons simulated were taken from real reconstructed data cosimc ray muons, and all other elements of the simulated muons come from CORSKIA [12].

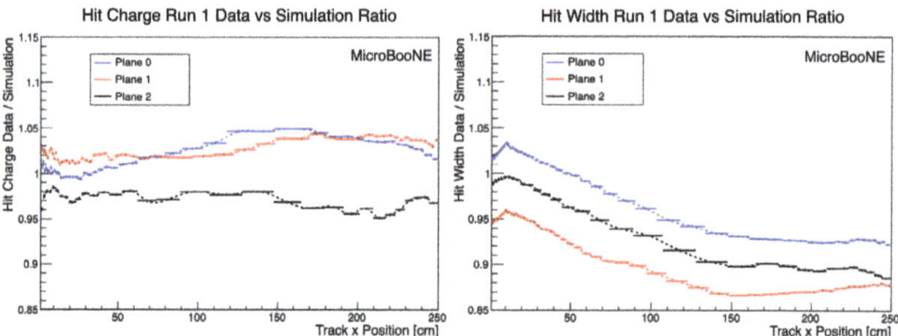

Figure 2. Plots of the ratios between data and simulation as a function of drift distance (X). The anode is approximately at x = 0. Images come from [16].

Second, these ratios go into a function that creates variation simulation samples, one made for each trajectory variable (X, YZ, θ_{XZ}, θ_{YZ}) for a total of four samples. The varied samples change the simulated waveforms associated with simulated neutrino interactions, which will be independent of the cosmic ray muon samples used to generate the ratios. The waveforms in these varied simulation samples are modulated by the hit charge and hit width ratios measured between data and simulation for the various track variables. In the case of overlapping hits, only the portion not related to the cosmic ray muon is modulated. The process begins by creating a scale factor. The scale factor is the ratio as a function of track position weighted by the amount of energy deposited in the simulation. This new weighted ratio value for the associated hit width and hit charge rescales the varied hit width and hit charge. These new, altered hit widths and hit charges are then fed into a Gaussian function, and the waveform associated with the hit is scaled relative to the unaltered simulated hit width and hit charge. Equation (1) shows the reweighting function as a function of the drift time in terms of the mean time (t), width (σ), and charge (Q) between the original hit (σ, Q) and the reweighted hit (σ', Q') for times associated with the waveform being varied (t_j).

$$w = \frac{\sum_j \frac{Q'}{\sqrt{2\pi\sigma'^2}} \exp\left(-\frac{(t-t_j)^2}{2\sigma'^2}\right)}{\sum_j \frac{Q}{\sqrt{2\pi\sigma^2}} \exp\left(-\frac{(t-t_j)^2}{2\sigma^2}\right)} \tag{1}$$

Figure 3 reveals two examples of the wire modification on both a single hit and overlapping hits [16]. These varied simulation samples, four wire modification samples in total, were then used to evaluate the detector-related systematic uncertainties for analyses. Examples include the following publications [17–20].

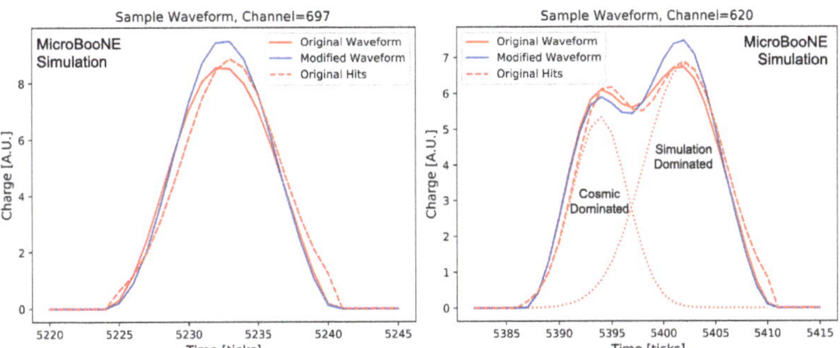

Figure 3. Examples of the original and modified waveforms for a single hit (**left**) and two concurrent hits from a cosmic ray muon overlaid in simulation and a simulated physics interaction (**right**). Only the waveform from the simulated physics event was modulated to form the new hit in the varied sample. The images come from [16].

3. Calibration of TPC Calorimetry

The precision of the energy reconstructed in data and simulation depends on a thorough and robust calibration of the amount of energy deposited per unit length. The need for detailed calorimetry calibration is especially true for LArTPCs, since the ability to measure individual energy deposits at sub-millimeter resolution is a proposed advantage of using this detector technology over others. In the case of MicroBooNE, the scheme calibrates the hit charge measured per unit length (dQ/dx) as a function of position and time. It then calibrates the energy scale using calorimetry from selected stopping muon candidates [21]. The two-step calibration has been adapted from the calibration schemes in MINOS [22]. In MicroBooNE's calibration process, each correction measured is used to generate the next set of dQ/dx corrections.

Since MicroBooNE operates on the surface, the cosmic ray muon flux is so significant that abundances of ionized electrons and positively charged argon ions perpetually exist in the detector. These positively charged argon ions move more slowly through the argon, and therefore build up on the edges of the TPC, pushing the ionized electron away from the detector faces [23]. This distortion is known as the space charge effect and leads to the stretching and squeezing of signals. This effect has been measured by studying the positional offset of cosmic ray muons entering and exiting the detector [23]. These measurements were then verified using a UV-laser [23].

It is essential to have a consistent standard candle of tracks, so the corrections of dQ/dx address detector effects and not the physics of interactions and stopping particles. For MicroBooNE, non-stopping cosmic ray muons are used that are anode-cathode crossing. These muons cross the whole drift volume. The arrival time can be ascertained by the hit closest to the anode. These tracks were identified in simulations and data with the Pandora reconstruction package [24,25].

A selected cathode–anode crossing muon track for dQ/dx calibration must have a track length between 250 and 270 cm. It must also have an angle relative to the drift distance and detector length (θ_{XZ}) of less than 75 degrees and an angle relative to the detector height and length (θ_{YZ}) less than 80 degrees. These cathode–anode crossing track samples calibrate each day of data-taking, and therefore are organized by the day the detector collected the event.

These cosmic ray muons samples are used to smear the dQ/dx calibration as a function of position. The smearing function (C) for a generic position variable (i), such as YZ or X, is seen in Equation (2) in terms of the global median dQ/dx and the local dQ/dx.

$$C_i = \frac{dQ/dx_{i,\text{global}}}{dQ/dx_{i,\text{local}}} \qquad (2)$$

Equation (2) is first used in terms of the detector height and length (YZ) to form C_{YZ}. The event sample statistics in data for YZ are shown in Figure 4. The corrections in YZ aim to address effects, such as unresponsive channels, space charge effect distortions in YZ, and electronics response. Then, the same process is used for C_X to calibrates as a function of drift distance (X), which corrects for attenuation due to electronegative impurities, diffusion effects, and remaining space charge effect corrections in the horizontal direction. The final calibration of dQ/dx aims to fix time-dependent differences between data-taking days [21] (Equation (3)).

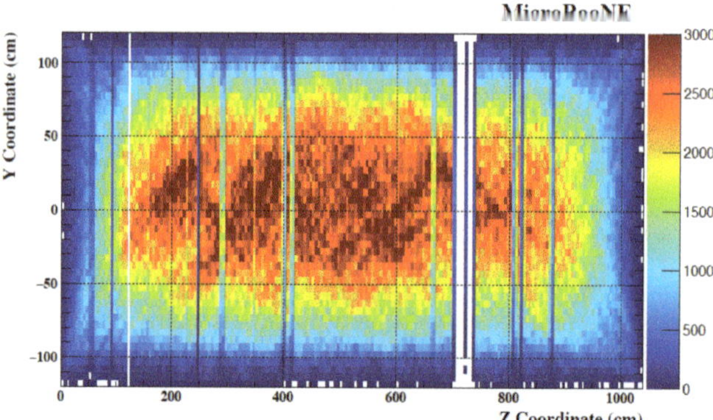

Figure 4. Display from data of number of hits measured as a function of position of the detector height (Y) and length (Z). The image is from [21].

$$C(t) = \frac{dQ/dx_{\text{ref.}}}{dQ/dx_{\text{global median xyzcorr.}}(t)} \qquad (3)$$

The calibrated dQ/dx is finally shown in Equation (4).

$$dQ/dx_{\text{calib.}} = dQ/dx \cdot C_X \cdot C_{YZ} \cdot C(t) \tag{4}$$

With the dQ/dx calibrated, the next step is to measure the energy scale. The energy scale measurement starts by selecting a dE/dx model for liquid argon and then measuring a conversion scale, or gain, from ADC values from the TPC electronics to the number of electrons ionized in the hit. MicroBooNE, for calibration, uses the modified Box model [26]. The gain is measured using a sample of neutrino-induced stopping muons with a track length of at least 150 cm and angular cuts identical to those used for the dQ/dx calibration. Like with dQ/dx, Pandora is used via the reconstruction package [24,25]. Most probable values are found for bins of residual range using a Landau–Gaussian fitter [27]. The calorimetric most probable value for dQ/dx in a bin is compared to the Landau–Vavilov predicted value in the region between 250 and 450 MeV. The gain value is found by minimizing the χ^2 between a sample of stopping cosmic ray muons and the expectation from the Landau–Vavilov theory [21,28]. Figure 5 shows the dE/dx of a stopping muon after measuring the gain value. Even outside the kinetic energy range used for calibration, there appears good agreement within uncertainties between the fitted data and the predicted values from Landau–Vavilov. The conversion from residual range to kinetic energy is accomplished using the continuous slowing down approximation table for muons [29]. The gain value measured for the modified Box model extracted from the χ^2 fit is then used as a global value for the data set and works as the final calibration step to convert electronics response to units of energy deposited [21].

Figure 6 evaluates the difference between two methods of assessing the total energy of neutrino-induced stopping muons, the hit-by-hit calibrated calorimetry from the collection plane, and the track length of the muon. The difference between the two methods in data is around 2%, which is near the predicted difference from simulation (1%) and considered a sufficient closure test for calibration [21].

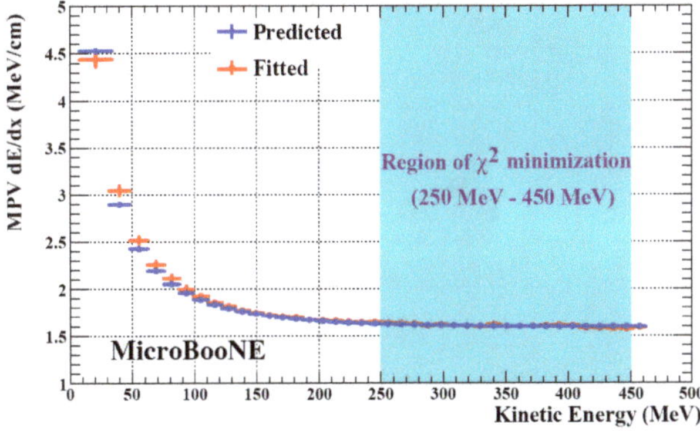

Figure 5. Energy deposited (dE/dx) as a function of the cosmic ray muon's kinetic energy for the collection plane calorimetry. The red represents data from 2016 with the best fit of the gain value, and the blue represents the expectation from [28]. The figure was originally seen in [21].

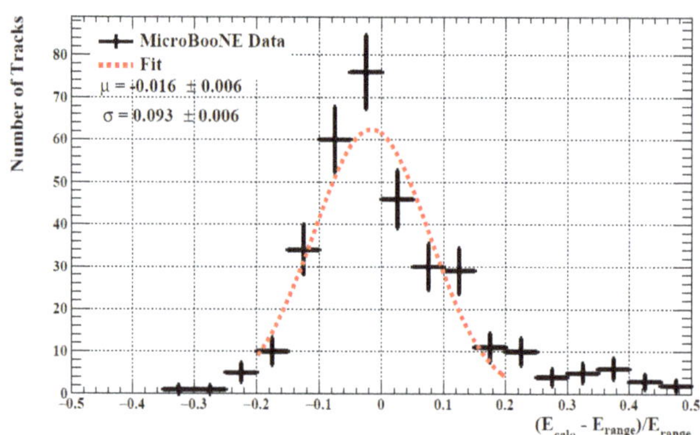

Figure 6. Comparison between the total energy of stopping muons measured by hit-by-hit calorimetry and the energy calculated from the track range. Taken from [21].

4. Example Applications of Higher-Level Reconstruction Using Calorimetry

MicroBooNE has developed a wide range of reconstruction techniques using calorimetry information. A log-likelihood ratio metric designed to separate reconstructed muons from protons serves as an example. In this method, a log-likelihood metric is from probing template probability density functions of dE/dx in slices of residual range of a muon or proton for each hit in each plane in the last 30 cm of the track [30]. Figure 7 shows the data to simulation comparison as separated by particle type in the simulation, with -1 hypothesizing a proton and 1 hypothesizing a muon.

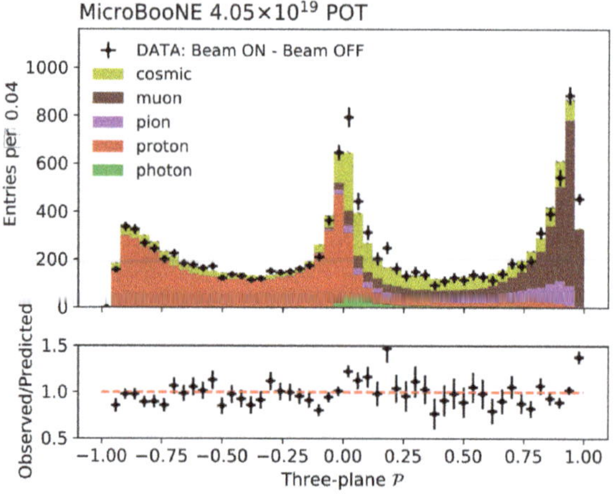

Figure 7. Distribution of data and simulation log-likelihood ratio measured as the difference between beam on samples and beam off samples to eliminate the cosmic background. With the Poisson uncertainties on the data distribution, the simulation and data appear within agreement. The image is from [30].

Another recent highlight is the publication of shower reconstruction using deep-learning methods with a SparseSSNet [31]. As a LArTPC, MicroBooNE has the unique capability of identifying photon-induced showers from electron-induced showers. Some analyses used a Kalman filter to accomplish this [17,32]. In tandem, a deep-learning selection using SparseSSNet can identify and isolate electromagnetic showers. The isolated hits of the shower form the total calorimetric energy measured for the event. Figures 8 and 9 reveal the total energy reconstructed with Michel electrons and neutral pions. There is good agreement between data and simulation with $\chi^2/d.o.f$ of 0.61 and 0.98, respectively.

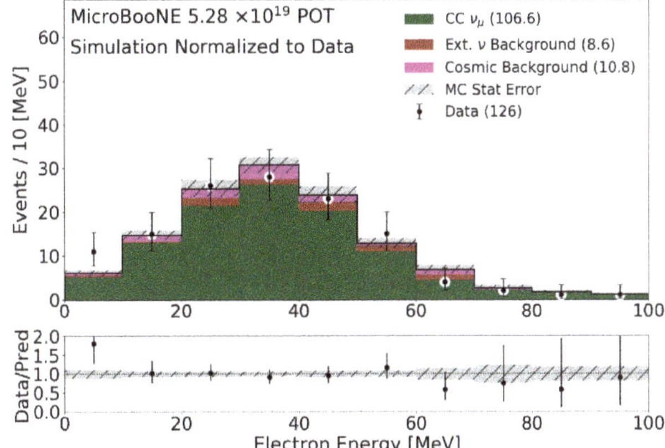

Figure 8. Total electron energy reconstructed from selected stopping muons using charged current (CC) muon neutrino events. Statistics for the simulation are scaled to the beam protons on target used in the data distribution. Plot from [31].

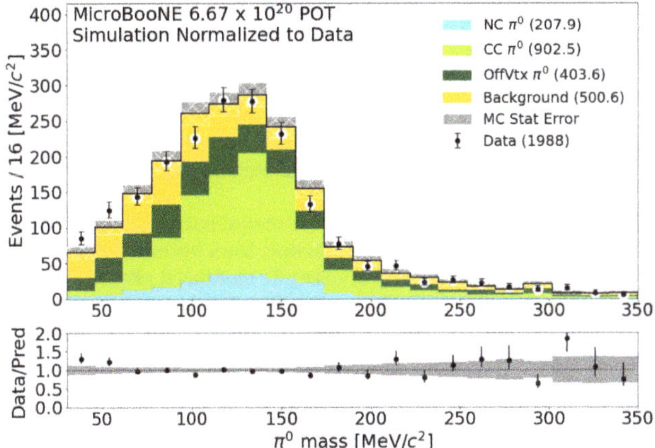

Figure 9. Total reconstructed neutral pion mass from charged current (CC) events, neutral current (NC) events, and neutrino interaction events off vertex (OffVtx). Statistics in simulation scaled to beam protons on target used for data distribution. Figure originally from [31].

5. Conclusions

This paper summarized recent publications from MicroBooNE. Highlights focused on extracting charge from waveforms, calibrating dE/dx using TPC hits, and applying the calorimetry to reconstruct shower energies and separate proton tracks from muon tracks. MicroBooNE has collected nearly six years of neutrino data and aims to continue developing techniques and applications for LArTPC reconstruction. As an example, techniques discussed were used in publications of cross section results [33,34] and publications searching for anomalous excesses of electron-like neutrino events [17–20].

Funding: This document was prepared by the MicroBooNE collaboration using the resources of the Fermi National Accelerator Laboratory (Fermilab), a U.S. Department of Energy, Office of Science, HEP User Facility. Fermilab is managed by Fermi Research Alliance, LLC (FRA), acting under contract number DE-AC02-07CH11359. MicroBooNE is supported by the following: the U.S. Department

 *instruments*

Article

FASER's Electromagnetic Calorimeter Test Beam Studies

Charlotte Cavanagh on behalf of the FASER Collaboration

Department of Physics, University of Liverpool, Liverpool L69 3BX, UK; charlotte.cavanagh@cern.ch

Abstract: FASER, or the Forward Search Experiment, is a new experiment at CERN designed to complement the LHC's ongoing physics program, extending its discovery potential to light and weakly interacting particles that may be produced copiously at the LHC in the far-forward region. New particles targeted by FASER, such as long-lived dark photons or axion-like particles, are characterised by a signature with two oppositely charged tracks or two photons in the multi-TeV range that emanate from a common vertex inside the detector. The full detector was successfully installed in March 2021 in an LHC side tunnel 480 m downstream from the interaction point in the ATLAS detector. FASER is planned to be operational for LHC Run 3. The experiment is composed of a silicon-strip tracking-based spectrometer using three dipole magnets with a 20 cm aperture, supplemented by four scintillator stations and an electromagnetic calorimeter. The FASER electromagnetic calorimeter is constructed from four spare LHCb calorimeter modules. The modules are of the Shashlik type with interleaved scintillator and lead plates that result in 25 radiation lengths and 1% energy resolution for TeV electromagnetic showers. In 2021, a test beam campaign was carried out using one of the CERN SPS beam lines to set up the calibration of the FASER calorimeter system in preparation for physics data taking. The relative calorimeter response to electrons with energies between 10 and 300 GeV, as well as high energy muons and pions, has been measured under various high voltage settings and beam positions. The measured calorimeter resolution, energy calibration, and particle identification capabilities are presented.

Keywords: resolution; calorimeter; pre-shower

Citation: Cavanagh, C., on behalf of the FASER Collaboration. FASER's Electromagnetic Calorimeter Test Beam Studies. *Instruments* **2022**, *6*, 31. https://doi.org/10.3390/instruments6030031

Academic Editors: Fabrizio Salvatore, Alessandro Cerri, Antonella De Santo and Iacopo Vivarelli

Received: 29 July 2022
Accepted: 26 August 2022
Published: 31 August 2022

Publisher's Note: MDPI stays neutral with regard to jurisdictional claims in published maps and institutional affiliations.

1. Introduction

There are many models to suggest new physics beyond the standard model (SM), for example, the possible existence of dark sectors (DS) that may contain new, light, weakly coupled particles that interact only very weakly with ordinary matter. FASER [1–3] is a new experiment designed to detect potentially long-lived particles (LLPs) produced at the ATLAS interaction point (IP1) in the forward region. It is located 480 m downstream of IP1 in the TI12 service tunnel. These particles are highly collimated, and their decay products have around TeV-scale energies.

The physics models targeted by FASER are characterised by the presence of LLPs, such as dark photons (A') and axion-like particles (ALPs, a). Dark photons [4] are hypothetical particles that belong to a DS and form a portal to the SM. This leads to a coupling between the SM and the DS, the strength of which is governed by a mixing parameter ϵ. The size of ϵ determines the strength of the interaction, hence the lifetime of the dark photon. Dark photons that have a mass $m_{A'}$ below a few hundred MeV predominantly decay into e^+e^- and $\mu^+\mu^-$ pairs. ALPs [4] are weakly interacting, pseudoscalar particles. In the photon-dominant case, there is a coupling $g_{a\gamma\gamma}$ between the SM and the ALP, a. The smaller this coupling, the more long-lived the particle. Identification of electrons, and photons in the case of ALPs, relies on energy deposits in the ECAL, hence FASER's main sensitivity in the low mass region is to e^+e^- pairs and $\gamma\gamma$ pairs resulting from dark photon decay and ALP decay, respectively. Figure 1 shows the decay mode, and the area of parameter space that FASER and the possible FASER2 upgrade will explore in the case of dark photons. Figure 2 shows the same for ALPs, in the photon-dominant case.

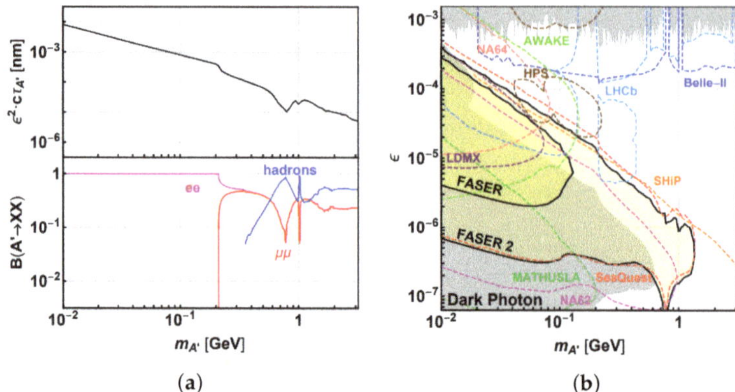

Figure 1. (**a**) Dark photon decay modes according to mass range. (**b**) FASER and FASER2 (proposed, enlarged successor of FASER) accessible parameter space compared with current and proposed experiments [3].

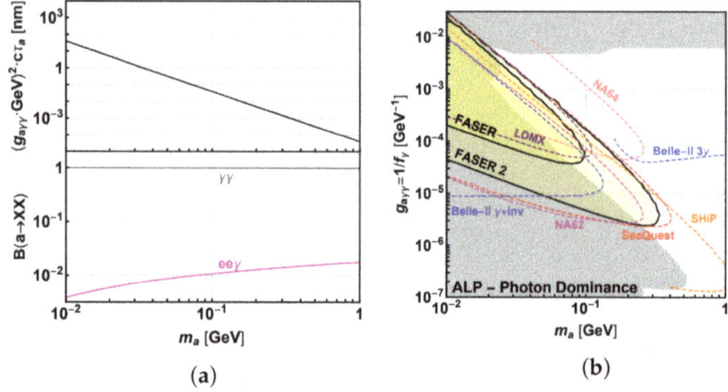

Figure 2. (**a**) ALP decay modes according to mass range. (**b**) FASER and FASER2 accessible parameter space compared with current and proposed experiments [3].

2. The FASER Experiment

The main components of the FASER experiment are shown in Figure 3 [1]. A particle produced at IP1 enters FASERν, an emulsion detector made of tungsten plates that act as a target for neutrinos, interleaved with emulsion films to record the trajectories of charged particles. FASERν is followed by two scintillator veto stations that veto charged particles coming through the tunnel walls from IP1, primarily muons. The veto stations are followed by a 0.6 T permanent dipole magnet that also acts as a decay volume for incoming LLPs. It has a 10 cm aperture radius and is 1.5 m long. Next is the spectrometer which consists of two 1 m long 0.6 T dipole magnets. FASER has the IFT (interface tracker) in addition to three tracking stations, each made up of three layers of silicon strip detectors. These are located at either end of the dipole magnets, and one is in between. FASER uses ATLAS silicon trackers (SCT) modules [5]. The pre-shower station, the details of which are shown in Figure 4, is made up of two scintillators, each preceded by a 3 mm thick tungsten radiator in addition to 5 cm of graphite. The primary role of the pre-shower station is for particle identification (PID). Finally, FASER's electromagnetic calorimeter (ECAL), shown in Figure 5, is made of 4 LHCb ECAL modules [6]. These Shashlik-type calorimeter modules contain 66 alternating layers of 2 mm lead and 4 mm plastic scintillator plates, with a total of 25 radiation lengths. Between the lead and scintillator plates, there is a layer of TYVEC paper. A 10 dynode-stage head-on photo-multiplier tube (PMT) module provides a readout signal in the form of PMT pulses, as with the pre-shower station. The role of the calorimeter is to measure the energy deposits made by the electrons and photons that result from the decay of the dark photons and ALPs.

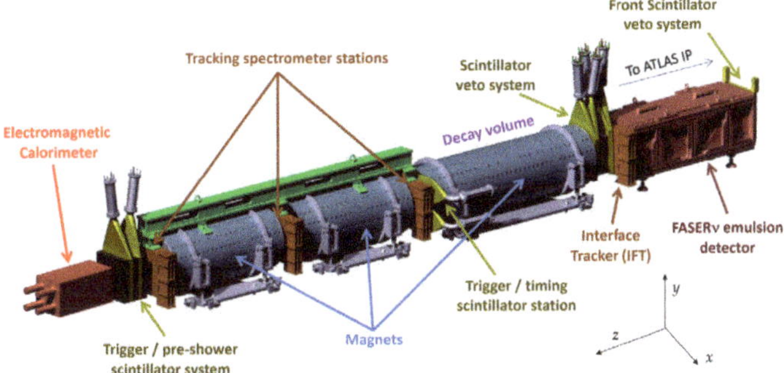

Figure 3. Diagram of FASER and FASERν components [1].

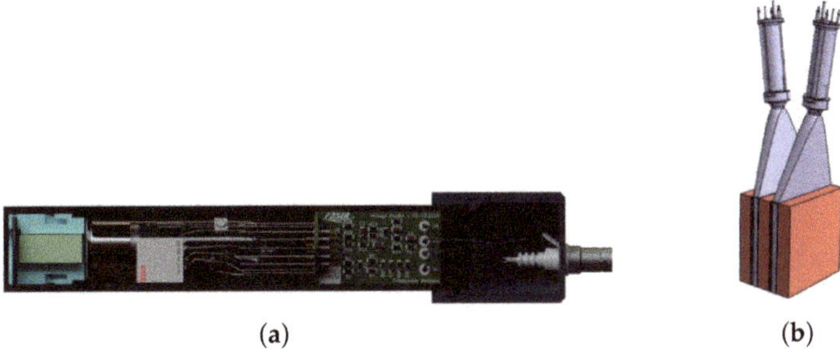

(a)   (b)

Figure 4. (**a**) A diagram of a PMT module that provides readout pulses to the pre-shower station and calorimeter modules. (**b**) A sketch of the pre-shower station showing the two scintillators (grey), each preceded by 3 mm of tungsten and 5 cm of graphite (red).

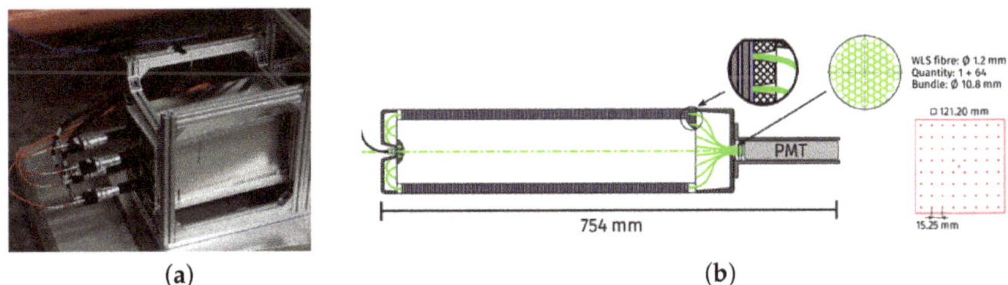

(a)   (b)

Figure 5. (**a**) A photograph taken of the FASER ECAL in the TI12 tunnel, showing the 2 × 2 layout. (**b**) A diagram of the LHCb outer ECAL modules [6] used in FASER.

3. Results of the 2021 FASER Calorimeter Test Beam

3.1. Aims and Overview

The aims of the test beam were to calibrate the calorimeter modules using electron beams with energy between 10 and 300 GeV. Twenty-four positions across the modules' surface were scanned. In addition, the uniformity of the muon response was measured at 150 GeV, and a pion scan was performed at 200 GeV to study the hadronic response. Over 150 million events were recorded over the course of the test beam, primarily from the centre of the upper middle ECAL module. There were a number of runs performed under special conditions with the removal of the pre-shower material to study the resulting effects on energy (charge) deposited in the calorimeter.

The test beam was carried out at the CERN H2 beam line [7]. The test beam detector setup shown in Figure 6 consists of the trigger scintillator counter, the IFT station, the pre-shower station, and the calorimeter. In the test beam, six calorimeter modules were used: the four intended to be installed in the TI12 tunnel, plus two spare modules arranged in a 3×2 configuration.

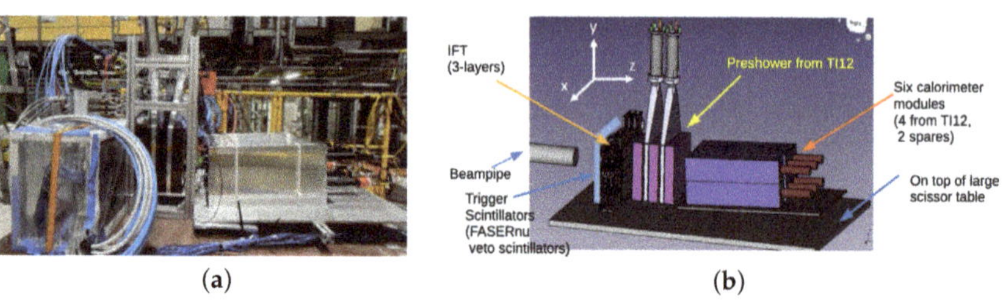

(**a**)   (**b**)

Figure 6. (**a**) A photograph and (**b**) a diagram of the test beam setup, carried out in EHN1 (Experimental Hall North) at CERN. The coordinate system is defined in this figure.

3.2. Data Analysis

The PMT signals from the ECAL, pre-shower, and trigger scintillators are digitised at 500 MHz by 14-bit ADCs and read out in a wide window (1.2 μs), and the integrated charge is summed in a window around the expected peak signal. The readout for most events is triggered by signals in both trigger scintillators exceeding a predefined threshold at the same time. Besides the digitiser, hits in the tracker stations are read out in a 75 ns window and used to reconstruct tracks. The response of the calorimeter modules is studied using events selected as follows:

- The event trigger bit must indicate that the front two trigger scintillators were hit.
- Only one track must be found in the event, with tracks reconstructed according to a dedicated tracking algorithm.
- Tracks must be relatively straight with angle cuts such that the angular spread in the x and y plane is $| \theta_x |$ and $| \theta_y | < 2$ degrees.
- The track position must be within a 20 mm $\times$ 20 mm square area surrounding the beam position, obtained from extrapolating the track to the face of the calorimeter.

For each beam energy E, the resulting charge distribution deposited in the calorimeter is converted in terms of energy deposits and fitted to a crystal ball such that the energy resolution $\frac{\sigma_E}{E}$ can be obtained. Three terms contributed to the resolution according to the relation below:

$$\frac{\sigma_E}{E} = \frac{a}{\sqrt{E}} \oplus \frac{b}{E} \oplus c$$

where $\oplus$ indicates the quadratic sum. The $\frac{a}{\sqrt{E}}$ term is the stochastic term, $\frac{b}{E}$ is the noise term, and c is the constant term.

3.3. Test Beam Simulation

FASER's simulation is based on the Geant4 package [8]. LHCb test beam results using the same ECAL modules were used for comparison when building the simulation and studying the energy response and resolution, before it could be validated by FASER's own test beam data. At this stage, the simulation does not include any digitisation. Digitisation is a step which mimics the detector electrons by converting the simulation output into an output similar to the PMT pulses of real data. A dedicated geometry was developed for the test beam simulation. An example of an event display, produced based on ATLAS VP1 software [9], is shown in Figure 7 with the full test beam setup.

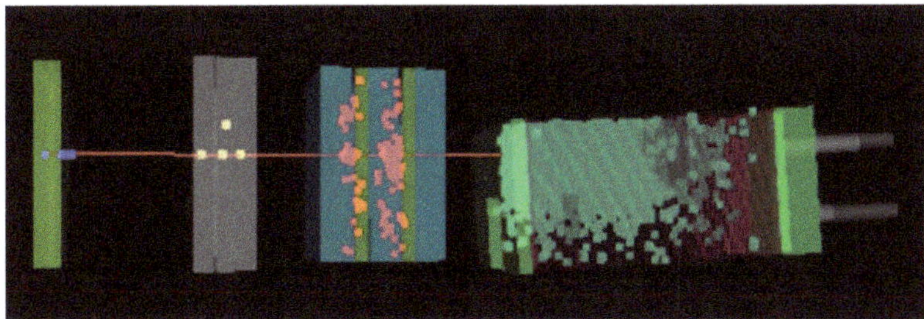

Figure 7. An event display of a simulated 100 GeV electron passing through the tracker station and the pre-shower station, before showering in the ECAL.

3.4. Pre-Shower Correction

The pre-shower "steals" a portion of the EM shower from the calorimeter, as a direct result of the two radiation lengths of the tungsten radiator. This effect varies on an event-by-event basis and thus degrades the energy resolution. This is corrected for in order to obtain the best energy resolution measurement. The total deposited charge in the pre-shower station compared to the total deposited charge in the calorimeter can be seen in Figure 8a for a 100 GeV electron.

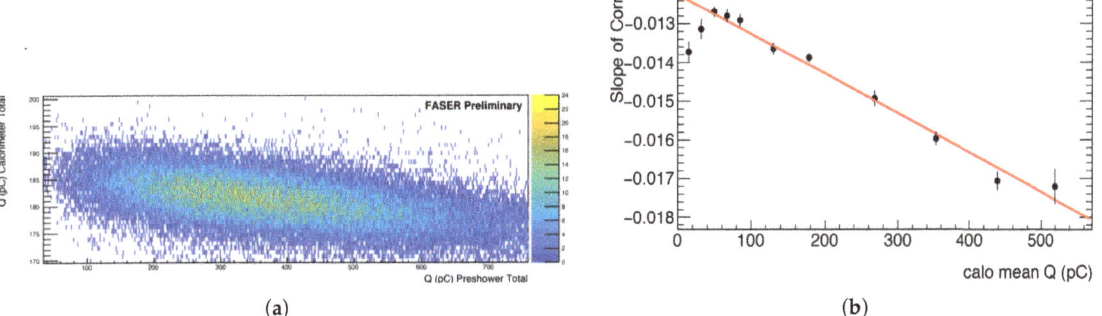

(**a**)   (**b**)

Figure 8. (**a**) The deposited charge in the pre-shower versus calorimeter for 100 GeV electron. (**b**) The slope of correlation fit between charge deposits in the pre-shower and calorimeter, as a function of the mean deposited charge in the calorimeter for electrons with an incident beam energy of 10–300 GeV.

Studying the gradient of Figure 8a, compared to similar plots over the full electron energy range, the test beam data show a dependence on beam energy in terms of the fraction deposited in the pre-shower relative to the calorimeter. This dependence on beam energy is shown in Figure 8b, which demonstrates the correlation of charge deposits in the pre-shower versus the calorimeter for 10 GeV–300 GeV electrons, as a function of the mean charge deposited in the calorimeter. A pre-shower correction was derived to mimic the absence of a pre-shower station, taking into account the deposited charge in the calorimeter and pre-shower station:

$$Q_{corrected} = Q_{calo} + \left(m * Q_{pre-shower} \right),$$

where Q is the total deposited charge and m is the gradient derived from the fit in Figure 8b. The pre-shower correction aims to mimic the distribution of the deposited charge, seen if there was no pre-shower station in the test beam setup, isolating the calorimeter response. This results in an increased energy response and a reduced energy resolution, shown in Figure 9.

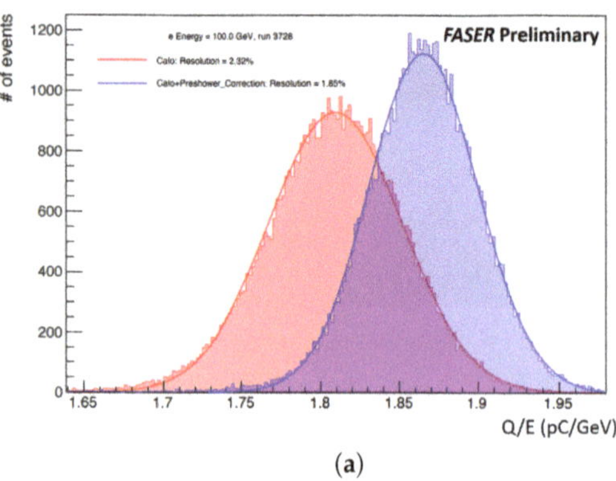

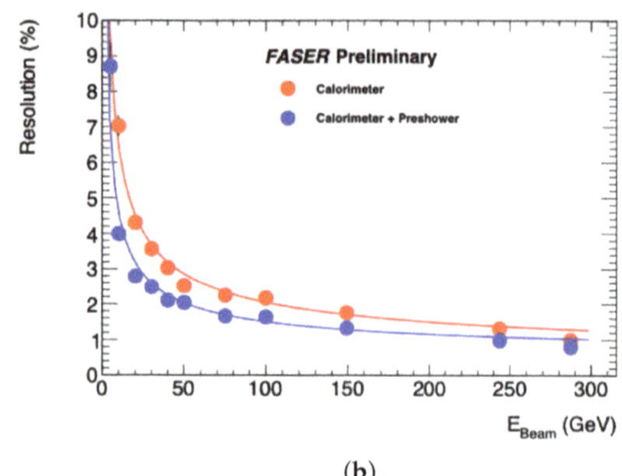

(a)   (b)

Figure 9. (**a**) The change in deposited charge as a result of the pre-shower correction. (**b**) The improvement in energy resolution as a result of the pre-shower correction.

A version of the pre-shower correction must also be applied to the simulation, in order to correct the amount of deposited energy seen in the calorimeter. The deposited energy in the pre-shower versus calorimeter for a 100 GeV electron simulation is shown in Figure 10, as well as the equivalent plot with the correction applied. As with the test beam data, the correction gives an idea of the calorimeter response in the absence of a pre-shower station.

Data with the pre-shower material removed were taken to mimic a setup without the pre-shower station and evaluate as closely as possible the isolated calorimeter response. This effect was also studied in the simulation; the change in energy response from the removal of the tungsten and graphite material from the pre-shower station is shown in Figure 11.

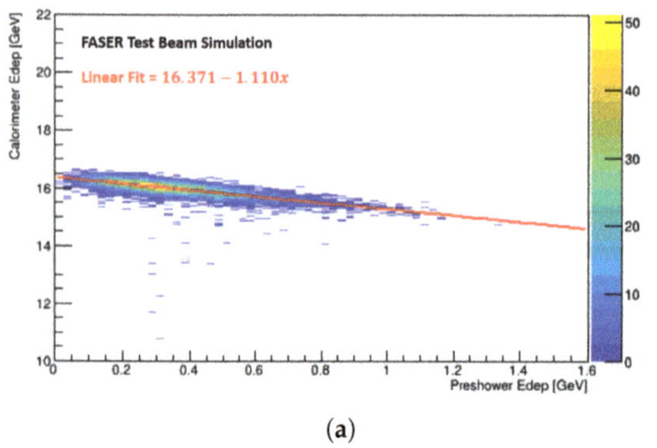

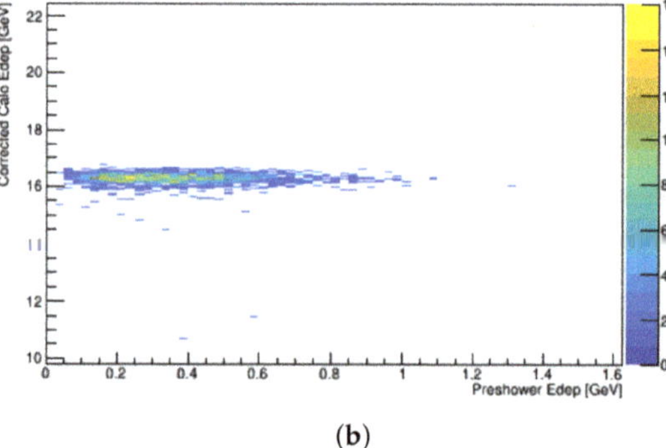

(a)   (b)

Figure 10. The distribution of deposited energy in the pre-shower versus the calorimeter for a simulated 100 GeV electron (**a**) before and (**b**) after the pre-shower correction has been applied.

The simulation software was modified to compare with these data, removing the tungsten radiator and graphite from the simulation. The fraction of total energy deposited in the calorimeter increases from 15.5% to 16.4%. The change in the shape of the distribution was reflected by an improvement in energy resolution, shown in Table 1. There is good agreement between the quantative change in the energy resolution measurements with the pre-shower in place in the data, as summarised in Table 2. This effect is comparable to the application of the pre-shower correction in the data and simulation.

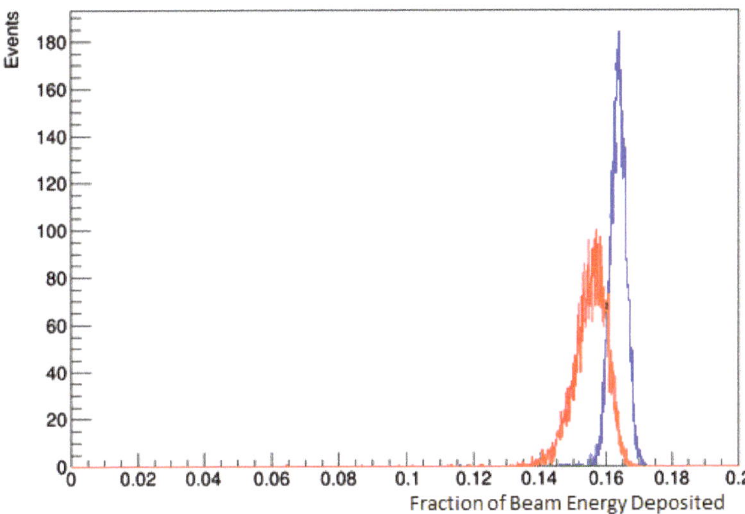

Figure 11. The fraction of beam energy deposited in the calorimeter for a simulated 30 GeV electron, with the tungsten/graphite material removed from the pre-shower station (blue), compared to the default simulation setup (red).

Table 1. Energy resolution measurements in test beam simulation at 2 different electron energies, with and without the pre-shower material.

Energy	With Pre-Shower	Tungsten/Graphite Removed
30 GeV	$2.72 \pm 0.04\%$	$1.58 \pm 0.02\%$
200 GeV	$0.9 \pm 0.01\%$	$0.71 \pm 0.01\%$

Table 2. Energy resolution measurements in test beam data at 2 different electron energies, with and without the pre-shower material.

Energy	With Pre-Shower	Tungsten/Graphite Removed
30 GeV	$3.76 \pm 0.03\%$	$2.84 \pm 0.02\%$
200 GeV	$1.89 \pm 0.01\%$	$1.67 \pm 0.01\%$

3.5. Energy Resolution

With the application of the pre-shower correction in both the test beam data and simulation, it is possible to compare the two, along with a parameterisation of the LHCb test beam results [10]. The calorimeter resolution as a function of electron energy is shown for 10 GeV–300 GeV in Figure 12.

Table 3 displays the energy resolution fit parameters. There is a difference in the a term in the data and simulation; the corrected simulation shows a lower value closer to that seen in the LHCb test beam. However, the addition of a noise term ($\frac{b}{E}$) vastly improves the fit in the case of data. This term was calculated from the measured noise of the digitiser signal. Since this term is related to the electronic noise of the readout chain, it was not included in the fit for the simulation. The simulation is also not able to accurately account for the constant term. Choosing a value of around 1%, in line with what LHCb measured, brings the higher energy end of the distribution upwards towards the test beam data; this is shown by the red fit in Figure 12. There are a number of factors that need to be considered before the differences in the energy response and energy resolution of the data and simulation can be fully understood, for example, the secondary effects discussed in Section 3.8. However, a resolution of 1–2% is more than sufficient for the next step of data analysis.

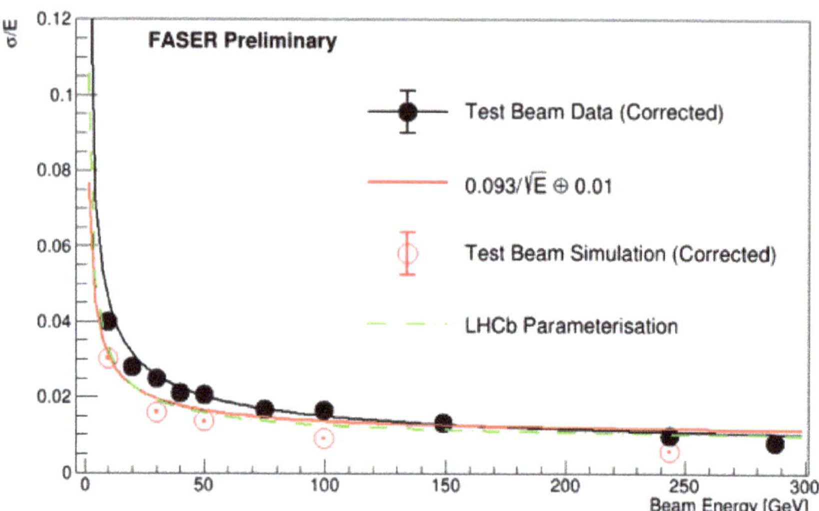

Figure 12. Energy resolution of the pre-shower corrected test beam data (black), the pre-shower corrected test beam simulation (red), and a parameterisation of LHCb test beam results using the same ECAL modules (green). The red line is a fit of the simulation points, with the addition of a 0.01 constant term.

Table 3. Calorimeter energy resolution values seen in Figure 12, including the fit used by LHCb, the corrected and uncorrected test beam data, and the corrected and uncorrected test beam simulation. Note that the test beam simulation does not include a noise term.

	$\sigma_E/E = a/\sqrt{E} \oplus b/E \oplus c$		
	a	b	c
Data (Corrected)	0.134	0.151	0.0065
Data	0.196	0.151	0.0057
Simulation (Corrected)	0.093 ± 0.003	-	0.0000 ± 0.0004
Simulation	0.135 ± 0.001	-	0.0000 ± 0.0017
LHCb	0.094 ± 0.004	0.108 ± 0.029	0.0083 ± 0.0002

3.6. Energy Calibration

The data are in terms of the charge in pC divided by the beam energy, whereas the simulation is shown as a fraction of the beam energy. The deposits in the calorimeter from a 100 GeV electron are shown, in both the data and simulation, in Figure 13. The data and simulation from minimum ionising particles (MIPs) can be used to calibrate the response and convert from pC to GeV. The expected energy from the simulation can be estimated using the below relation:

$$\frac{Q(e^-)}{Q(\mu^-)} = \frac{E(e^-)}{E(\mu^-)}$$

where Q is the mean deposited charge in the calorimeter (from data), and E is the simulated mean deposited energy in the calorimeter, derived from crystal ball fits of the distribution in the case of electrons. For MIPs, the most probable value is derived from Landau fits of the distribution to account for the large tails. The pC to GeV conversion can be applied by considering the ratio of the charge deposited by an electron and the charge deposited by a MIP. Taking this ratio and scaling it according to the simulated energy deposition of an equivalent MIP signal gives an approximation of the energy, in GeV, deposited by such an electron. Comparison with Monte Carlo simulations allows for the extrapolation of other signal types. The resulting calculation showed that roughly 16.5% of beam energy is deposited in the calorimeter; this can be seen in Figure 13b and agrees with predictions from simulations run prior to the test beam.

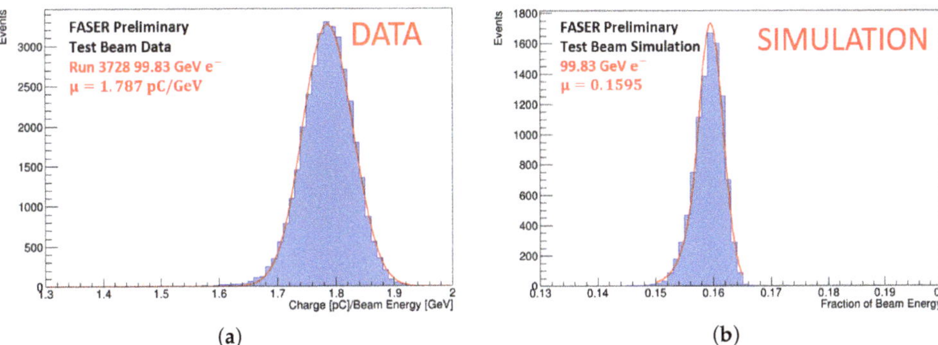

Figure 13. (**a**) Deposited charge in the calorimeter for a 100 GeV test beam electron. (**b**) Deposited energy in the calorimeter for a simulated 100 GeV electron.

3.7. PID Capabilities

With the GeV conversion in place, it is possible to directly compare the data and simulation plots for the purpose of particle identification (PID). Here are example plots that show the signals of muons, pions, and electrons in the data and simulation in the pre-shower and in the calorimeter. The simulated signal from a 30 GeV electron, a 150 GeV muon, and a 200 GeV pion are shown in Figure 14. Plots derived from the test beam data for a 200 GeV electron, 150 GeV muon, and 200 GeV pion are shown in Figure 15. Exploring the PID capabilities of the pre-shower and calorimeter is important for understanding the potential to distinguish a signal from the background. Although the electron signal, particularly in the calorimeter, is distinct, it is likely to become more difficult to distinguish particle types at higher energies. When overlaid, there is a shift in the response between the data and simulation; further investigation is needed into the PID studies and energy calibration.

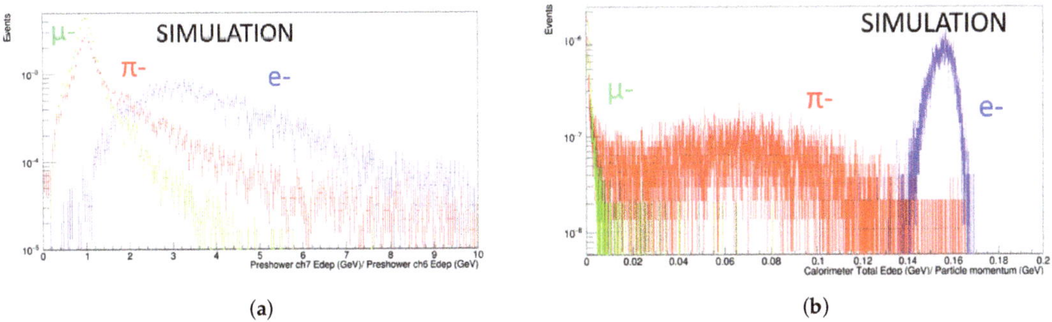

Figure 14. Test beam simulation PID plots showing 200 GeV pion (green), 150 GeV muon (red), and 30 GeV electron (blue) simulated signals. (**a**) The ratio of deposited energy (GeV) in the two pre-shower station scintillator layers. (**b**) The ratio of deposited energy in the calorimeter and particle momentum (GeV).

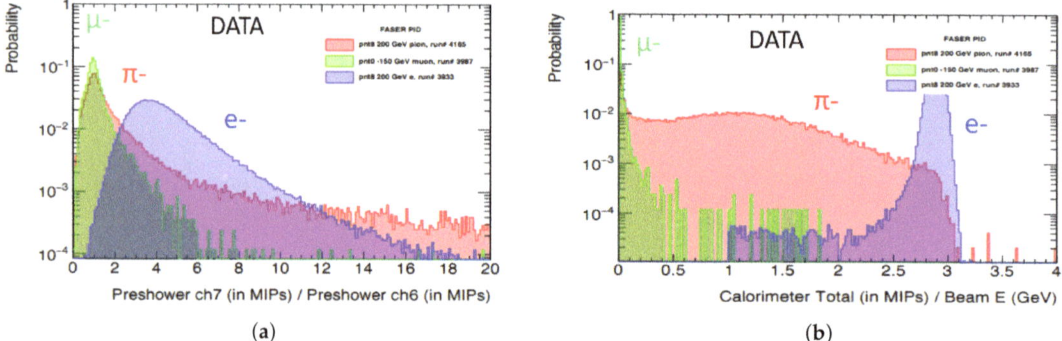

Figure 15. Test beam data PID plots showing 200 GeV pion (green), 150 GeV muon (red), and 200 GeV electron (blue) simulated signal. (**a**) The ratio of deposited energy (GeV) in the two pre-shower station scintillator layers. (**b**) The ratio of deposited energy in the calorimeter and beam energy (GeV).

3.8. Local Calorimeter Effects

It was demonstrated in Section 3.5 that there are differences in the energy resolution fit between the simulation and data; the test beam was an opportunity to study effects in the calorimeter that may play a part in explaining these differences. One such observed effect was improvement in the light collection near wavelength shifting (WLS) optical fibres in the ECAL modules. LHCb set the amplitude of this effect to zero, and this had been coded into the simulation, but now it is possible to tune the specific amplitude based on the FASER test beam results. The calorimeter response as a function of the track position, fitted with a cosine function to extract an amplitude, is shown in Figure 16a. Another effect observed in the data was the increase in the generation of additional light, likely Cherenkov light, in the plastic light mixer in front of the PMT. This is seen when an MIP passes directly through a PMT in the calorimeter. This effect is illustrated in the heat map in Figure 16b.

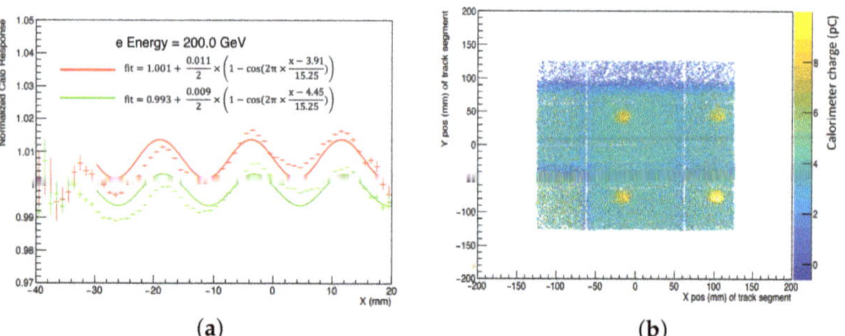

Figure 16. (**a**) Measurement of improvement in light collection near WLS fibres made in the FASER test beam. The normalised calorimeter response as a function of the x position of the track segment is fit with a cosine function in order to extract the amplitude. (**b**) A heat map of the calorimeter, showing an increase in light deposits when an MIP travels through a PMT region.

4. Conclusions

In conclusion, the test beam saw efficient data taking with good overall beam quality and purity. The test beam results have been compared with the simulation, and energy calibration efforts are underway. In addition to this, the PID capabilities of the pre-shower and calorimeter are being studied. The energy response and resolution show generally good agreement between the data and simulation with some differences that are being investigated, and the energy resolution is sufficient for the next step of data analysis. The detector is once again situated in the TI12 tunnel, and the data taking for Run 3 has begun.

Funding: This research received no external funding.

Data Availability Statement: Not applicable

Acknowledgments: FASER is supported by CERN, the Simons Foundation, and the Heising-Simons Foundation.

Conflicts of Interest: The authors declare no conflict of interest.

Abbreviations

The following abbreviations are used in this manuscript:

FASER	Forward Search Experiment at the LHC	IFT	Interface Tracker
ALP	Axion-like Particle	ECAL	Electromagetic Calorimeter
LLP	Long-lived Particle	xAOD	Analysis Object Data
SM	Standard Model	ADC	Analogue to Digital Conversion
DS	Dark Sector	WLS	Wavelength Shifting Fibre
IP1	ATLAS Interaction Point	MIP	Minimum Ionising Particle
SCT	ATLAS Semiconductor Tracker	PMT	Photomultiplier Tube
PID	Particle Identification		

References

1. Ariga, A.; Abreu, H.; Mansour, E.A.; Antel, C.; Ariga, T.; Bernlochner, F.; Boeckh, T.; Boyd, J.; Brenner, L.; Cadoux, F.; et al. The FASER Detector. CERN-FASER-2022-001. *arXiv* **2022**, arXiv:2207.11427.
2. Ariga, A.; Ariga, T.; Boyd, J.; Cadoux, F.; Casper, D.W.; Favre, Y.; Feng, J.L.; Ferrere, D.; Galon, I.; Gonzalez-Sevilla, S.; et al. *Technical Proposal: FASER, the Forward Search Experiment at the LHC*; Tech. Rep. CERN-LHCC-2018-036. LHCC-P-013; CERN: Geneva, Switzerland, 14 December 2018. Available online: http://cds.cern.ch/record/2651328 (accessed on 24 July 2022).
3. Ariga, A.; Ariga, T.; Boyd, J.; Cadoux, F.; Casper, D.W.; Favre, Y.; Feng, J.L.; Ferrere, D.; Galon, I.; Gonzalez-Sevilla, S.; et al. FASER's Physics Reach for Long-lived Particles. *Phys. Rev. D* **2019**, *99*, 095011. [CrossRef]
4. Battaglieri, M.; Belloni, A.; Chou, A.; Cushman, P.; Echenard, B.; Essig, R.; Estrada, J.; Feng, J.L.; Flaugher, B.; Fox, P.J.; et al. US Cosmic Visions: New Ideas in Dark Matter 2017: Community Report. *arXiv* **2017**, arXiv:1707.04591.
5. Abdesselam, A.; Akimoto, T.; Allport, P.P.; Alonso, J.; Anderson, B.; Andricek, L.; Anghinolfi, F.; Apsimon, R.J.; Barbier, G.; MacWaters, C.; et al. The barrel modules of the ATLAS semiconductor tracker. *Nucl. Instrum. Meth. A* **2006**, *568*, 642–671. [CrossRef]
6. LHCb Collaboration. *LHCb Calorimeters: Technical Design Report*; CERN: Geneva, Switzerland, 2000. Available online: http://cds.cern.ch/record/494264 (accessed on 24 July 2022).
7. H2 Beam Line. Available online: http://sba.web.cern.ch/sba/BeamsAndAreas/H2/H2_presentation.html (accessed on 24 July 2022).
8. Geant4 Collaboration. CERN. Available online: https://geant4.web.cern.ch/ (accessed on 24 July 2022).
9. ATLAS Collaboration. The VP1 ATLAS 3D Event Display. Available online: http://atlas-vp1.web.cern.ch/ (accessed on 24 July 2022).
10. Arefev, A.; Kvaratskheliia, T.; Morozov, A.; Melnikov, E.; Voronchev, K.; Roussinov, D.; Barsuk, S.; Belyaev, I.; Bobchencko, B.; Martemyanov, M.; et al. *Beam Test Results of the LHCb Electromagnetic Calorimeter*; CERN-LHCB-2007-149; CERN: Geneva, Switzerland, 2008.

Article

Development of a Novel Highly Granular Hadronic Calorimeter with Scintillating Glass Tiles

Dejing Du [1,2,3] **on behalf of the CEPC Calorimeter Working Group and Yong Liu** [1,2,3,*] **on behalf of the Scintillating Glass R&D Collaboration**

1   Institute of High Energy Physics, Chinese Academy of Sciences, Yuquan Road 19B, Beijing 100049, China
2   University of Chinese Academy of Sciences, Yuquan Road 19A, Beijing 100049, China
3   State Key Laboratory of Particle Detection and Electronics, Yuquan Road 19B, Beijing 100049, China
*   Correspondence: liuyong@ihep.ac.cn; Tel.: +86-10-88236066

Abstract: Based on the particle-flow paradigm, a new hadronic calorimeter (HCAL) with scintillating glass tiles is proposed to address major challenges from precision measurements of jets at the future lepton colliders, such as the Circular Electron Positron Collider (CEPC). Tiles of high-density scintillating glass, with a high-energy sampling fraction, can significantly improve the hadronic energy resolution in the low-energy region (typically below 10 GeV for major jet components at Higgs factories). The hadronic energy resolution of single hadrons and the effects of key parameters of scintillating glass have been evaluated in the Geant4 full simulation, followed by the physics benchmark studies on the Higgs boson with jets in the final state. R&D efforts of scintillating glass materials are ongoing within a dedicated collaboration since 2021 with the aim to achieve a high light yield, a high density, and a low cost. Measurements have been performed for the first batches of scintillating glass samples including the light yield, emission and scintillation spectra, scintillation decay times, and cosmic responses. An optical simulation model of a single scintillating glass tile has been established to provide guidance in the development of scintillating glass. Highlights of the expected detector performance and the latest scintillating glass developments are presented in this contribution.

Keywords: scintillating glass; hadronic calorimeter; high granularity calorimetry; silicon photomultiplier

Citation: Du, D., on behalf of the CEPC Calorimeter Working Group; Liu, Y., on behalf of the Scintillating Glass R&D Collaboration. Development of a Novel Highly Granular Hadronic Calorimeter with Scintillating Glass Tiles. *Instruments* **2022**, *6*, 32. https://doi.org/10.3390/instruments6030032

Academic Editors: Fabrizio Salvatore, Alessandro Cerri, Antonella De Santo and Iacopo Vivarelli

Received: 31 July 2022
Accepted: 25 August 2022
Published: 2 September 2022

Publisher's Note: MDPI stays neutral with regard to jurisdictional claims in published maps and institutional affiliations.

1. Introduction

High-energy electron–positron collider experiments have been proposed for precision measurements of the Higgs boson, which was discovered at the Large Hadron Collider (LHC) in 2012 [1,2], and to explore new physics beyond the Standard Model. the Circular Electron Positron Collider (CEPC), as one option among next-generation colliders as Higgs factories, requires accurate identification and reconstruction of all final states from Higgs, W, and Z bosons. Therefore, the jet energy resolution of the CEPC detector needs to achieve $\sim 30\%/\sqrt{E_{jet}(GeV)}$ [3], which poses challenges for the calorimetry system. A feasible paradigm to achieve this goal is the high granular calorimetry based on the particle flow algorithm (PFA) [4], which makes use of the optimal sub-detector accordingly to determine the energy-momentum of each particle within a jet. An essential prerequisite for calorimeters is to distinguish clusters of nearby individual particles in order to match the tracking system for charged particles and identify clusters originating from neutral particles, which can only be measured in calorimeters. PFA-oriented calorimeters with various technical options featuring high granularity to achieve an excellent three-dimensional spatial resolution are being developed and extensively studied within the CALICE collaboration [5].

As the majority of jet components at Higgs factories with a center-of-mass energy of 240 GeV are relatively low energy (mostly below 10 GeV, as shown in Figure 1), a better hadronic energy resolution would be useful for better PFA performance and jet measurement precision. Hereby, we propose a new design for a highly granular HCAL with high-density scintillating glass tiles, with a higher-energy sampling fraction and PFA compatibility, to further improve the hadronic energy resolution. Its detector layout generally is similar to the CALICE scintillator-steel hadronic calorimetry (AHCAL) technique, proposed in the CEPC Conceptual Design Report [3], but instead of a plastic scintillator, scintillating glass tiles are instrumented.

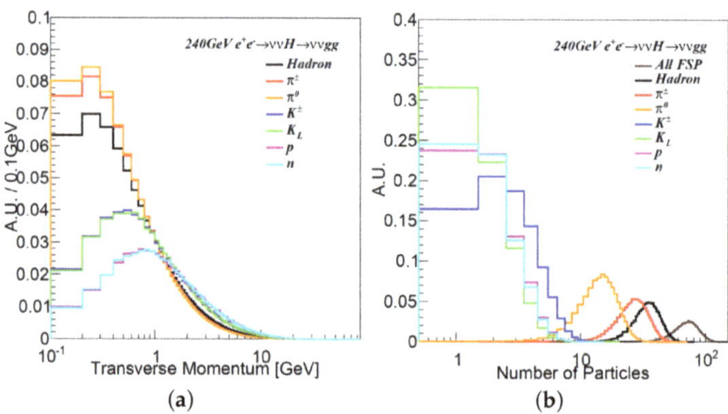

Figure 1. Jet components of the $e^+e^- \to \nu\nu H \to \bar{\nu}\nu gg$ at 240 GeV with distributions of (**a**) transverse momenta and (**b**) the number of different particles within a jet.

In this proceeding, Section 2 introduces the performance studies and physics potentials with this HCAL design. Recent progress of high-density scintillating glass R&D activities and characterization results of glass samples are covered in Section 3, followed by simulation studies, as well as measurements in Section 4 for an HCAL detector unit and a summary in Section 5.

2. Performance Studies of the Scintillating Glass HCAL

Dense scintillating glass with a moderate light yield and adjustable ingredients is usually considered as a promising option for calorimetry applications and more cost effective compared with scintillating crystals. Traditional calorimetry designs used crystals or glass in the form factor of large-volume blocks and, thus, required considerably high intrinsic light yield and transmittance. On the other hand, small-sized tiles with SiPM readout for PFA calorimetry would collect scintillation light more efficiently and, thus, significantly loosen the requirements on light yield and transmittance, which makes scintillating glass a promising option for the applications in high-granularity calorimetry.

The scintillating glass HCAL is designed as a sampling calorimeter, which consists of 40 longitudinal layers with around 4.8 λ_I (λ_I as the nuclear interaction length). Each layer with 0.12 λ_I contains a steel plate as the absorber and a sensitive layer with scintillating glass tiles read out individually by silicon photomultipliers. Geant4 [6] full simulation (with version 10.7.4 and the physics list "QGSP_BERT") has been established, including all relevant physics processes for EM and hadronic showers, and its general setup is shown in Figure 2 with an HCAL module and the layer structure. Selected results on the performance evaluation and optimizations of the HCAL design are presented in the following.

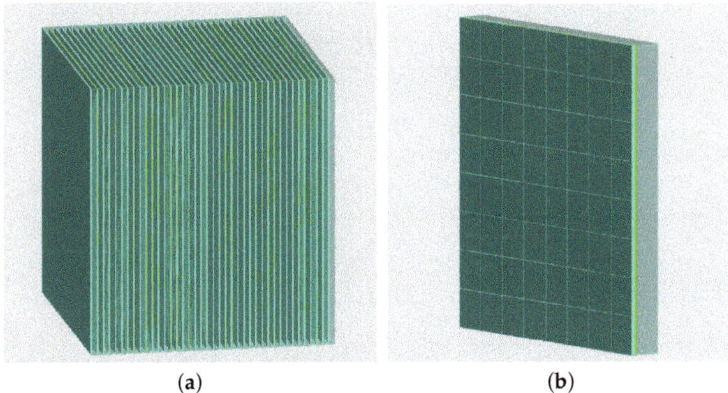

(a)   (b)

Figure 2. The HCAL structure visualization using Geant4: (**a**) an HCAL module with a transverse size of 108×108 mm^2 and 40 longitudinal layers; (**b**) a longitudinal layer including a steel plate, scintillator tiles, and 2 mm-thick readout PCB. The transverse size of the scintillator tiles is set to 3×3 mm^2. The thickness of a steel plate and a layer of scintillator tiles can be tuned, but with a fixed value of 0.12 λ_I to meet the requirement of the CEPC CDR.

2.1. Hadronic Energy Resolution

Compared with the plastic scintillator, high-density scintillating glass can significantly increase the energy sampling fraction, which is beneficial to improve the energy resolution. Based on PFA fast simulation to factorize the jet energy resolution or the boson mass resolution, the hadronic energy resolution (obtained with single hadrons), among many key factors, ranks the second-most important factor for the PFA's performance [7].

In order to evaluate the performance potential, hadronic energy resolutions of the HCAL with plastic and glass were compared with the Geant4 simulation. The properties of the scintillating glass in the simulation corresponds to the glass sample #7 in Table 1. Three scenarios are compared and shown in Figure 3: each layer with: (1) a 3 mm-thick plastic scintillator (negligible in terms of λ_I) and a 20 mm steel plate (blue); (2) a 3 mm (0.011 λ_I)-thick scintillating glass and a 20 mm steel plate (red); (3) a 23 mm (0.084 λ_I)-thick scintillating glass (green), all with the ideal energy threshold of 0 MIP, so that all hits are effectively collected. It shows that scintillating glass HCAL is expected to have a better hadronic energy resolution especially with incident kinetic energies below 30 GeV. It should be fair to state that the plastic scintillator options provide an acceptable energy resolution, but also that scintillating glass offers substantially better performance.

For further detailed studies presented in the following, the thickness of the scintillating glass varies from 0.01 λ_I to 0.12 λ_I, while the steel thickness is changed accordingly, so that each layer is fixed at 0.12 λ_I in all scenarios. The λ_I of the scintillating glass (with the constituent recipe described in Section 3) and steel are 22.4 cm and 16.8 cm, respectively. Figure 4 shows the impact of the scintillating glass thickness on the hadronic energy resolution, using neutral kaon MC samples, including two sets of energy thresholds per channel. It can be found that the thickness of the scintillating glass and the energy threshold can significantly affect the energy resolution. A lower threshold would always be desirable for better energy resolution. It should be pointed out that the glass density is another crucial factor. Figure 3 represents an earlier study with the density being 4.94 g/cm^3, corresponding to scintillating glass samples. In the rest of the studies, for example in Figure 3, the density used in the simulation was set to 6 g/cm^3, which corresponds to the goal of scintillating glass R&D.

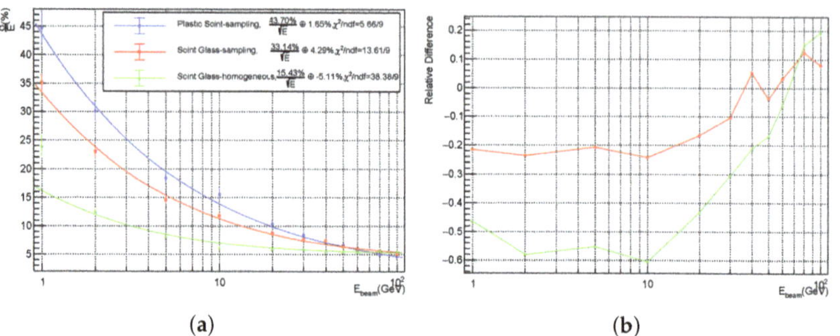

(a)　　　　　　　　　　　　　　　　**(b)**

Figure 3. Performance with single neutral kaons (K_L^0) in the kinetic energy range from 1 GeV to 100 GeV perpendicular to the incidence of the calorimeter surface: (**a**) hadronic energy resolutions with different sensitive materials; (**b**) relative differences of hadronic energy resolutions with scintillating glass compared with with 3 mm plastic scintillator and 20 mm steel.

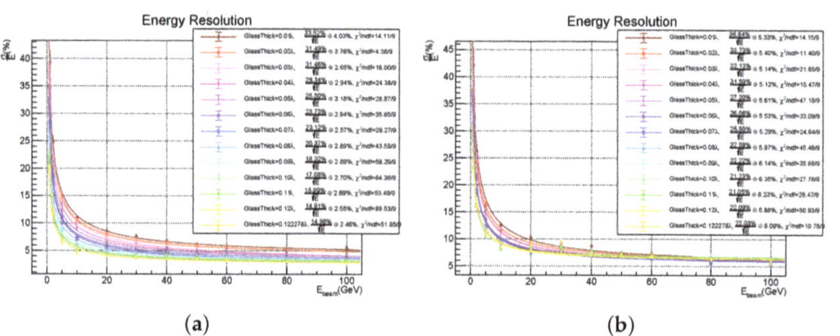

(a)　　　　　　　　　　　　　　　　**(b)**

Figure 4. Energy resolutions of the HCAL with different thicknesses of scintillating glass for K_L^0 with the kinetic energy ranging from 1 GeV to 100 GeV. The glass thickness varies from 0.01 λ_I to 0.12 λ_I and corresponds to solid points in different colors (dark red and orange for the minimum and maximum sampling fraction, respectively) with different energy thresholds per channel of 0–0.3 MIP. The sampling fraction uses the energy deposition obtained directly from Geant4, without considering the readout implementation or corrections. (**a**) Threshold = 0 MIP per channel; (**b**) threshold = 0.3 MIP per channel.

The stochastic and constant terms of the energy resolution are extracted from each scenario in Figure 4 and shown in Figure 5. The two sets of energy thresholds of 0 and 0.3 MIP per readout channel correspond to an ideal configuration and a realistic one, respectively. Generally, with a given threshold, the stochastic term is improved with thicker glass. With the energy threshold above 0.3 MIP, the stochastic term remains almost constant when the glass becomes thicker than 0.08 λ_I. As shown in Figure 4c,d, a higher threshold significantly degrades the constant term.

It should be noted that the HCAL design is non-compensated in general, i.e., the responses to the hadronic components (denoted as h) and electromagnetic (EM) ones (denoted as e) in hadronic showers are not equal ($h/e < 1$), and normally, the h/e ratio increases along with the incident particle energy [8]. Therefore, the energy resolution degrades in the high-energy region when the glass becomes thicker, as it is more sensitive to the EM components. The software compensation technique [9] in high-granularity calorimeters is a feasible option to assign different energy density weights, determined by the energy deposition per tile, to equalize responses to EM and hadronic components, which can significantly improve the energy resolution, and simulation studies are ongoing for the scintillating glass option.

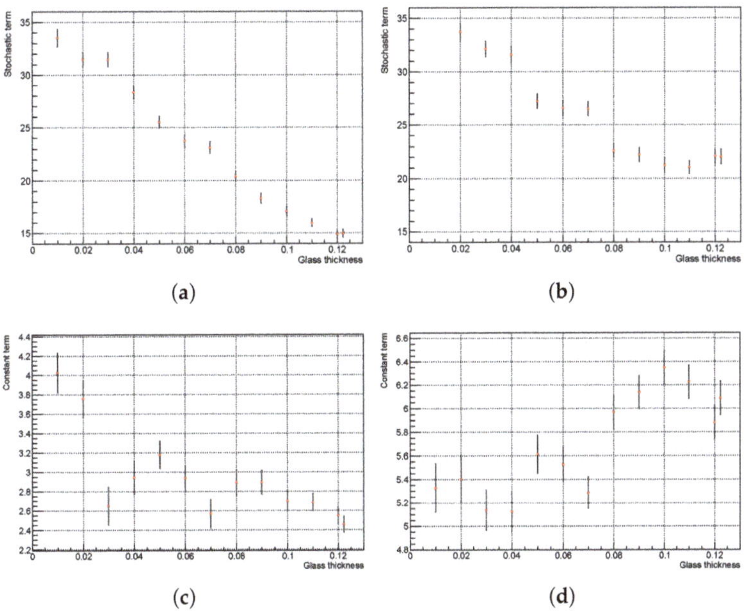

Figure 5. Stochastic and constant terms of hadronic energy resolution versus the thickness of scintillating glass with different energy thresholds. (**a**) Stochastic term with the threshold = 0 MIP; (**b**) stochastic term with the threshold = 0.3 MIP; (**c**) constant term with the threshold = 0 MIP; (**d**) constant term with the threshold = 0.3 MIP.

2.2. Boson Mass Resolution

As the majority of hadrons in jets at the CEPC are low energy, the scintillating glass HCAL has great potential for improving the jet energy resolution. Jet performance with Higgs hadronic decays has been evaluated with the scintillating glass HCAL implemented in the full CEPC detector, where the other sub-detectors are kept the same as the CEPC CDR baseline.

The boson mass resolution (BMR) is hereby used as a key parameter to quantify the physics performance. In this study of $ZH \rightarrow \nu\nu gg$ at 240 GeV, the BMR is the resolution of the Higgs invariant mass reconstructed from two gluon jets. As shown in Figure 6, the BMR with the CEPC CDR baseline detector design is around 3.8%. In the scenario of a homogeneous HCAL with scintillating glass tiles to replace the CDR baseline HCAL, the BMR is improved by around 10% to be 3.45%. A particle flow algorithm named "ArborPFA" [10] was used in the study, and the PFA-related parameters were those tuned with the CDR baseline HCAL. It is expected that the BMR can be further improved by optimizing the PFA for the scintillating glass HCAL.

Given that the stochastic energy resolution term is strongly affected by the per-channel energy threshold, it should be pointed out that a sufficiently low energy threshold (around 2.5% MIP) was implemented in the BMR simulation at this stage to illustrate the physics potentials. A range of realistic threshold values, considering possible constraints from photosensors, front-end electronics, the trigger, the DAQ, etc., will be further studied to evaluate the impacts on the BMR performance.

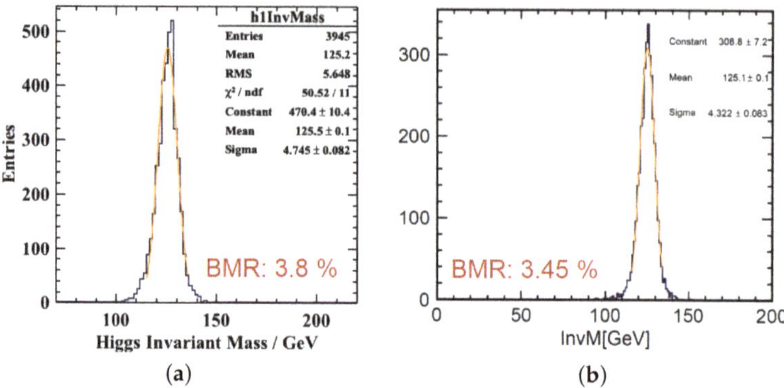

Figure 6. BMR of $ZH \rightarrow \nu\nu gg$ at 240 GeV with a low energy threshold around 2.5% MIP. (**a**) CEPC CDR baseline detector; (**b**) CEPC CDR baseline detector with the baseline SiW ECAL and the HCAL replaced by a homogeneous HCAL, which is 40 layers of $40 \times 40 \times 40$ mm^3 glass tiles (included with readout PCB and without any absorber). It needs to be noted that this setup configuration is not meant to be the final HCAL design, nor to meet the CEPC CDR requirement, but only to illustrate the physics potentials with the sufficient depth in the HCAL. Ongoing studies are being carried out with the exact same depth as the CDR requirement of around 4.8 λ_I as the total depth.

3. R&D of Scintillating Glass

The R&D activities of scintillating glass materials for the CEPC-PFA-oriented hadronic calorimeter were initiated in 2021, and a scintillating glass collaboration was established in China. The collaboration aims to synthesize high-density, transparent, high-light-yield, and cost-effective glass materials and has developed several sample batches. To evaluate the glass performance, dedicated setups have been developed to measure the optical and scintillating characteristics of scintillating glass samples in the mm scale (the transverse size around 5×5 mm^2, the thickness around 3 mm): (1) a setup with radioactive sources (Cs-137 and Na-22) for the intrinsic light yields, energy resolutions, and decay times; (2) an ultraviolet–visible spectrometer (Lambda 650, PerkinElmer, Waltham, MA, USA) for transmission spectra; (3) X-ray sources with a tungsten target (Moxtek, MAGPRO) and a spectrometer (Omni Lambda 300i, Zolix) for X-ray excited luminescence (XEL) spectra. More details about the instrumentations and methods can be found in [11]. Over 30 pieces of samples have been measured, among which the glass sample with the best performance was aluminoborosilicate glass, with the ingredients of $B_2O_3 - SiO_2 - Al_2O_3 - Gd_2O_3 - Ce_2O_3$.

The transmittance, defined as the ratio of the light passing through to the incident light on the samples [12], is a key parameter to quantify the glass transparency and to affect its light output. The transmission spectra of scintillating glass samples are shown in Figure 7a. The absorption edge of all samples is located near 360 nm. When the wavelength is longer than 400 nm, the transmittance of sample #4 is higher than 72%. Relatively high transmittance (>75%) is required for scintillating glass.

The XEL emission spectra of seven samples (#1 to #7) are shown in Figure 7b. The measured results show that scintillating glass has broadband emissions in the range of 300 to 600 nm. The gap at around 365 nm is due to the switching of the filter in the instrument [11]. The peak of the emission spectra of all samples is around 393 nm, which matches the photon detection efficiency (PDE) spectrum of most common SiPMs.

The full-energy peak from radioactive sources is used for energy calibration. The energy resolution is defined as the FWHM (namely $2.355 \times \frac{\sigma}{mean}$) of the full-energy peak. Figure 7c shows the energy spectra of sample #7 with a ^{137}Cs source (662 keV gammas); its energy resolution is 27.5% at 662 keV, and the light yield was measured to be about 881 photons/MeV [11]. The intrinsic light yield of scintillating glass is aimed to be in the range of 1000–2000 photons/MeV.

Figure 7d shows the decay times of sample #7, consisting of a fast component and a slow one, which is 329 ns (20%) and 839 ns (80%), respectively. For the foreseen high-luminosity Z-pole operation at the CEPC, the decay time of scintillating glass needs to be reduced significantly and is in general required to be on the order of 100 ns.

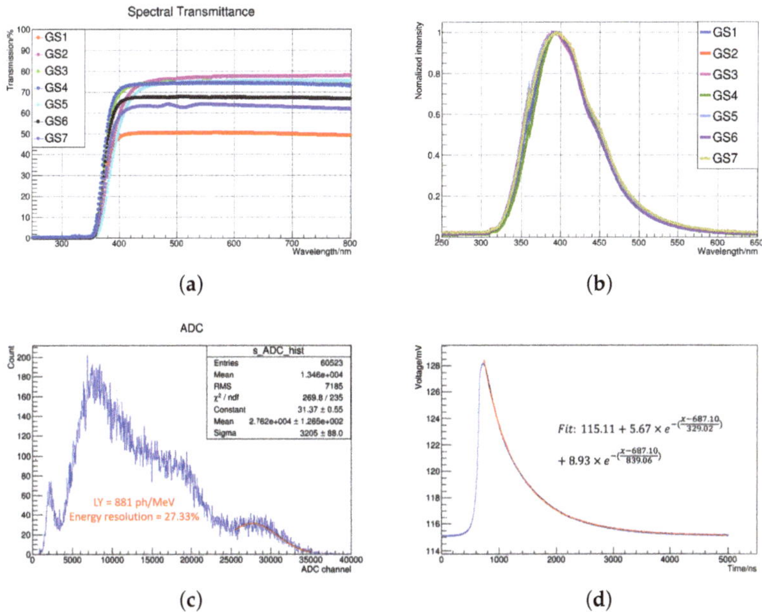

Figure 7. The partial measured results of aluminoborosilicate glass samples. (**a**) Transmission spectra of samples #1–#7; (**b**) normalized X-ray-excited luminescence emission spectra for samples #1–#7; (**c**) the energy spectra of sample #7 under a ^{137}Cs (662 keV) radioactive source; (**d**) scintillating decay time of sample #7.

Table 1 summarizes the characterization results of the optical and scintillating properties of the glass samples (all aluminoborosilicate glass). In general, glass sample #7 with a composition of $25B_2O_3 - 30SiO_2 - 10Al_2O_3 - 34Gd_2O_3 - 1Ce_2O_3$ shows the best performance, with a transmittance at visible wavelengths around 64%, a light yield of 881 photons/MeV, and a density of about 5 g/cm^3. The targeted values of the properties of scintillating glass are summarized as follows: a density around 6 g/cm^3, a light yield in the range of 1000–2000 photons/MeV, a transmittance around 75%, a decay time on the order of 100 ns. Till now, all glass samples were in the millimeter scale, and the current R&D efforts focus on the development of centimeter-scale scintillating glass samples.

Table 1. Characterization results of samples #1–#7 of scintillating glass.

Sample	Density (g/cm^3)	Transmittance (%)	Emission Peak (nm)	Light Yield (ph/MeV)	Energy Resolution (%)	Decay Time (ns)
#1	~4.5	50	394	546	31.04	273, 1007
#2	~4.5	78	392	536	36.47	334, 939
#3	~4.5	75	393	680	29.02	351, 1123
#4	4.65	74	396	660	30.46	308, 1363
#5	4.94	74	392	705	27.84	354, 760
#6	4.53	67	393	802	26.77	318, 1380
#7	4.94	64	394	881	27.33	329, 839

4. Simulation Studies and Measurements of an HCAL Detector Unit

The HCAL detector unit consists of a scintillating glass tile and a silicon photomultiplier (SiPM). Glass sample #7 ($4.5 \times 4.5 \times 3.5$ mm^3 after cutting and polishing) in Table 1, which was tested to have the best performance, was selected for detailed studies as the basis to extrapolate to the cm-sized tiles. The SiPM-type with the following studies was selected with the Hamamatsu S13360-6025PE [13].

4.1. MIP Response

The minimum ionizing particle (MIP) response of an individual detector unit provides the energy scale for the energy reconstruction of the highly granular HCAL. Muons in cosmic rays on the ground are good MIP candidates and were used for MIP calibration. Hereby, the MIP response is defined as the number of photons detected at the SiPM placed in the tile center of the transverse plane.

4.1.1. Cosmic Ray Experiment

As shown in Figure 8, a dedicated cosmic ray experiment was developed, using plastic scintillator tiles (as the top and bottom triggers) and the scintillating glass sample placed in between. The glass sample was wrapped with an ESR foil (with an air gap between the glass surface and the ESR foil) and directly air-coupled with a SiPM. Two trigger tiles were used for the coincidence to make sure cosmic muons pass the glass sample and constrain the incident angle range. However, as the size of the trigger tiles is larger than the sample, there was still a part of the cosmic muons with an incident angle to the glass surface normal. Figure 9a shows the MIP response of a scintillating glass tile measured by cosmic ray muons with the most probable value (MPV) of 277 detected photons at the SiPM.

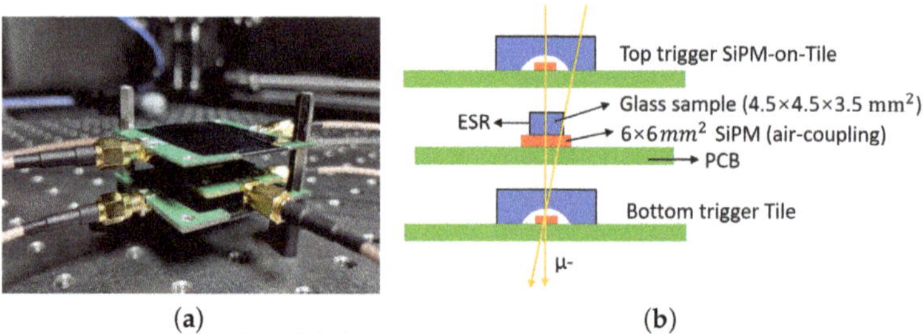

(a)   (b)

Figure 8. The experimental setup used for the cosmic ray test. (**a**) is the real experimental setup and (**b**) is the schematic of the experimental setup.

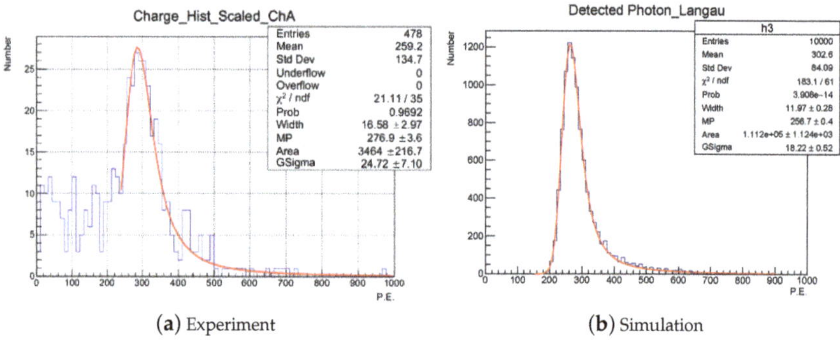

(**a**) Experiment   (**b**) Simulation

Figure 9. The MIP response of a scintillating glass tile: (**a**) cosmic ray test and (**b**) optical simulation.

4.1.2. Optical Simulation

Based on the cosmic ray experiment, the measurement data were used to validate the Geant4 optical simulation for an HCAL detector unit. As shown in Figure 10a, muons pass vertically through the tile. The scintillating glass sample contains many small bubbles, which were taken into account in the simulation, as shown in Figure 10b. Figure 9b shows the MIP response of a scintillating glass tile with the Geant4 optical simulation, with the most probable value (MPV) of 257 detected photons. This study demonstrated that the Geant4 optical simulation can well reproduce the measurements. As the muon's incidence in the simulation is exactly perpendicular to the tile surface, it is reasonable that the simulation expects a slightly smaller MIP response than the measurements.

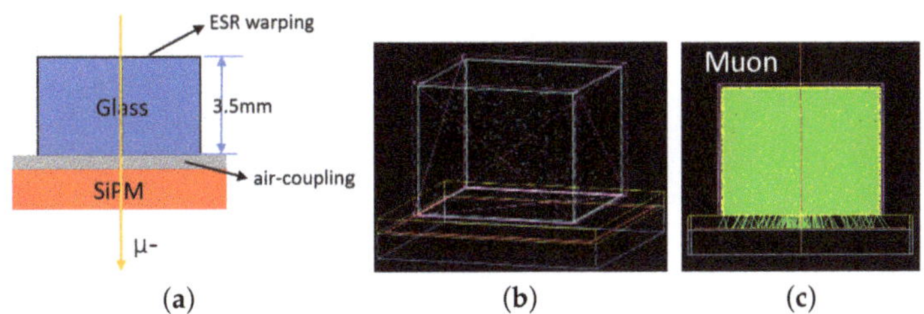

Figure 10. (**a**) The schematic of the geometry setup; (**b**) geometry display in the Geant4 simulation; (**c**) event display of a 1 GeV muon in the Geant4 simulation.

4.2. Projected Performance

At present, as the size of all samples is in the millimeter scale, the performance of the detector unit with realistic transverse size (30×30 mm^2) tiles can only be obtained through simulation. Assuming that the properties of larger glass tiles remain the same as small glass samples, the response uniformity of a scintillating glass tile was studied in the Geant4 optical simulation (after validation with cosmic ray tests) by changing the incident position with a step size of 0.5 mm. The response refers to the number of photons detected at the SiPM when the incident muons pass perpendicularly through the position (X-Y in Figure 11) of a tile.

As shown in Figure 11a, the center of the tile has the highest response, as the SiPM is coupled in the tile center and has a much higher light collection efficiency. The response uniformity of the tile is represented by $\frac{max-min}{average}$, where max, min, and average are the maximum, minimum, and average values of the response. The average response and uniformity of scintillating glass tiles of different thicknesses are shown in Table 2. It can be found that the optimal thickness appears to be around 10 mm. Nevertheless, the response uniformity of the tile is far from optimal, and the uniformity needs to be improved by increasing the transmittance of the scintillating glass and optimizing the tile design.

Due to the limited transmittance of scintillation glass, a significant part o the scintillation photons will be self-absorbed by the glass. As the SiPM is located in the geometric center of the glass tile, the tile response depends on the scintillation location, determined by the position of incident particles. The farther the distance between the incident position from the SiPM, the more photons will be absorbed in the propagation process. When the incident position is within the SiPM sensitive area (hereby, 6×6 mm^2), the light collection efficiency is much higher, as most photons are directly detected without many reflections, leading to a much higher response in this region. When the glass becomes thicker than 3 mm, more scintillation photons are generated due to the higher energy deposition, leading to the higher response in general. On the other hand, when the glass becomes even thicker, the self-absorption effect will start to dominate and more photons will be absorbed in the glass, before being detected by the SiPM. The tendency near the tile central region with the SiPM shows that the tile response becomes lower (dimmer in in color scale) when the tile becomes thicker (e.g., thicker than 10 mm). In general, Geant4 also very much preserves the detailed information of the optical processes and can be extracted for further quantitative studies, which would be essential for optimizing the tile design and improving the response non-uniformity.

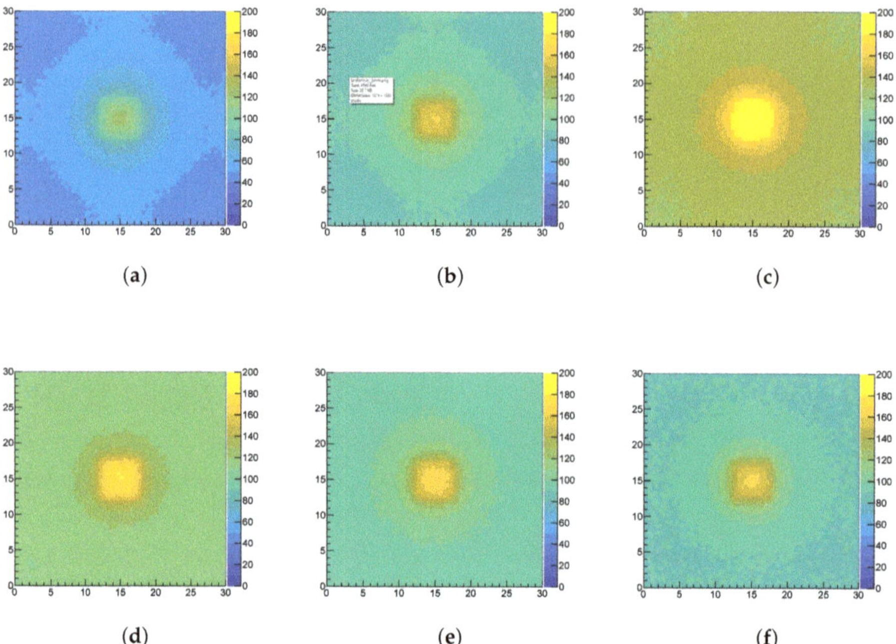

Figure 11. Response uniformity of different thicknesses of scintillating glass tile. Transverse size is fixed with 30×30 mm^2: (**a**) $30 \times 30 \times 3$ mm^3; (**b**) $30 \times 30 \times 5$ mm^3; (**c**) $30 \times 30 \times 10$ mm^3; (**d**) $30 \times 30 \times 15$ mm^3; (**e**) $30 \times 30 \times 20$ mm^3; (**f**) $30 \times 30 \times 25$ mm^3.

Table 2. Average response and non-uniformity with different thicknesses.

Thickness (mm)	3	5	10	15	20	23
Average Response	65	106	149	127	112	104
Non-Uniformity	1.1	0.67	0.47	0.53	0.63	0.71

5. Summary and Prospects

A new high-granularity HCAL concept with high-density scintillating glass tiles was proposed to further improve the energy resolution and the BMR. Compared with the plastic scintillator, the scintillating glass HCAL with a higher energy sampling fraction has a better hadronic energy resolution, especially with incident kinetic energies below 30 GeV. The software compensation technique [9] in high-granularity calorimeters is expected to significantly improve the energy resolution, and simulation studies are ongoing for the scintillating glass option. In addition, the scintillating glass HCAL has great potential to improve the BMR. The R&D of centimeter-scale scintillating glass with high density, transmittance, and light yield is ongoing. The cosmic ray experiment and simulation of the detector unit were carried, and the results were used as a guide for the detector design and the R&D of scintillating glass materials.

Funding: This project received funding support from the Chinese Academy of Sciences (CAS) Talent Program (Category B for young professionals) and the CAS Center for Excellence in Particle Physics (CCEPP).

Data Availability Statement: Not applicable.

Acknowledgments: The authors would like to express their gratitude for the fruitful discussions with many of the colleagues in the CEPC calorimeter working group and the CALICE collaboration.

Conflicts of Interest: The authors declare no conflict of interest.

References

1. Collaboration, C. Observation of a new particle in the search for the Standard Model Higgs boson with the ATLAS detector at the LHC. *Phys. Lett. B* **2012**, *716*, 1–29. [CrossRef]
2. Collaboration, A. Observation of a new boson at a mass of 125 GeV with the CMS experiment at the LHC. *Phys. Lett. B* **2012**, *716*, 30–61. [CrossRef]
3. The CEPC Study Group. CEPC Conceptual Design Report: Volume 2—Physics & Detector. *arXiv* **2018**, arXiv:1811.10545.
4. Brient, J.C. Improving the Jet Reconstruction with the Particle Flow Method: An Introduction. In *Calorimetry in Particle Physics*; World Scientific Publishing Co.: Singapore, 2004. [CrossRef]
5. CALICE Website. Available online: https://twiki.cern.ch/twiki/bin/view/CALICE/WebHome (accessed on 30 July 2022).
6. Agostinelli, S.; Allison, J.; Amako, K.; Apostolakis, J.; Araujo, H.; Arce, P.; Asai, M.; Axen, D.; Banerjee, S.; Barrand, G.; et al. Geant4—A simulation toolkit. *Nucl. Instrum. Methods Phys. Res. Sect. Accel. Spectrom. Detect. Assoc. Equip.* **2003**, *506*, 250–303. [CrossRef]
7. Ruan, M. Performance Requirement from the Hadronic event/jet. LCWS2019. Available online: https://agenda.linearcollider.org/event/8217/contributions/44771/attachments/34967/54047/Jet_Requirement-LCWS.pdf (accessed on 30 July 2022).
8. Eigen, G.; Price, T.; Watson, N.K.; Marshall, J.S.; Thomson, M.A.; Ward, D.R.; Benchekroun, D.; Hoummada, A.; Khoulaki, Y.; Apostolakis, J.; et al. Hadron shower decomposition in the highly granular CALICE analogue hadron calorimeter. *J. Instrum.* **2016**, *11*, P06013. [CrossRef]
9. Adloff, C.; Blaha, J.; Blaising, J.J.; Drancourt, A.O. Hadronic energy resolution of a highly granular scintillator-steel hadron calorimeter using software compensation techniques. *JINST* **2012**, *7*, P09017. [CrossRef]
10. Ruan, M.; Zhao, H.; Li, G.; Fu, C.; Wang, Z.; Lou, X.; Yu, D.; Boudry, V.; Videau, H.; Balagura, V.; et al. Reconstruction of physics objects at the Circular Electron Positron Collider with Arbor. *Eur. Phys. J. C* **2018**, *78*, 1–14. [CrossRef]
11. Tang, G.; Hua, Z.; Qian, S.; Sun, X.; Ban, H.; Cai, H.; Li, S.; Liu, H.; Liu, S.; Ma, L.; et al. Optical and scintillation properties of aluminoborosilicate glass. *Opt. Mater.* **2022**, *130*, 112585. [CrossRef]
12. Raymond, C.; Ronca, S. Chapter 6—Relation of Structure to Electrical and Optical Properties. In *Brydson's Plastics Materials*, 8th ed.; Gilbert, M., Ed.; Butterworth-Heinemann: Oxford, UK, 2017; pp. 103–125. [CrossRef]
13. Hamamatsu S13360 Series. Available online: http://www.hamamatsu.com.cn/UserFiles/upload/file/20190728/1.pdf (accessed on 30 July 2022).

Article

Photodiode Read-Out System for the Calorimeter of the Herd Experiment

Pietro Betti [1,2,*], Oscar Adriani [1,2], Matias Antonelli [3], Yonglin Bai [4], Xiaohong Bai [4], Tianwei Bao [4], Eugenio Berti [1,2], Lorenzo Bonechi [2], Massimo Bongi [1,2], Valter Bonvicini [3], Sergio Bottai [2], Weiwei Cao [4], Jorge Casaus [5], Zhen Chen [4], Xingzhu Cui [4], Raffaello D'Alessandro [1,2], Sebastiano Detti [2], Yongwei Dong [4], Noemi Finetti [2,6], Valerio Formato [7], Miguel Angel Velasco Frutos [5], Jiarui Gao [4], Xiaozhen Liang [4], Ran Li [4], Xin Liu [4], Linwei Lyu [4], Gustavo Martinez [5], Nicola Mori [2], Jesus Marin Munoz [5], Lorenzo Pacini [2], Paolo Papini [2], Cecilia Pizzolotto [3], Zheng Quan [4], JunJun Qin [4], Dalian Shi [4], Oleksandr Starodubtsev [2], Zhicheng Tang [4], Alessio Tiberio [2], Valerio Vagelli [8,9], Elena Vannuccini [2], Bo Wang [4], Junjing Wang [4], Le Wang [4], Ruijie Wang [4], Gianluigi Zampa [3], Nicola Zampa [3], Zhigang Wang [4], Ming Xu [4], Li Zhang [4] and Jinkun Zheng [4]

1   Dipartimento di Fisica e Astronomia, Università degli Studi di Firenze, 50019 Sesto Fiorentino, Italy
2   INFN Firenze, 50019 Sesto Fiorentino, Italy
3   INFN Trieste, 34149 Trieste, Italy
4   Chinese Academy of Sciences, Beijing 100049, China
5   CIEMAT, 28040 Madrid, Spain
6   Dipartimento di Scienze Fisiche e Chimiche, Università degli Studi dell'Aquila, 67100 L'Aquila, Italy
7   INFN—Sezione di Roma Tor Vergata, 00133 Roma, Italy
8   Agenzia Spaziale Italiana, 00133 Roma, Italy
9   INFN Perugia, 06100 Perugia, Italy
*   Correspondence: betti@fi.infn.it

Citation: Betti, P.; Adriani, O.; Antonelli, M.; Bai, Y.; Bai, X.; Bao, T.; Berti, E.; Bonechi, L.; Bongi, M.; Bonvicini, V.; et al. Photodiode Read-Out System for the Calorimeter of the Herd Experiment. *Instruments* **2022**, *6*, 33. https://doi.org/10.3390/instruments6030033

Academic Editors: Fabrizio Salvatore, Alessandro Cerri, Antonella De Santo and Iacopo Vivarelli

Received: 30 July 2022
Accepted: 31 August 2022
Published: 2 September 2022

Publisher's Note: MDPI stays neutral with regard to jurisdictional claims in published maps and institutional affiliations.

Abstract: HERD is a future experiment for the direct detection of high energy cosmic rays. The instrument is based on a calorimeter optimized not only for a good energy resolution but also for a large acceptance. Each crystal composing the calorimeter is equipped with two read-out systems: one based on wavelength-shifting fibers and the other based on two photodiodes with different active areas assembled in a monolithic package. In this paper, we describe the photodiode read-out system, focusing on experimental requirements, design and estimated performances. Finally, we show how these features lead to the flight model project of the photodiode read-out system.

Keywords: cosmic rays; calorimeters; space instrumentation; large detector systems for particle and astroparticle physics

1. Introduction

The direct measurements of high-energy cosmic rays are limited by the geometrical acceptance of space experiments, since at high energy the flux is described by a power law $J^{-\gamma}$ with a spectral index γ of about 2.7, strongly limiting the number of particles at these energies. The HERD (*High Energy cosmic-Radiation Detector*) [1,2] experiment is a new direct experiment planned to be installed on the *Chinese Space Station* in 2027, whose aim is to improve and extend the current measurements at high energies. Indeed, the experiment has been designed in order to directly measure protons and nuclei up to the cosmic ray *knee* region at about 1 PeV, and *electron+positron* flux up to tens of TeV, at least one order of magnitude higher than the current experiments. In addition, HERD will also detect high-energy photons in order to look for cosmic ray sources. Furthermore, both measurements of *electron+positron* flux and high energy photons are a valuable tool in searching for indirect evidence of dark matter.

The main component of the HERD detector is a homogeneous, isotropic, 3-D, finely segmented and deep (about 55 X_0 and 3 λ_I) calorimeter. As shown on the left side of

Figure 1, the calorimeter will be composed of about 7500 LYSO cubic scintillating crystals (with a side of 3 cm) assembled in an octagonal-based prism.

Figure 1. On the left: scheme of the structure of the calorimeter, about 7500 LYSO cubic crystals are assembled in an octagonal-based prism. In the center: picture of a LYSO crystal with WLSF and PD read-out systems installed; the crystal is covered with a reflective coating. On the front side of the crystal a monolithic package with photodiodes is glued to the crystal. The WLSFs are glued on the top face of the crystal, but are covered by the reflective coating; however, we can see the fibers coming outside of the reflective coating in the upper right corner of the image. On the right: an illustration of an in-house-built prototype of a monolithic package for the PD read-out system.

Each crystal of the calorimeter is equipped with two different read-out systems: one based on *Wave Length Shifting Fibers* (WLSF) coupled with Intensified scientific CMOS cameras, the other one based on the use of two photodiodes with different active areas (PD system). A picture of a crystal equipped with both read-out systems is shown in the central panel of Figure 1. The use of two independent read-out systems is very important in order to have strong control on the energy scale (two independent calibrations), two independent triggers and redundancy. In this proceeding, the system requirements, the design and the performances of the PD system are analyzed. Before going on, we briefly describe the WLSF read-out system in order to give the reader a general view of the read-out system alternative to the PD system. The WLSF read-out system consists of two fibers (with a diameter of 300 μm each) for each crystal, that collect light and then emit it at their extremities in the green wavelength region. The fibers are connected via a thin optical guide to a surface of the crystal in order to collect LYSO scintillating photons. One extremity of each fiber is connected to a CMOS camera: one high-gain CMOS and one low-gain CMOS. The use of two CMOS with different gains is necessary in order to reach a very high dynamic range. In addition, the remaining extremities of the fibers are connected to photomultiplier tubes that collect and sum signals from different crystals in order to provide fast information on the energy deposit in a certain region of the calorimeter that can be used for trigger purposes.

The HERD calorimeter is surrounded by other subdetectors for tracking, charge measurement and anticoincidence purposes. In this way, HERD is able to detect particles entering the detector not only from the zenith but also from the lateral sides. This design, together with the calorimeter features, achieves an acceptance which is about three times larger than that of typical calorimeters with the same volume and mass, as was demonstrated by the CaloCube collaboration [3–8]. Indeed, HERD will have an effective geometric factor (convolution of geometric acceptance and detection efficiency) about one order of magnitude larger than that of current space experiments: about 2–3 m^2sr for electrons and about 1 m^2sr for protons, instead of the 0.3 m^2sr and 0.1 m^2sr of DAMPE experiment [9], the largest effective geometric factors of currently in-orbit experiments. This is a key factor for extending cosmic rays fluxes measurements at higher energies.

2. Design of the Photodiode Read-Out System

The main requirements for both WLSF and PD systems are low power consumption and an exceptionally large dynamic range. The first is constrained by the limited power availability on the space station, while the second is necessary to measure the deposit by high energy showers and, at the same time, to calibrate the detector with MIP (*minimum ionizing particles*) protons and nuclei. Indeed simulations show that a shower induced by a PeV proton can release up to 250 TeV in a single crystal. On the other hand, in-flight calibration with MIPs requires measuring energy deposits of about 30 MeV, i.e., the energy deposit of a minimum ionizing proton in a 3 cm LYSO cube. Thus, a very high dynamic range of about 10^7 is needed for the read-out systems.

The PD system is based on pairs of photodiodes with different active areas: the large photodiode (LPD, model VTH2110 produced by Excelitas) and the small photodiode (SPD, model VTP9412 produced by Excelitas) with active areas, respectively, of 25 mm^2 and 1.6 mm^2. LPD and SPD are assembled in a custom monolithic package, as shown in the right panel of Figure 1 (a new monolithic package developed with Excelitas is currently under production as we will show in Section 5). Using two photodiodes with different active areas allows for an increase in the dynamic range: LPD can detect signals smaller than the SPD thanks to the larger active area, while the SPD signals saturate the front-end electronics at higher energies than LPD ones. Multiple measurements of the ratio between LPD and SPD gains at different test beams will be discussed in the next sections.

To satisfy the requirements on power consumption, noise and dynamic range, a dedicated front-end chip has been developed specifically for this system. This chip is called HiDRA2, based on the CASIS ASIC [10], and is the principal component of the front-end-electronics (FEE) for the PD system. The main characteristics of this chip are: a low power consumption of about 3.75 mW per channel, a low noise with an *Equivalent Noise Charge* (ENC) of about 2500 *equivalent electrons* and a high dynamic range from few fC to 52.6 pC. The single read-out channel of the HiDRA2 chip is composed of two main parts: a *Charge Sensitive Amplifier* (CSA) followed by a *Correlated Double Sampling* (CDS). The CSA has two different gains. The gain is automatically selected for each channel on an event-by-event basis; this allows it to reach a high dynamic range of the FEE without doubling the number of channels. The ratio between *high-gain* and *low-gain* is about 20, and laboratory measurements show that the uncertainty on this value is smaller than 1%.

Photodiodes are connected to the FEE via specifically designed Kapton cables. Every cable can simultaneously connect up to 10 crystals (10 LPD-SPD couples) and is long about 26 cm. In the flight model Kapton cables will be longer and will connect up to 21 crystals. A long Kapton cable extension (about 68 cm) has been produced in order to extend the length of the current one and was used to estimate the noise for the flight model.

3. Performance of the Photodiode Read-Out System

In this section, we show the main performances of the PD read-out system measured with several laboratory tests and test beams. In particular, we will focus on the noise of the system, the measurement of cosmic muon MIP, and the ratio between LPD and SPD gains. The noise of the system was estimated using the standard deviation of the Gaussian fit performed on the pedestal distribution. These measurements were performed in two different configurations: with the normal Kapton cable (as shown in the left peak of Figure 2 left panel) and with the addition of the Kapton extension to simulate the flight model length.

Furthermore, the noise was measured connecting photodiodes of several instrumented crystals: the typical noise values were about 22.5 *ADC channels* without the extension and 27.5 *ADC* with the extension. Hence, in the flight model, an increase of noise of at least 20% with respect to the current configuration is expected due to the longer cables.

The energy released by an MIP is the minimum signal that we want to be able to detect in a single crystal. We measured the typical MIP distribution by using cosmic muons, as shown in the right peak of Figure 2 central panel for an LPD. LPD muon signal distributions were fitted with the convolution of Gaussian and Landau distributions, while for SPD signal distributions simple Gaussian distributions were used since the MIP values for SPD were very small and the noise dominated the signal. MIP measurements were performed with a few crystals. From these measurements we determined an average

$MIP_{LPD} = (110 \pm 10)\ ADC$, and $MIP_{SPD} = (6 \pm 2)\ ADC$. Even considering a noise of about 30 ADC (larger than the measured one due to possible degradation in the final detector) the *Signal to Noise Ratio* is expected to be $SNR_{LPD} > 3.5$, while $SNR_{SPD} < 1$. Thus, LPD can be calibrated with MIP signals, whereas SPD can be calibrated by exploiting the correlation with LPD.

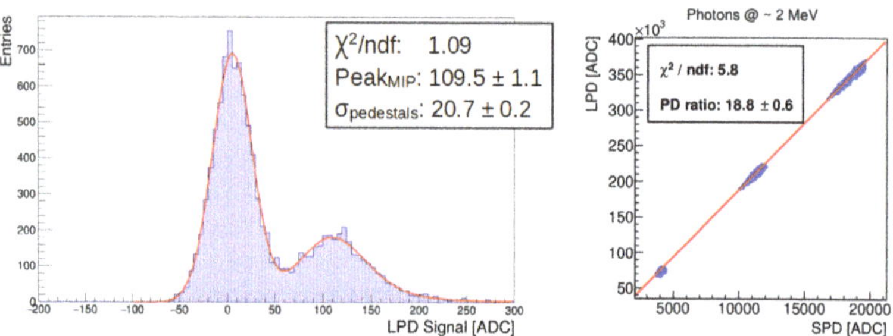

Figure 2. Left panel: MIP distribution of a LPD; the left peak is the pedestal peak while the one on the right is the MIP events peak. Right panel: correlation plot between LPD and SPD measured with a multiparticle beam of photons with a mean energy of 2 MeV.

The LPD/SPD gain ratio was measured at two test beams with multiparticle beams of 2 MeV protons at the Labec facility [11] and photons with a mean energy of 2 MeV with a medical radiotherapy accelerator. In particular, beam multiplicities were varied in order to scan a sufficiently large region of the dynamic range of the system. Then, correlation plots between LPD and SPD signals were built and a linear fit was performed. The angular coefficient resulting from the fit was the ratio between LPD and SPD gains. This is shown in Figure 2 right for one of the few crystals tested. The resulting average LPD/SPD gain ratio was 19.0 ± 1.5. Further measurements with different beams are foreseen to extend the range of the energy scan and confirm these values.

4. Calibration of the Prototype Tested at Sps

After developing and testing the sensor system for a single crystal, a prototype of the calorimeter was assembled. The prototype was composed of 525 crystals all equipped with WLSF, with only 63 of them equipped with photodiodes due to mechanical and procurement constraints. The prototype was tested at the CERN SPS in October 2021. In this section we show the calibration procedure of the PD system of the prototype as it was performed with the data acquired at the test beam; other results of the ongoing analysis will be the subject of a forthcoming paper.

The noise was estimated as in the previous section, performing Gaussian fits on pedestal histograms. The noise mean value was about 18.5 ADC. This value was lower than the one measured in laboratory tests (shown in Section 3).

LPDs were calibrated by measuring energy releases of MIP. In this case, a muon beam with energy of 250 GeV was used. A typical histogram for MIP signals is shown on the left side of Figure 3.

The distribution showed two peaks: the right peak was the main peak of MIP signals and was fitted with the convolution of Gaussian and Landau distributions. The left peak was due to triggered events in which the beam did not hit the crystal (i.e., pedestal events) and was fitted with a Gaussian distribution. As we can see, the pedestal distribution was not centered on zero because of a baseline recovery problem in case of high energy release. Thus, the simultaneous fit of the two peaks was important to properly estimate the energy deposit. This problem is currently being solved thanks to a firmware and electronic update, and it will not be present in future tests. The mean value of the MIP for 250 GeV muons for all crystals was equal to $(126 \pm 4)\ ADC$. This value was slightly larger than that measured with cosmic muons. This was mainly due to the different energy of the muons considered, indeed cosmic muons have an energy distribution centered at a few GeV, while muons at the test beam had 250 GeV energy.

Then the ratio between LPD and SPD gains has been measured with a correlation plot and linear fit, in the same way, that was described in the previous section. A correlation plot for a crystal is shown on the right side of Figure 3. Considering all the crystals, the mean value of the LPD/SPD gain ratio was equal to 18.7 ± 0.9, compatible with the value reported in the previous section. In the next section, these parameters will be used to determine the dynamic range of the PD read-out system.

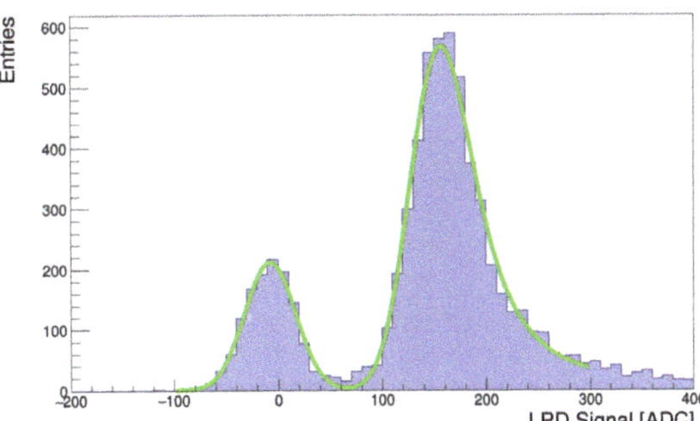
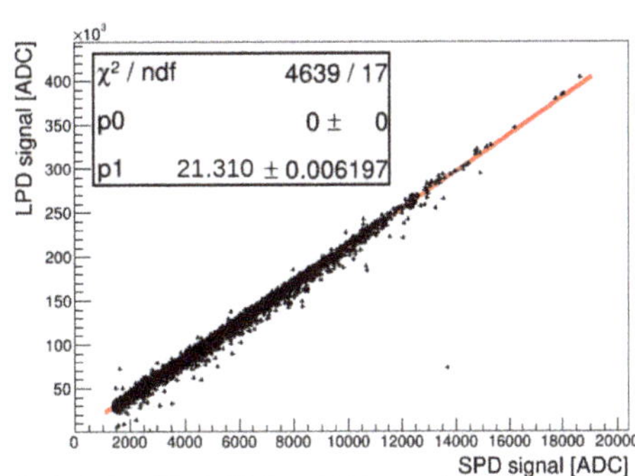

Figure 3. (**Left** side): MIP distribution of an LPD. (**Right** side): correlation plot between LPD and SPD signals for a crystal.

5. Results

In the previous section, the MIP values and LPD/SPD gain ratio values for multiple crystals have been measured. In addition, from simulations, we know that the muon MIP releases are about 30 MeV in the single crystal. From the saturation level of the HiDRA2 chip it is possible to estimate the saturation levels for LPD and SPD: about 190 GeV for LPD and 3.5 TeV for SPD. However, the system should be able to measure up to 250 TeV energy releases in a single crystal, thus the SPD input signal should be attenuated in order to shift the saturation level to higher energies. In doing this we must be careful in keeping an overlapping region between the operative ranges of LPD and SPD since their correlation is the only possible method to calibrate the SPD gain.

It has been decided to cover the SPD surface with an optical filter with a transmittance of 1.5%. In this way, the new saturation level is about 250 TeV, while the SNR of SPD at the LPD saturation level is bigger than 15, which permits the SPD calibration. In Figure 4 a scheme of the dynamic range of the system with both normal and optically filtered SPD is shown.

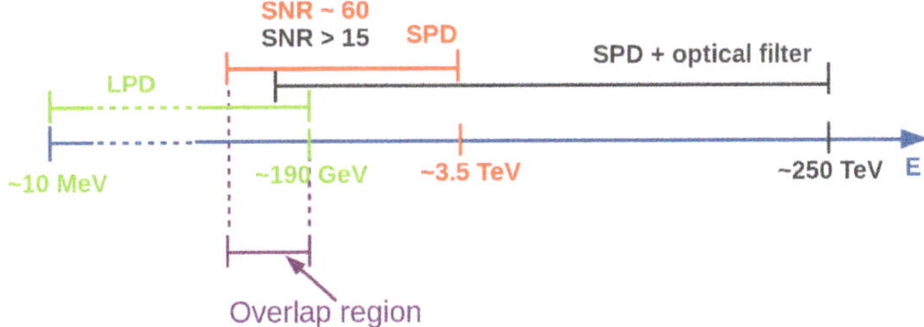

Figure 4. Scheme of the dynamic range of the system. The absolute energy scale is depicted in blue, the LPD dynamic range in green, the SPD dynamic range in red, the SPD optically filtered dynamic range in grey, and the overlapping region between LPD and SPD in purple.

The new project of the monolithic package with the optical filter on the SPD surface has already been developed in collaboration with Excelitas. The new packages are in production and the first 1000 packages will be ready at the beginning of 2023. They will be characterized and mounted in a new calorimeter prototype of 1000 crystals, that will be tested at SPS at the end of 2023.

6. Discussion

In this paper, we have shown the developments and the tests of the photodiode read-out system of the calorimeter of the HERD experiment. The performances of the system and the developments of the monolithic package for the flight model, which is now under production, have been illustrated. Finally, it is planned to equip a new calorimeter prototype with 1000 new monolithic packages and to test it at SPS at the end of 2023.

Author Contributions: Conceptualization, O.A., R.D. and J.C.; methodology, P.B., E.B., L.P.; software, P.B., E.B., L.P. and A.T.; validation, P.B., E.B. and L.P.; formal analysis, P.B., E.B. and L.P.; investigation, O.S., P.B., E.B., L.P.; resources, Y.B., T.B., W.C., X.C., J.G., R.L., X.L. (Xiaozhen Liang), L.L., Z.Q., J.Q., D.S., Z.T., B.W., J.W., R.W., Z.W., M.X., L.Z., J.Z., S.D., G.M., J.M.M., V.F., M.A.V.F., C.P., V.V., S.B., L.B., M.B., N.F., P.P., E.V., M.A., V.B., G.Z., N.Z., X.L. (Xin Liu), X.B., Z.C., L.W., M.A.V.F. and J.M.M.; data curation, P.B., E.B., L.P.; writing—original draft preparation, P.B.; writing—review and editing, P.B. and N.M.; visualization, P.B. and L.P.; supervision, R.D.; project administration, O.A. and Y.D.; funding acquisition, N.M. All authors have read and agreed to the published version of the manuscript.

Funding: This research received no external funding.

Institutional Review Board Statement: Not applicable.

Data Availability Statement: Not applicable.

Acknowledgments: We would like to thanks all the HERD collaboration for its support and for making this experiment possible.

Conflicts of Interest: The authors declare no conflict of interest.

References

1. HERD The High Energy Cosmic Radiation Detection Facility. Available online: http://herd.ihep.ac.cn/ (accessed on 27 July 2022).
2. Pacini, L.; Adriani, O.; Bai, Y.l.; Bao, T.W.; Berti, E.; Bottai, S.; Cao, W.W.; Casaus, J.; Cui, X.Z.; D'Alessandro, R.; et al. Design and expected performances of the large acceptance calorimeter for the HERD space mission. In Proceedings of the 37th International Cosmic Ray Conference—PoS(ICRC2021), Berlin, Germany, 12–23 July 2021
3. Berti, E.; Adriani, O.; Albergo, S.; Ambrosi, G.; Auditore, L.; Basti, A.; Bigongiari, G.; Bonechi, L.; Bonechi, S.; Bongi, M.; et al. CaloCube: A new-concept calorimeter for the detection of high-energy cosmic rays in space. *Nucl. Instrum.* **2017**, *845*, 421–424. [CrossRef]
4. Adriani, O.; Albergo, S.; Auditore, L.; Basti, A.; Berti, E.; Bigongiari, G.; Bonechi, L.; Bongi, M.; Bonvicini, V.; Bottai, S.; et al. Calocube—A highly segmented calorimeter for a space based experiment. *Nucl. Instrum.* **2016**, *824*, 609–613.
5. Bongi, M.; Adriani, O.; Albergo, S.; Auditore, L.; Bagliesi, M.G.; Berti, E.; Bigongiari, G.; Boezio, M.; Bonechi, L.; Bonechi, S.; et al. CALOCUBE: An approach to high-granularity and homogeneous calorimetry for space based detectors. *J. Phys. Conf. Ser.* **2015**, *587*, 012029. [CrossRef]
6. Pacini, L.; Adriani, O.; Agnesi, A.; Albergo, S.; Auditore, L.; Basti, A.; Berti, E.; Bigongiari, G.; Bonechi, L.; Bonechi, S.; et al. CaloCube: An innovative homogeneous calorimeter for the next-generation space experiments. *J. Phys. Conf. Ser.* **2017**, *928*, 012013. [CrossRef]
7. Adriani1, O.; Agnesi, A.; Albergo, S.; Antonelli, M.; Auditore, L.; Basti, A.; Berti, E.; Bigongiari, G.; Bonechi, L.; Bongi, M.; et al. CaloCube: An isotropic spaceborne calorimeter for high-energy cosmic rays. Optimization of the detector performance for protons and nuclei. *Astropart. Phys.* **2017**, *96*, 11–17. [CrossRef]
8. Berti, E.; Adriani, O.; Albergo, S.; Ambrosi, G.; Auditore, L.; Basti, A.; Bigongiari, G.; Bonechi, L.; Bonechi, S.; Bongi, M.; et al. CaloCube: A new concept calorimeter for the detection of high energy cosmic rays in space. *J. Phys. Conf. Ser.* **2019**, *1162*, 012042. [CrossRef]
9. Chang, J.; Ambrosi, G.; An, Q.; Asfiyarov, R.; Azzarello, P.; Bernardini, P.; Bertucci, B.; Cai, M.S.; Caragiulo, M.; Chen, D.Y.; et al. The DArk Matter Particle Explorer mission. *Astropart. Phys.* **2007**, *95*, 6–24. [CrossRef]

10. Bonvicini, V.; Orzan, G.; Zampa, G.; Zampa, N. A Double-Gain, Large Dynamic Range Front-end ASIC with A/D Conversion for Silicon Detectors Read-Out. *IEEE Trans. Nucl. Sci.* **2010**, *57*, 2963–2970. [CrossRef]
11. Chiari, M.; Barone, S.; Bombini, A.; Calzolai, G.; Carraresi, L.; Castelli, L.; Czelusniak, C.; Fedi, M.E.; Gelli, N.; Giambi, F.; et al. LABEC the INFN ion beam laboratory of nuclear techniques for environment and cultural heritage. *Eur. Phys. J. Plus* **2021**, *136*, 472. [CrossRef] [PubMed]

 *instruments*

Article

Performance Study of a New Cluster Splitting Algorithm for the Reconstruction of PANDA EMC Data

Ziyu Zhang [1], Guang Zhao [2,*], Shengsen Sun [2,3], Qing Pu [1], Chunxiu Liu [2], Chunxu Yu [1], Dong Liu [4], Hang Qi [4], Guangshun Huang [4], Tobias Stockmanns [5], Beijiang Liu [2], Fei Wang [6], Yitong Zhang [7] and Xiaoyan Shen [2,3]

[1] School of Physics, Nankai University, Tianjin 300071, China
[2] Institute of High Energy Physics, Beijing 100049, China
[3] School of Physical Sciences, University of Chinese Academy of Sciences, Beijing 100049, China
[4] Department of Modern Physics, University of Science and Technology of China, Hefei 230026, China
[5] Forschungszentrum Jülich, Institut für Kernphysik, 52428 Jülich, Germany
[6] School of Nuclear Science and Technology, University of South China, Hengyang 421001, China
[7] School of Physics, Liaoning University, Shenyang 110036, China
* Correspondence: zhaog@ihep.ac.cn

Abstract: For high-energy π^0 mesons, the angle between the two final-state photons decreases with the increase in the energy of the π^0, which enhances the probability of overlapping electromagnetic showers. The performance of the cluster splitting algorithm in the EMC reconstruction is crucial for the mass resolution measurement of π^0 with high energy. The cluster splitting algorithm is based on the theoretical lateral distribution of the electromagnetic showers. A simple implementation of the lateral distribution can be described as a (multi-)exponential function. In a realistic electromagnetic calorimeter, considering the granularity of the detector, the measured energy in a cell is actually the integral of the theoretical energy deposition, which deviates from the exponential function. Based on the simulation of the barrel EMC in the $\overline{\text{P}}$ANDA experiment, a cluster splitting algorithm with a new lateral energy development function is developed. The energy resolution of overlapping showers with high energy has been improved.

Keywords: calorimeter; energy reconstruction; cluster splitting algorithm

Citation: Zhang, Z.; Zhao, G.; Sun, S.; Pu, Q.; Liu, C.; Yu, C.; Liu, D.; Qi, H.; Huang, G.; Stockmanns, T.; et al. Performance Study of a New Cluster Splitting Algorithm for the Reconstruction of PANDA EMC Data. *Instruments* **2022**, *6*, 34. https://doi.org/10.3390/instruments6030034

Academic Editor: Fabrizio Salvatore, Alessandro Cerri, Antonella De Santo, Iacopo Vivarelli

Received: 31 July 2022
Accepted: 30 August 2022
Published: 5 September 2022

Publisher's Note: MDPI stays neutral with regard to jurisdictional claims in published maps and institutional affiliations.

1. Introduction

The antiProton ANnihilations at DArmstadt ($\overline{\text{P}}$ANDA) experiment [1,2] is planned to operate in 2026 at the Facility for Antiproton and Ion Research (FAIR). $\overline{\text{P}}$ANDA aims to perform the studies of charmonium spectroscopy, exotic states, charmed hadrons in nuclear matter, and the γ-ray spectroscopy of excited states in doubly strange $\Lambda\Lambda$ hypernuclei with the beam momentum in the range from 1.5 GeV/c to 15 GeV/c.

In order to ensure a geometrical acceptance close to 4π, the $\overline{\text{P}}$ANDA detector consists of two spectrometers: the target spectrometer (TS) and the forward spectrometer (FS), as shown in Figure 1 [1]. The TS is arranged in a barrel part for angles larger than $22°$ and an end cap part for the forward range down to $5°$ in the vertical and $10°$ in the horizontal plane, while the FS covers the very forward angles. Both the TS and FS contain detectors for tracking, charged particle identification, electromagnetic calorimetry, and muon identification.

The electromagnetic calorimeter (EMC) is an essential detector in $\overline{\text{P}}$ANDA to measure the energy of the photons and electrons. The EMC in the TS consists of one barrel and two end caps. To achieve the desired very low detection threshold, the improved $PbWO_4$ scintillator was chosen, providing a small radiation length of 0.89 cm and a short decay time of 6.5 ns. The calorimeter will be operated at a temperature of $-25\,°C$ for an increased light yield [3]. There are 11200 crystals in total for the barrel EMC, with an average lateral size of 21.3 mm. The barrel is approximately projective. The crystals almost focus on the collision vertex with a $4°$ tilt in both the azimuthal and longitudinal directions.

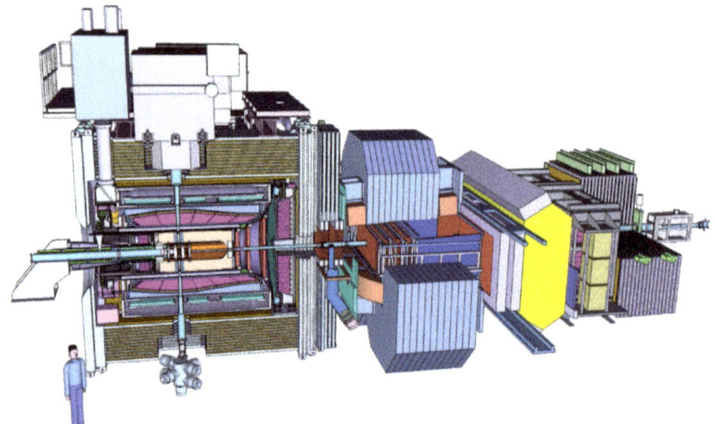

Figure 1. Side view of $\overline{\text{P}}$ANDA with the target spectrometer (TS) on the left side and the forward spectrometer (FS), starting with the dipole magnet center, on the right side. The antiproton beam enters from the left.

The main task of the EMC is to provide the four-vector momentum of a photon by measuring its energy and position. When photons overlap, the EMC should also be able to isolate the overlapped photons, especially in the case of a high-energy π^0. Figure 2 shows the angle between the two final-state photons from π^0 decays as a function of the π^0 energy. The overlapping is much more severe with the increase in energy. Therefore, any significant improvement in cluster splitting can improve the π^0 reconstruction at high energy.

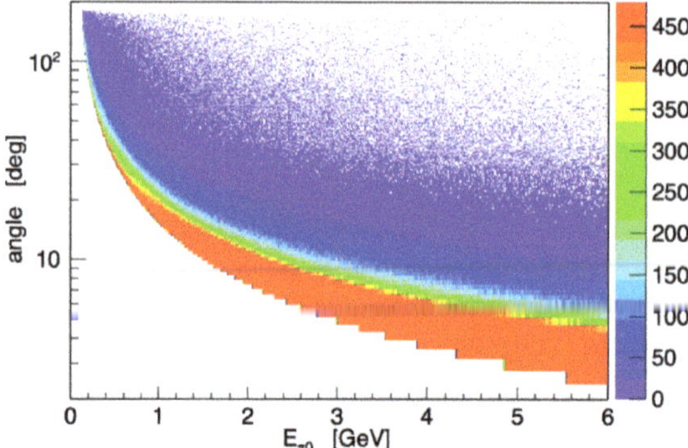

Figure 2. The angle between the two final-state photons with the increase in the energy of π^0.

In this paper, an optimized cluster splitting algorithm is presented, applying an updated description of the lateral development formula of electromagnetic showers that takes the detector granularity into account.

2. Cluster Splitting Algorithm

The two major procedures of the EMC reconstruction are cluster finding and cluster splitting. The cluster finding algorithm attains the objective of finding a cluster formed by a series of crystals with an energy deposition higher than the threshold in a continuous region. In a cluster, a seed crystal is defined as the crystal with local maximum energy. If there are multiple seeds in a cluster, it means shower overlaps exist, and it should be further separated. The cluster splitting algorithm is intended to separate and precisely assign the energy and position of each overlapped shower.

The cluster splitting algorithm is based on the knowledge of the lateral development of electromagnetic showers, which can be expressed as:

$$\frac{E_{\text{target}}}{E_{\text{seed}}} = \text{LAT}(r),$$ (1)

where E_{target} and E_{seed} are energy depositions of the target crystal and seed crystal, r is the distance from shower center to the target crystal, and LAT is a function of r describing the lateral development. As shown in Figure 3, for a target crystal, the energy deposition fraction from the i-th shower can be calculated as:

$$f_i = \frac{(E_{\text{seed}})_i \cdot \text{LAT}(r_i)}{\Sigma_j (E_{\text{seed}})_j \cdot \text{LAT}(r_j)},$$ (2)

where the sum in the denominator runs over all showers in the cluster. With the energy deposition fraction f_i, the energy deposition from different showers in a crystal can be separated, and the energy and position of the individual shower can be calculated. The energy fraction will be updated iteratively according to the energy and position of the showers in the previous step until the results converge.

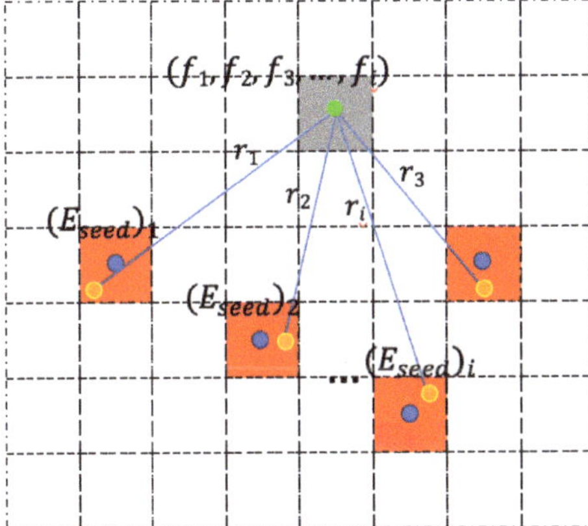

Figure 3. Demonstration of calculating the energy deposition fraction from seeds (red) for a target crystal (grey). f_1, f_2, f_3, and f_4 are energy fractions; $(E_{\text{seed}})_1$, $(E_{\text{seed}})_2$, $(E_{\text{seed}})_3$, and $(E_{\text{seed}})_4$ are different seed crystal energies; and r_1, r_2, r_3, and r_4 are the distances from the center of showers to center of target crystals.

3. Lateral Energy Distribution of Electromagnetic Shower Considering the Detector Granularity

The lateral development distribution of an electromagnetic shower can be empirically described as an exponential function (shown by the black dashed line in Figure 4):

$$\text{LAT}^{(\text{E})}(r) = \exp(-c \cdot r / R_{\text{M}}),$$ (3)

where R_{M} is the Molière radius of $PbWO_4$, and the c is a constant. For a realistic electromagnetic calorimeter geometry, the energy deposition in a crystal is the integral of Equation (3) over the crystal area, which deviates from the exponential function. Based on the $\overline{P}$ANDA barrel EMC, a measurement of the lateral development is performed using MC-simulated events. The results of the simulation are shown by the scattered points in Figure 4. An updated parameterization for the lateral development that considers the detector granularity is proposed as:

$$\mathrm{LAT}^{(\mathrm{G})}(r) = \exp\left(-\frac{p_1}{R_\mathrm{M}} \cdot \left(r - p_2 \cdot r \cdot \exp\left(-\left(\frac{r}{p_3 \cdot R_\mathrm{M}}\right)^{p_4}\right)\right)\right), \tag{4}$$

where p_1, p_2, p_3 and p_4 are free parameters. Equation (4) can be viewed as adding a correction term for the small r region to the original exponential Equation (3).

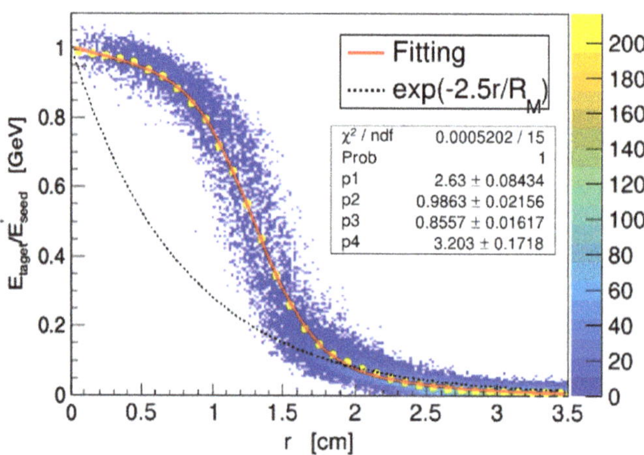

Figure 4. The $E_\mathrm{target}/E_\mathrm{seed}$ varies with the r. The r is the distance from the shower center to the target crystal. Each point in the scatter plot represents a target crystal in the showers with the energy E_target. The black dashed line is the empirical exponential function; the red line represents the fitting analytic function.

With the updated lateral development model, a correction on the E_seed is further applied. When the shower center does not coincide with the center of the seed crystal, the actual seed is a virtual seed that slightly deviates from the physical crystal. As shown in Figure 5, if we consider a virtual seed as the yellow point, the virtual seed energy (E'_seed) can relate to the seed crystal energy (E_seed) as:

$$E_\mathrm{seed} = E'_\mathrm{seed} \cdot \mathrm{LAT}^{(\mathrm{G})}(r_\mathrm{seed}), \tag{5}$$

where r_seed is the distance from the center of the shower to the geometric center of the seed crystal. The lateral development distribution formula is updated with the seed energy correction:

$$\mathrm{LAT}^{(\mathrm{G})}_\mathrm{corr}(r) = \exp\left(-\frac{p_1}{R_\mathrm{M}} \cdot (\xi(r, p_2, p_3, p_4) - \xi(r_\mathrm{seed}, p_2, p_3, p_4)),\right.$$
$$\left.\text{where } \xi(r) = r - p_2 \cdot r \cdot \exp\left(-\left(\frac{r}{p_3 \cdot R_\mathrm{M}}\right)^{p_4}\right).\right) \tag{6}$$

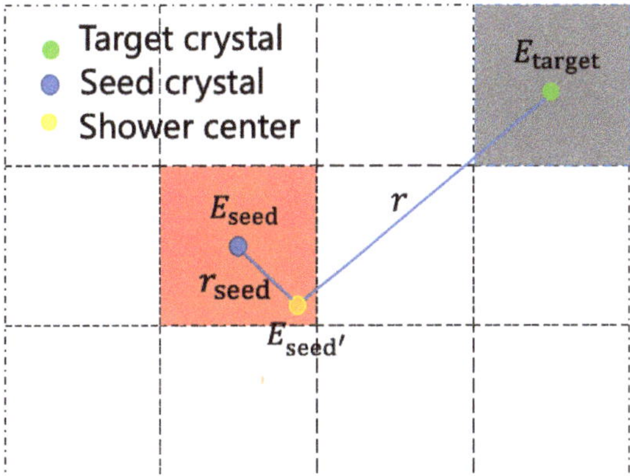

Figure 5. This figure shows a case in which the center of the shower is inconsistent with the center of the seed crystal. The green point is the center of the target crystal, and the blue point and yellow point represent the geometric center of the seed crystal and the center of the shower, respectively.

4. Performance Results

The performance of the updated cluster splitting algorithm that considers the detector granularity has been checked using small cross-angle photon samples. Samples of two photons with the same energy (6 GeV) and small opening angle at which the photons are emitted ($<6.75°$) are simulated. The small open angle ensures overlapping between the two electromagnetic showers on the $\overline{\text{P}}$ANDA EMC. Figures 6 and 7 show the reconstructed energy distributions. The black and red histograms represent the energy of photons reconstructed by the default algorithm and the updated algorithm, respectively. The comparison of two histograms demonstrates intuitively that the energy reconstruction of photons is better using the updated algorithm. For a more precise comparison, the histograms have been fitted by double-Gaussian functions. From the fitting results, the energy resolutions are obtained as 199.1 ± 2.6 MeV for the default algorithm and 156.2 ± 1.3 MeV for the updated algorithm. For 6 GeV overlapping showers, a roughly 20% improvement is achieved on the energy resolution with the updated cluster splitting algorithm.

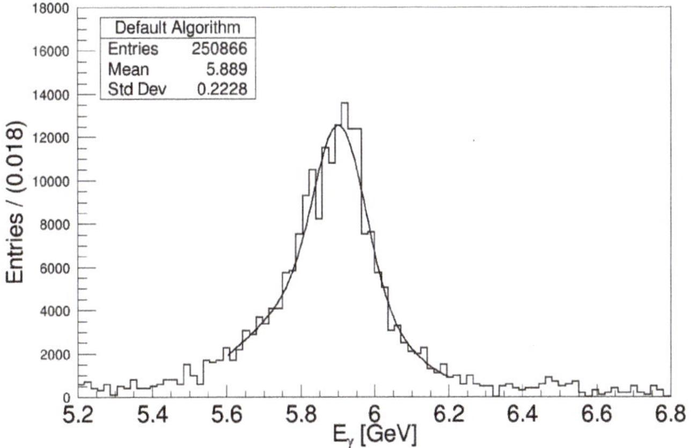

Figure 6. Reconstructed photon energy distribution with the default algorithm.

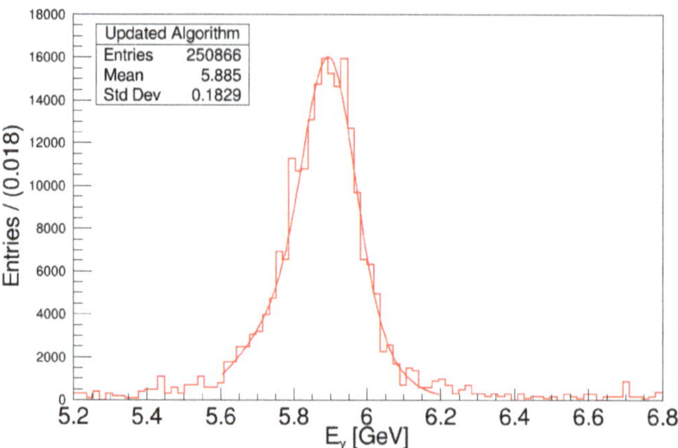

Figure 7. Reconstructed photon energy distribution with the updated algorithm.

5. Summary

The cluster splitting algorithm has been updated by establishing a new description of the lateral development of an electromagnetic shower for the barrel $\overline{\mathrm{P}}$ANDA EMC. The performance is studied in a small cross-angle photon MC-simulated sample. A clear improvement in the energy resolution is achieved for high-energy photons, which may lead to improvement in the π^0 mass resolution. More validation of the updated cluster splitting algorithm will be carried out in the future. The algorithm can also easily be applied to other calorimeter sub-detectors in $\overline{\mathrm{P}}$ANDA.

Author Contributions: Conceptualization, G.Z. and S.S.; methodology, Z.Z., Q.P., G.Z., S.S., and T.S.; software, Z.Z., Q.P., C.L., D.L., H.Q., and T.S.; validation, Z.Z., F.W., and Y.Z.; formal analysis, Z.Z., Q.P., and G.Z.; investigation, Z.Z., S.S., G.H., and B.L.; writing—original draft preparation, Z.Z. and G.Z.; writing—review and editing, G.Z. and S.S.; supervision, C.Y., G.H., and X.S.; project administration, S.S. and X.S. All authors have read and agreed to the published version of the manuscript.

Funding: This research was funded by National Natural Science Foundation of China (NSFC) under contract nos. 11875277 and 12061131003 and the National Key R&D Program of China under contract no. 2020YFA0406300.

Data Availability Statement: Not applicable.

Conflicts of Interest: The authors declare no conflict of interest.

References

1. PANDA Collaboration. Technical Design Report for PANDA Electromagnetic Calorimeter (EMC). *arXiv* **2008**, arXiv:0810.1216.
2. Panda Collaboration. Update to the Technical Design Report for the PANDA Electromagnetic Calorimeter. 2020. Available online: https://panda.gsi.de/system/files/user_uploads/heinsius%40ep1.rub.de/RE-TDR-2020-007.pdf (accessed on 30 July 2022)
3. Eissner, T. The new PWO Crystal Generation and Concepts for the Performance Optimisation of the PANDA EMC. Available online: http://geb.uni-giessen.de/geb/volltexte/2013/10382 (accessed on 30 July 2022).

Article

Development of Novel Designs of Resistive Plate Chambers

Burak Bilki [1,2,3,*], **Yasar Onel** [2], **Jose Repond** [2], **Kutlu Kagan Sahbaz** [1,3,4], **Mehmet Tosun** [1,3,4] **and Lei Xia** [5] **on behalf of the CALICE Collaboration**

[1] Department of Mathematics, Beykent University, Istanbul 34500, Turkey
[2] Department of Physics and Astronomy, University of Iowa, Iowa City, IA 52242, USA
[3] Turkish Accelerator and Radiation Laboratory, Ankara 06830, Turkey
[4] AInstitute of Nuclear Sciences, Ankara University, Ankara 06100, Turkey
[5] Argonne National Laboratory, Lemont, IL 60439, USA
*   Correspondence: burak.bilki@cern.ch

Abstract: Resistive Plate Chambers (RPCs) are a key active media of the muon systems of current and future collider experiments as well as the CALICE (semi-)digital hadron calorimeter. The outstanding issues with RPCs can be listed as the loss of efficiency for the detection of particles when subjected to high particle fluxes and the limitations associated with the common RPC gases. We developed novel RPC designs with: low resistivity glass plates; a single resistive plate; and a single resistive plate and a special anode plate coated with high secondary electron emission yield material. The cosmic and beam tests confirmed the viability of these new approaches for calorimetric applications. The chambers also have improved single-particle response, such as a pad multiplicity close to unity. Here, we report on the construction of various different glass RPC designs and their performance measurements in laboratory tests and with particle beams. We also discuss future test plans, which include the long-term performance tests of the newly developed RPCs, investigation of minimal gas flow chambers, and feasibility study for the large-size chambers.

Keywords: resistive plate chambers; gaseous imaging and tracking detectors; hadron calorimetry

Citation: Bilki, B.; Onel, Y.; Repond, J.; Sahbaz, K.K.; Tosun, M.; Xia, L., on behalf of the CALICE Collaboration Development of Novel Designs of Resistive Plate Chambers. *Instruments* **2022**, *6*, 35. https://doi.org/10.3390/instruments6030035

Academic Editors: Fabrizio Salvatore, Alessandro Cerri, Antonella De Santo and Iacopo Vivarelli

Received: 15 August 2022
Accepted: 29 August 2022
Published: 7 September 2022

Publisher's Note: MDPI stays neutral with regard to jurisdictional claims in published maps and institutional affiliations.

1. Introduction

Resistive Plate Chambers (RPCs) are particle detectors which were introduced in the 1980s [1] and have been widely used in High Energy Physics experiments since then. The experimental implementations are mostly on triggering and precision timing. They consist of two or more resistive plates of high resistance (glass or Bakelite) that are separated by thin gas gaps. The readout is provided either by strips or pads, which are placed on the outside of the chambers.

Under high particle fluxes, RPCs exhibit a significant loss of efficiency due to the high resistance of the resistive plates. Lower-resistivity glass samples are produced and purchased, and various size RPCs were constructed with them. As expected, lower-resistivity glasses offer higher rate capability. Although other types of low-resistivity planar materials are being probed as resistive plates, low-resistivity glass option remains viable due to large flexibility in tuning the glass composition and production parameters for optimal long-term performance.

In the context of studies of imaging calorimetry for a future lepton collider, as carried out by the CALICE Collaboration [2], we developed a novel design of RPC based on a single resistive plate. The RPCs were read out using the standard Digital Hadron Calorimeter (DHCAL) [3] electronic readout system featuring 1×1 cm^2 signal pads. Tests were performed with both cosmic rays and particle beams, and the results point towards an improvement in pad multiplicity and rate capability.

Recently, we developed 1-glass RPCs with anode planes coated with a thin layer of high secondary electron yield material. The purpose is to relax the requirements on the RPC gases, both in terms of type and operational parameters, in order to mitigate the limitations of their use due to green house effects, which are present in many of them. A number of 1-glass RPC samples were produced and tested with both cosmic rays and particle beams resulting in promising first results.

Here, we report on the construction of various different glass RPC designs and their performance measurements in laboratory tests and with particle beams.

2. Development of Semi-Conductive Glass

One of the major areas that the RPCs require R&D for future implementations is the improvement of their rate capability. This has been a long-standing problem with limited solution thus far. The rate limitation is related to the usually high resistivity of the resistive plates used in their construction [4]. A simple approach to handle the limitation is to increase the electrical conductivity of the glass to allow the resistive plates to restabilize faster. Soda–lime silicate glasses in the current RPCs are well-known, inexpensive, and easily manufactured, on the other hand, they come in a bulk electrical conductivity at the order of 10^{-15} S/cm. For high-rate implementations, the target conductivity is two to five orders of magnitude higher. In addition to the conductivity requirements, the RPC glass must be homogenous, radiation-hard, and it must be easily manufactured and must not be ionically conductive to provide long-term stability.

In this context, we developed low resistivity vanadate-based glasses. The lead vanadates ended up with conductivities of the order of 10^{-10} S/cm, and the samples were not able to hold the high voltage with sparking across the plates. The target conductivity range was obtained with tellurium vanadates doped with zinc oxide (ZnO). Figure 1 (left) shows the electronic conductivity of the binary tellurium vanadate glass as a function of the vanadium oxide mole fraction. The conductivity of the binary system was between 10^{-6} and 10^{-10} S/cm and ZnO doping was an economical modifier to reduce the conductivity to within the target range. From 25% to 55% ZnO, conductivity ranged four orders of magnitude from around 10^{-11} to around 10^{-15} S/cm.

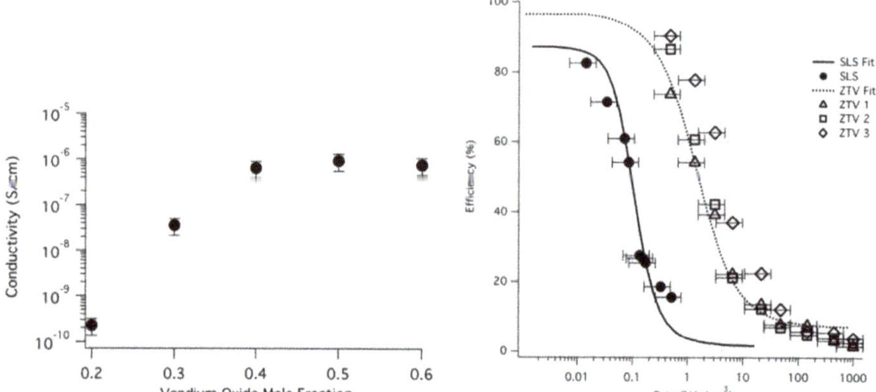

Figure 1. The electronic conductivity of the binary tellurium vanadate glass as a function of the vanadium oxide mole fraction (**left**) and the efficiency vs. particle rate for the RPCs made with zinc-tellurium vanadate glasses (**right**).

The glass composition of $0.40ZnO–0.40TeO_2–0.20V_2O_5$ was used to make three 5 cm × 5 cm two-glass RPCs. The RPCs were tested at the Fermilab Test Beam Facility (FTBF) [5] with 120 GeV proton beam. Figure 1 (right) shows the efficiency vs. particle rate for the RPC made with soda–lime silicate glasses (denoted as SLS) and the three RPCs made with the zinc tellurium vanadate glasses (denoted ZTV1, ZTV2 and ZTV3). The improvement in the rate capability of the RPCs at a given efficiency is more than an order of magnitude [6]. Since the production is made in house, the conductivity can be tuned to very fine levels. The performance of the RPCs is consistent across different glass production campaigns and is also consistent with the theoretical calculations [4] denoted with the solid curves. The final composition can be easily transferred to larger production facilities. On the other hand, the mechanical properties of several square-meter-size glasses with this composition should be studied in detail once available.

3. Development of 1-Glass RPC

The novel 1-glass RPC design offers a number of distinct advantages including an average pad multiplicity close to unity, indicating better position resolution and easier calibration; less strict parameters on the resistive layer of the cathode plate, as the performance of the two- (or multi-) glass chamber mostly depends on the quality of the coating on the anode plate; lower thickness, which is highly desirable for calorimetry; and an improved rate capability [7].

Chambers with a lateral size of 32×48 cm^2 were built with a single soda–lime float glass and were read out with standard DHCAL electronics. Figure 2 shows pictures of the readout (left) and front (right) side of the 1-glass RPC. The average efficiency of the chambers was 95%, and the pad multiplicity was close to unity as measured in the cosmic ray test stand [7].

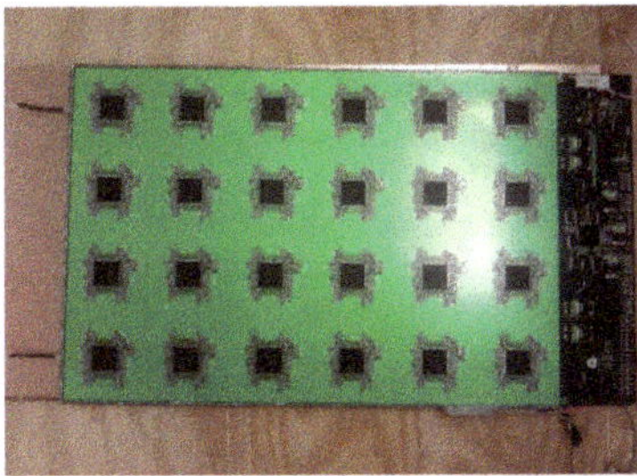

Figure 2. Pictures of the readout (**left**) and front (**right**) side of the 1-glass RPC. The readout side picture depicts the array of the DCAL chips and the data concentrator circuitry of the digital readout. The front side picture depicts the cathode glass covering the entire area of the 1 cm $\times$ 1 cm pads.

4. Measurement of Rate Capability

Three different RPC designs were tested for rate capability at FTBF:

- 2-glass RPCs with standard glass: The chambers were built with two standard soda–lime float glass plates with a thickness of 1.1 mm each. The gas gap was 1.2 mm. The chambers were 20×20 cm^2 in size.
- 1-glass RPCs with standard glass: The chambers were built with one standard soda–lime float glass plate with a thickness of 1.15 mm. The gas gap was also 1.15 mm. The lateral size of the chamber was dictated by the size of the readout board, i.e., 32×48 cm^2, as described in the previous section.
- 2-glass RPCs with semi-conductive glass: These chambers utilize 1.4 mm semi-conductive glass with a bulk resistivity several orders of magnitude smaller than standard soda–lime float glass, obtained from Schott Glass Technologies Inc. [8]. The gas gap of these chambers was also 1.15 mm and the area of the chambers measured 20×20 cm^2.

The usually pencil-like 120 GeV primary proton beam of FTBF was defocused upstream of the experimental hall, and the Gaussian beam profile was measured with the wire chambers. The widths in the x and y directions were measured to be $\sigma_x = 1.0$ cm and $\sigma_y = 0.8$ cm. In the calculation of the beam intensity, in units of Hz/cm^2, the size of the beam spot was taken to be $2\sigma_x \times 2\sigma_y$, with an error derived from the measurement error of the widths of the Gaussians.

Figure 3 shows the efficiency (left) and average pad multiplicity (center) as a function of beam rate for six different RPCs [9]. The performance across different construction campaigns is consistent, and the rate at which the chambers retain high efficiency is observed to increase with decreasing overall resistance of the chambers. The 1-glass RPC is approximately a factor of two better than the two-glass RPC in terms of rate capability. The average pad multiplicity is close to unity for the entire range of particle rates for the 1-glass chambers. For the 2-glass chambers, the average pad multiplicity remains below two. Figure 3 (right) shows the rate at 50% efficiency as a function of conductance per area of the glass plates. The points are fit empirically to the following function which was drawn as a red curve: $I_{50\%} = 1.7 \times 10^5 + 3.2 \times 10^6 H - 1.7 \times 10^8 H^3$ where $H = 1/\log_{10} G$, G being the conductance per unit area of the glass plates. There is indication that higher rate capability can be achieved with further R&D.

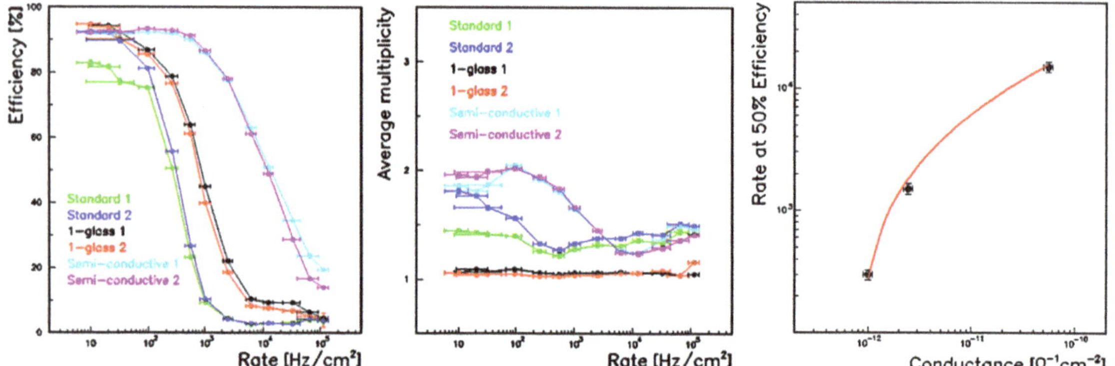

Figure 3. The efficiency (**left**) and the average pad multiplicity (**center**) as a function of beam rate for six different RPCs, and the rate at 50% efficiency as a function of conductance per area of the glass plates (**right**).

5. Development of Hybrid RPC

In order to mitigate the performance limitations associated with the alternative RPC gas mixtures, reduce the overall fresh gas needs also investigating the possibility of sealed chambers, and study the alternative multiplication mechanisms, we probed a new technique based on further electron multiplication in a thin layer of high secondary electron yield material. The first set of materials probed were Al_2O_3 and TiO_2. The material is applied as a coating on the anode plane, which is inside the chamber for the 1-glass RPC designs. The coating of Al_2O_3 was carried out with magnetron sputtering, and the coating of TiO_2 was applied with an airbrush. We constructed several 10 cm × 10 cm chambers, also standard 1- and 2-glass RPCs. Following the laboratory tests, the RPCs were tested with FTBF muons. The list of the RPCs tested is as follows: one standard 1-glass RPC; one standard 2-glass RPC; two 1-glass RPCs with anodes coated with 500 nm and 350 nm Al_2O_3 (Al_2O_3_v1 and Al_2O_3_v2); and three 1-glass RPCs with anodes coated with 1 mg/cm^2, 0.5 mg/cm^2, and 0.15 mg/cm^2 TiO_2 (TiO_2_v1, TiO_2_v2, and TiO_2_v3). The gas mixture was the standard DHCAL mixture: R134a (94.5%), Isobutane (5.0%), and SF_6 (0.5%); the gas flow rate was 2–3 cc/min, roughly half of the rate for the DHCAL.

Figure 4 shows the efficiency as a function of the applied high voltage. The standard 2-glass RPC becomes 90% efficient around 8.5 kV. Similarly, if one considers the high voltage value at the 90% efficiency crossing as a measure, for the standard 1-glass RPC and TiO_2_v3, it is around 7.5 kV. The advantage of the 1-glass RPC over the 2-glass RPC is manifest, and there is no measurable effect on the performance with 0.15 mg/cm^2 TiO_2 coating applied on the anode plate.

The major improvement on the performance is obtained with the RPCs which have anode plates coated with Al_2O_3 and thicker TiO_2. The 90% efficiency crossing voltage setting is around 6.5 kV, clearly indicating the contribution of the electron multiplication in the coating since the standard 1-glass RPC efficiency is negligible at these voltages. Following the promising results with these first-generation hybrid RPCs, data analysis and simulation studies to quantify the effect of the coating as a function of the material and its thickness are underway, and further tests with segmented anode and different gases are planned.

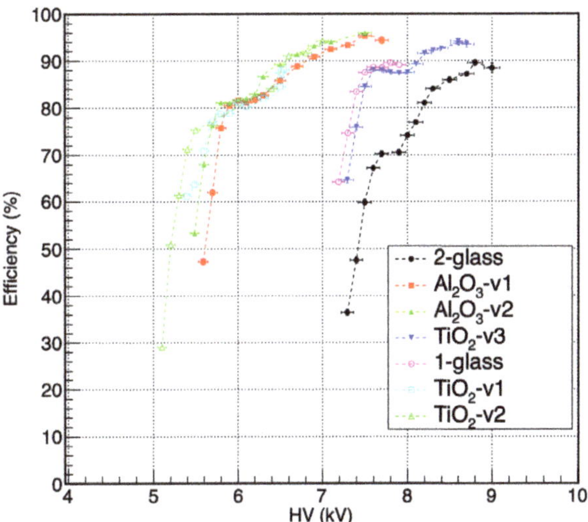

Figure 4. The efficiency as a function of applied high voltage for seven RPCs, including the five hybrid RPCs.

6. Conclusions

Novel designs of Resistive Plate Chambers were developed in order to respond to several outstanding issues.

A dedicated R&D was performed to develop semi-conductive glasses with flexible design parameters. It was found that the $0.40ZnO$–$0.40TeO_2$–$0.20V_2O_5$ glass provides a feasible starting point with a large phase space of options of constituents and processes to pursue further R&D.

The 1-glass RPC design offers several advantages over 2-glass RPCs, such as an average pad multiplicity around one and an increased rate capability. It also starts a new chapter where the in-chamber anode plate can be made more functional. By coating the anode plate with high secondary electron yield materials, electron multiplication in the chamber can be enhanced considerably. R&D is underway to fully characterize the newly developed, so-called hybrid RPCs. The hybrid RPCs have the potential to mitigate the limitations associated with RPC gases and to relax the overall operating conditions.

Several RPCs, including 1-glass and 2-glass RPCs made with the standard soda–lime float glass, and 2-glass RPCs made with semiconductive glasses, were tested for their rate capabilities in a 120 GeV proton beam at FTBF. The results indicate that with increasing conductance per area of the glass plates, the rate capability of RPCs increases. In addition, the range of particle rates for which the chambers retain their full particle detection efficiency also increases. An empirical relation for the dependence of the rate at 50% efficiency on the conductance per unit area of the glass plates was obtained.

Author Contributions: Investigation, K.K.S. and M.T.; Project administration, B.B. and J.R. and L.X.; Writing–review & editing, Y.O. All authors have read and agreed to the published version of the manuscript.

Funding: This research received no external funding.

Data Availability Statement: Not applicable.

Acknowledgments: B. Bilki, K. K. Sahbaz and M. Tosun acknowledge support under Tübitak grant no 118C224.

Conflicts of Interest: The authors declare no conflict of interest.

References and Note

1. Santonico, R.; Cardarelli, R. Development of resistive plate counters. *Nucl. Instr. Meth.* **1981**, *187*, 377. [CrossRef]
2. Available online: https://twiki.cern.ch/twiki/bin/view/CALICE/WebHome (accessed on 31 August 2022).
3. Adams, C.; Bambaugh, A.; Bilki, B.; Butler, J.; Corriveau, F.; Cundiff, T.; Drake, G.; Francis, K.; Furst, B.; Guarino, V.; et al. Design, construction and commissioning of the Digital Hadron Calorimeter—DHCAL. *J. Instrum.* **2016**, *11*, P07007. [CrossRef]
4. Bilki, B.; Butler, J.; May, E.; Mavromanolakis, G.; Norbeck, E.; Repond, J.; Underwood, D.; Xia, L.; Zhang, Q. Measurement of the rate capability of Resistive Plate Chambers. *J. Instrum.* **2009**, *4*, P06003. [CrossRef]
5. Available online: https://ftbf.fnal.gov/ (accessed on 31 August 2022).
6. Johnson, N.; Wehr, G.; Hoar, E.; Xian, S.; Akgun, U.; Feller, S.; Affatigato, M.; Repond, J.; Xia, L.; Bilki, B. Electronically Conductive Vanadate Glasses for Resistive Plate Chamber Particle Detectors. *Int. J. Appl. Glass Sci.* **2015**, *6*, 26. [CrossRef]
7. Bilki, B.; Corriveau, F.; Freund, B.; Neubüser, C.; Onel, Y.; Repond, J.; Schlereth, J.; Xia, L. Tests of a novel design of Resistive Plate Chambers. *J. Instrum.* **2015**, *10*, P05003. [CrossRef]
8. Schott GlassTechnologies Inc. 400 York Ave, Duryea, PA 18642, USA.
9. Affatigato, M.; Akgun, U.; Bilki, B.; Corriveau, F.; Freund, B.; Johnson, N.; Neubüser, C.; Onel, Y.; Repond, J.; Xia, L. Measurements of the rate capability of various Resistive Plate Chambers. *J. Instrum.* **2015**, *10*, P10037. [CrossRef]

 *instruments*

Editorial

25 Years of Dual-Readout Calorimetry

Richard Wigmans

Department of Physics, Texas Tech University, Lubbock, TX 79409-1051, USA; richard.wigmans@ttu.edu

Abstract: Twenty-five years ago, at the CALOR1997 conference in Tucson, the idea of dual-readout calorimetry was first presented. In this talk, I discuss the considerations that led to that proposal, and describe the developments that have since taken place, to the point where dual-readout calorimetry is now considered a major candidate for experiments at future colliders.

Keywords: dual-readout calorimetry; calorimeter calibration; calorimeter performance; future colliders

---

In 1997, at the seventh edition of the CALOR conference series, I introduced what was to become dual-readout calorimetry. After describing the potential advantages of simultaneous detection of dE/dx and the Čerenkov light generated in the absorption of high-energy hadrons, I ended my presentation with the following sentence (Paper 1, Table 1):

> *"I am convinced that resources for a dedicated R&D program to investigate these possibilities may turn out to be extremely well spent".*

In my talk today, I will demonstrate that this prediction has turned out to be correct. I will, in a chronological way, describe the R&D efforts that have been carried out in the past 25 years. These efforts have led to the point that today dual-readout calorimetry is considered a major candidate for future experiments in particle physics. However, let me start by providing some context by summarizing the developments in our understanding of calorimetry that took place in the decade preceding the 1997 conference:

- Hadron showers consist of an electromagnetic (em) and a non-em component.
- The non-em component involves nuclear reactions; the nuclear binding energy of nucleons released in these reactions does *not* contribute to the calorimeter signals. This is what is usually referred to as **invisible energy**.
- The hadronic energy resolution of a calorimeter is usually determined by fluctuations in invisible energy.
- The relative effect of such fluctuations does *not* become smaller as $1/\sqrt{E}$ at increasing energy, unlike in em calorimeters, where sampling fluctuations and fluctuations in the number of signal quanta dominate.
- The average value of the em shower fraction increases with energy. This is the reason for the **non-linearity** typical for most hadron calorimeters.
- A crucial calorimeter performance parameter is e/h, the ratio of the calorimeter response (i.e., average signal/GeV) to the em and non-em shower components. Typically, $e/h > 1$, because of invisible energy. If $e/h = 1$, a major performance improvement can be obtained. Such calorimeters are called **compensating**.

So how can these facts be used to improve hadron calorimeter performance?

- By designing a calorimeter in such a way that $e/h = 1$. This works **only** for sampling calorimeters.
- In sampling calorimeters, different classes of shower particles may be sampled very differently. The electrons and positrons that make up the em shower component

Citation: Wigmans, R. 25 Years of Dual-Readout Calorimetry. *Instruments* **2022**, *6*, 36. https://doi.org/10.3390/instruments6030036

Received: 4 August 2022

Accepted: 22 August 2022

Published: 7 September 2022

Publisher's Note: MDPI stays neutral with regard to jurisdictional claims in published maps and institutional affiliations.

are sampled according to dE/dx[1]. In the non-em component, **neutrons** produced in nuclear breakup may be sampled *much* (10–100 times) more efficiently, when the active medium contains hydrogen. There is little or no competition for np elastic scattering in that case.

- The total kinetic neutron energy is correlated with the invisible energy loss, especially in high-Z materials.
- The amplification factor for neutron signals should be chosen such that it **compensates** for the invisible energy loss: $e/h = 1$. This amplification factor is determined by the sampling fraction for charged shower particles in the calorimeter, e.g., $\sim$2% for Pb/plastic scintillator, $\sim$6% for U/plastic scintillator structures.

Figure 1 illustrates the importance of the e/h value for the performance of hadron calorimeters. Diagram c shows that the compensating calorimeter is nicely linear, while the calorimeters with $e/h \neq 1$ exhibit substantial non-linearities. A comparison of diagrams Figure 1a,b illustrates the importance for the hadronic energy resolution. At high energy, the resolution of the 2% sampling device (Figure 1b) was measured to be 3 times better than that of the homogeneous (i.e., sampling fraction 100%) calorimeter (Figure 1a), and the difference in energy dependence is also very striking.

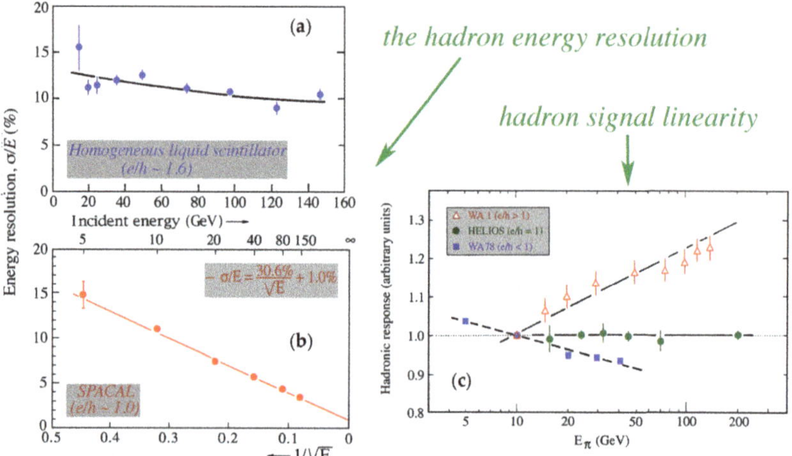

Figure 1. The effects of compensation for the performance of hadron calorimeters.

Despite the obvious advantages offered by compensating calorimeters (hadronic energy resolution, signal linearity, Gaussian response functions, as well as the fact that calibration becomes very easy since there are no more differences between electrons and hadrons) there are also some disadvantages. The small sampling fraction that is needed limits the em energy resolution, and the crucial reliance on detecting neutrons requires larger integration volumes and integration times than may be practical. However, the most important disadvantage may concern the jet performance. Typically, a substantial energy fraction of jets comes in the form of relatively low-energy particles. As illustrated by Figure 2a, the response to hadrons gradually decreases for kinetic energies below 5 GeV, since an increasing fraction of the hadrons range out before initiating a nuclear interaction, and thus only lose energy by ionizing the absorber medium, just like muons. Furthermore, since the e/mip value may be quite different from 1.0 (e.g., 0.6 in the case of the uranium calorimeter shown in this figure), response non-linearity is thus also an issue for jet detection in this compensating calorimeter. Figure 2b shows that this problem could be mitigated

---

[1] in the last stages of an EM shower, sampling of soft photons depends on the Z value of the absorber medium, which may lead to $e/mip \neq 1$.

by using a lower-Z absorber medium. Dual-readout calorimetry offers that option. It is also not bound by the small sampling fractions required for compensation. These were important considerations for the proposal to study the possibilities of this alternative method to improve the performance of hadron calorimeters.

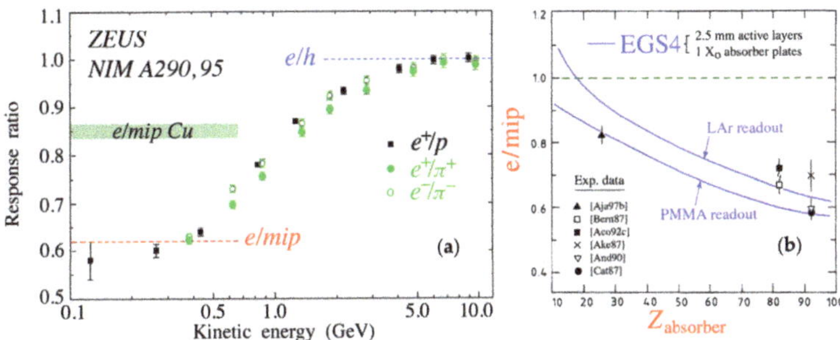

Figure 2. The low-energy response of the compensating ZEUS calorimeter (**a**) and the e/mip values measured for sampling calorimeters with different Z absorber material (**b**).

Having a good idea is one thing, finding the money to test it experimentally is a completely different issue, especially when there is no immediate application, e.g., in an approved future experiment. However, in this case NASA came to the rescue, thanks to a former postdoc of mine who was working on a project to detect ultra-high cosmic hadrons outside the Earth's atmosphere. NASA was considering an experiment at the International Space Station, called ACCESS, and had issued a call for proposals for a suitable detector. In the TeV-PeV energy region, calorimetry was one of very few viable options. However, mass restrictions (<2000 kg) and the required large aperture made this an extremely challenging proposition, since how could one expect to do any meaningful measurements with an instrument that was only 2 nuclear interactions lengths deep (or even less) on a particle that needed at least 10 interaction lengths to be absorbed?

The properties of such a thin calorimeter would be completely determined by leakage fluctuations. Furthermore, unless one could get a handle on these fluctuations, *event by event*, no acceptable performance should be expected. We argued (successfully) that dual-readout calorimetry could provide such a handle. The argument went as follows. At a depth of $2\,\lambda_{\text{int}}$, the overwhelming majority of the hadrons will initiate a nuclear interaction in the calorimeter. In this interaction, some fraction of the energy will be used for π^0 production. If that fraction is large, there will, on average, be relatively little energy leakage, since the em showers developed by the π^0s may be contained, to a large extent, in the absorber, especially if this is made of high-Z material ($\lambda_{\text{int}}/X_0 \approx 30$ in lead). On the other hand, if the em fraction is low, the energy that leaks out of the thin calorimeter is relatively large. It was already known from the prototype studies for the CMS very-forward calorimeter, which uses quartz fibers as active material, that the Čerenkov light produced in these fibers is overwhelmingly generated by the em components of the hadron showers. For these reasons, simultaneous detection of Čerenkov light and scintillation light (a measure for dE/dx) would provide not only information about the energy deposited in the calorimeter, but also on the relative fraction of em shower energy and, therefore, about the undetected energy leaking out.

We tested the validity of these ideas (Paper 2, Table 1) with the calorimeter depicted in Figure 3. The absorber material was lead, 39 plates, each 6.4 mm thick, for a total depth of $1.4\,\lambda_{\text{int}}$. In between these plates, layers of ribbons of fibers were inserted. These fibers were alternatingly made of plastic scintillator and quartz. The fibers from each layer were read out by small photomultiplier tubes. As shown in the figure, these PMTs were arranged in such a way that x-y granularity was achieved for both types of readout. Essentially, in this way we constructed two calorimeters that provided completely independent scintillator (S) and Čerenkov (Q) signals from the same events.

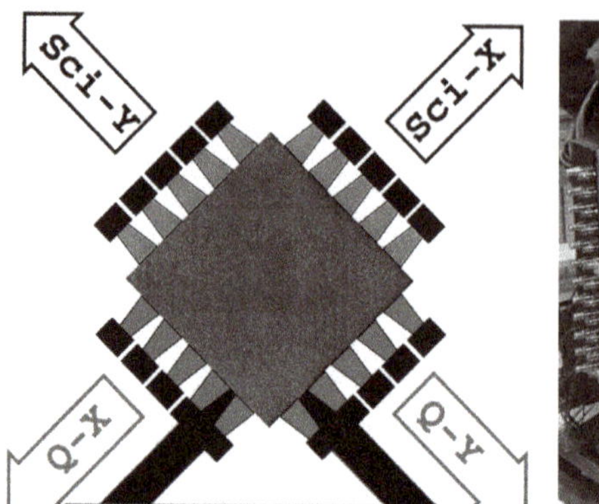

Figure 3. The first dual-readout calorimeter, a prototype for the ACCESS experiment at the International Space Station.

In order to study whether the ideas described above worked in practice, this instrument was exposed to beams of high-energy pions at the CERN SPS. Figure 4 shows some results from measurements with 375 GeV π^-. The energy scale for both types of signals was set with a beam of electrons. Figure 4a shows a scatter plot of the signals from the pions recorded in the quartz (vertical scale) and scintillation (horizontal scale) fibers. If there was no extra information in the Čerenkov signals, the data points would cluster around the diagonal in this plot. However, the observed banana shape indicates otherwise. A straight line through the origin of this plot describes data with a fixed ratio of the two types of signals. In Figure 4c,d, the scintillation signal distributions are shown for cuts $Q/S < 0.45$ and $0.75 < Q/S < 0.85$, respectively. These distributions are subsets of the *total* scintillation signal distribution, which is shown in Figure 4b. Clearly, small Q/S values select events with relatively little π^0 content, and thus large shower energy leakage and a relatively small total signal. On the other hand, events with relatively large Q/S values are indicative for showers with a relatively large em fraction, and hence relatively little energy leakage and a correspondingly large calorimeter signal. This is precisely what we hoped to achieve with this dual-readout calorimeter.

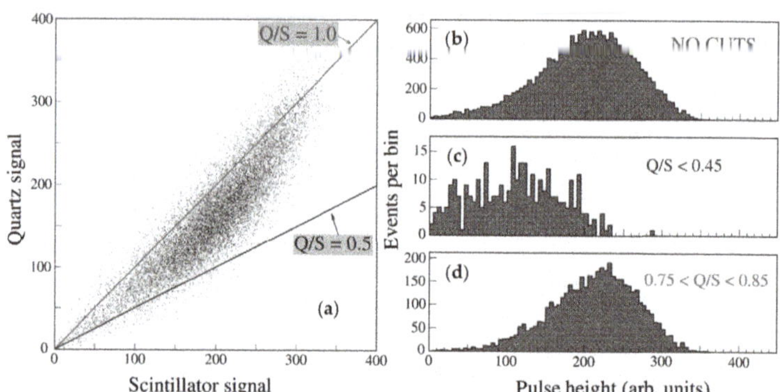

Figure 4. Results from measurements of the signals from 375 GeV π^- sent into the 1.4 λ_{int} thick ACCESS dual-readout calorimeter. See text for details.

Encouraged by the fact that the dual-readout principle worked already so well in this extremely thin calorimeter[2], we started to plan for a much larger instrument for particle

---

[2] Unfortunately, NASA cancelled the ACCESS project after the accident with the Columbia Space Shuttle (2003).

physics experiments, here on Earth where the mass limitations imposed by NASA do not apply. To contain high-energy hadron showers, such a detector needed to be at least 10 λ_{int} deep. We chose copper absorber, which has many advantages over lead (weight, machinability, e/mip ratio, ...). Based on the mentioned results, we convinced US-DOE (our funding agency) to support us financially (160 k\$). To save money, we used as much material from previous projects as possible (e.g., quartz fibers from CMS-HFCAL, PMTs, etc.). The DREAM calorimeter, shown in Figure 5, was built at TTU in 2003 and tested at the CERN SPS with beams of high-energy electrons, pions and muons (Paper 3, Paper 4, Table 1), by a small group of TTU people, with help from some friends (Hans Paar, John Hauptman, Aldo Penzo).

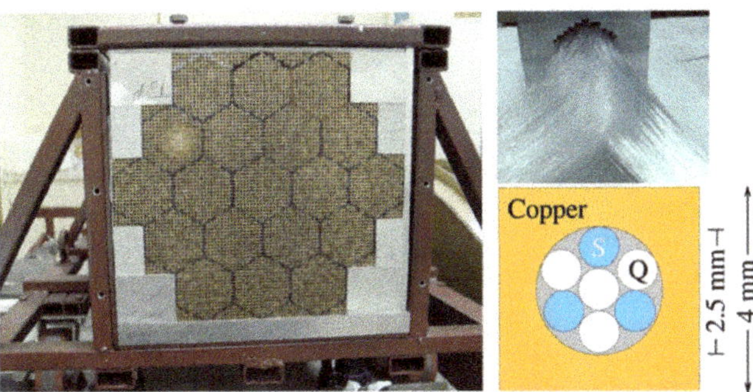

Figure 5. The DREAM calorimeter. See text for details.

The basic building block was a 2 m long copper rod with a central hole in it. In this hole were inserted seven optical fibers, three scintillating and four undoped ones (quartz in the central region, PMMA in the periphery). The calorimeter consisted of 5150 such rods, arranged in a pattern of 19 hexagonal cells. The fibers from each cell were split into two bunches, one for each type of fiber. Each bunch was connected to a PMT, so that there were thus 19 + 19 = 38 signals recorded for every shower developing in this instrument, which had a total fiducial mass of 1030 kg.

Figure 6b shows the first surprise encountered when we tested this detector with muons of different energies (Paper 3, Table 1). Even though the calorimeter was calibrated with a beam of high-energy electrons, the signals for muons, which also exclusively lose energy through the em interaction, were different for the two types of signals provided by the calorimeter. The S signals were, on average, larger than the C ones, *by a constant amount*. The reason for this is the fact that the direct Čerenkov light emitted as the muons travel in the direction of the fibers through the calorimeter, falls outside the numerical aperture of the fibers, and thus does not contribute to the signals. For the S signals, this is inconsequential, but the C fibers only produce a signal from the radiative energy losses. Since the relative importance of these processes increases with the muon energy, so do the signals, both in the S and the C channels. The constant difference between these two signals thus makes it possible to measure (event by event) the fractions of the measured scintillation signal due to direct ionization and to higher-order em processes: A unique feature.

Measurement of the e/mip ratio (Figure 6a) confirmed that the value for this copper calorimeter (0.82) was indeed considerably closer to 1 than the values typical for calorimeters based on lead or uranium absorber.

Additionally, the results obtained for the hadronic performance (Paper 4, Table 1) confirmed the beneficial effects of the dual-readout method. Figure 7a shows that the response became linear and equal to that of electrons when the information from both types of signals was used to reconstruct the particle energy. The hadronic energy resolution also improved in this process, especially at high energies, because the deviation from $1\sqrt{E}$ scaling (almost) disappeared (Figure 7b).

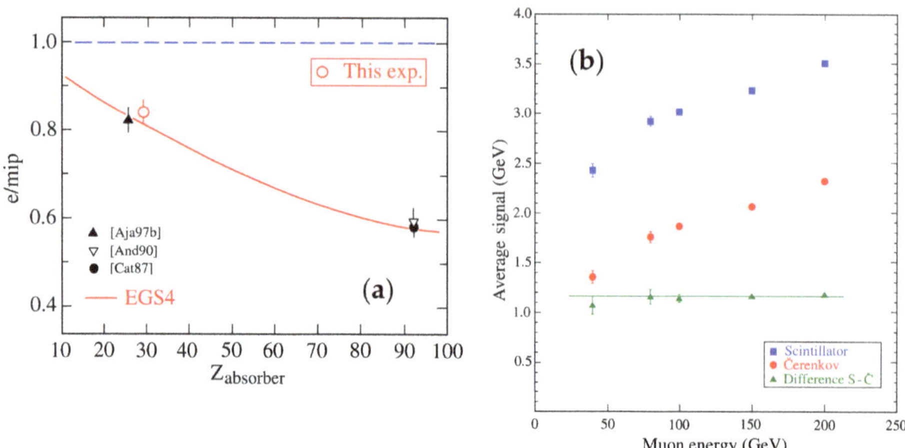

Figure 6. Results of muon detection with the DREAM calorimeter.

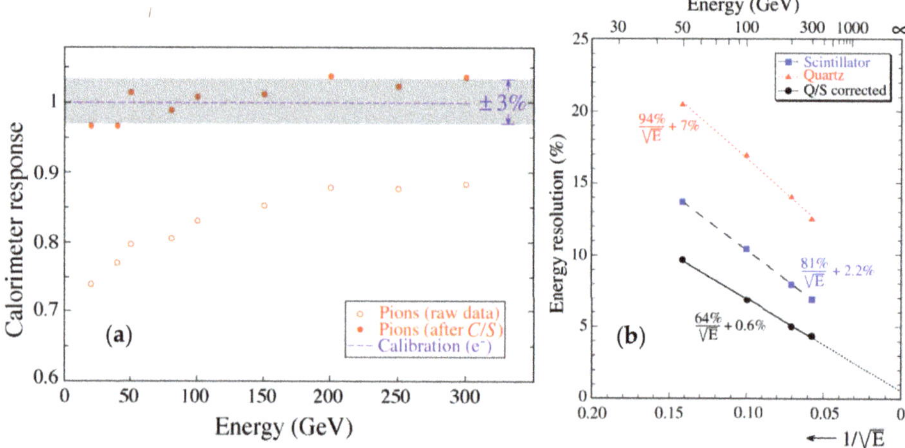

Figure 7. Results of pion detection with the DREAM calorimeter.

The method we used to determine the em shower fraction and the energy of the hadron is illustrated in Figure 8a, which shows the scatter plot of the two types of signals recorded for each event. The two signals, S and C, can each be described in terms of an em plus a non-em component. The latter is weighed by a factor $\frac{1}{e/h}$. The crucial point of the method is that these factors (h/e) are *different* for the S and C components, otherwise all data points would be scattered around the $C = S$ diagonal, and the method to determine the em fraction and the hadron energy from the C/S signal ratio, as described in this figure, would not work. It turns out that the total signal distribution, which may be obtained by projecting the scatter plot on either the horizontal (S) or vertical axis (C), is non-Gaussian, which is typical for calorimeters with $e/h \neq 1$ (Figure 8b). However, is turns out that this distribution is in fact a superposition of many distributions with different fixed em fractions (Figure 8c), and the overall signal distribution just reflects the extent to which these different f_{em} fractions occur in practice. The dual-readout method eliminates these differences and results in Gaussian signal distributions with the correct average value, i.e., the same as the value for electrons of the same energy.

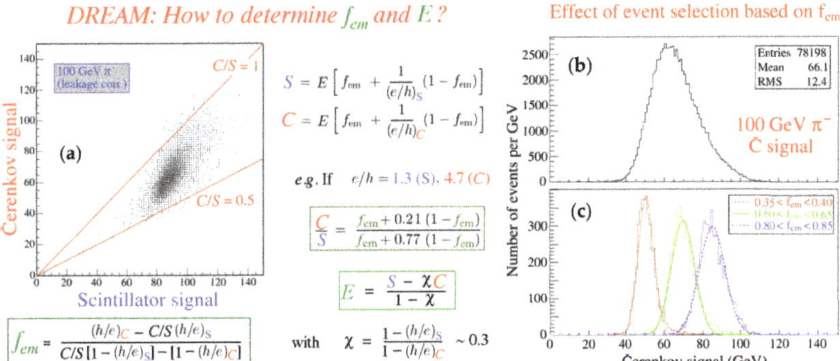

Figure 8. The dual-readout method used to determine the em shower fraction and the hadron energy, based on the S and C signals (**a**). The non-Gaussian 100 GeV signal distribution (**b**) is a superposition of Gaussian distributions with different f_{em} values (**c**).

A closer look at the (energy normalized) $[S/E, C/E]$ scatter plot reveals several other interesting features (Figure 9). All experimental data points are located on a straight line that connects the points $[(h/e)_S, (h/e)_C]$ and $[1,1]$. The larger the f_{em} value, the closer the data point is to $[1,1]$, i.e., the point where the electron data congregate ($f_{em} = 1$). The distribution of the hadron data points on this line reflects the distribution of the f_{em} values. The f_{em} distributions may be different for different types of hadrons (e.g., for pions the average f_{em} value is larger than for protons of the same energy, see Figure 9), but the data points cluster around the same line. The same is true for hadrons of different energy, where the average f_{em} increases with energy. The parameter χ, which was introduced in Figure 8, is equivalent to the cotangent of the angle θ that defines the direction of the line around which the data points cluster in Figure 9. This leads to the important conclusion that both χ and θ, which form the essence of the dual-readout method, are **independent of the hadron energy and the type of hadron** (Paper 10, Table 1).

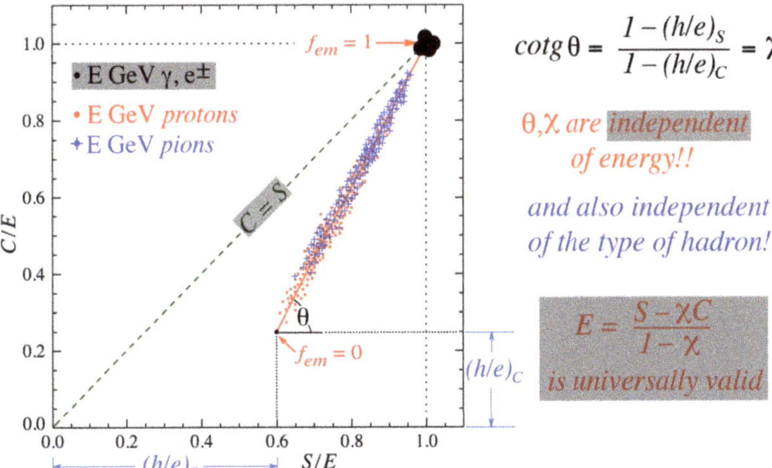

Figure 9. The dual-readout method is independent of the energy and the type of hadron. It also works well for jets.

After the successes obtained with the DREAM calorimeter, a new collaboration was formed that further pursued the possibilities of this novel concept in the context of CERN's R&D program. This RD52 Collaboration consisted of American, Italian and South Korean scientists. The dual-readout technique does not necessarily require a sampling calorimeter. However, it is essential that the signals provided by the detector can be separated into Čerenkov and scintillation components. RD52 demonstrated the possibility to achieve that in a number of high-density crystals (Paper 5, Table 1), e.g., $PbWO_4$, which is being used in CMS and other hep experiments. It turned out that a substantial fraction of the signals produced by this type of crystal is actually the result of Čerenkov light, and that it can be distinguished from the scintillation component in a variety of ways, including the time structure, the spectral properties and the polarization characteristics of the signals.

The Collaboration also built a substantial em calorimeter section consisting of BGO crystals and used this in conjunction with the DREAM detector, which served as the hadronic section. Figure 10 shows some results of tests of this configuration. By using UV filters, the signals from the BGO section exhibited both a Čerenkov and a scintillation component, which could be extracted with two separated time gates (Figure 10a). Figure 10b shows that it even turned out to be possible to obtain information on the production of neutrons in the shower development, which manifested itself as a slow tail in the scintillation signals. This tail was absent for tests with electrons and in all Čerenkov signals. It was demonstrated that by using the neutron information event-by-event a similar unraveling of the overall signal distribution could be obtained as on the basis of the f_{em} value (Figure 8b,c).

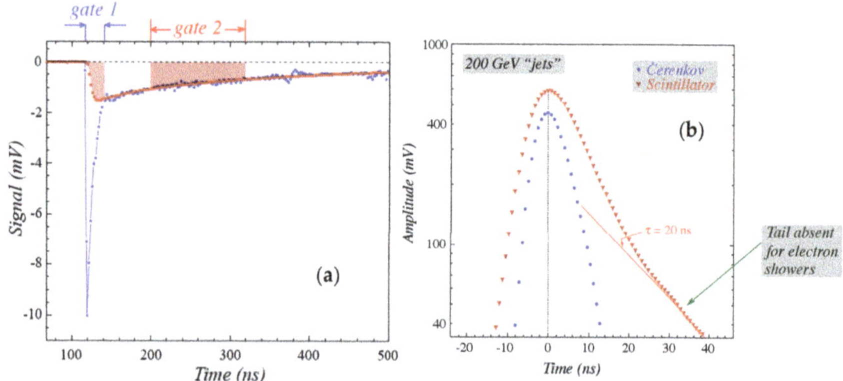

Figure 10. RD52 results on tests of the DREAM calorimeter preceded by a BGO em section. See text for details.

The RD52 Collaboration also built and tested fiber sampling calorimeters with a much finer sampling than the DREAM one. Figure 11 shows a copper based instrument that was primarily tested with electrons (Paper 7, Table 1). A larger, 1350 kg calorimeter using lead as absorber material was built by the Pavia group, and was tested with beams of high-energy electrons, muons, pions and protons at CERN.

These measurements were also used to test the predictions depicted in Figure 9, i.e., the fact that the dual-readout method yields results that are independent on the type of hadron. Figure 12a shows excellent signal linearity, both for pions and protons, over the full energy range of the measurements (Paper 9, Table 1). The average signals per GeV were within ∼2% equal for electrons, protons and pions. Figure 12b shows that the response functions were extremely well described by a Gaussian function, and the width of this function scaled within experimental errors with $30\%/\sqrt{E}$, without any need for additional resolution terms, both for pions and for protons. To obtain these results, the calorimeter was surrounded by large slabs of plastic scintillator, whose signals were used to determine the side leakage of the hadron showers, event by event.

Figure 11. A fine-sampling copper-fiber calorimeter built by the Pisa group of the RD52 Collaboration.

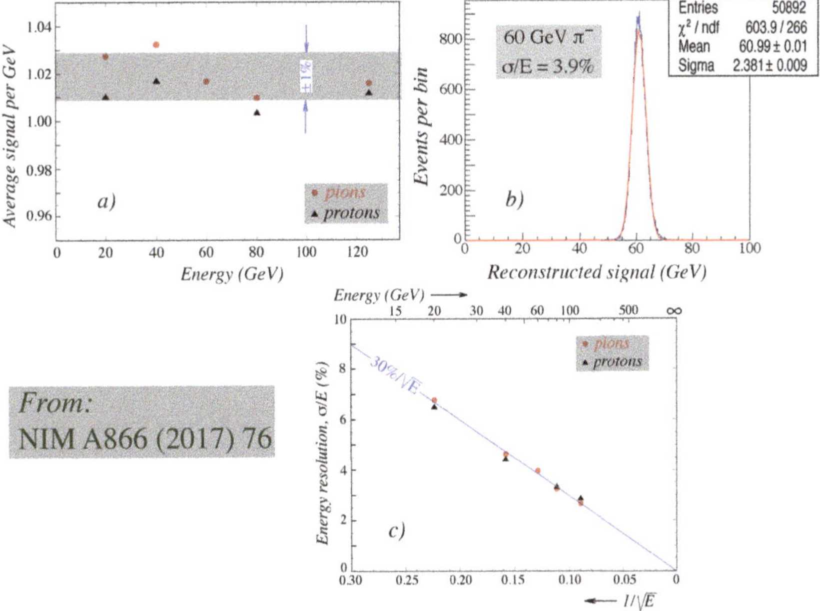

Figure 12. Results on pion and proton detection with a fine-sampling lead based dual-readout calorimeter built by the RD52 Collaboration. Shown are results on hadron signal linearity (**a**), response function (**b**) and energy resolution (**c**).

Side leakage fluctuations were still a significant remaining contribution to this record setting resolution performance. Other factors were the Čerenkov light yield and sampling fluctuations. However, the ultimate limit that can be achieved is determined by the correlation between the reconstructed hadron energy and the nuclear binding energy loss in the shower development. Our Monte Carlo studies (Paper 8, Table 1) revealed that this ultimate limit is actually better than that achievable with compensation techniques, both for copper and lead absorber (Figure 13).

In summary, dual-readout fiber calorimetry offers potentially a spectacular performance improvement over the calorimeters used in past and present hep experiments. It is often assumed that the fact that the fiber structure prevents longitudinal segmentation is a significant disadvantage, especially when it comes to electron identification. Actually, this is a **myth**. Not only does the very fine lateral segmentation allowed by the fibers offer fantastic alternative options, but the time structure of the signals also turns out to make it possible to determine the depth of the light production in the fibers with a precision of $\sim$1 λ_{int}. We have demonstrated (Paper 6, Table 1) that a combination of the various available options made it possible to identify electrons with >99% efficiency, with <0.2% pion misidentification, in a mixed electron/pion beam. Apart from that, the fiber geometry allows for a very compact structure without significant dead space, while the absence of longitudinal segmentation avoids the huge and often unsolvable problems encountered when intercalibrating the different longitudinal sections of a segmented device. This fiber calorimeter can be calibrated with electrons and that is all that is needed to obtain the correct energy scale for all hadrons, and jets.

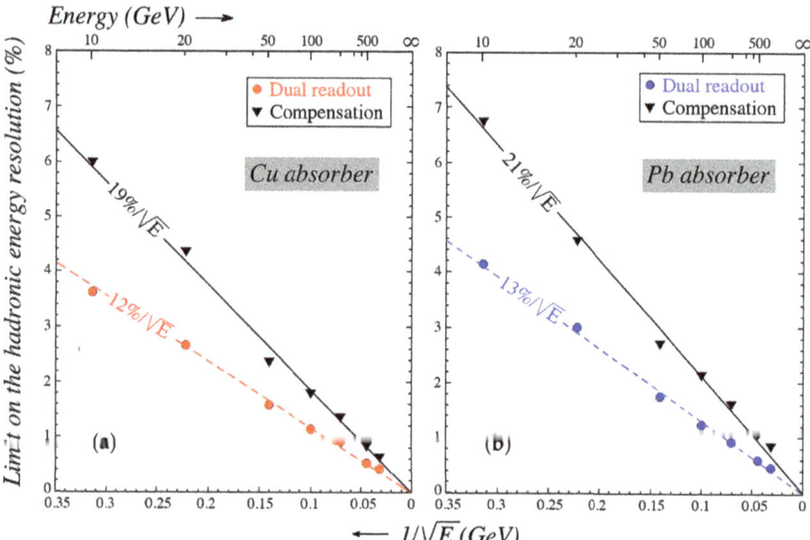

Figure 13. The limit on the hadronic energy resolution derived from the correlation between nuclear binding energy losses and the parameters measured in dual-readout or compensating calorimeters, as a function of the particle energy. Results from GEANT Monte Carlo simulations of pion showers developing in a massive block of copper (**a**) or lead (**b**).

Table 1 lists some significant publications from 25 years of dual-readout calorimetry, in chronological order. A complete list can be found in the Rev. Mod. Phys. article (Paper 11, Table 1), indicated by the red arrows.

Table 1. A chronological selection of publications on dual-readout calorimetry.

The beginning of Dual-Readout Calorimetry:

1) *Quartz Fibers and the Prospects for Hadron Calorimetry at the 1% Resolution Level,*
R. Wigmans, Proceedings of the 7th International Conference on Calorimetry in
High Energy Physics, Tucson (AZ), Nov. 9-14, 1997.

Selected papers in the refereed literature:

2) *Beam Tests of a Thin Dual-Readout Calorimeter for Detecting Cosmic Rays Outside the Earth's Atmosphere,*
V. Nagaslaev, A. Sill and R. Wigmans, Nucl. Instr. and Meth. **A462** (2001) 411–425.

3) *Muon Detection with a Dual-Readout Calorimeter,*
N. Akchurin *et al.,* Nucl. Instr. and Meth. **A533** (2004) 305–321.

4) *Hadron and Jet Detection with a Dual-Readout Calorimeter,*
N. Akchurin *et al.,* Nucl. Instr. and Meth. **A537** (2005) 537 – 561.

5) *Dual-Readout Calorimetry with Crystal Calorimeters,*
N. Akchurin *et al.,* Nucl. Instr. and Meth. **A598** (2009) 710 - 721.

6) *Particle identification in the longitudinally unsegmented RD52 calorimeter,*
N. Akchurin *et al.,* Nucl. Instr. and Meth. **A735** (2014) 120 - 129.

7) *The electromagnetic performance of the RD52 fiber calorimeter,*
N. Akchurin *et al.,* Nucl. Instr. and Meth. **A735** (2014) 130 - 144.

8) *Lessons from Monte Carlo simulations of a dual-readout fiber calorimeter,*
N. Akchurin *et al.,* Nucl. Instr. and Meth. **A762** (2014) 100 - 118.

9) *Hadron detection with a dual-readout fiber calorimeter,*
S. Lee *et al.,* Nucl. Instr. and Meth. **A866** (2017) 76 - 90.

10) *On the limit of the hadronic energy resolution of calorimeters,*
S. Lee, M. Livan and R. Wigmans, Nucl. Instr. and Meth. **A882** (2018) 148 - 157.

11) *Dual-readout calorimetry,*
S. Lee, M. Livan and R. Wigmans, Rev. Mod. Phys. **90** (2018) 025002. 

12) *New Developments in Calorimetric Particle Detection,*
R. Wigmans, J. Progr. Part. Nucl. Phys. **103** (2018) 109 - 161

Conflicts of Interest: The author declares no conflict of interest.

instruments

MDPI

Article

ATLAS LAr Calorimeter Commissioning for the LHC Run 3

Alessandra Betti [1,2] on behalf of the ATLAS Liquid Argon Calorimeter Group

1   Istituto Nazionale di Fisica Nucleare Sezione di Roma (IT), Sapienza Università di Roma (IT), 00185 Roma, Italy; alessandra.betti@cern.ch
2   Department of Physics, Southern Methodist University, Dallas, TX 75205, USA

Abstract: The Liquid Argon Calorimeters are employed by ATLAS for all electromagnetic calorimetry in the pseudo-rapidity region $|\eta| < 3.2$, and for hadronic and forward calorimetry in the region from $|\eta| = 1.5$ to $|\eta| = 4.9$. They also provide inputs to the first level of the ATLAS trigger. After a successful period of data taking during the LHC Run 2 between 2015 and 2018, the ATLAS detector entered into a long period of shutdown. In 2022, the LHC will restart and the Run 3 period should see an increase of luminosity and pile-up of up to 80 interactions per bunch crossing. To cope with these harsher conditions, a new trigger readout path has been installed during the long shutdown. This new path should significantly improve the triggering performance on electromagnetic objects. This will be achieved by increasing the granularity of the objects available at trigger level by up to a factor of ten. The installation of this new trigger readout chain also required the update of the legacy system. More than 1500 boards of the precision readout have been extracted from the ATLAS pit, refurbished and re-installed. The legacy analog trigger readout, which will remain during the LHC Run 3 as a backup of the new digital trigger system, has also been updated. For the new system, 124 new on-detector boards have been added. Those boards that are operating in a radiative environment are digitizing the calorimeter trigger signals at 40 MHz. The digital signal is sent to the off-detector system and processed online to provide the measured energy value for each unit of readout. In total up to 31 Tbps are analyzed by the processing system and more than 62Tbps are generated for downstream reconstruction. To minimize the triggering latency the processing system had to be installed underground. The limited available space imposed a very compact hardware structure. To achieve a compact system, large FPGAs with high throughput have been mounted on ATCA mezzanines cards. In total, no more than three ATCA shelves are used to process the signal from approximately 34,000 channels. Given that modern technologies have been used compared to the previous system, all the monitoring and control infrastructure is being adapted and commissioned as well. This contribution presents the challenges of the installation, the commissioning and the milestones still to be completed towards the full operation of both the legacy and the new readout paths for the LHC Run 3.

Keywords: ATLAS; LAr; calorimeter; Phase-1; upgrade; commissioning

Citation: Betti, A., on behalf of the ATLAS Liquid Argon Calorimeter Group. ATLAS LAr Calorimeter Commissioning for the LHC Run 3. *Instruments* **2022**, *6*, 37. https://doi.org/10.3390/instruments6030037

Academic Editors: Fabrizio Salvatore, Alessandro Cerri, Antonella De Santo and Iacopo Vivarelli

Received: 25 July 2022
Accepted: 5 September 2022
Published: 8 September 2022

1. The ATLAS Liquid Argon (LAr) Calorimeter

The ATLAS Liquid Argon (LAr) Calorimeter [1,2] is a system of sampling calorimeters with full azimuthal coverage. Lead, copper and tungsten are used as absorbers in different parts of the detector and Liquid Argon (LAr) is used as active material. The calorimeter is divided in different sub-detectors as represented in Figure 1: a high granularity Electromagnetic Barrel (EMB) with accordion geometry that covers the pseudo-rapidity region $|\eta| < 1.475$, two Electromagnetic EndCaps (EMEC) that cover the region $1.375 < |\eta| < 3.2$, two Hadronic EndCaps (HEC) that cover the region $1.5 < |\eta| < 3.2$ and Forward Calorimeters (FCal) that can detect particles up to $|\eta| = 4.9$. These components are divided into two sides, side-A and side-C, oriented respectively along the positive and negative z-axis of the experiment. Each sub-detector is longitudinally segmented in three layers, which are referred as "front", "middle" and "back" layers. The subdetectors covering the regions with $|\eta| < 1.8$ also have an additional thin layer called "presampler," which is used to recover upstream energy loss. Three different cryostats enclose the Barrel and the two EndCap parts in order to maintain the Liquid Argon at a temperature of 88 K.

Incoming particles produced in the LHC proton–proton collisions shower in the absorbers and ionize the Liquid Argon. In order to collect the ionization signal produced in the LAr, a high electric field is applied across the gap and the current produced by the drift of the electrons is read out by electrodes. A triangular-shaped pulse is produced and it is then amplified and shaped into a bipolar pulse and digitized at the LHC bunch-crossing frequency of 40 MHz. The geometry of the LAr gap, the readout electrodes with the high voltage lines and the pulse shape are shown in Figure 2.

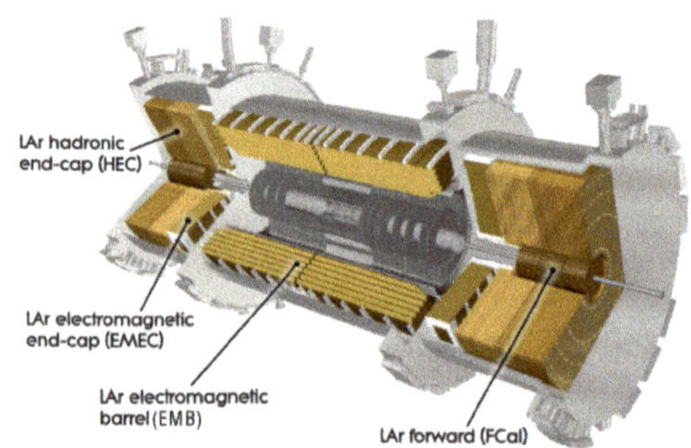

Figure 1. Schematic view of the different sub-detectors composing the ATLAS Liquid Argon (LAr) Calorimeter system [1,2].

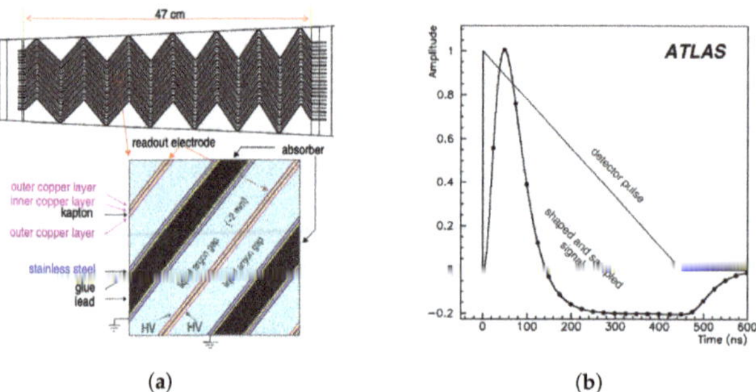

Figure 2. Accordion-shaped LAr gap of the barrel with readout electrodes and high voltage lines (**a**) and triangular pulse produced in the LAr gap compared to the shaped and sampled signal (**b**) [2].

Lar Calorimeter Main Readout and Legacy Analog Trigger Electronics

The signals from the detector are read out by more than 180,000 channels and are routed towards the Front End Crates (FECs). The Front End Boards (FEBs) [3] receive up to 128 analogue channel signals from each layer of the calorimeter. The signals are amplified with three different gains and then shaped with a bipolar analogue filter, producing the signal shape shown in Figure 2. The shaped signal is then sampled at the LHC bunch-crossing frequency of 40 MHz and stored in an analog memory buffer, while awaiting the Level-1 (L1) trigger accept. Upon a L1 trigger accept, the signal is digitized and transmitted to the back-end Readout Drivers (RODs) for the main readout energy computation. In addition to this, the FEBs also contain the Layer Summing Boards (LSBs) that send the analog sums of signals within one layer through the baseplane to the Tower Builder Board (TBB). The TBB sums the signals in the Trigger Towers (TTs), corresponding to detector cells of $\Delta\eta \times \Delta\Phi = 0.1 \times 0.1$ summed over the different layers and then sends the sums to the back-end receivers for the L1 trigger system that computes the L1 accepts, and sends them back to the ATLAS sub-detectors including the LAr system.

2. Atlas LAr Calorimeter Phase-1 Upgrade

After a successful period of data-taking during the LHC Run 2 between 2015 and 2018, the ATLAS detector entered into a long period of shutdown. In 2022, the LHC Run 3 is starting with running conditions characterised by higher instantaneous luminosity and higher average pile-up compared to the LHC Run 2, which results in an overall increased occupancy of the ATLAS detector compared to the Run 2 conditions. Specifically, the instantaneous luminosity and average pile-up in Run 3 will reach 3×10^{34} cm^{-2}s^{-1} and 80 respectively, compared to 2×10^{34} cm^{-2}s^{-1} and 40 of Run 2. During the Run 3 data-taking, the sustainable ATLAS level-1 (L1) trigger and High Level Trigger (HLT) rates remain at the same Run 2 levels of 100 kHz and 1 kHz, respectively. Given the increased rate of particles and the unchanged limitations on the rates, an upgrade of the L1 trigger is required in order to improve the discrimination power between different physics objects at trigger level and keep the same rates of accepted events without degrading the physics performance. Particularly, in the so-called Phase-1 upgrade, the LAr calorimeter trigger readout electronics are upgraded in order to improve the discrimination between electrons, photons, jets and τ-leptons at trigger level [4,5]. The new LAr calorimeter digital trigger readout system installed for the Run 3 data-taking increases the readout granularity by up to a factor of ten: instead of summing the energies from the calorimeter cells in regions of $\Delta\eta \times \Delta\Phi = 0.1 \times 0.1$ and over the different layers of the calorimeter to form the Trigger Towers, the energies are summed into smaller clusters called Super Cells (SCs) in each layer. The SCs provide higher cells granularity and the longitudinal information coming from the four layers of the calorimeter. Figure 3 shows an image of the expected energy deposit for a 70 GeV electron as seen by the old analog trigger system compared to the same deposit measured by the new digital system. Specifically, the SCs provide the same $\Delta\eta \times \Delta\Phi$ information of the TTs for the presampler and back layers but a $\Delta\eta \times \Delta\Phi = 0.025 \times 0.1$ information for the front and middle layers. With the higher granularity information provided by the SCs, the new L1 trigger algorithm has improved discrimination between different physics objects thanks to more detailed information about the shower shape development. This allows us to keep, during Run 3, the same trigger rate as in Run 2 with the same energy thresholds even in the more challenging Run 3 environment, as shown in Figure 4. As an example, the maximum trigger rate allowed for electrons is 20 kHz. This rate was corresponding in Run 2 to a transverse energy (E_T) threshold of 20 GeV for 95% electron efficiency and in Run 3 the rate can be maintained at the same level, with the same efficiency and E_T threshold, by using the additional shower information from the new digital trigger readout electronics.

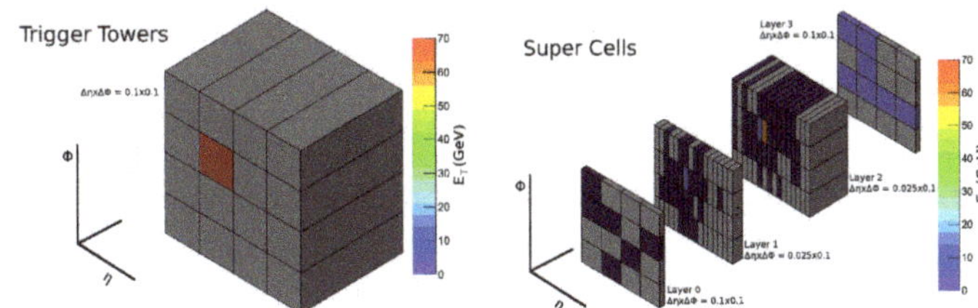

Figure 3. Simulation of the energy deposits of a 70 GeV electron reconstructed with the Run 2 Trigger Towers (**left**) and with the new Super Cells that will be used during Run 3 (**right**) [4].

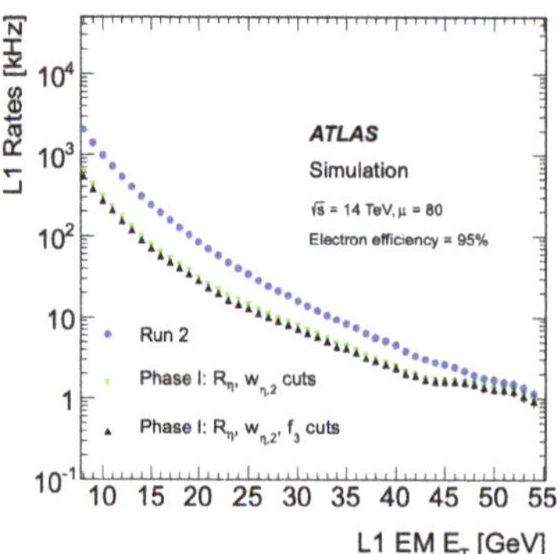

Figure 4. Expected L1 trigger rates in Run 3 as a function of the EM transverse energy threshold (E_T) for the case of the Run 2 trigger electronics (blue) and the new upgraded digital trigger electronics with different selections of shower shape variables (black and green) [4].

2.1. New LAr Calorimeter Digital Trigger Readout Electronics

The LAr readout electronics Phase-1 upgrade, installed before the start of the LHC Run 3, extends the existing readout system with new front-end and back-end components [4,5]. The new front-end boards send the Super Cells digital data to the new back-end components that compute the energies and transmit them to the new digital trigger system. An overview of the new LAr calorimeter readout electronics architecture for Run 3 is shown in Figure 5. The legacy analog trigger readout system will be maintained and it will be used for triggering at the beginning of Run 3 until the new Phase-1 digital trigger readout system is fully commissioned. The new LAr digital trigger readout system is composed of new Layer Sum Boards providing the analog sums of the signals from the calorimeter cells within one layer for the higher trigger readout granularity of the SCs, new baseplanes to route the increased number of analog signals and host the new LAr Trigger Digitizer Boards (LTDBs) that digitize the SC analog signals and provide the sums for the TBBs of the legacy analog trigger, and new Back-End boards, i.e., the LAr Digital Processing Blades (LDPBs), which read the SC signals from the LTDBs, compute the energies and send them to the new L1 trigger system. The LDPBs are built in Advanced Telecommunications Computing Architecture (ATCA) format and are composed of one ATCA carrier named LAr Carrier (LArC) equipped with four LAr Trigger prOcessing MEzzanines (LATOMEs) each and are controlled and monitored via an Intelligent Platform Management Controller (IPMC) plugged into the LArCs. The LATOMEs receive the SC ADC samples from the LTDBs via optical links for each brunch crossing and then transmit the elaborated data to the L1Calo trigger. To minimize the triggering latency, the LDPB boards are equipped with high-performance FPGAs. The total latency for the new LAr trigger path starting from the proton–proton collision time up to the computation of the energy amounts to 43.8 proton bunch-crossings (BCs), where one BC corresponds to a 25 ns time interval. In addition, the processors require 14 BCs to extract the trigger primitives and transmits them to the ATLAS Topological Trigger processors. The overall 57.8 BCs latency of the calorimeter trigger system conforms with the maximum (65 BCs) value allowed by ATLAS at the input of the Topological Trigger processors where data from both the calorimeter and the muon trigger modules are combined, and it is reduced compared to the latency of the legacy analog trigger system.

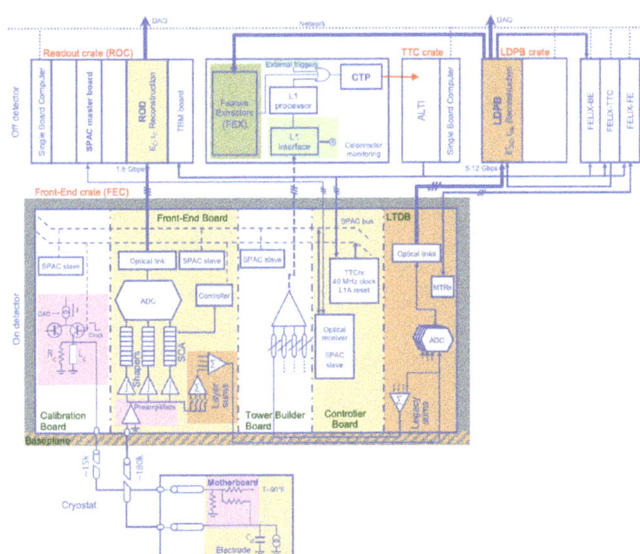

Figure 5. The ATLAS LAr calorimeter readout electronics architecture for the LHC Run 3. The new LAr boards are highlighted in orange. This diagram depicts specifically the electronics of the EM calorimeters [5].

2.2. Installation

The Phase-1 upgrade of the LAr trigger readout electronics was installed between 2019 and 2021 during the period of LHC shutdown for machine and detector upgrades before the start of the LHC Run 3. The installation of the front-end and back-end electronics was fully completed in October 2021. On the detector, 1524 FEBs have been extracted, refurbished with new LSBs and installed back in the FECs, 114 new baseplanes have been installed to host the refurbished FEBs and the 124 LTDBs, and the cooling plates and hoses were replaced. Simultaneously, 30 LDPBs, corresponding to 30 LArCs and 116 LATOMEs, have been installed in the ATLAS service cavern, with data fibers connected to the LTDBs. Monitoring and control systems for both new front-end and back-end boards have been put in place as well.

2.3. Validation and Commissioning

The newly installed readout electronics needed to be integrated in the ATLAS readout system; its performance had to be validated and the full new readout system needed to be commissioned for operations before the start of the Run 3 data-taking. The integration, validation and commissioning efforts started already during the installation period on subsets of the LAr detector after each half-FEC was refurbished and the corresponding digital trigger readout path was connected. The validation of the readout system is performed both by using injected pulses from the calibration boards into the front-end electronics and real data from the LAr detector signals [4,5].

For the main readout path, it had to be checked that, after the installation of the Phase-1 upgrade, the FEBs have a similar level of noise and calibration coefficients as before because, although the Phase-1 upgrade did not involve an upgrade of this readout path, the FEBs were extracted and re-inserted with refurbished electronics and additional routing. Calibration runs are taken and compared to the reference runs from the end of the LHC Run 2. These scans provide detailed information on the noise levels and on the calibration coefficients which result unchanged after the refurbishment, as can be seen in Figure 6. The figure shows an example of the comparison of the results from the new calibration runs to the Run 2 reference for the mean pedestal values in ADC counts, mean value of the

RMS in ADC counts and mean value of the gain over the LAr Calorimeter cells in a given pseudorapidity (η) range.

The new digital trigger readout is a completely new system installed for the Phase-1 upgrade that had to be validated. For this purpose, the full chain of the LAr digital trigger data acquisition, including the front-end and the back-end electronics, is tested after the installation. A set of input signal scans is defined and performed in order to validate the new system: mapping scans to check the connectivity of all channels, timing scans to align the various components in time, and calibration scans to validate the pedestal values, the pulse shape and the the gain value and linearity. The pulse shape collected by the LATOME can be verified by performing the so-called "delay runs", consisting of a series of injected calibration pulses with a single input signal current with an increasing delay, used for reconstructing the pulse shape with high granularity of readout points (each every 1.04 ns) as shown in Figure 7 for different energy regimes. Distortion in the pulse shape can be seen at high energy due to saturation effects. The linearity of the response of the new digital trigger readout electronics can be measured using the so-called "ramp runs", consisting of a series of injected calibration pulses with different amplitudes and measuring the peak ADC value with respect to the pedestal as a function of the E_T corresponding to the injected pulse, as shown in Figure 7. The ADC values are linearly increasing with the deposited E_T up to about 800 GeV, where saturation of the SC pulse occurs. There is no reference from Run 2 available for the new digital trigger readout, but the performance of this new readout path can be compared to the legacy analog trigger readout and to the main readout. Particularly, the E_T deposited in the SCs can be compared to the E_T deposited in the TTs and to the sum of the E_T in the corresponding cells of the LAr calorimeter for injected signals. The SC data are collected by the new LATOME boards while the energy deposited in the cells and in the TTs is collected by the legacy main and trigger readout systems. As shown in Figure 7, the deposited energy measured by the LATOME of the new digital trigger system corresponds well to the one measured by the TTs of the legacy trigger system and by the main readout system up to the level where the signal on the legacy TTs or on the SCs is saturated.

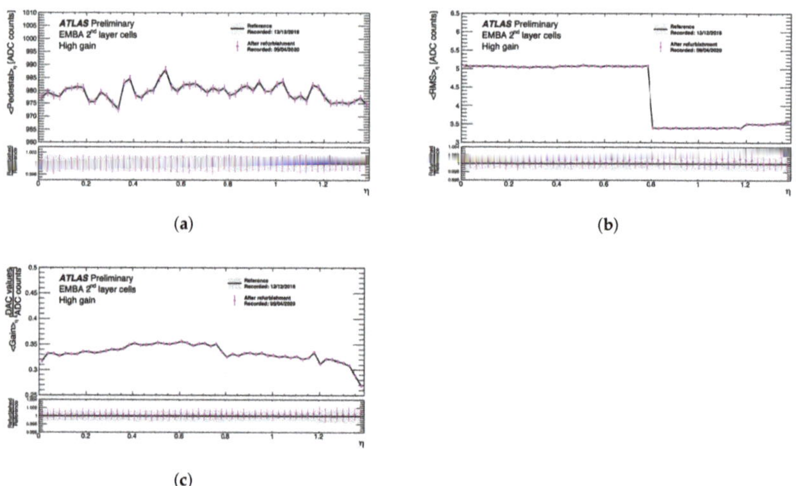

(a)

(b)

(c)

Figure 6. Mean pedestal values in ADC counts (**a**), mean value of the RMS in ADC counts (**b**) and mean value of the gain over the LAr Calorimeter cells (**c**), in a given pseudorapidity (η) range. Only the cells of the second layer of the electromagnetic barrel on side A (EMBA) are included. The black line shows the values measured at the end of Run 2. The purple dots show the values measured after the refurbishment of the front-end crates and front-end boards.

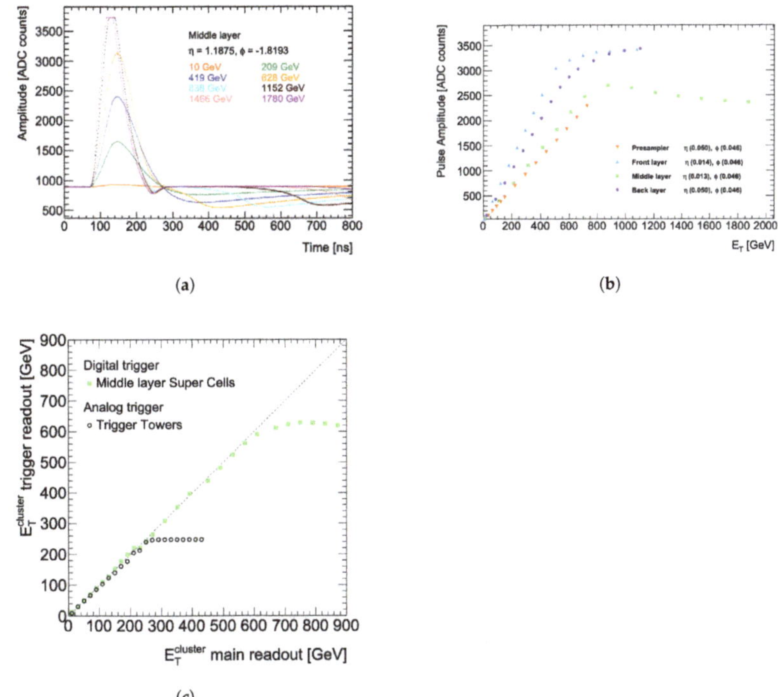

Figure 7. Pulse shape collected by the LATOME board for one SC in the barrel middle layer for several injected current pulses corresponding to different E_T values (**a**), peak ADC value with respect to the pedestal as function of the E_T corresponding to the injected pulse as seen on a LATOME board by four channels of the different calorimeter layers (**b**), comparison between the E_T deposited in the middle layer SCs or TTs and the sum of the E_T in the corresponding cells of the LAr barrel calorimeter (**c**) [5].

For the initial commissioning and validation of the upgraded detector with real data from the LAr detector system, data events from the first LHC proton beams are triggered using the legacy trigger system and the response of the new trigger system is read-out as well to check its performance.

In October 2021, the LHC Run 3 pilot run took place, with the first proton beams circulating again in the LHC after 3 years of shutdown. During this run, the full ATLAS detector was operational and the LAr calorimeter system was already including both the legacy trigger readout system and the new digital trigger readout system. The LHC pilot run gave the opportunity to further test the LAr readout system after the Phase-1 upgrade with real data, checking both legacy and new trigger readouts, as well as the main readout, with the first observations of particle collisions after the upgrade, in preparation for the start of the LHC Run 3 data-taking. Figures 8 and 9 illustrate two events recorded by the ATLAS detector from the LHC beam splashes during the pilot run: the proton beam was accelerated to 450 GeV and focused to hit collimators placed in the beamline before the detector, such that the particles created in the interaction moved on, along the beamline and outwards, passing through the detector. The data visualised in the pictures have been recorded using the legacy trigger based on energy deposits in the LAr electromagnetic calorimeter on the C-Side of the detector. By looking at images like those, it was possible to check that all the sub-detectors were working and confirm the validity of the data-taking. This was a very important step in testing the whole data workflow, in anticipation of the LHC Run 3 physics program.

Figure 8. Illustration of a data event recorded by the ATLAS detector from the LHC beam splashes during the LHC pilot run in October 2021. The image shows a cutout view of the ATLAS detector for an event where the proton test beam from the LHC is coming into the ATLAS detector from the left of the picture (which shows the A-Side of the detector) and travelling to the right (showing the C-Side). The red stripes are used to visualise the particle-matter interactions in the inner layers, the green boxes show the energy deposits in the LAr calorimeters, the yellow boxes visualise energy deposits in the TileCal hadronic calorimeter and the blue boxes surrounding the central part of the image are part of the ATLAS muon spectrometer, shown here for context.

Figure 9. Illustrations of a data event recorded by the ATLAS detector from the LHC beam splashes during the LHC pilot run in October 2021. The proton test beam from the LHC is coming into the ATLAS detector from the right of the picture (showing the A-Side of the detector) travelling to the left (the C-Side). In the image, the yellow boxes show energy deposits in all layers of the ATLAS detector. From the centre moving outwards, the image shows particle interactions in the inner tracking detectors, in the electromagnetic and hadronic calorimeters, and in the muon detectors.

The data recorded by ATLAS from the beam splashes during the LHC pilot beam tests in October 2021 were also used to specifically check the performance of the LAr calorimeters after the Phase-1 upgrade. Figure 10a shows the LAr cell energy sums, distributed in a hypothetical tower grid with $\Delta\eta \times \Delta\Phi = 0.025 \times 0.025$, in the endcap C, in the barrel and in the endcap A. Figure 10b shows the measured SC E_T from all layers of the LAr electromagnetic barrel (EMB) and electromagnetic endcaps (EMEC) compared to the summed transverse energies from their constituent calorimeter cells obtained through the main readout path. All these results are obtained using data from a single event of a beam splash run in October 2021. These first data from the LHC gave a confirmation of the good coverage of the detector readout after the upgrade and a confirmation of the good agreement between the energies measured by the new digital trigger readout and by the main readout even with preliminary calibration constants.

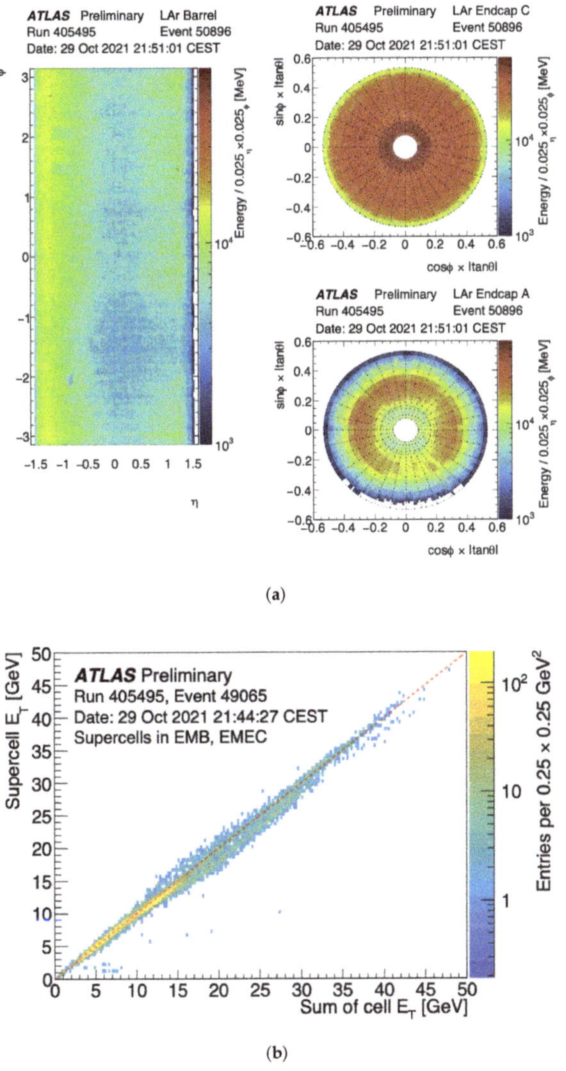

Figure 10. LAr cell energy sums distributed in a hypothetical tower grid with $\Delta\eta \times \Delta\Phi = 0.025 \times 0.025$ in the endcap C, in the barrel and in the endcap A (**a**), and measured SC E_T from all layers of the LAr Electromagnetic Barrel (EMB) and Electromagnetic Endcaps (EMEC), compared to the summed transverse energies from their constituent calorimeter cells, obtained through the main readout path (**b**) with data from a single event of a beam splash run in October 2021.

After the October 2021 LHC pilot run, the next milestone of the commissioning of the upgraded detector happened in April 2022, when the proton beams started circulating again in the LHC for the start of the LHC Run 3. Figure 11 illustrates two events recorded by the ATLAS detector from the LHC beam splashes on the first day of LHC Run 3 operations in April 2022. The data visualised in the picture have been recorded using the legacy trigger based on energy deposits in the LAr electromagnetic calorimeter on the C-Side of the detector. The analysis of these first LHC Run 3 beam splashes confirmed that the LAr calorimeter is ready for the start of the LHC Run 3 stable beams and physics data-taking runs in Summer 2022.

The strategy of triggering the events with the legacy analog trigger and reading out in parallel the response of the new digital trigger system is going to be kept for the first initial part of the Run 3 data-taking until the performance of the new digital trigger system will be fully checked and validated. After the full validation and confirmation that the performance of the new trigger system are satisfactory, the new digital trigger system will be used for triggering for the later main part of the Run 3 physics data-taking.

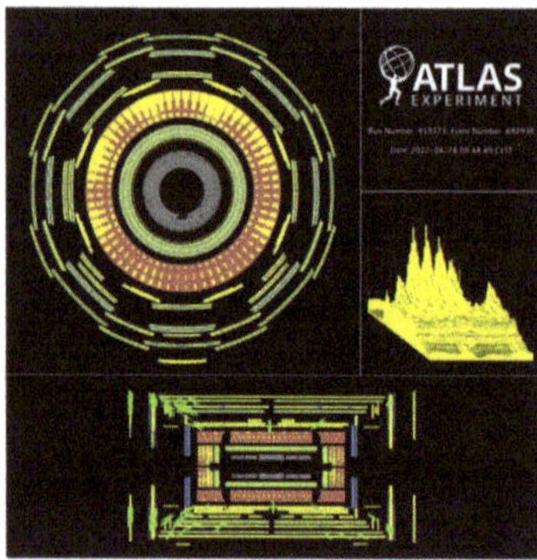
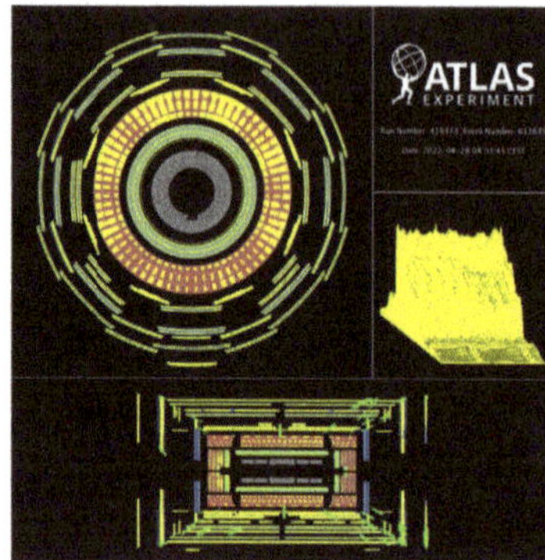

Figure 11. Illustrations of data events recorded by the ATLAS detector from the LHC Run 3 beam splashes tests in April 2022. In the event shown on the left, the spray of particles enters ATLAS from the right hand side of the picture (the A-Side of the detector) travelling to the left (the C-Side), while in the event shown on the right the particle enter ATLAS from the lefthand side of the picture (the C-Side of the detector) and travel to the right (the A-Side). In both images, the yellow boxes show energy deposits in all layers of the ATLAS detector. From the centre moving outwards, the image shows particle interactions in the inner tracking detectors, in the electromagnetic and hadronic calorimeters, and in the muon detectors.

3. Conclusions

The installation of the ATLAS LAr calorimeter Phase-1 upgrade front-end and back-end readout electronics for the LHC Run 3 was completed successfully in October 2021. The newly installed components were integrated in the ATLAS detector readout system and their performance were validated already during the installation period using injected pulses into the front-end electronics. In October 2021, the full ATLAS detector was already operational during the LHC pilot beam splashes and was operational again during the first LHC Run 3 beam splashes tests in April 2022. The LHC beam splashes data confirmed good performance of the LAr calorimeters after the Phase-1 upgrade, with the main readout, legacy analog trigger and new digital trigger readout systems. The ATLAS LAr calorimeters are ready for the start of the LHC Run 3 physics data-taking in Summer 2022.

Author Contributions: LAr phase-1 upgrade validation and commissioning and calibration software development, Alessandra Betti.

Funding: This research was supported in part by the US Department of Energy Award N. DE-SC0010129.

Institutional Review Board Statement: Not applicable.

Informed Consent Statement: Not applicable.

Data Availability Statement: Not applicable.

Conflicts of Interest: The authors declare no conflict of interest.

References

1. ATLAS Collaboration. *The ATLAS Experiment at the CERN Large Hadron Collider. J. Instrum.* **2008** 3, S08003.
2. ATLAS Collaboration. *ATLAS Liquid-Argon Calorimeter: Technical Design Report*; CERN-LHCC-96-041; CERN, Geneva, Switzerland, 1996.
3. Buchanan, N.J.; Chen, L.; Gingrich, D.M.; Liu, S.; Chen, H.; Damazio, D.; Densing, F.; Duffin, S.; Farrell, J.; Kasamy, S.; et al. ATLAS liquid argon calorimeter front end electronics. *J. Instrum.* **2008**, 3, P09003. [CrossRef]
4. Aleksa, M.; Hervas, L.; Fincke-Keeler, M.; Wingerter-Seez, I.; Marino, C.; Enari, Y.; Majewski, S.; Lanni, F.; Clel, W. *ATLAS Liquid Argon Calorimeter Phase-I Upgrade: Technical Design Report*; CERN-LHCC-2013-017; CERN: Geneva, Switzerland, 2013.
5. Aad, G.; Akimov, A.V.; Al Khoury, K.; Aleksa, M.; Andeen, T.; Anelli, C.; Aranzabal, N.; Armijo, C.; Bagulia, A.; Ban, J.; et al. The Phase-I trigger readout electronics upgrade of the ATLAS Liquid Argon calorimeters. *J. Instrum.* **2022**, *17*, P05024. [CrossRef]

Article

Reconstruction of 3D Shower Shape with the Dual-Readout Calorimeter

Sanghyun Ko [1],*, Hwidong Yoo [2] and Seungkyu Ha [2] on behalf of the IDEA Dual-Readout Group

1   Department of Physics and Astronomy, Seoul National University, Seoul 08826, Korea
2   Department of Physics, Yonsei University, Seoul 03722, Korea
*   Correspondence: sanghyun.ko@cern.ch

Abstract: The dual-readout calorimeter has two channels, Cherenkov and scintillation, that measure the fraction of an electromagnetic (EM) component within a shower by using different responses of each channel to the EM and hadronic component. It can measure the energy of EM and hadronic shower simultaneously—its concept inspired the integrated design for measuring both EM and hadronic showers, which left the task of reconstructing longitudinal shower shapes to the utilization of timing. We explore the possibility of longitudinal shower shape reconstruction using signal processing on silicon photomultiplier timing, and 3D shower shape by combining lateral and longitudinal information. We present a comparison between Monte Carlo (MC) and reconstructed longitudinal shower shapes from the simulation, and the application of 3D shower shapes associated with the dual nature of the calorimeter to identify electrons, hadrons, and hadronic punch-thru or muons.

Keywords: calorimetry; particle detectors; photomultipliers; particle identification

Citation: Ko, S.; Yoo, H.; Ha, S., on behalf of the IDEA Dual-Readout Group. Reconstruction of 3D Shower Shape with the Dual-Readout Calorimeter. *Instruments* **2022**, *6*, 39. https://doi.org/10.3390/instruments6030039

Academic Editors: Antonella De Santo, Alessandro Cerri, Fabrizio Salvatore and Iacopo Vivarelli

Received: 31 July 2022
Accepted: 5 September 2022
Published: 13 September 2022

Publisher's Note: MDPI stays neutral with regard to jurisdictional claims in published maps and institutional affiliations.

1. Introduction

In high-energy physics experiments, the energy of incident particles can be measured through a destructive interaction with absorbers that generates subsequent showers. A hadronic particle can develop not only the hadronic component that mainly consists of charged mesons and protons, but also the EM component from the production and decay of neutral pions. However, the detector's response to the hadronic component is significantly lower than that of the EM component for most calorimeters. Hence, the fluctuation of EM fraction within the shower initiated by hadronic particles limits the equivalent measurement of hadronic and EM components, hindering measuring energy.

The dual-readout calorimeter [1,2] is one of the proposed solutions to counter the fluctuation of EM fraction within the hadronic shower, simultaneously measuring the EM and hadronic components by utilizing different responses of Cherenkov and scintillation channels to relativistic and nonrelativistic particles. Each channel has a distinct response ratio for the hadronic component to the EM component, and the ratio of Cherenkov to scintillation channel (C/S) allows for estimating the EM fraction within the shower.

The ability to measure the energy of EM and hadronic showers simultaneously has led to contemporary designs of the dual-readout calorimeter that have no longitudinal segmentation. However, the longitudinal profile of a hadronic shower carries certain information that may improve the particle identification and energy reconstruction performance of hadronic showers. Studies with 3D-segmented particle flow calorimeters [3] suggest that details of shower shapes can be used for the software compensation technique. For instance, EM parts of the shower are more compact and denser compared to hadronic parts of the shower due to different scales between the radiation length and the nuclear interaction length. Therefore, we try to exploit timing information to reconstruct longitudinal and 3D shower shapes for a dual-readout calorimeter without physical segmentation.

2. Simulation Setup

The 4π projective geometry of the dual-readout calorimeter was implemented in the simulation with DD4hep [4]. Figure 1 illustrates the arrangement of Cherenkov and scintillation fibers with a 3D-printed projective module. A unit module is a trapezoidal tower consisting of a copper absorber with a 2 m length in the longitudinal direction. Inside the tower, scintillation and Cherenkov fibers were inserted in a checkerboard pattern at a 1.5 mm distance between fibers.

In the dual-readout calorimeter, fibers' optical properties determine the timing characteristics. Therefore, detailed descriptions of optical properties are essential. The Cherenkov fiber implemented in the simulation was a Mitsubishi Eska SK-40 clear fiber with a polymethylmethacrylate (PMMA) core and fluorinated polymer cladding. The scintillation fiber was Kuraray SCSF-78, consisting of PMMA cladding and polystyrene-based scintillating core. We emulated the refractive index [5–7], attenuation length [8,9], light yield, emission spectra, and decay time in the detector descriptions [10].

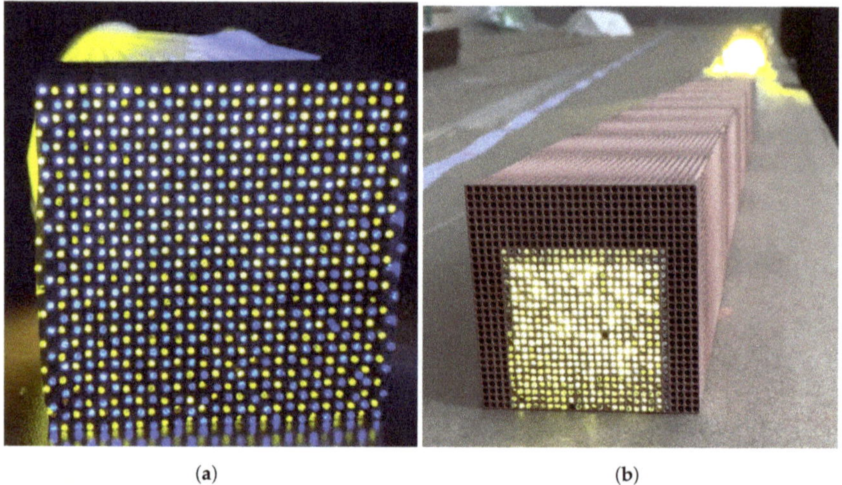

(a)   (b)

Figure 1. (**a**) Dual-readout calorimeter with Cherenkov fibers (blue light) and scintillation fibers (yellow light). (**b**) Projective geometry of dual-readout calorimeter with a copper absorber where only lit fibers on the rear side had full length that reached the front of the tower.

The above detector descriptions in DD4hep were interfaced to a GEANT4 [11–13] MC simulation. Figure 2 shows optical physics within the fibers simulated with GEANT4, where we can observe the unique behaviors of Cherenkov and scintillation light emission, and the propagation of optical photons via total internal reflection.

Generated optical photons are detected at the rear end of the tower. The collected number of photons, time of arrival, and wavelength information are plugged into silicon photomultiplier (SiPM) emulation software library SimSiPM [14]. This describes the response of SiPMs driven by parameters obtained from either lab measurements or data sheets from manufacturers. In the simulation, the data sheet of Hamamatsu S14160-1310PS SiPM [15] was used to describe SiPM behaviors, including dark count rate, afterpulse, cross-talk, and pulse shape as a function of time. Between the rear ends of scintillation fibers and SiPMs, Kodak Wratten number 9 yellow filters were inserted, which prevents the saturation of SiPMs from the high light yield of the scintillation channel, and absorbs the spatially dependent short-attenuation-length blue light.

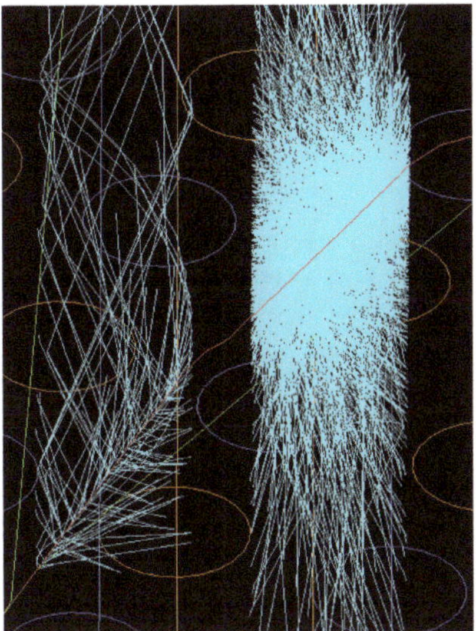

Figure 2. Optical physics simulated with GEANT4. The red line from the bottom left to the upper right indicates the electron from the shower fragment. The green line is a low-energy photon from the radiation, and cyan lines are optical photons generated from Cherenkov and scintillation process on the left and right sides, respectively. Blue and orange lines represent a sectional view of Cherenkov and scintillation fibers.

3. Longitudinal and 3D Shower Shape Reconstruction

3.1. Removing the Exponential Decay Signature

For fiber-sampling calorimeters such as the dual-readout calorimeter, the conventional approach to obtain longitudinal information regarding the shower is to estimate it using the time typically taken from the signal's peak or time of arrival. Setting the impact point and the moment of collision as $\vec{x} = 0$, $t = 0$, the observed time can be expressed as the sum of a high-energy particle's time of flight (ToF) and the propagation time of an optical photon within the fiber with group velocity v. SiPM's position $\vec{l}$ can also be described as a vector sum of the flight path of the high-energy particle and the distance that the optical photon propagated.

$$t = \frac{|\vec{x}|}{c} + \frac{|\vec{k}|}{v} \qquad \vec{l} = \vec{x} + \vec{k} \tag{1}$$

Here, we can benefit from the projective geometry to reconstruct the position of energy deposits by approximating that the three vectors are almost parallel.

$$|\vec{k}| = \frac{t - |\vec{l}|/c}{1/v - 1/c} \qquad \vec{x} = \vec{l} - \frac{t - |\vec{l}|/c}{1/v - 1/c} \frac{\vec{k}}{|\vec{k}|} \tag{2}$$

However, using only the time of a peak or arrival yields only a single number per shower, eventually ignoring details aside from the depth of the shower maximum or the tail. Therefore, understanding the full details of longitudinal shower shapes require utilizing the entire timing structure.

Without longitudinal segmentation, the calorimeter solely depends on the timing to reconstruct longitudinal shower shapes. Unfortunately, interpreting an electronic pulse shape into a physical shower shape is very challenging due to the many hidden layers between the two. For instance, a SiPM does not show a narrow pulse from a single photon. Instead, it returns an exponentially decaying pulse with a relatively short rise time compared to the decay time. Moreover, the number of photons follows the exponential decay by the scintillation process even emitted at the same depth.

Fortunately, the common nature of exponential decay allows for us to establish the energy density contributed to the pulse shape we observed from the SiPM by using the classic Fourier transform technique. For example, a pulse shape can be modeled as a convolution of exponential decay with time translation.

$$f(t) = \Theta(t - t_0)e^{-k(t-t_0)} \tag{3}$$

The exponential decay is described as a Lorentzian function in the frequency domain, while time translation becomes an oscillating component. Δt is a unit time of the discrete Fourier transform, corresponding to the sampling time of electronics.

$$F(\omega) = \frac{1}{1 - e^{-(k+i\omega)\Delta t}} \tag{4}$$

Provided that the time translation (time of arrival) given by the time window is not too large compared to the decay time of SiPMs, we can roughly interpret the full-width half maximum (FWHM) $\Delta\omega$ as an effective decay time.

$$cosh(k\Delta t) = 2 - cos(\Delta\omega\Delta t/2) \tag{5}$$

Hence, we can remove the decay term while leaving the oscillating component untouched, yielding time translation information solely without the exponential decay.

$$\frac{F(\omega)}{1 - e^{-(k+i\omega)\Delta t}} \rightarrow F(\omega) \tag{6}$$

Figure 3 is a signal-processing example simulated with SimSiPM: 1 ns rise time and 6.5 ns decay time based on the Hamamatsu S14160-1310PS data sheet but with higher signal-to-noise (SNR) ratio for the clear demonstration. It represents an analog signal consisting of four pulses—a two-photon equivalent pulse from 13 to 14 ns, the main five-photon equivalent pulse from 15 to 16 ns, followed by two late single photon contributions at 18 and 25 ns. As photons are collected at the rear side of the tower, understanding the shape of early contributions is essential for reconstructing the tail part of the shower. However, it is challenging to discriminate them from the primary pulse, as shown in Figure 3a. Figure 3c demonstrates that signal processing significantly reduces decay structures for each pulse; hence, we can recognize individual contributions within the signal.

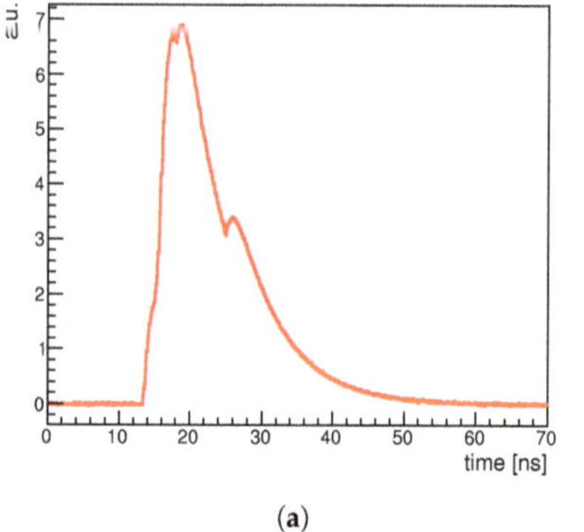

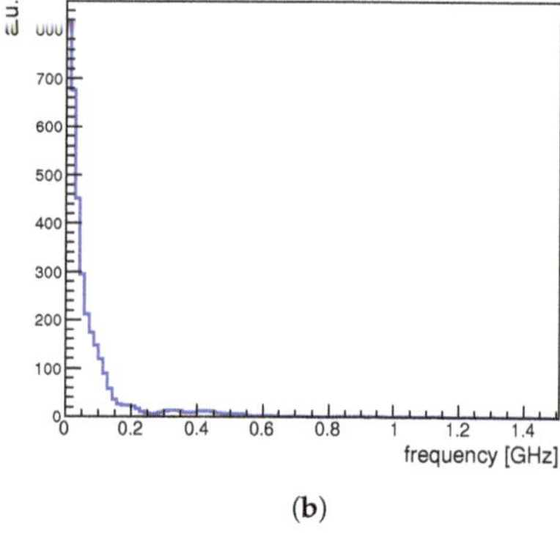

(a)

(b)

Figure 3. *Cont.*

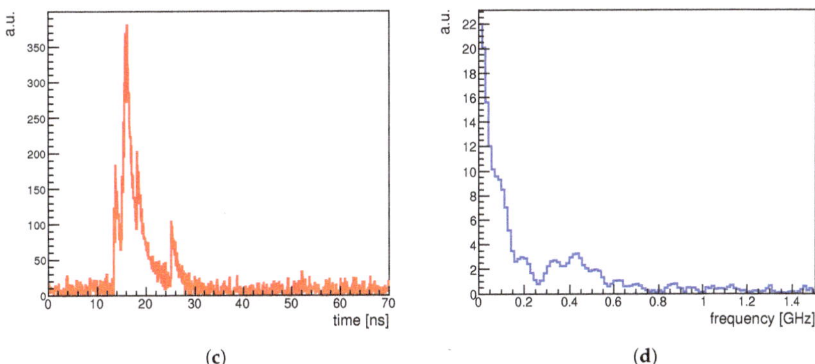

(c)   (d)

Figure 3. An example of a simulated signal in (**a**) time and (**b**) frequency domains. Signal after decay removal in the (**c**) time and (**d**) frequency domains, respectively.

3.2. Mitigating Modal Dispersion of Optical Fibers

However, the signal pulse shape is still far from the physical shower shape even after removing the exponential decay signature due to the substantial effect caused by the modal dispersion of optical fibers. In a step-index multimode fiber, the group velocity of the signal pulse is slower if the number of modes is higher, causing the dispersion of the pulse shape due to the different group velocities. The intrinsic approach to resolve modal dispersion is using a graded-index multimode fiber. It uses a relatively higher refractive index at the core and a lower one at the outer region, compensating for the group velocity with the refractive index.

Unfortunately, the market situation does not allow it as a viable solution because of expensive clear fibers. Furthermore, no graded-index scintillating fiber is available commercially. Therefore, we took the software compensation to tackle this issue by assigning faster group velocity for early components of the pulse shape, and a slower one for late components after decay removal.

The group velocity is profiled as a function of ΔT—the time passed from the time of arrival t_0. We used the well-understood longitudinal profile of the EM shower from the EGS4 simulation [16], equalizing the relative area of the integrated pulse from t_0 to the energy contained from the tail of the shower and the depth x, which corresponds to time $t_0 + \Delta T$.

$$\frac{\int_{t_0}^{t_0+\Delta T} f(t)\, dt}{\int_{t_0}^{\infty} f(t)\, dt} = \frac{\int_{x}^{\infty} \frac{dE(x)}{dx}\, dx}{\int_{0}^{\infty} \frac{dE(x)}{dx}\, dx} \tag{7}$$

Then, we can express the group velocity by using $t = t_0 + \Delta T$ and x given by Equation (7).

$$v_{group} = \frac{|\vec{l}| - |\vec{x}|}{t - |\vec{x}|/c} \tag{8}$$

Figure 4 shows the profiled group velocity for scintillation and Cherenkov channel as a function of ΔT. As intended, the group velocity had a slightly lower value than the speed of light within the medium at $\Delta T = 0$, and gradually decreased as ΔT increases, compensating for the mode increment.

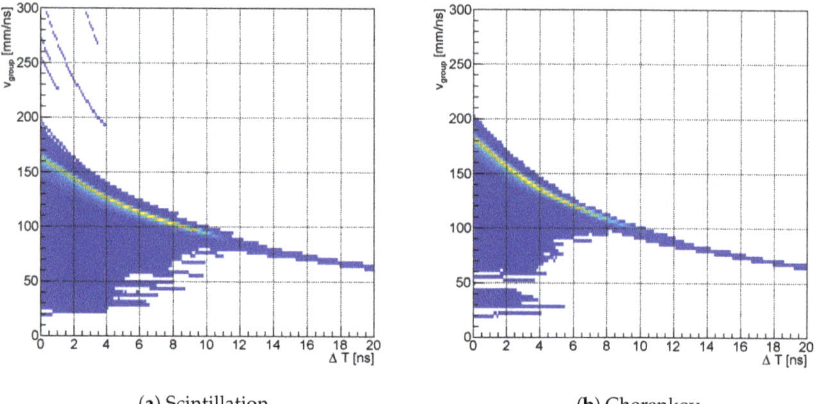

(**a**) Scintillation. (**b**) Cherenkov.

Figure 4. Group velocity for (**a**) scintillation and (**b**) Cherenkov channel as a function of time ΔT passed since arrival t_0, profiled using the signal pulse amplitude per sampling time per SiPM.

3.3. Longitudinal Shower Depth, Length, and 3D Shower Shapes

After mitigating modal dispersion within optical fibers, we can describe longitudinal shower shapes using the timing pulses. To test the reliability of reconstructed longitudinal shower shapes, we compared them with the MC truth energy deposits retrieved from GEANT4 steps. We describe the shape with two parameters, depth and length. The depthlike observable is defined as the distance from the tower's front end to the shower maximum, and the lengthlike observable represents the distance between the two positions where the local energy density exceeds 10% of the shower maximum.

Figure 5 shows the depth- and lengthlike observables for 20 GeV electrons and pions. Here, the sampling time of electronics was assumed to be 100 ps, and timing resolution in the order of 10 ps, so that the jitter had a negligible effect. The comparison shows a decent correlation between the reconstructed and MC truth observables. However, the group velocity profiling on the EM shower could not completely correct the dispersion effect and it affected the late arrival photons more strongly. Figure 5a reveals the asymmetrical impact that deviated the head of the reconstructed shower further frontward.

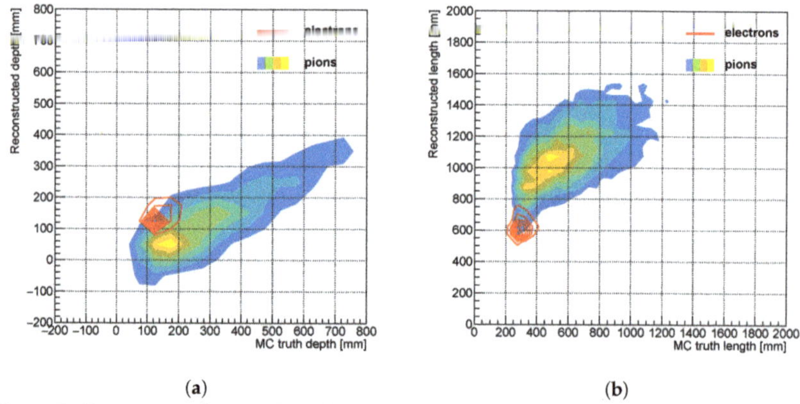

(**a**) (**b**)

Figure 5. Reconstructed vs. MC truth shower (**a**) depthlike and (**b**) lengthlike observables for 10^4 events of 20 GeV electrons (lined contour) and pions (filled contour), respectively. Contour lines indicate the density of events.

Having descriptions of the longitudinal shower shape, we can illustrate the 3D shower shape reconstructed with the dual-readout calorimeter by mixing it with the lateral shower shape. Figure 6 renders several event displays of reconstructed 3D shower shape on the left and MC truth on the right.

The reconstructed 3D shower shape of the 20 GeV pion in Figure 6b illustrates typical hadronic shower characteristics that consist of the EM component mainly represented by the Cherenkov hits and the non-EM component by the scintillation hits. The event display shows that the EM component was densely deposited along the center of the shower, while the hadronic component tended to reach deeper and be located away from the center. Furthermore, Figure 6c,d suggest that 3D reconstruction reveals the unique shape of hadronic punch-thru and minimal ionizing particle (MIP), allowing for us to identify them using the shower substructure analysis that was not possible beforehand for particles with arbitrary incident energy.

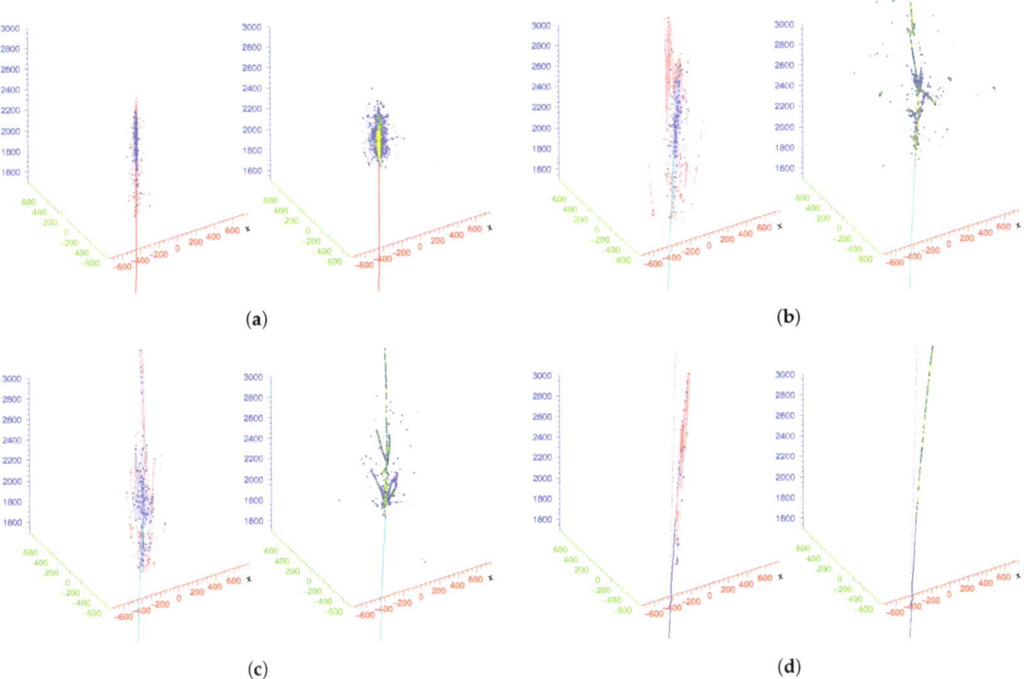

(a)   (b)

(c)   (d)

Figure 6. (**left**) Reconstructed and (**right**) MC truth 3D shower shapes of 20 GeV particles. For reconstructed cases, red dots represent the 3D hits from the scintillation channel and blue dots from the Cherenkov channel. The color code portrays the relative energy density for the MC truth case. E_S, E_C, and E_{DR} indicate the reconstructed energy in GeV from the scintillation, Cherenkov channel, and the dual-readout corrected energy. The 2 T magnetic field was applied to all four cases. (**a**) Electron ($E_S = 19.96, E_C = 20.84$). (**b**) Pion ($E_S = 17.29, E_C = 10.22, E_{DR} = 20.04$). (**c**) Punch-thru ($E_S = 13.05, E_C = 8.410, E_{DR} = 14.85$). (**d**) Muon ($E_S = 1.550, E_C = 1.243$).

4. 3D Shower Substructure with Density-Based Clustering

We attempted to take advantage of the novel 3D shower shape reconstruction with the dual-readout calorimeter by looking into the properties of its shape. Counting the number of substructures is the simplest way to define the characteristics of a given shower. The DBSCAN algorithm [17] is used to cluster hits from 3D reconstruction, and it has several handy features to cluster shower substructures.

The nature of hadronic shower fluctuation forbids us from knowing how many substructures there are within the shower or where particles head after the scattering process. The DBSCAN allows for us to cluster shower substructures under these circumstances due to its feature that does not require the number of clusters a priori and works on arbitrarily shaped clusters. Furthermore, it can weigh each point by the 3D hit's energy, which equals the amplitude of the pulse shape after the Fourier transformation at the corresponding time.

The DBSCAN has two input parameters—the maximal distance between the neighboring points within the same cluster (*eps*) and the minimal number of weighted points to create a separate cluster (*minPts*). Considering the lateral granularity of the dual-readout calorimeter (1.5 mm), the *eps* parameter was set to 7.5 mm, and the *minPts* parameter was set to 0.1% of the total number of 3D hits to incorporate as many shower fragments as possible. We equalized each point's weight to the amplitude of the pulse at the corresponding time after the Fourier transformation so that the DBSCAN took the position and energy of 3D hits for the clustering.

However, the DBSCAN did not consider the different lateral and longitudinal accuracy. In the longitudinal direction, it did not have any physical segmentation as in the lateral case. Instead, it relied on the sampling time for determining the depth of each point, where 100 ps of sampling time coincided with a 4 cm bin along the longitudinal axis. Therefore, we scaled the longitudinal axis by a factor of 20 to match the lateral and longitudinal direction accuracy.

Figure 7 depicts the clustered shower substructure for 20 GeV particles. The number of colors in Figure 7b shows that the DBSCAN separately distinguished substructures within the hadronic shower. However, in the attempt to discriminate two photons from the neutral pion decay in Figure 7c, the *eps* parameter had to be reduced to 5 mm. This possibly indicates that the appropriate DBSCAN parameter is a subject of optimization based on the overall size of the shower, since different scales between the radiation length and the nuclear interaction length may require different scopes for substructural clustering.

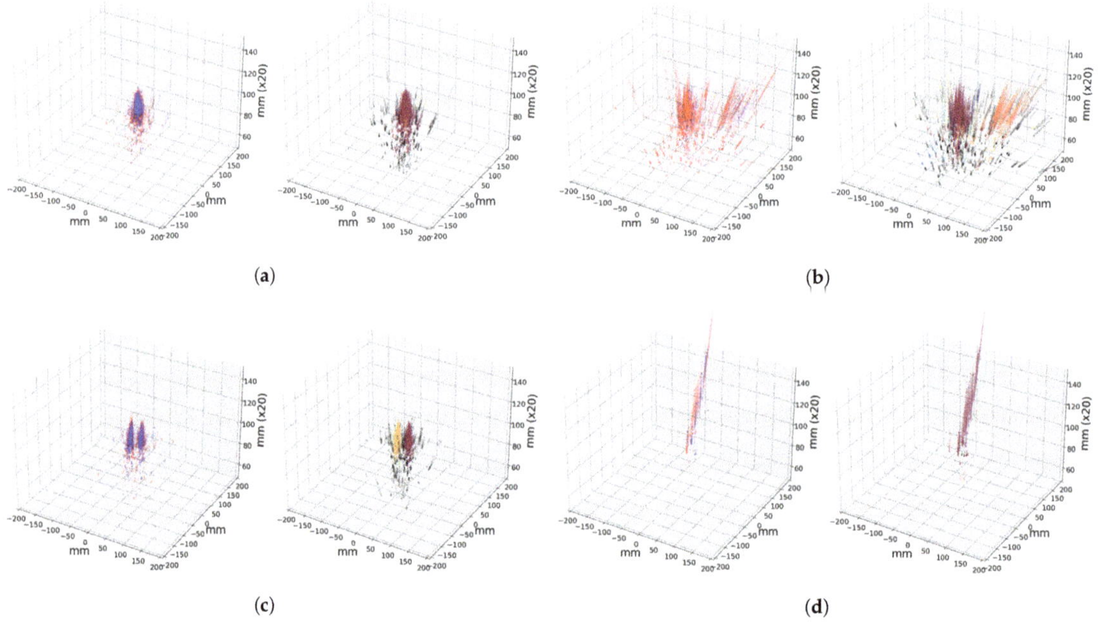

Figure 7. (**left**) Reconstructed 3D hits and (**right**) 3D hits after clustering shower substructures. On the left pad of each figure, the red dots represent 3D hits from the scintillation channel and blue dots from the Cherenkov channel. Color codes on the right pad illustrate different substructures clustered by the DBSCAN. The longitudinal (vertical) axis is scaled by 20, where a unit length equals 20 mm. (**a**) Electron. (**b**) Pion. (**c**) Neutral pion. (**d**) Muon.

The estimated number of clusters provides an additional source of information from the C/S variable of the dual-readout calorimeter. Thus, mixing the number of clusters with the C/S can bring insights into the behavior of rare showered particles. For instance, as seen in Figure 8, it allows for us to distinguish hadronic showers not only from EM showers but also showers initiated by particles acting like MIPs, such as punch-thru and muons.

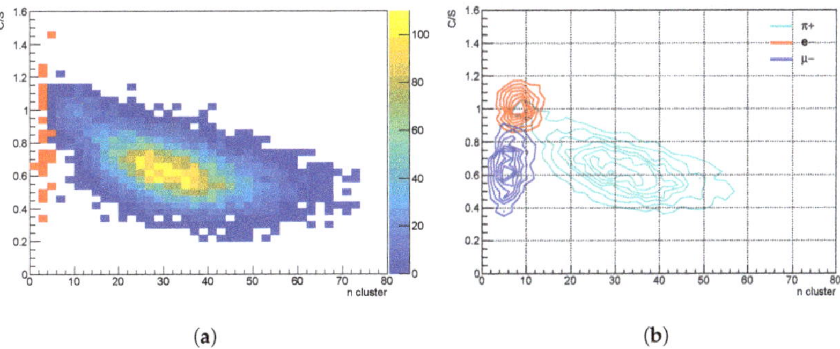

(a)   (b)

Figure 8. The number of clusters vs C/S for (**a**) 20 GeV pions (palette) overlain by the ones with the dual-readout corrected energy less than 5 GeV (red), (**b**) 20 GeV pions (cyan), electrons (red), and muons (blue).

5. Summary and Future Developments

Longitudinal and 3D shower shape reconstruction with a dual-readout calorimeter was presented. Despite no physical longitudinal segmentation, the comparison with reconstructed and MC truth shower shapes suggests that exploiting timing information using signal processing may allow for us to reconstruct shower substructures without losing details. Moreover, clustering with DBSCAN reveals that the amount of information contained in the reconstructed shower substructures is substantial to perform basic particle identification by using only the number of clusters mixed with the dual-readout calorimeter's C/S variable.

In this study, we strictly relied on the software-based approach to demonstrate physics and detector response via the simulation and emulation of SiPMs. The simulation study shows that the detector's fine lateral granularity and excellent timing characteristics are essential for reconstructing 3D shower shapes. We plan to perform a beam test at the CERN SPS facility heading towards complete proof of concept by collecting some waveforms of each fiber using Hamamatsu S14160-1310PS SiPMs with a fast decay time and the DRS4 digitizer [18] with 200 ps sampling time.

Author Contributions: Conceptualization, S.K.; methodology, S.K.; software, S.K.; validation, S.K.; formal analysis, S.K.; investigation, S.K.; resources, S.K.; data curation, S.K.; writing—original draft preparation, S.K.; writing—review and editing, S.K. and H.Y.; visualization, S.K.; supervision, H.Y. and S.H.; project administration, H.Y.; funding acquisition, H.Y. All authors have read and agreed to the published version of the manuscript.

Funding: This work was supported by the Yonsei University Research Fund (Post Doc. Researcher Supporting Program) of 2022 (project no.: 2021-12-0147) and the National Research Foundation of Korea (NRF) grants NRF-2020R1A2C3013540 and NRF-2021K1A3A1A79097711.

Data Availability Statement: Not applicable.

Conflicts of Interest: The authors declare no conflict of interest.

References

1. Lee, S.; Livan, M.; Wigmans, R. Dual-readout calorimetry. *Rev. Mod. Phys.* **2018**, *90*, 025002. [CrossRef]
2. Akchurin, N.; Carrell, K.; Hauptman, J.; Kim, H.; Paar, H.; Penzo, A.; Thomas, R.; Wigmans, R. Hadron and jet detection with a dual-readout calorimeter. *Nucl. Instrum. Methods Phys. Res. Sect. Accel. Spectrometers Detect. Assoc. Equip.* **2005**, *537*, 537–561. [CrossRef]
3. Sefkow, F.; White, A.; Kawagoe, K.; Pöschl, R.; Repond, J. Experimental tests of particle flow calorimetry. *Rev. Mod. Phys.* **2016**, *88*, 015003. [CrossRef]
4. Frank, M.; Gaede, F.; Petric, M.; Sailer, A. AIDASoft/DD4hep, 2018. Available online: http://dd4hep.cern.ch/ (accessed on 20 July 2022). [CrossRef]
5. Akchurin, N.; Bedeschi, F.; Cardini, A.; Cascella, M.; De Pedis, D.; Ferrari, R.; Fracchia, S.; Franchino, S.; Fraternali, M.; Gaudio, G.; et al. Lessons from Monte Carlo simulations of the performance of a dual-readout fiber calorimeter. *Nucl. Instrum. Methods Phys. Res. A* **2014**, *762*, 100–118. [CrossRef]
6. Beadie, G.; Brindza, M.; Flynn, R.A.; Rosenberg, A.; Shirk, J.S. Refractive index measurements of poly(methyl methacrylate) (PMMA) from 0.4–1.6 µm. *Appl. Opt.* **2015**, *54*, F139–F143. [CrossRef] [PubMed]
7. Sultanova, N.; Kasarova, S.; Nikolov, I. Dispersion Properties of Optical Polymers. *Acta Phys. Pol. A* **2009**, *116*, 585–587. [CrossRef]
8. Abrate, S. Chapter 3—Plastic Optical Fibers for Data Communications. In *Handbook of Fiber Optic Data Communication*, 4th ed.; De Cusatis, C., Ed.; Academic Press: Oxford, UK, 2013; pp. 37–54. [CrossRef]
9. Kaino, T.; Fujiki, M.; Nara, S. Low-loss polystyrene core-optical fibers. *J. Appl. Phys.* **1981**, *52*, 7061–7063. [CrossRef]
10. Alfieri, C.; Rodrigues Cavalcante, A.B.; Joram, C. *A Set-Up to Measure the Optical Attenuation Length of Scintillating Fibres*; Technical Report; CERN: Geneva, Switzerland, 2015.
11. Agostinelli, S.; Allison, J.; Amako, K.; Apostolakis, J.; Araujo, H.; Arce, P.; Asai, M.; Axen, D.; Banerjee, S.; Barrand, G.; et al. Geant4—A simulation toolkit. *Nucl. Instrum. Methods Phys. Res. Sect. A Accel. Spectrometers Detect. Assoc. Equip.* **2003**, *506*, 250–303. [CrossRef]
12. Allison, J.; Amako, K.; Apostolakis, J.; Araujo, H.; Arce Dubois, P.; Asai, M.; Barrand, G.; Capra, R.; Chauvie, S.; Chytracek, R.; et al. Geant4 developments and applications. *IEEE Trans. Nucl. Sci.* **2006**, *53*, 270–278. [CrossRef]
13. Allison, J.; Amako, K.; Apostolakis, J.; Arce, P.; Asai, M.; Aso, T.; Bagli, E.; Bagulya, A.; Banerjee, S.; Barrand, G.; et al. Recent developments in Geant4. *Nucl. Instrum. Methods Phys. Res. Sect. A Accel. Spectrometers Detect. Assoc. Equip.* **2016**, *835*, 186–225. [CrossRef]
14. Edoardo, P.; Romualdo, S. *SimSiPM: A Library for SiPM Simulation*; GitHub: Como, Italy, 2021. Available online: https://github.com/EdoPro98/SimSiPM (accessed on 20 July 2022).
15. Hamamatsu Photonics, K.K. MPPC S14160-1310PS Low Breakdown Voltage, Wide Dynamic Range Type MPPC with Small Pixels, 2021. Available online: https://www.hamamatsu.com/content/dam/hamamatsu-photonics/sites/documents/99_SALES_LIBRARY/ssd/s14160-1310ps_etc_kapd1070e.pdf (accessed on 20 July 2022).
16. Nelson, W.R.; Hirayama, H.; Rogers, D.W.O. *The Egs4 Code System*; Stanford Linear Accelerator Center: Menlo Park, CA, USA, 1985.
17. Ester, M.; Kriegel, H.P.; Sander, J.; Xu, X. A Density-Based Algorithm for Discovering Clusters in Large Spatial Databases with Noise. In Proceedings of the Second International Conference on Knowledge Discovery and Data Mining (KDD'96), Portland, OR, USA, 2–4 August 1996; AAAI Press: California, CA, USA, 1996; pp. 226–231.
18. Ritt, S. Design and performance of the 6 GHz waveform digitizing chip DRS4. In Proceedings of the 2000 IEEE Nuclear Science Symposium Conference Record, Dresden, Germany, 19–25 October 2008; pp. 1512–1515.

instruments

Article

R&D of a Novel High Granularity Crystal Electromagnetic Calorimeter

Baohua Qi [1,2,3] **and Yong Liu** [1,2,3,*] **on behalf of the CEPC Calorimeter Working Group**

1   Institute of High Energy Physics, Chinese Academy of Sciences, Yuquan Road 19B, Beijing 100049, China
2   University of Chinese Academy of Sciences, Yuquan Road 19A, Beijing 100049, China
3   State Key Laboratory of Particle Detection and Electronics, Yuquan Road 19B, Beijing 100049, China
*   Correspondence: liuyong@ihep.ac.cn; Tel.: +86-010-88236066

Abstract: Future electron-positron collider experiments aim at the precise measurement of the Higgs boson, electroweak physics and the top quark. Based on the particle-flow paradigm, a novel highly granular crystal electromagnetic calorimeter (ECAL) is proposed to address major challenges from jet reconstruction and to achieve the optimal EM energy resolution of around 2–3%/$\sqrt{E(\text{GeV})}$ with the homogeneous structure. Extensive R&D efforts have been carried out to evaluate the requirements and potentials of the crystal calorimeter concept from sensitive detection units to a full sub-detector system. The requirements on crystal candidates, photon sensors as well as readout electronics are parameterized and quantified in Geant4 full simulation. Experiments including characterizations of crystals and silicon photomultipliers (SiPMs) are performed to validate and improve the simulation results. The physics performance of the crystal ECAL is been studied with the particle flow algorithm "ArborPFA" which is also being optimized. Furthermore, a small-scale detector module with a crystal matrix and SiPM arrays is under development for future beam tests to study the performance for EM showers.

Keywords: Higgs factory; calorimeter; crystal; SiPM; high granularity

Citation: Qi, B.; Liu, Y., on behalf of the CEPC Calorimeter Working Group. R&D of a Novel High Granularity Crystal Electromagnetic Calorimeter. *Instruments* **2022**, *6*, 40. https://doi.org/10.3390/instruments6030040

Academic Editors: Fabrizio Salvatore, Alessandro Cerri, Antonella De Santo and Iacopo Vivarelli

Received: 31 July 2022
Accepted: 10 September 2022
Published: 15 September 2022

Publisher's Note: MDPI stays neutral with regard to jurisdictional claims in published maps and institutional affiliations.

1. Introduction

The discovery of the Higgs boson in 2012 by the ATLAS [1] and CMS [2] Collaborations at the Large Hadron Collider (LHC) offers exciting opportunities for future particle physics in the coming decades. Precise measurements of properties of the Higgs boson, the W and Z bosons, as well as the top quark, will provide crucial tests of the standard model (SM) and are promising in searching for new physics beyond the SM (BSM). Following the demands from physics, future electron–positron colliders including the CEPC [3], FCC-ee [4], ILC [5] and CLIC [6] are among options planned to devote most of their operation hours as Higgs factories, to produce millions of Higgs bosons within a clean collision environment.

To fully exploit the physics potentials of a future lepton collider, the detectors should achieve an unprecedented jet energy resolution, which poses stringent requirements on the detector design. The invariant mass resolution of the Higgs boson aims to reach 3–4% at the CEPC. A jet is typically composed of ~65% charged particles, ~15% photons, and ~10% neutral hadrons. Particle flow algorithms [7] (PFA) aim to measure each of the final-state particles with one of the optimal sub-detectors and require high granularity in calorimeters to separate close-by particle showers and try to match with the tracking system for charged particles. A PFA-oriented electromagnetic calorimeter (ECAL) should be precise enough for photon energy reconstruction and particle identification (PID). Therefore, a highly segmented ECAL with excellent three-dimensional (3D) spatial resolution as well as good energy and time resolution is desired. Among the PFA-ECAL options that are being developed within the CALICE [8] Collaboration, e.g., the silicon-tungsten ECAL [9] and scintillator-tungsten ECAL [10], the energy resolution would be limited by the sampling structures (on the order of $\sim 10\%/\sqrt{E(\text{GeV})}$). On the other hand, legendary crystal-based calorimeters have demonstrated excellent EM resolution of 2–$3\%/\sqrt{E(\text{GeV})}$ (for instance, the BGO calorimeter in the LEP-L3 experiment [11] and the PWO calorimeter in the LHC-CMS experiment [12]) while there was no fine 3D segmentation in crystals. Hereby, we propose a new conceptual design of the high granularity crystal ECAL with fine 3D segmentation to be compatible with the PFA, to reach an optimal EM resolution, and to significantly improve the sensitivity to low-energy particles. The precision measurements of γ and π^0 can also be portals to further explore the new physics beyond the standard model [13].

There are two major designs of the high granularity crystal ECAL in the proposal, as shown in Figure 1. Design 1, with finely segmented crystals in both longitudinal and transversal dimensions, would be naturally compatible with PFA. Initial PFA studies have been performed using this design to demonstrate the potentials in physics, which will be presented in Section 2. It also sets a stringent requirement on the total material budget of readout boards, cooling services and mechanical structures between longitudinal layers, which must be strictly controlled to avoid significant performance degradation [14].

To address this challenge, a new detector layout is proposed (Design 2 in Figure 1) with long crystal bars arranged to be orthogonal to each other in two neighboring layers for a maximum longitudinal segmentation, minimum inactive materials in between and a significant reduction in readout channels. High transverse granularity is expected to be realized by combining the information of every two adjacent layers. The basic module of a "supercell" is 40×40 cm^2 in the transverse plane with, typically, $24X_0$ in depth. Each crystal bar is read out by 2 SiPMs at two ends, which can also provide timing information from the two sides for positioning reconstruction, clustering and particle identification. Nevertheless, this layout also poses a challenge for 3D pattern recognition and requires the development of dedicated reconstruction software, which will be briefly introduced in Section 3.

Besides the highlights of physics performance studies and software development, preliminary results from hardware R&D activities on the key aspects of the calorimeter design are presented in Section 4, followed by the summary and prospects in Section 5. It should be noted that most results presented in this proceeding are selected from active working progress.

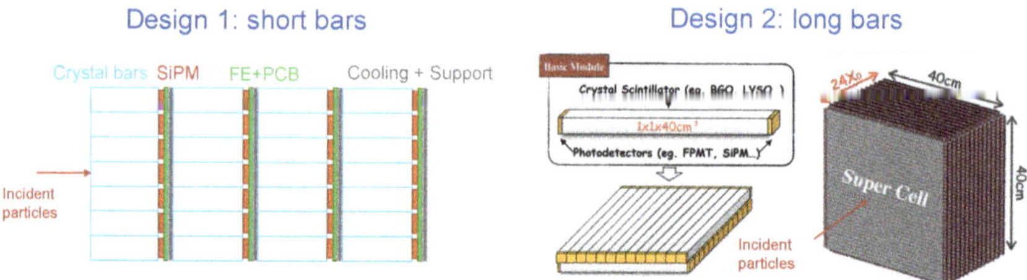

Figure 1. Two designs proposed for a high granularity crystal calorimeter: Design 1 with finely segmented crystals and single-ended readout; Design 2 with long crystal bars and two-ended readout.

2. PFA Performance

The CEPC baseline detector ("CEPCv4") in the CEPC Conceptual Design Report [3] and the PFA software "ArborPFA" [15] were implemented in the CEPC software framework [16] for the PFA performance studies and PFA optimizations. Baseline sub-systems from the inside out include silicon vertex detector, Time Projection Chamber (TPC), silicon-tungsten (SiW) ECAL, RPC-based HCAL, magnet coils and the return yoke as a muon detector, as shown in Figure 2a. The SiW-ECAL was replaced by finely segmented crystal bars with two sets of granularity, i.e., $1 \times 1 \times 1$ cm^3 and $1 \times 1 \times 2$ cm^3. The granularity of $1 \times 1 \times 1$ cm^3 corresponds to an ideal setup to fully exploit 3D information to evaluate the potentials and optimize the PFA parameters for crystals. The slightly coarser granularity of $1 \times 1 \times 2$ cm^3 is the goal for the layout of long crystal bars (Design 2), assuming that the dedicated reconstruction software can finally resolve ambiguities of pattern recognition in shower clustering due to the special geometry arrangement. The crystal option implemented for the crystal ECAL was selected to be the Bismuth Germanate (BGO) at this stage.

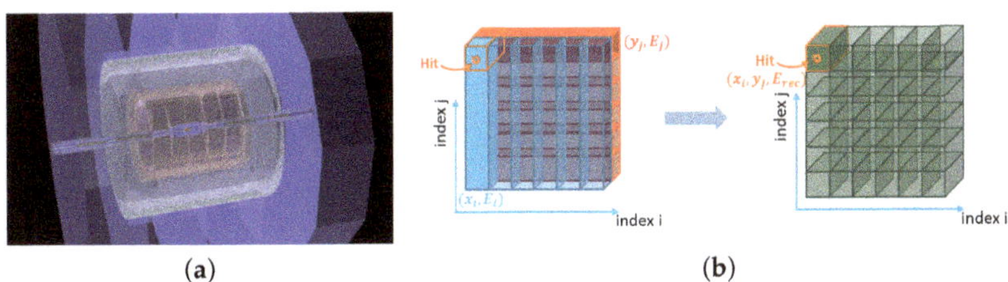

(a)   (b)

Figure 2. The CEPC CDR detector layout ("CEPCv4") for PFA performance studies with implementations of the SiW-ECAL (as the CDR baseline) and crystal ECAL in two scenarios. (**a**) CEPC CDR baseline detector. (**b**) Reconstruction with long crystal bars with an aim of fine transverse granularity.

2.1. Separation Power of Close-By Showers

The separation capability of showers initiated by close-by particles is an essential requirement for PFA calorimetry. Considering typical jet components, key scenarios are: (1) the separation of two photons (denoted as $\gamma's$) and (2) the separation of a charged pion ($\pi^{\pm}$) and a γ [17].

Compared with the SiW-ECAL, EM showers in crystals tend to have a larger number of hits and a wider lateral distribution, since crystals are more sensitive to low-energy particles in secondary cascades and the Molière radius (R_M) for crystals ($R_M = 2.26$ cm for BGO) is significantly larger than tungsten ($R_M = 0.93$ cm) [18]. As the ArborPFA was optimized for the CEPC CDR with sampling calorimeters, it needs further optimizations to take into account different shower features in a homogeneous calorimeter.

Major updates implemented in the ArborPFA algorithm for the crystal ECAL include: (1) to search for the cluster skeleton using a relatively high energy threshold and by temporarily ignoring low-energy hits; and (2) to re-cluster low-energy hits near the shower skeleton. Simulation studies so far show that these two updates can significantly improve the separation capability (presented below) and energy linearity and resolution (not shown due to page constraints), respectively.

The separation of near-by particles is based on the overlay of events of single particles generated by a particle gun with varying positioning offsets. Showering events of γs and charged pions before entering the ECAL are excluded for further analysis. The separation efficiency is defined as the ratio of the events with two correctly reconstructed physics objects from ArborPFA algorithm (PFA Objects, or PFOs) and the total events. A correct reconstruction is defined as the PFO energy being in the range of 3.3–6.6 GeV for incident 5 GeV photons, and of 9.9–10.1 GeV for incident 10 GeV charged pions. The latter scenario corresponds to an essential part of the ArborPFA, i.e., cluster matching between trackers and calorimeters for charged particles, where trackers can provide the momentum information with much higher accuracy than calorimeters.

As shown in Figure 3, the separation efficiency of two $\gamma's$ increases along with a higher energy threshold (implemented for each crystal cell) in the updated ArborPFA with the multi-threshold reconstruction scheme. The performance of the granularity of $1 \times 1 \times 2$ cm^3 is similar to the ideal granularity of $1 \times 1 \times 1$ cm^3. It would be more challenging to separate hadronic clusters from EM ones, due to the complicated hadronic shower profiles leading to issues of matching the charged clusters in calorimeters with trackers. Clusters of the hadrons will be much more widely distributed in 3D space, and fluctuations in hadronic shower profiles and energy depositions are much more significant, compared with EM showers. Parameters concerning the distance between clusters and tracks can hardly be effective enough. For the particle-flow algorithms, there is always a significant chance that clusters originally from a hadron will not be correctly associated with its track.

It should be noted that the crystal ECAL geometry in simulation was kept the same as the SiW-ECAL layout, where gaps between modules are based on the SiW option. This study shows these gaps (e.g., around 160 mm in the plots) will degrade the separation efficiency of the crystal ECAL. A new general layout dedicated to crystal modules will be studied to include realistic estimates and minimize the impacts from gaps.

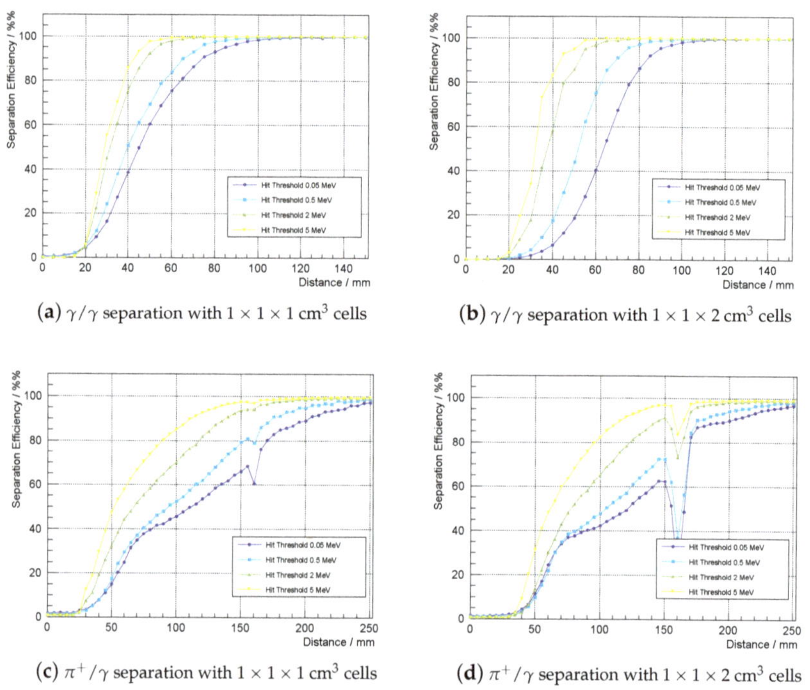

(**a**) γ/γ separation with $1 \times 1 \times 1$ cm^3 cells

(**b**) γ/γ separation with $1 \times 1 \times 2$ cm^3 cells

(**c**) π^+/γ separation with $1 \times 1 \times 1$ cm^3 cells

(**d**) π^+/γ separation with $1 \times 1 \times 2$ cm^3 cells

Figure 3. Separation efficiency with varying distances between two incident particles using two sets of crystal granularity.

2.2. Higgs Benchmark with Jets

The boson mass resolution (BMR) is a key parameter to evaluate the PFA performance of jets in Z/W/Higgs hadronic decays. At the CEPC, the Higgs BMR needs to be <4% in order to reach the 2σ separation power of the Higgs and Z/W bosons. The MC samples of $e^+e^- \rightarrow ZH \rightarrow \nu\nu gg$ including initial-state radiation (ISR) photons at 240 GeV were used for the studies with two gluon jets of the Higgs final state, where the BMR is defined as the invariant mass with two jets reconstructed by the ArborPFA. The simulation setup remains the same as in Section 2.1, i.e., the CEPC CDR baseline detector with the SiW-ECAL superseded by the crystal ECAL using the granularity of 1 cm^3. Since the default ArborPFA was optimized for the CEPC CDR baseline detector, as shown in Figure 4a, it cannot deliver the required performance with the crystal ECAL even with the fine granularity. After implementing the updates in ArborPFA as discussed in Section 2.1 for separation of close-by particles in crystals, there is a significant improvement in BMR by 3.7%, as shown in Figure 4b. Further optimizations and studies of impacts from granularity are ongoing.

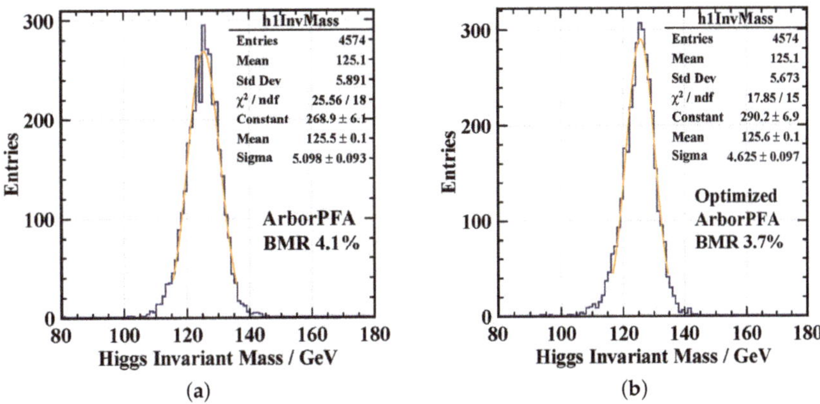

Figure 4. Higgs boson mass resolution with the ArborPFA algorithm. (**a**) BMR with the crystal ECAL using the default ArborPFA. (**b**) BMR with the crystal ECAL using the optimized ArborPFA.

3. Reconstruction Algorithm Dedicated to New Geometry Design

For the long bar layout, the geometry was implemented using DD4hep [19] and a dedicated reconstruction algorithm is under development in the new software framework named "CEPCSW" [20]. The idea is to start reconstruction from raw hits for each longitudinal layer to form 1D clusters and obtain 2D clusters by combining information from every two adjacent layers. Then, 2D clusters are linked in the longitudinal direction into 3D clusters required by the PFA. The basic reconstruction flow is shown in Figure 5. To reach its full physics potential, key questions including pattern recognition and cluster matching with the tracking system need to be addressed. In general, the new algorithm aims to achieve the granularity of $1 \times 1 \times 2$ cm^3, and minimize the impact from the ambiguity of pattern recognitions. In Figure 6, preliminary performance studies of the crystal ECAL with the long bar layout show promising results in particles reconstruction and separation [21]. Further development and validation works are still ongoing.

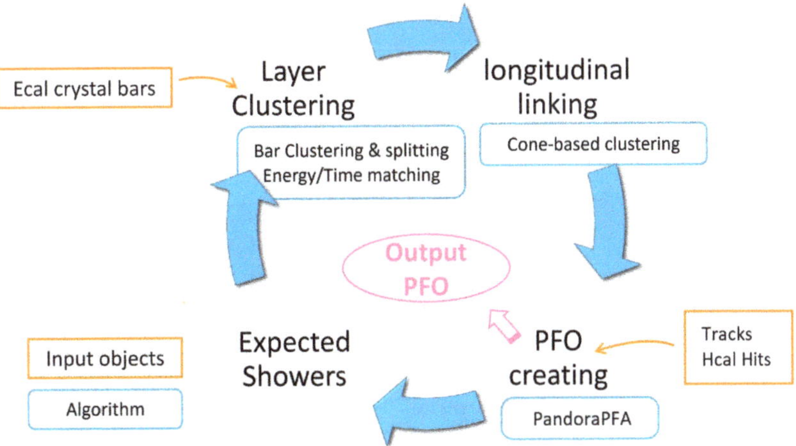

Figure 5. Dedicated reconstruction flow for the crossed long crystal bar ECAL.

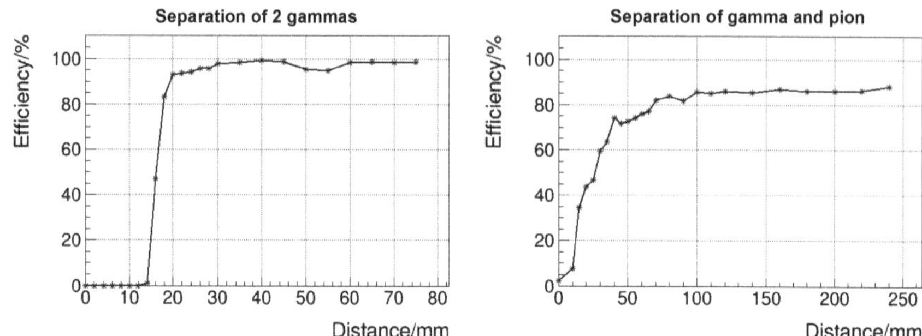

Figure 6. Preliminary study of separation efficiency with the dedicated reconstruction algorithm.

4. Detector Design and Characterizations

The quantitative requirements on the crystal and readout unit are presented in this section, including the EM energy resolution, photon statistics, energy threshold, and response uniformity, followed by measurements of a typical crystal-SiPM unit. Parameters in the digitization tool used by the Geant4 [22] simulation were tuned based on the cosmic-ray and radioactive test results.

4.1. Energy Resolution and Requirements

The EM energy resolution of the crystal calorimeter was studied with the Geant4 simulation (without the optical photon simulation). To take into account the important factors related to photon statistics, we developed a digitization tool to convert the raw energy deposition into the number of photons detected by a SiPM. The digitization tool includes the energy threshold per readout channel, fluctuations in photons detected at the SiPM, the time window for signal integration, etc. The photon statistics are primarily evaluated in terms of the MIP (minimum ionising particle) response, which corresponds to the most probable energy deposition of high energy muons in 1 cm BGO crystal around 8.9 MeV [18].

As shown in Figure 7a, the EM energy resolution for a given MIP response of 100 p.e. is dominated by the energy threshold per readout channel and the threshold should be lower than 0.5 MIP to achieve the aim of $<3\%/\sqrt{E(\mathrm{GeV})}$. The MIP light yield per channel, i.e., photo-statistics, significantly impacts the stochastic term extracted from the EM energy resolution, as shown in Figure 7b, and the MIP light yield is required to be more than 100 p.e. A sufficiently low energy threshold is always desirable to further improve the resolution. SiPMs with a low inter-pixel crosstalk level are promising to achieve a reasonably low energy threshold. For a good balance between the MIP light yield and the dynamic range, the appropriate crystal candidate needs to have a moderate intrinsic light yield. The crystal candidate is required to be friendly for mechanical processing (especially important for the long crystal bar design) and not prohibitively costly for the large volume usage required by the final detector (unit price on the order of a few USD per c.c.).

Crystal options under study primarily focus on the bismuth germanate (BGO). BGO has a high intrinsic light yield, in the range of 8000–10,000 photons/MeV, and its mechanical stability is good for cutting and polishing, but with a typical long scintillation decay time of 300 ns. The lead tungstate (PWO) has fast scintillation components (the fast and slow components are 10 ns and 30 ns, respectively), but its low intrinsic light yield, in the range of 100–200 photons/MeV, makes it difficult to meet the requirement of more than 100 p.e./MIP. Furthermore, it is quite brittle and, thus, further mechanical processing is very challenging, especially for long crystal bars.

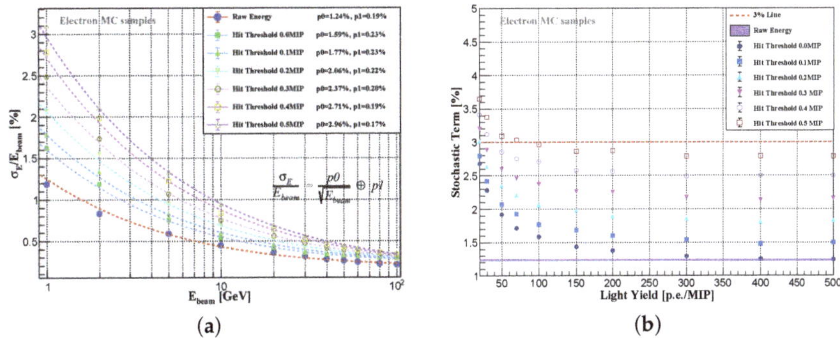

Figure 7. Energy resolution under different energy thresholds and crystal light yields. (**a**) The EM energy resolution when different energy thresholds are implemented on calorimeter hits. The light yield is set to 100 detected photons per MIP. (**b**) The stochastic term of the EM energy resolution with varying light yields (number of detected photons per MIP) and energy thresholds for hits.

4.2. Laser Calibration of SiPMs

The SiPMs for the crystal ECAL readout need to cover a large dynamic range (on the order of 10^3 MIPs), which requires a high pixel density (on the order of 10^5 pixels for the total sensitive area). A laser test stand was built with a picosecond laser source (405 nm, NKT Photonics [23]) for SiPM characterizations including the single photon calibration and the response linearity. The SiPMs under test are listed in Table 1. In Figure 8, single photons can be well-separated for the two types of SiPMs, and the NDL-EQR06 SiPM is expected with better timing performance due to the fast-rising edge. The narrow pulse shape (~10 ns) is helpful to achieve an effectively larger dynamic range for crystals with a long decay time (e.g., BGO) due to the short recovery time of pixels. Results of ongoing measurements of the response linearity and the dynamic range will be presented in future conferences or papers.

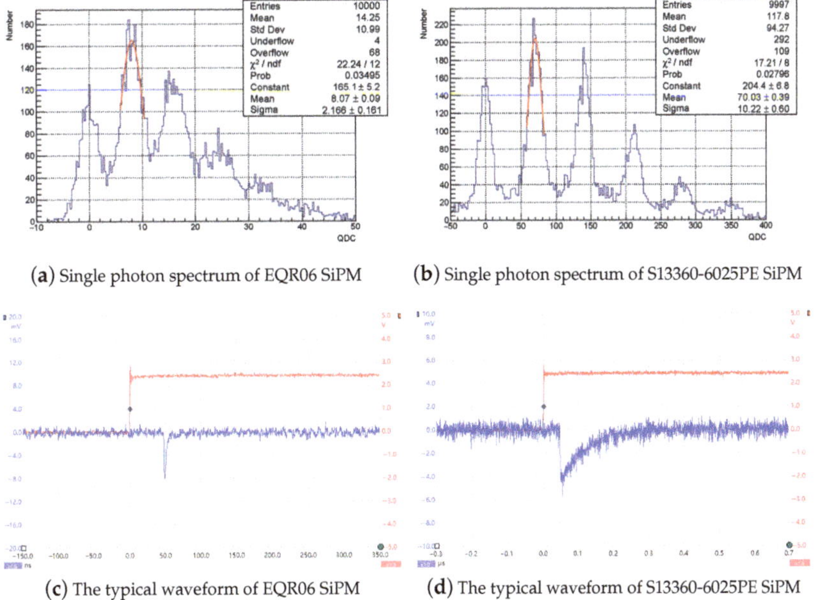

(**a**) Single photon spectrum of EQR06 SiPM

(**b**) Single photon spectrum of S13360-6025PE SiPM

(**c**) The typical waveform of EQR06 SiPM

(**d**) The typical waveform of S13360-6025PE SiPM

Figure 8. Single photon spectrum and typical waveforms in the oscilloscope of DUTs. Red pulses in (**c**,**d**) correspond to trigger signals.

Table 1. Devices under test and their basic parameters.

Type	Pixel Pitch (μm)	Sensitive Area (mm)	Number of Pixels	Typical PDE	Typical Gain
Hamamatsu S13360-6025PE [24]	25	6×6	57,600	25%	7×10^5
NDL EQR06 11-3030D-S [25]	6	3×3	244,720	30%	8×10^4

4.3. Characterization of Long BGO Crystal Bar

The characterizations of a 40 cm long BGO crystal bar (manufactured by the Shanghai Institute of Ceramics [26]) were carried out to evaluate the MIP response performance and the results are also used to validate the digitization tool.

4.3.1. Cosmic-Ray Test

A $1 \times 1 \times 40$ cm^3 BGO crystal was tested with cosmic-ray muons to evaluate the MIP response. As shown in Figure 9a, the crystal bar was wrapped with ESR film and placed onto a rail. Two apertures were cut out through the wrapping of the two ends for the light detection at SiPMs. Hamamatsu C13365-3050SA [27] SiPM modules (3×3 mm^2 active area) were coupled to the two sides of the crystal bar with silicone grease (index of refraction 1.465). In addition, two 3×3 cm^2 triggers were used to select coincidence muon events. The most probable value of the MIP response is 2028 p.e./MIP, as shown in Figure 9b. The fact that the trigger size is larger than the crystal cross section leads to a structure at the left side of the MIP signal, i.e., muons penetrates only partially but not the full crystal. It shows that the MIP response can meet the requirement, but we also need to consider the dynamic range as another critical issue, which remains under study.

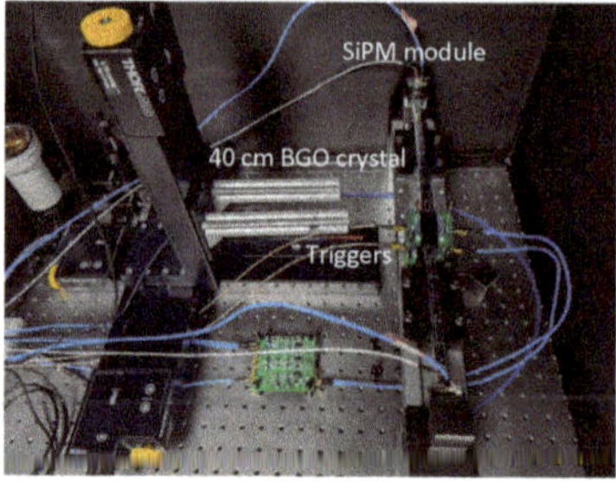

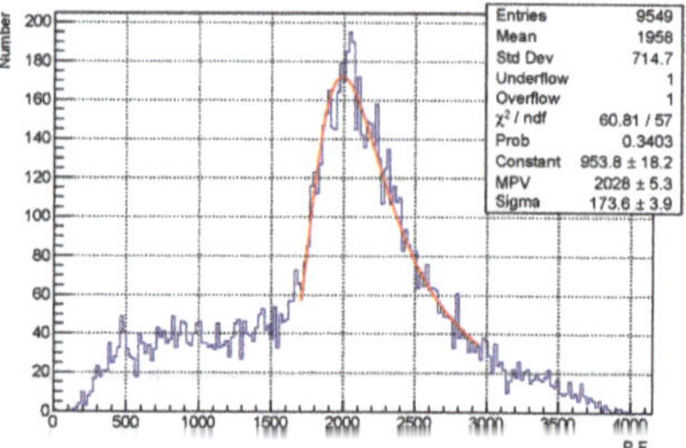

(**a**) A photo of the setup of the cosmic-ray test   (**b**) Measured MIP response with BGO crystal

Figure 9. Setup and result of the cosmic-ray test of the BGO crystal.

4.3.2. Energy Calibration with a Radioactive Source

Measurements with radioactive sources are important to evaluate the energy resolution and to validate the digitization tool. Tests were performed with the same BGO crystal bar of $1 \times 1 \times 40$ cm^3 as mentioned above. Signals were read out by two SiPMs (Hamamatsu S13360-6025PE) directly air-coupled with each end. A slide rail system was used to improve the stability for better SiPM–crystal coupling. A radioactive source of Cs-137 was placed on the 1D movable rail to scan along the crystal length direction, as shown in Figure 10a. It should be noted that the source collimation diameter is around 8 mm. Figure 10b shows the energy resolution is 11.2% for 662 keV gammas, and the full-energy peak corresponds to around 134 p.e. at the SiPMs.

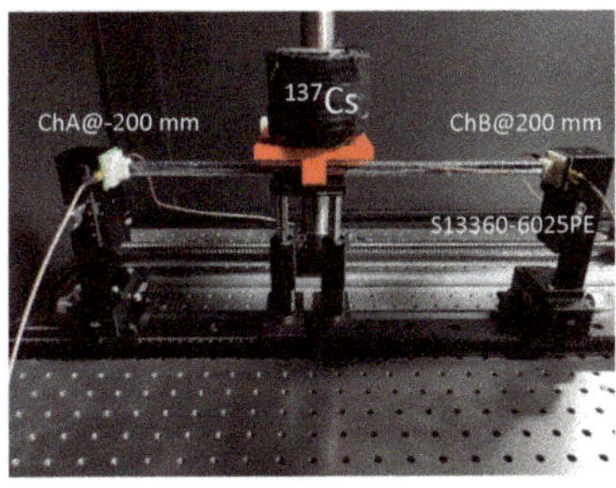

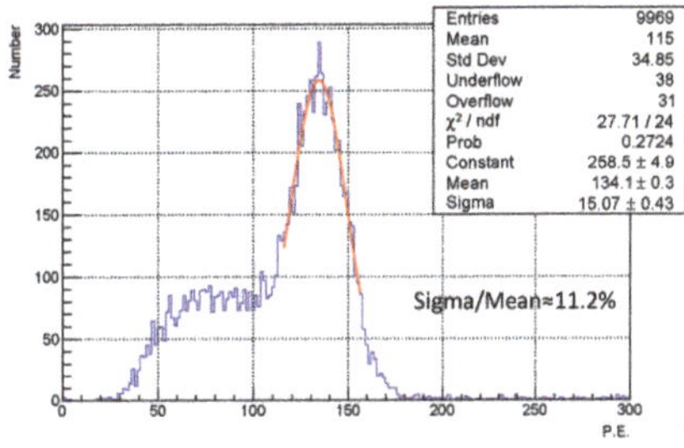

(**a**) A photo of the radioactive source test stand (**b**) Measured energy spectrum of Cs-137

Figure 10. Radioactive source test of a 40 cm long BGO crystal bar.

4.3.3. Response Uniformity Studies

The response uniformity along the crystal bar was measured with the same radioactive test-stand mentioned above by moving the radioactive source along the crystal bar length direction. The results from measurements and Geant4 optical simulation are shown in Figure 11a,b. The simulation assumed the intrinsic light yield of BGO crystals is 8200 p.e./MeV, and the crystal surface roughness was considered. Here, the incident-angle distribution of gamma-rays was not implemented and the simulation will be improved in the future.

Preliminary results show that the response uniformity in measurements is, in general, better than the simulation, but only features an asymmetrical pattern. In general, the results in data and simulation are reasonably consistent at a 10% level, which is sufficient for the validation of the optical simulation and the digitization tool at the first order. For better consistency of 1%, there are several subtle parts to be studied, such as optical model parameters and the modeling of defects on the crystal surface, guided by the measurements. On the other hand, more crystals will be tested to evaluate repetitive precision.

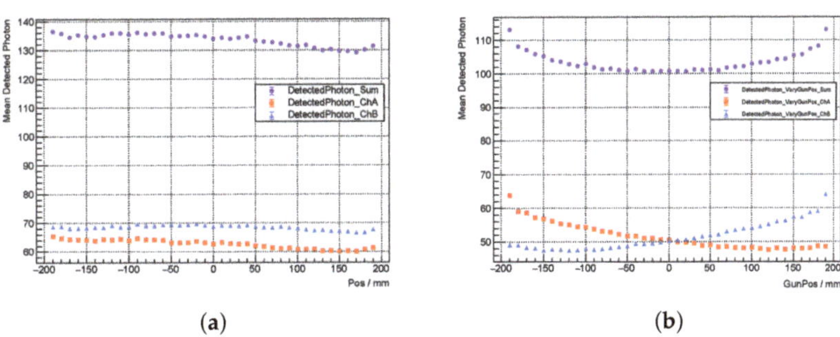

(**a**) (**b**)

Figure 11. Uniformity scan for a 40 cm long BGO crystal bar. (**a**) Measured response uniformity with Cs-137. (**b**) Simulated response uniformity with 662 keV photons.

A 2D uniformity map is obtained in simulation with muons perpendicularly passing through the module, as shown in Figure 12a, and was implemented in the digitization tool for an ECAL module with the transverse size of 40×40 cm^2 with the detector layout of crossed long crystal bars. The 2D non-uniformity effect will lead to position dependence for the energy reconstruction. Monte Carlo samples of high-energy electrons were used to evaluate the impacts on the energy reconstruction of EM showers, dependent on different levels of non-uniformity, as shown in Figure 12b. Therefore, the crystal-SiPM response uniformity needs to be well-controlled and carefully calibrated.

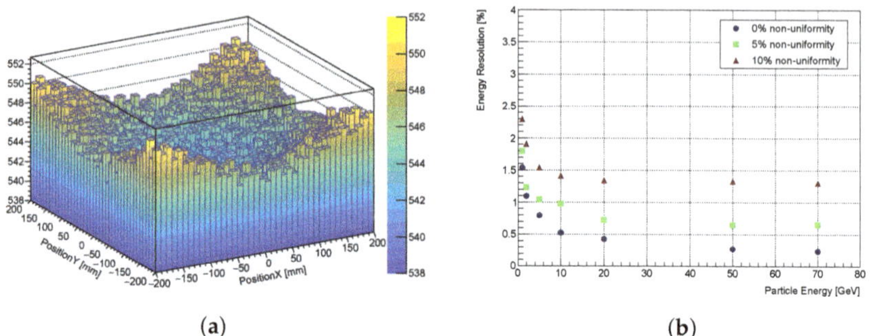

Figure 12. Simulations of uniformity of the crystal ECAL module. (**a**) 2D response uniformity of the ECAL module, 1 GeV muons are used for scanning. (**b**) Energy resolution under certain non-uniformity of long crystal bars. Nine modules are placed in the simulation to prevent energy leakage.

5. Summary and Prospects

High-granularity calorimetry options enable an excellent jet reconstruction capability for future high-energy experiments. A highly granular crystal calorimeter was proposed to aim at a superior EM energy resolution and PFA performance for future Higgs factories. Physics potentials were presented using the CEPC detector with a high-granularity crystal calorimeter, including the PFA performance on separation power and the Higgs benchmark with two jets. The optimization of ArborPFA for crystals is ongoing. A dedicated reconstruction algorithm is under development for the detector layout with long crystal bars arranged to be orthogonal in adjacent layers. Hardware activities focus on the crystal–SiPM readout unit to address key questions of the detector requirements were studied. Characterizations of BGO crystals and SiPMs were carried out and the results were used to validate the simulation. In the near future, small-scale ECAL modules will be developed to evaluate the EM shower performance in beam tests, to gain experience in the large-scale module design, and to deliver reliable inputs to evaluate the whole detector performance.

Author Contributions: Conceptualization, Y.L.; methodology, B.Q. and Y.L.; software, B.Q.; validation, B.Q. and Y.L.; formal analysis, B.Q.; investigation, B.Q.; resources, Y.L.; data curation, B.Q. and Y.L.; writing—original draft preparation, B.Q.; writing—review and editing, Y.L.; visualization, B.Q.; supervision, Y.L.; project administration, Y.L.; funding acquisition, Y.L. All authors have read and agreed to the published version of the manuscript.

Funding: This project has received funding support from the Chinese Academy of Sciences (CAS) for young professionals and the CAS Center for Excellence in Particle Physics (CCEPP).

Acknowledgments: The authors would like to express their gratitude for the fruitful discussions with many of the colleagues in the CEPC calorimeter working group and the CALICE collaboration.

References

1. The ATLAS Collaboration. Observation of a new particle in the search for the Standard Model Higgs boson with the ATLAS detector at the LHC. *Phys. Lett. B* **2012**, *716*, 1–29. [CrossRef]
2. The CMS Collaboration. Observation of a new boson at a mass of 125 GeV with the CMS experiment at the LHC. *Phys. Lett. B* **2012**, *716*, 30–61. [CrossRef]
3. The CEPC Study Group. CEPC Conceptual Design Report Volume II—Physics & Detector. *arXiv* **2018**, arXiv:1811.10545.
4. The FCC Collaboration. FCC-ee: The Lepton Collider - Future Circular Collider Conceptual Design Report Volume 2. *Eur. Phys. J. Spec. Top.* **2019**, *228*, 261–623. [CrossRef]
5. The ILC Collaboration. The International Linear Collider Technical Design Report Volume 4: Detectors. *arXiv* **2013**, arXiv:1306.6329.

6.   Linssen, L.; Miyamoto, A.; Stanitzki, M.; Weerts, H. Physics and Detectors at CLIC: CLIC Conceptual Design Report. *arXiv* **2012**, arXiv:1202.5940.

7.   Thomson, M.A. Particle flow calorimetry and the PandoraPFA algorithm. *Nucl. Instrum. Meth. A* **2009**, *611*, 25–40. [CrossRef]

8.   CALICE Collaboration Home Page. Available online: https://twiki.cern.ch/twiki/bin/view/CALICE/WebHome (accessed on 20 July 2022).

9.   The CALICE Collaboration. Design and electronics commissioning of the physics prototype of a Si-W electromagnetic calorimeter for the International Linear Collider. *J. Instrum.* **2008**, *3*, P08001.

10. The CALICE Collaboration. Construction and response of a highly granular scintillator-based electromagnetic calorimeter. *Nucl. Instrum. Meth. A* **2018**, *887*, 150–168. [CrossRef]

11. Adeva, B.; Aguilar-Benitez, M.; Akbari, H.; Alcaraz, J.; Aloisio, A.; Alvarez-Taviel, J.; Alverson, G.; Alviggi, M.G.; Anderhub, H.; Anderson, A.L.; et al. The construction of the L3 experiment. *Nucl. Instrum. Meth. A* **1990**, *289*, 35–102. [CrossRef]

12. The CMS Collaboration. The CMS Electromagnetic Calorimeter Project: Technical Design Report. 1997. Available online: https://cds.cern.ch/record/349375 (accessed on 20 July 2022).

13. Escrihuela, F.J.; Flores, L.J.; Miranda, O.G. Neutrino counting experiments and non-unitarity from LEP and future experiments. *Phys. Lett. B* **2020**, *802*, 135241. [CrossRef]

14. Liu, Y. High-Granularity Crystal Calorimeter: R&D Status. Online Mini-Workshop on a Detector Concept with a Crystal ECAL, 22–23 July 2020. Available online: https://indico.ihep.ac.cn/event/11938/contributions/14784/attachments/6980/7889/2020_0722_Crystal_ECAL_Status_YL.pdf (accessed on 21 August 2022).

15. Ruan, M.; Videau, H. Arbor, a new approach of the Particle Flow Algorithm. In Proceedings of the International Conference on Calorimetry for the High Energy Frontier, Paris, France, 22–25 April 2013.

16. CEPC Software Home Page. Available online: http://cepcsoft.ihep.ac.cn/ (accessed on 21 August 2022).

17. Yuexin, W. Crystal ECAL Design for CEPC. Online Mini-Workshop on a Detector Concept with a Crystal ECAL, 22–23 July 2020. Available online: https://indico.ihep.ac.cn/event/11938/contributions/14786/attachments/6967/7875/Yuexin_CrystalECALWS_20200722.pdf (accessed on 21 August 2022).

18. PDG: Atomic and Nuclear Properties of Materials (2022). Available online: https://pdg.lbl.gov/2022/AtomicNuclearProperties/index.html (accessed on 21 August 2022).

19. DD4hep (Detector Description for High Energy Physics). Available online: https://github.com/AIDASoft/DD4hep (accessed on 20 July 2022).

20. CEPCSW (CEPC Offline Software Prototype Based on Key4hep). Available online: https://github.com/cepc/CEPCSW (accessed on 20 July 2022).

21. Fangyi, G. High-granularity Crystal ECAL for CEPC. The 2021 International Workshop on the High Energy Circular Electron Positron Collider, 8–12 November 2021. Available online: https://indico.ihep.ac.cn/event/14938/contributions/35587/attachments/17472/19907/CrystalECAL_CEPCWorkshop2021.pdf (accessed on 21 August 2022).

22. The GEANT4 Collaboration. Geant4—A simulation toolkit. *Nucl. Instrum. Meth. A* **2003**, *506*, 250–303. [CrossRef]

23. NKT PILAS—Picosecond Pulsed Diode Lasers Datasheet. Available online: https://www.nktphotonics.com/products/pulsed-diode-lasers/pilas/ (accessed on 20 July 2022).

24. Hamamatsu Photonics S13360 Series MPPC Datasheet. Available online: https://www.hamamatsu.com/content/dam/hamamatsu-photonics/sites/documents/99_SALES_LIBRARY/ssd/s13360_series_kapd1052e.pdf (accessed on 20 July 2022).

25. NDL (Novel Device Laboratory) EQR06 11-3030D-S SiPM Datasheet. Available online: http://www.ndl-sipm.net/PDF/Datasheet-EQR06.pdf (accessed on 20 July 2022).

26. Shanghai Institute of Ceramics, Chinese Academy of Sciences Home Page. Available online: http://english.sic.cas.cn/ (accessed on 20 July 2022).

27. Hamamatsu Photonics C13365 Series MPPC Module Datasheet. Available online: https://www.hamamatsu.com/content/dam/hamamatsu-photonics/sites/documents/99_SALES_LIBRARY/ssd/c13365_series_kacc1227e.pdf (accessed on 21 August 2022).

Article

Including Calorimeter Test Beams in Geant-val—The Physics Validation Testing Suite of Geant4

Lorenzo Pezzotti [1,*], Andrey Kiryunin [2], Dmitri Konstantinov [3], Alberto Ribon [1], Pavol Strizenec [4] and on behalf of the Geant4 Collaboration

[1] European Organization for Nuclear Research, CERN, Esplanade des Particules 1, CH-1217 Geneva, Switzerland

[2] Max Planck Institute for Physics, D-80805 Munich, Germany

[3] NRC Kurchatov Institute—IHEP, 142280 Protvino, Russia

[4] Institute of Experimental Physics, Slovak Academy of Sciences (SAS), 04001 Kosice, Slovakia

* Correspondence: lorenzo.pezzotti@cern.ch

Abstract: The Geant4 simulation toolkit is currently adopted by many particle physics experiments, including those at the Large Hadron Collider and the ones proposed for future lepton and hadron colliders. In the present era of precision tests for the Standard Model and increasingly detailed detectors proposed for the future colliders scenario, Geant4 plays a key role. It is required to remain a reliable and stable toolkit for detector simulations and at the same time undergo major improvements in both physics accuracy and computational performance. Calorimeter beam tests involve various particles at different energy scales and represent ideal benchmarks for the physics modeling and assessment of Monte Carlo tools for radiation–matter simulation. We present the first results of a broad validation campaign on test beam data targeting data deployment and preservation with geant-val, the Geant4 validation and testing suite. We investigate the Geant4 capability to model the calorimeter response, energy fluctuations, and shower shapes using data from the ATLAS hadronic end-cap calorimeter and the CALICE silicon-tungsten calorimeter. The evolution over the recent years of the recommended set of physics processes for high-energy physics applications is outlined and compared to alternative models for hadronic interactions.

Keywords: Geant4; geant-val; simulation; hadronic interaction; calorimeter; test-beam

Citation: Pezzotti, L.; Kiryunin, A.; Konstantinov, D.; Ribon, A.; Strizenec, P.; on behalf of the Geant4 Collaboration. Including Calorimeter Test Beams in Geant-val—The Physics Validation Testing Suite of Geant4. *Instruments* **2022**, *6*, 41. https://doi.org/10.3390/instruments6030041

Academic Editors: Fabrizio Salvatore, Alessandro Cerri, Antonella De Santo and Iacopo Vivarelli

Received: 31 July 2022
Accepted: 30 August 2022
Published: 15 September 2022

Publisher's Note: MDPI stays neutral with regard to jurisdictional claims in published maps and institutional affiliations.

1. The Geant4 Toolkit

The Geant4 simulation toolkit [1–3] is a general purpose Monte Carlo (MC) code for radiation–matter interaction simulation. It consists of nearly two million lines of code written in object-oriented C++ that have been developed over three decades by an international collaboration of physicists, computer scientists, mathematicians, and engineers. It currently supports both high-energy and low-energy particle physics experiments, neutrino experiments, and detector design studies for post-Large-Hadron-Collider (LHC) experiments as well as medical, space, and atmosphere applications.

One of the key goals of Geant4 is to provide the main LHC experiments with a reliable and stable MC tool to simulate the response of complex detectors to the passage of the large variety of particles produced during beam collisions. The most challenging part of these simulations, both in terms of simulation speed and physics accuracy, comes from the particles multiplication mechanisms responsible for the creation of showers in calorimeters. From the computing budget point of view, it requires tracking thousands of particles within each shower, with the actual number heavily depending on the production cuts applied. The MC simulation currently accounts for the largest contribution to the computing time of big experiments, with this contribution being dominated by the simulation of the calorimetric component. For instance, it amounted to 38% of the entire computing time in the case of the ATLAS experiment in 2018 [4]. From the physics modeling point of view, showers in calorimeters involve several particle types, processes, and energy scales.

Hadronic showers are a good example of the complexity involved. Their first stage is governed by the occurrence of the first inelastic nuclear interaction of the primary projectile, while most of the calorimeter signal is carried by relatively low-energy charged particles arising from nuclear breakups. These particles are a mixture of purely electromagnetic and hadronic interacting particles. Therefore, for a reliable description of both the shower shape fluctuations and the calorimeter response, the MC engine must provide excellent representations of hadronic processes happening at high-energy scales as well as electromagnetic and hadronic low-energy ones.

A smart solution is offered by Geant4 via the Physics Lists (PLs). PLs are easy-to-use descriptions of consistent sets of particles and processes to be simulated. Usually, the more physically accurate a PL is, the larger the computing time needed for a given event to be simulated. Each PL represents a meeting point between physical accuracy and computational cost; it is the responsibility of the user to pick the most suitable one for their application. Within Geant4, each user is allowed to create their own PL; however, it is worth noting that the four largest LHC experiments, for their Run2 simulations (2015–2018), recently adopted the Geant4-recommended PL for the high-energy-physics application, FTFP_BERT, eventually with mild variants. The same choice is foreseen for the upcoming LHC Run3 simulations and will likely apply to the High-Luminosity-LHC simulations as well.

The Standard Model tests envisaged for future LHC runs will require outstanding descriptions of all the physics processes involved in calorimeters in order to limit the systematic errors driven by simulation as much as possible. At the same time, calorimeters envisaged for future colliders will improve the current shower descriptions both in terms of energy resolution and sampling granularity. The prototypes under construction offer a unique chance for superior Geant4 validation, which will open the possibility to provide a realistic description of complex conceptual detector designs, thus helping to save money and time. The Geant4 Collaboration recently started a validation campaign on calorimeter beam tests in close collaboration with ATLAS and CALICE Calorimetry Groups; extensions to other groups are under investigation. Each validation study targets its inclusion into geant-val, the Geant4 validation and testing suite, which is outlined in Section 2. The main Geant4 validation results related to the ATLAS hadronic end-cap calorimeter (HEC) and the CALICE silicon-tungsten (SiW) calorimeter are described, together with the detector features, in Sections 3 and 4, respectively. Conclusions are drawn in Section 5.

2. The Geant-val Project

Geant4 validation on beam tests is usually performed via regression testing and PL comparison. Regression testing consists of running the same simulation with different software versions while comparing the results with experimental data, thus finding indications of any temporal evolution of a single PL. PL comparison, on the other hand, exploits a fixed software version and compares the results for different PLs, thus providing indications of which model is more accurate with respect to experimental data. On top of that, the user might want to investigate the dependence of results on other parameters, e.g., the production cuts or the signal integration time. Large validation campaigns typically require the same MC data production and analysis to be performed over tens or hundreds of different combinations of beam particle type, beam particle energy, detector description, and physics list. They stand among the most time-consuming tasks that the Geant4 Collaboration undertakes. Data preservation and deployment to the entire Geant4 Community is another major task, as each validation test should be updatable and distributable. To facilitate these tasks, the geant-val team developed a validation and testing suite [5] to support both the validator and the end-user. For the benefit of the end-user, geant-val offers a web interface (https://geant-val.cern.ch/, (accessed on 29 August 2022)) that makes it possible to fetch data in the form of static images for every PL and software version desired.

For the benefit of the validator, geant-val offers a Python tool, the *mc-config-generator*, to encapsulate simulation job metadata (software version, compiler, physics list, primary particles, etc.) in the form of a JSON file. Geant-val also provides a uniform way of preparing and running the jobs in parallel on common batch systems, thus providing a consistent way of executing all the combinations at once. Currently, the geant-val database hosts results of about forty validation tests from different Geant4 domains. In addition to data visual inspection, geant-val performs χ^2 and Kolmogorov–Smirnov statistical tests for results comparison.

3. Geant4 Validation on ATLAS HEC Test Beam Data

The ATLAS HEC [6] is a sampling calorimeter that exploits liquid argon (LAr) gaps interspersed between parallel copper plates. Within the ATLAS Detector, it is devoted to (almost) fully absorb hadrons in the pseudorapidity range of $1.5 < |\eta| < 3.2$. The HEC adopts a wedge-shaped design with 32 azimuthal modules replicated around the beam axis; longitudinally, it is divided into two main wheels (HEC1 and HEC2). The absorber layers are 2.5 and 5.0 cm thick, respectively, for the HEC1 and the HEC2. Annular spacers are used to define an 8.5 mm-thick region for LAr gaps. The total thickness amounts to $\simeq 9.7\ \lambda_{int}$ ($\simeq 103\ X_0$). Each wedge module is read out by 88 channels, with the readout scheme being optimized in order to easily reveal the η coordinate of the impinging particles in the ATLAS experiment. The transverse size of the HEC readout cells is $\Delta\eta \times \Delta\phi = 0.1 \times 2\pi/64$ in the region $|\eta| < 2.5$, and $0.2 \times 2\pi/32$ for larger values of pseudorapidity. Longitudinally, modules are readout by 4 layers. Production modules were exposed to the CERN-SPS particle beam during the construction stage in 2000 and 2001. Secondary and tertiary beams from the H6 (beam) line were used to steer electrons, muons, and hadrons in the energy range of $6 \leq E_{Beam} \leq 200$ GeV on the front face of three ϕ-modules positioned inside a cryostat; see Figure 1 (left). For several years, related results [7,8] have represented a test bed for the ATLAS simulation framework [9–11]. In 2022, a summary of selected test-beam results was made public by the ATLAS Liquid Argon HEC Collaboration [12]. In the following, experimental results are extracted from the 2022 reference where details of the test-beam setup as well as of the treatment of the uncertainties are given.

Recently, in a close collaboration between the ATLAS HEC Group and the Geant4 Hadronic Working Group, the test beam simulation was refactored in a standalone Geant4-based code and included in geant-val for the benefit of data preservation and distribution. The outcome of this activity represents a valuable test for hadronic interaction models and will be exploited in future Geant4 validation studies and parameter tuning. Figure 1 (right) shows the simulated test beam geometry. Simulated results in the following are obtained with the Geant4 standalone simulation.

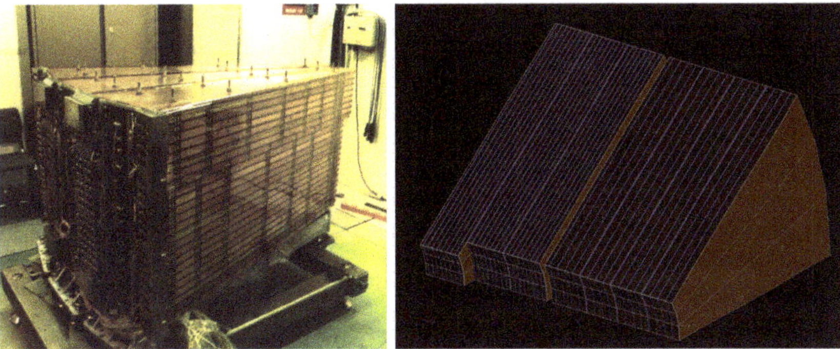

Figure 1. Left: Picture of three ϕ-modules from a section of the ATLAS HEC first wheel from [13]. **Right:** Graphical representation of the three ϕ-modules from a section of the ATLAS HEC, as simulated with Geant4.

Monte Carlo-to-Data Comparison

The Geant4 standalone simulation takes into account realistic materials and geometry descriptions of the calorimeter modules and the cryostat. For the sake of Geant4 validation, the impact of the beam-line auxiliary detectors is considered marginal and not reproduced in the simulation.

The seed for detector response simulation is provided by Geant4 in the form of ionizing energy deposition for every charged particle hit in the LAr gaps. The HEC signal is induced by the free electric charged particles on capacitively coupled copper boards that are immersed in between the LAr gaps. To model the ion recombination mechanism, an attenuation law for the ionizing energy deposition was included in the simulation, with the actual attenuation dependence on specific energy losses being parameterized with the same functional form used by the Birks Law to describe light emission in organic scintillators.

The gauge for energy reconstruction is offered by the sampling fraction estimated with e^- beams. According to Geant4 simulations, it amounts to 4.5% in the first wheel (HEC1) and is constant within 0.1% in the $20 \leq E_{Beam} \leq 150$ GeV energy range. Due to the different thickness of copper plates in the two wheels, the sampling fraction was divided by a factor of 2 when the energy depositions in the second wheel (HEC2) were calibrated. A signal integration time of 75 ns was also considered in the simulation, with the actual readout cut adjusted for the four longitudinal layers mimicking the test beam readout electronics performance. Signal integration over calorimetric cells depends on the nature of the impinging particle. Electron energies are reconstructed from the cumulated signal over the seven cells with the highest average signals; cell selection does not depend on the e^- energy and is kept identical for every event. Pion energies are reconstructed from the cumulated signals in the cells with a visible energy deposition greater than 2.1 MeV (corresponding to an integrated charge of 15 nA according to the simulation of the readout chain). This procedure leads to a selection of $\simeq$50 cells estimated with 180 GeV π^- events; this cell selection is kept fixed for every event regardless of the energy scale.

No sources of systematic uncertainties, as for instance the ones arising from a non-pure beam composition, are included in the Monte Carlo. The stochastic uncertainties for MC results are within 0.1% of the corresponding value. The π/e ratio, i.e., the ratio of the response to the charged pions and electrons, is directly estimated as the ratio of the average π^- reconstructed energy calibrated at the electromagnetic scale, divided by the beam energy. Figure 2 (left) shows the Geant4 FTFP_BERT PL prediction for recent releases (2017–2020) and compares it to experimental data. We observe a systematic increase over the years in the π/e value as described by Geant4. The best Monte Carlo-to-data agreement is offered by the 10.4.p01 version, while the 10.7.p01 one predicts hadronic response values that are 2% higher than the experimental measurements. This consideration is valid for the energy range of $20 \leq E_{Beam} \leq 120$ GeV, while at 150 GeV, the Monte Carlo estimation lies within the experimental uncertainties for every release.

At the time of writing, Geant4 10.7.p01 is the latest tested version. Figure 2 (right) shows the prediction for this software release, with four different PLs: FTFP_BERT, FTFP_BERT_ATL, QGSP_BERT, and FTFP_INCLXX. The FTFP_BERT_ATL PL adopts the Bertini intra-nuclear cascade model [14] and the Fritiof string model [15], with the two models overlapping in the range of 9–12 GeV. This is the only difference in comparison to FTFP_BERT, for which the overlap region is 3–6 GeV. This larger use of the Bertini model is responsible for the lowering of the calorimeter response to hadrons, and it leads to a better agreement with data. Currently, the ATLAS experiment adopts the FTFP_BERT_ATL PL. The INCL intra-nuclear cascade model is used in the (experimental) FTFP_INCLXX PL, which uses an overlapping range of 15–20 GeV for the transition with the Fritiof model; the overall prediction is 5–6% higher than the experimental reference, almost independently of the energy scale. The QGSP_BERT PL corresponds to the FTFP_BERT at low energies and introduces the QGSP string model within the overlapping range of 12–25 GeV, resulting in a hadronic response about 3–4% higher than the experimental data.

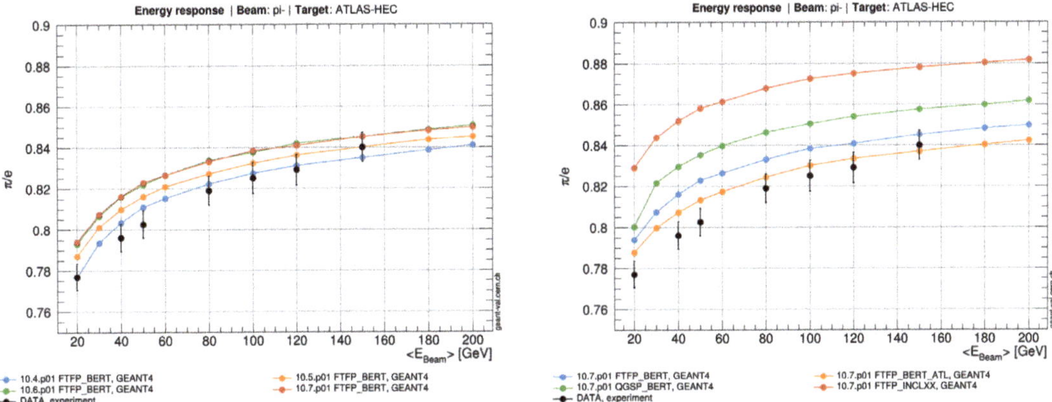

Figure 2. Left: π/e comparison of the Geant4 FTFP_BERT PL prediction for recent releases (2017–2020) with the experimental data. **Right:** Comparison of several PL predictions from Geant4 10.7.p01 for the same variable.

Important information also comes from the hadronic response fluctuations. They are measured as the σ/E value, with σ and E having been extracted from a Gaussian fit to the π^- energy distributions calibrated at the electromagnetic scale. Figure 3 (left) illustrates the FTFP_BERT estimation for four recent releases and compares them to the experimental reference. We observe a reduction of the hadronic response fluctuations when switching from Geant4 10.4.p01 to 10.5.p01. This discrepancy amounts to $\simeq 20\%$ regardless of the energy scale. Such a result is a great example of the importance of regular Geant4 validation on experimental data through realistic simulations. Foreseeing such a big change through code examination would be impossible, while simplified simulation tests would spot the difference without indicating whether it had improved the Monte Carlo-to-data agreement. Figure 3 (right) shows the 10.7.p01 comparison of the previously described PLs, indicating that the minimal fluctuations of energy response correspond to those of the FTFP_INCLXX PL.

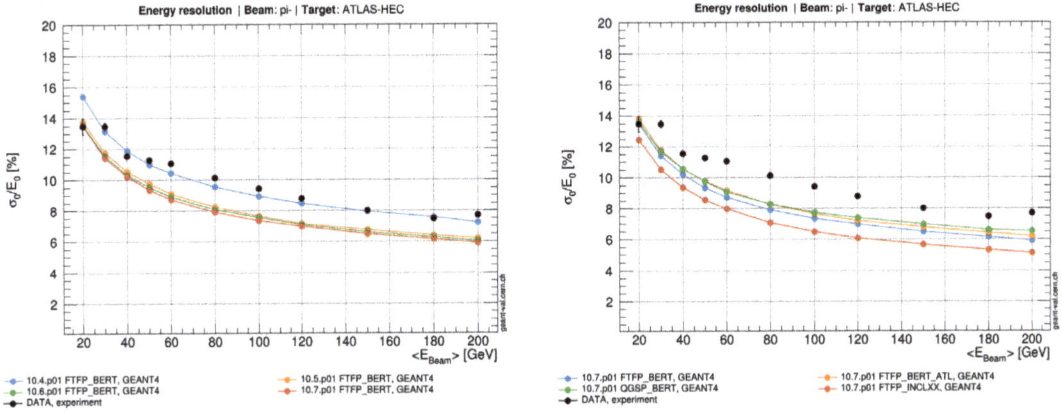

Figure 3. Left: σ/E comparison of the Geant4 FTFP_BERT PL prediction for recent releases (2017–2020) with the experimental data. **Right:** Comparison of several PL predictions from Geant4 10.7.p01 for the same variable.

The path of calorimetry to future lepton colliders is leading to detectors with higher granularity. Such detectors will allow for detailed shower shape reconstruction to be used for particle identification purposes in combination with the tracking information. It is therefore of paramount importance for Geant4 to provide an accurate description of shower shapes. The ATLAS HEC is a good benchmark for longitudinal shower shapes in copper-based sampling calorimeters. The shower profile is extracted from the fraction of the measured energy deposited in each layer, $F_i = E_i/E_{sum}$, with E_{sum} being the total measured energy, while E_i is the energy measured in the layer i. The mean of the profile (L_0) is a direct measurement of the shower barycenter longitudinal position. The L_0 evolution with the beam energy (E_{Beam}) is shown in Figure 4. Figure 4 (left) shows the FTFP_BERT evolution for the L_0 values in the energy range of $20 \leq E_{Beam} \leq 200$ GeV. It indicates the constant shortening of the barycenter longitudinal position from 2017 to 2020; currently the barycenter position can be constrained with a sub-percent precision for every energy point. Figure 4 (right) compares the same variable for several PLs of the 10.7.p01 Geant4 release, with the longitudinal barycenter position being 5% higher for the QGSP_BERT description with respect to the other PLs and the experimental data.

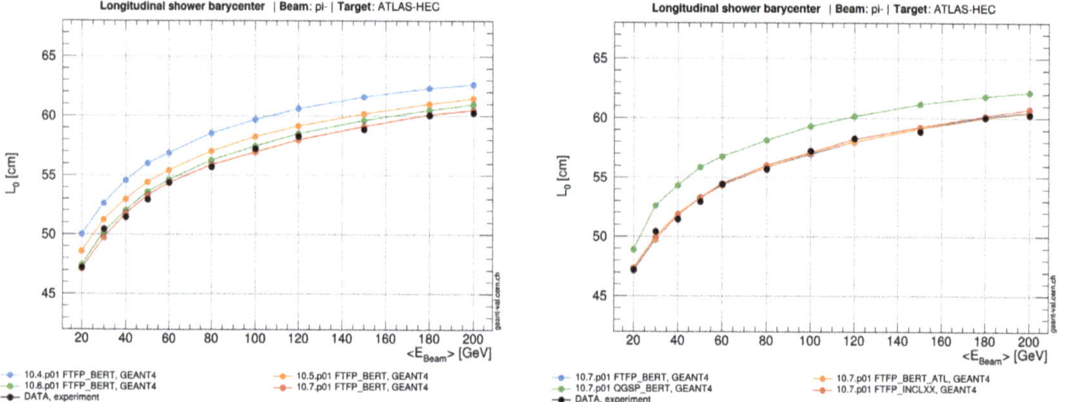

Figure 4. Left: L_0 comparison of the Geant4 FTFP_BERT PL prediction for recent releases (2017–2020) with the experimental data. **Right**: Comparison of several PL predictions from Geant4 10.7.p01 for the same variable.

Another key aspect is the Geant4 capability to reconstruct the hadronic shower length. An indirect measurement comes from the RMS (σ_L) of the longitudinal energy profile introduced above (the longer the shower, the higher the RMS value). Figure 5 (left) shows the FTFP_BERT evolution of the σ_L measurement and compares it to the experimental reference in the energy range of $20 \leq E_{Beam} \leq 200$ GeV. The FTFP_BERT PL recently evolved towards shorter π^- showers in the copper-based calorimeter, finding a recent Monte Carlo-to-data agreement of $\simeq$2%. A similar agreement for the 10.7.p01 version is provided by the FTFP_BERT_ATL and the FTFP_INCLXX PLs, while according to the QGSP_BERT PL, the hadronic showers σ_L is on average $\simeq$ 2% higher, see Figure 5 (right).

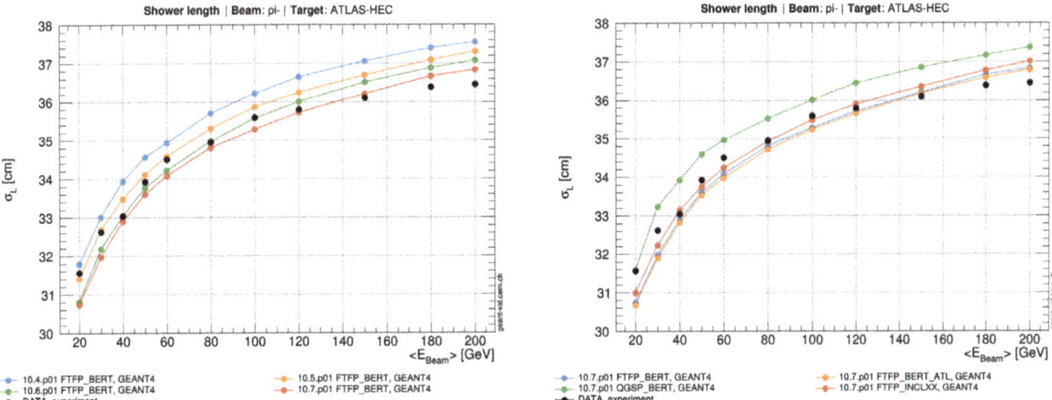

Figure 5. Left: σ_L comparison of the Geant4 FTFP_BERT PL prediction for recent releases (2017–2020) with experimental data. **Right**: Comparison of several PL predictions from Geant4 10.7.p01 for the same variable.

4. Geant4 Validation on CALICE SiW Test Beam Data

The CALICE SiW detector prototype [16] is a sampling calorimeter made of alternating layers of silicon and tungsten. Each of the 30 silicon layers has an active area of 18×18 cm^2 segmented into a 3×3 matrix of Si-wafers. Each wafer consists of 6×6 pixels, for a total of 9720 active elements. The pixel dimension is 1×1 cm^2. The first ten Si-layers (1–10) are interspersed between 1.4 mm-thick W-slabs. The absorber layer thickness changes to 2.8 mm for the following ten layers (11–20) and to 4.2 mm for the remaining layers (21–30). The effective length amounts to 24 X_0 ($\simeq 1$ λ_{int}), therefore more than half of the hadrons traversing it would undergo a nuclear interaction. Figure 6 (left) is a schematic reconstruction of the prototype. The prototype was tested at the Fermilab Test Beam Facility in 2008. Runs with π^- mesons in the energy range of 2–10 GeV were used to study the properties of the first stage of hadronic showers in a W-based calorimeter. The results were published in 2015 [17]. In 2022, the Geant4 Collaboration developed a Geant4-standalone simulation code, with the aim of ensuring regular Geant4 validation on these data as well as deployment to geant-val. Figure 6 (right) shows a simulated π^- interacting with the SiW prototype via a hadronic inelastic process. For clarity, only the first ten layers are displayed. Hits are marked with yellow dots, which correspond to tracks interacting in the SiW calorimeter and in the hadronic calorimeter placed downstream. The hadronic calorimeter is not shown, and no information from the latter is used in the analysis. The subsequent experimental results come from [17]. Beam purity studies, contamination removal, and data corrections together with the systematic and stochastic error treatment are discussed at length in [17] and thus are not repeated here. The following Monte Carlo results were obtained with the Geant4-standalone simulation. They correspond to pure π^- beams. No systematic error is considered in the simulation, and stochastic uncertainties are within 1% of the corresponding value.

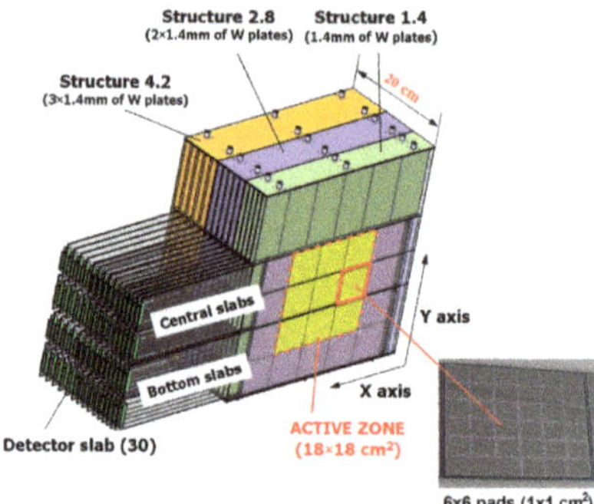

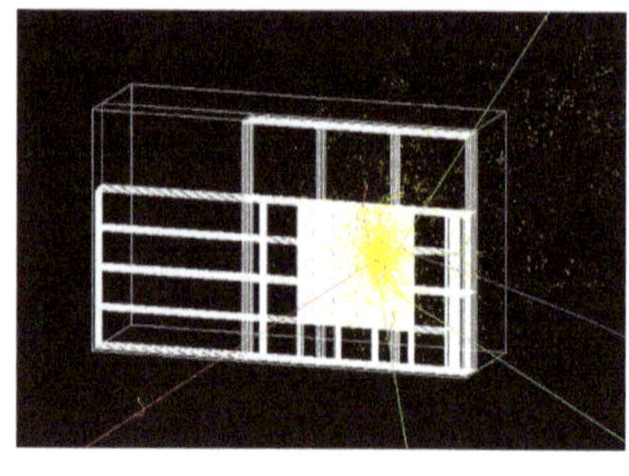

Figure 6. Left: Scheme of the CALICE SiW detector prototype from [17]. **Right**: Graphical representation of the Geant4 simulation of a π^- event interacting in the calorimeter via a nuclear inelastic process. See text for details.

Monte Carlo-to-Data Comparison

The calorimeter signals are calibrated with μ^- beams whose energy loss in an active Si-pixel defines the energy unit MIP. In the following, an energy threshold for a pixel selection of 0.6 MIP is considered, and only events with at least 25 fired pixels are retained. Pion events undergoing a nuclear inelastic reaction are tagged according to two selecting cuts:

- They correspond to events that have three consecutive layers with a measured energy (E_i) greater than 8 MIP. The first of the consecutive layers is considered as the one closer to the point where the nuclear breakup occurred.
- Alternatively, they are selected as the events with a relative increase in the layer energy above a certain threshold F_{cut}:

$$\frac{E_i + E_{i+1}}{E_{i-1} + E_{i-2}} > F_{cut} \text{ and } \frac{E_{i+1} + E_{i+2}}{E_{i-1} + E_{i-2}} > F_{cut} \tag{1}$$

with $F_{cut} = 6$.

To reduce the e^- contamination in π^- beams, events with an interacting layer number ≤ 7 are neglected. Figure 7 shows the longitudinal energy profiles in MIP units for π^- showers in the energy range of 2–10 GeV. The first layer corresponds to the identified interaction layer, so the x-axis represents the shower depth in layers. As most of the hadronic shower extends beyond the detector, the average value in a given bin is determined by considering only the events that contribute energy to the corresponding layer. To take into account the sampling fraction decrease with the layer number, pseudolayers are introduced. As explained in [17], the first ten layers correspond to the first ten pseudolayers, while each layer with a number within the ranges of 11–20 and 21–30 is assigned to two and three pseudolayers, respectively. Experimental data are compared to the Geant4 FTFP_BERT prediction for releases from 2017 to 2020. We observed a constant improvement in the Monte Carlo-to-data agreement over the years in the energy range of 6–10 GeV. For 10 GeV π^- events simulated with the 10.7.p03 PL, a residual tension between data and simulation has been observed to affect the longitudinal shower maximum; see Figure 7 (bottom-left). The same is not true for the description provided by the (experimental) FTFP_INCLXX PL; see Figure 7 (bottom-right).

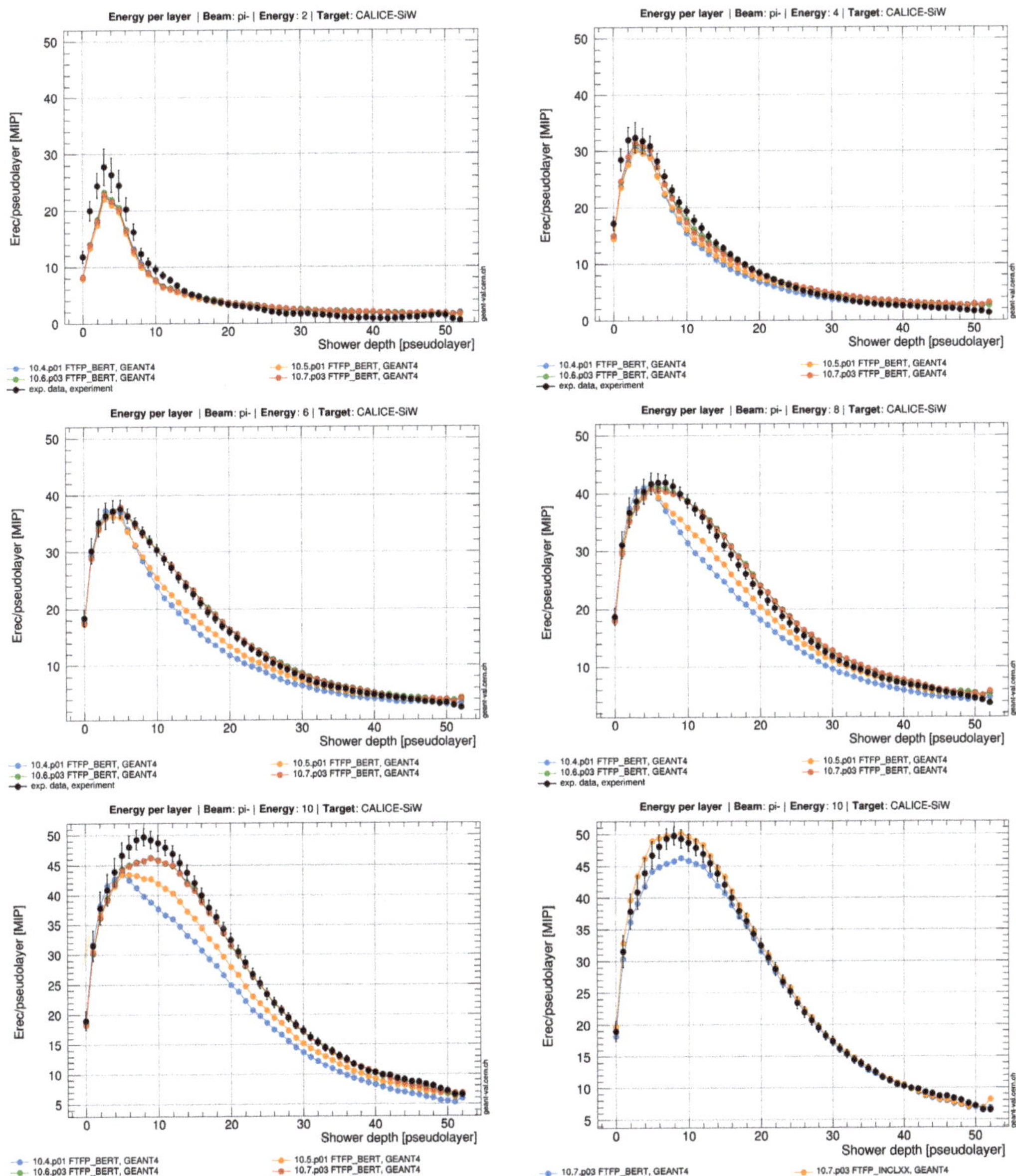

Figure 7. Longitudinal energy profiles for 2 (**top-left**), 4 (**top-right**), 6 (**center-left**), 8 (**center-right**), and 10 (**bottom-left**) GeV π^- events. Results from experimental measurements and Geant4 simulation with the FTFP_BERT PL and releases from 2017 to 2020. **Bottom-right**: Comparison of the same variable as simulated with the FTFP_BERT and the FTFP_INCLXX PLs from the Geant4 10.7.p03 release.

Another important source of information comes from the longitudinal hit distribution for π^- showers. Hits are individual pixels retained after the cleaning cuts. As stated before, the average value in a given bin is determined by considering only events which contribute in the corresponding layer and the first layer corresponds to the identified interaction layer. To compare longitudinal distribution shapes, they have been normalized to unity; see

Figure 8. Experimental data are compared to the FTFP_BERT prediction for Geant4 releases from 2017 to 2020. Simulated results are stable over the years and indicate a better capability to reproduce longitudinal hit distributions in the 6–10 GeV energy scale with respect to the 2–4 GeV one. It corresponds to a higher precision in the hit distribution description for the Fritiof string model as compared to the Bertini intra-nuclear cascade one. Similar results are obtained with the FTFP_INCLXX PL and are available on the geant-val website.

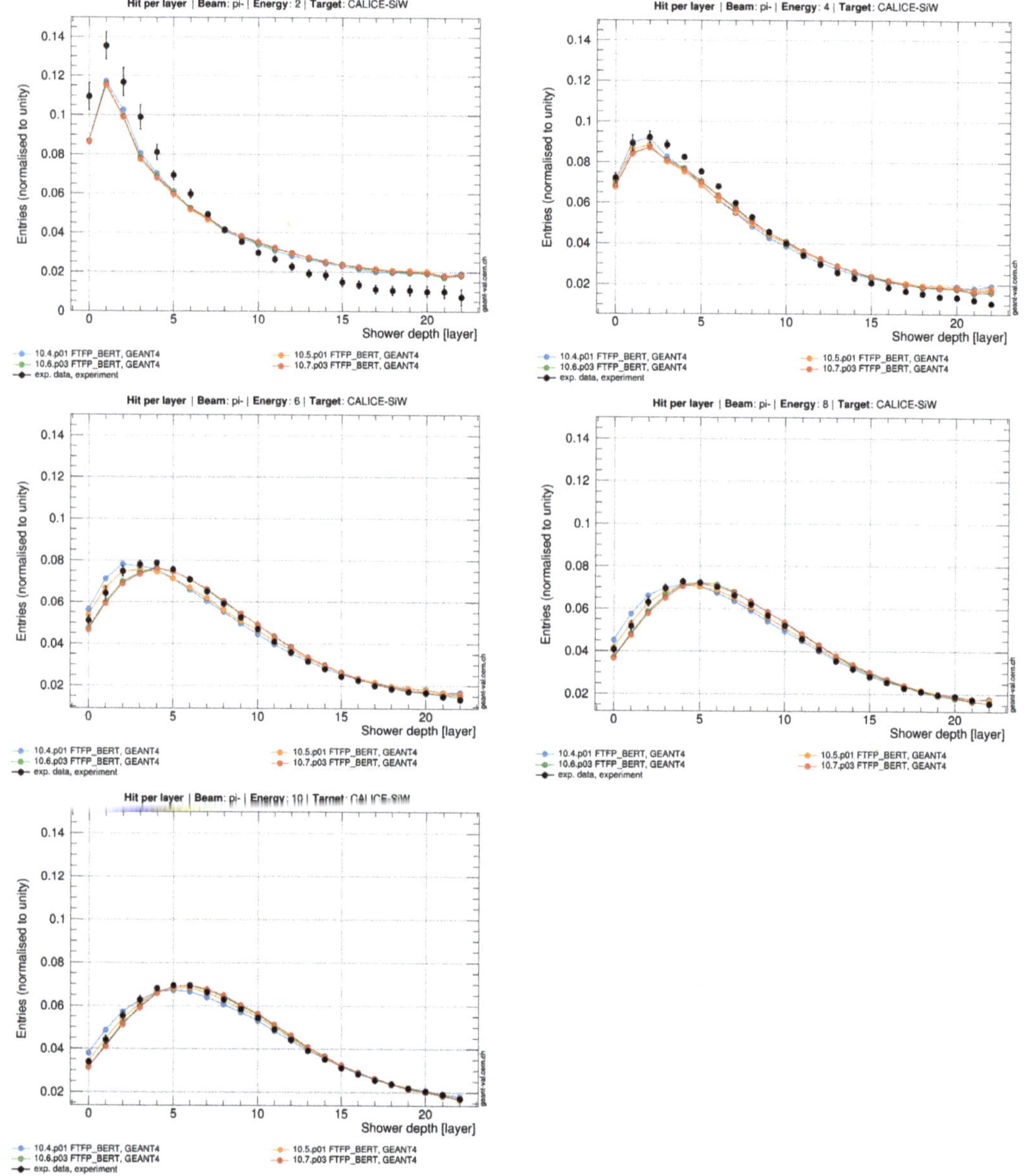

Figure 8. Longitudinal hit profiles for 2 (**top-left**), 4 (**top-right**), 6 (**center-left**), 8 (**center-right**), and 10 (**bottom-left**) GeV π^- events. Results from experimental measurements and Geant4 simulation with the FTFP_BERT PL and releases from 2017 to 2020.

5. Conclusions

The Geant4 simulation toolkit is widely adopted by nuclear and particle physics experiments for calorimetric simulation. The success of the Geant4 project stands on the continuous validation of its physics models as well as on the R&D targeting faster and better computing solutions. In the last years, most of the hadronic model development has been based on thin target experiments, with typical absorber dimensions of a few millimeters. This allowed for the precise modeling of nuclear interactions at fixed energy and projectile type. The possibility to achieve wider, more general validation is offered by test beam results on calorimeters, in which several processes and different particle types with wide energy ranges are involved. Such validation tests require strong collaborations between Geant4 developers and experimental groups in order to design accurate and realistic simulations. In this respect, both calorimeters from the past and prototypes designed for future experiments are equally suitable.

We showed an example from the first case, the ATLAS HEC beam test, and the latter case, the CALICE SiW calorimeter beam test. The HEC results clearly indicate a trend in the simulated response to pions in copper-based calorimeters towards higher values. They also provide a good example of the solution adopted by the ATLAS experiment, i.e., the FTFP_BERT_ATL PL, in which the Monte Carlo-to-data agreement in the calorimeter response simulation is largely improved with respect to the FTFP_BERT PL. Results from the HEC also show the great capability of Geant4 to reproduce the correct hadronic shower shape in terms of shower length and barycenter location in calorimeters with the typical segmentation of a few longitudinal layers as the ones currently adopted by the main LHC experiments.

The CALICE SiW calorimeter stands among the prototypes with the highest granularity envisaged for the post-LHC era. When tested with π^- showers, it provides extremely valuable information on the first stage of the hadronic shower development. Its results on longitudinal distributions have been compared with recent versions of the FTFP_BERT PL, observing a good improvement in the MC-to-data comparison for the visible energy depositions following a nuclear breakup in the 2–10 GeV energy range. It also shows how the INCL model provides a good alternative to the FTFP_BERT PL for the highest energy considered in that test beam.

We strongly believe the validation of Geant4 to be a collective effort shared among diverse experiments, and the most valuable drivers of physics model changes to be the ones arising from different sources of inputs. In this respect, the geant-val project represents the best effort to preserve experimental inputs, compare simulated results, and distribute the information to the broadest community possible. On top of that, geant-val offers solutions for standardized job preparation and submission on batch systems that help the Geant4 validators to perform extremely time- and CPU-consuming validation campaigns.

Author Contributions: Conceptualization, L.P., A.K., D.K., A.R. and P.S.; methodology, L.P.; software, L.P.; validation, L.P.; formal analysis, L.P.; investigation, L.P.; resources, L.P.; data curation, L.P.; writing—original draft preparation, L.P.; writing—review and editing, L.P.; visualization, L.P.; supervision, L.P., A.K., D.K., A.R. and P.S.; project administration, A.R. and the Geant4 Collaboration. All authors have read and agreed to the published version of the manuscript.

Funding: This research received no external funding.

Data Availability Statement: All results accomplished with the validation study described are available at the geant-val website (https://geant-val.cern.ch/, (accessed on 29 August 2022)).

Acknowledgments: We would like to thank Fabrizio Salvatore and the Organizing Committee of the beautiful CALOR2022 Conference.

Conflicts of Interest: The authors declare no conflict of interest.

References

1. Agostinelli, S.; Allison, J.; Amako, K.; Apostolakis, J.; Araujo, H.; Arce, P.; Asai, M.; Axen, D.; Banerjee, S.; Barrand, G.; et al. Geant4—A simulation toolkit. *Nucl. Instrum. Methods Phys. Res. A* **2003**, *506*, 250–303.
2. Allison, J.; Amako, K.; Apostolakis, J.; Araujo, H.; Arce Dubois, P.; Asai, M.; Barrand, G.; Capra, R.; Chauvie, S.; Chytracek, R.; et al. Geant4 Developments and Applications. *IEEE Trans. Nucl. Sci.* **2006**, *53*, 270–278. [CrossRef]
3. Allison, J.; Amako, K.; Apostolakis, J.; Arce, P.; Asai, M.; Aso, T.; Bagli, E.; Bagulya, A.; Banerjee, S.; Barrand, G.; et al. Recent developments in GEANT4. *Nucl. Instrum. Methods Phys. Res. A* **2016**, *835*, 186–225. [CrossRef]
4. Calafiura, P.; Catmore, J.; Costanzo, D.; Di Girolamo, A. *ATLAS HL-LHC Computing Conceptual Design Report*; CERN-LHCC-2020-015; LHCC-G-178; CERN: Geneva, Switzerland, 2020.
5. Freyermuth, L.; Konstantinov, D.; Latyshev, G.; Razumov, I.; Pokorski, W.; Ribon, A. Geant-val: A web application for validation of detector simulations. *EPJ Web Conf.* **2019**, *214*, 05002. [CrossRef]
6. Gingrich, D.M.; Lachat, G.; Pinfold, J.; Soukup, J.; Axen, D.; Cojocaru, C.; Oakham, G.; O'Neill, M.; Vincter, M.G.; Aleksa, M.; et al. Construction, assembly and testing of the ATLAS hadronic end-cap calorimeter. *J. Instrum.* **2007**, *2*, P05005. [CrossRef]
7. Dowler, B.; Pinfold, J.; Soukup, J.; Vincter, M.; Cheplakov, A.; Datskov, V.; Fedorov, A.; Javadov, N.; Kalinnikov, V.; Kakurin, S.; et al. Performance of the ATLAS hadronic end-cap calorimeter in beam tests. *Nucl. Instrum. Methods Phys. Res. Sect. A* **2002**, *482*, 94–124. [CrossRef]
8. Kiryunin, A.E.; on behalf of the ATLAS Liquid Argon HEC Collaboration. Performance of the ATLAS Hadronic End-Cap Calorimeter in Beam Tests. In Proceedings of the Tenth International Conference on Calorimetry in Particle Physics (CALOR 2002), Pasadena, CA, USA, 25–29 March 2002; p. 720.
9. Kiryunin, A.E.; Oberlack, H.; Salihagić, D.; Schacht, P.; Strizenec, P. GEANT4 physics evaluation with testbeam data of the ATLAS hadronic end-cap calorimeter. *Nucl. Instrum. Meth. A* **2006**, *560*, 278–290. [CrossRef]
10. Kiryunin, A.E.; Salihagić, D. Hadronic Shower Validation Experience for the ATLAS End-Cap Calorimeter. *AIP Conf. Proc.* **2007**, *896*, 205–214.
11. Kiryunin, A.E.; Oberlack, H.; Salihagić, D.; Schacht, P.; Strizenec, P. GEANT4 physics evaluation with testbeam data of the ATLAS hadronic end-cap calorimeter. *J. Phys. Conf. Ser.* **2009**, *160*, 012075. [CrossRef]
12. Dowler, B.; Pinfold, J.; Soukup, J.; Vincter, M.; Cheplakov, A.; Datskov, V.; Fedorov, A.; Javadov, N.; Kalinnikov, V.; Kakurin, S.; et al. Performance of the ATLAS Hadronic Endcap Calorimeter in Beam Tests: Selected Results. *ATL-LARG-PUB-2022-001*. June 2022. Available online: https://cds.cern.ch/record/2811731 (accessed on 29 August 2022).
13. The ATLAS Collaboration. Photos from MPI: Module Installation at CERN for 1999 Test Beam. *ATL-PHO-LARG-2001-013*. 1999. Available online: https://cds.cern.ch/record/42434 (accessed on 29 August 2022).
14. Heikkinen, A.; Stepanov, N.; Wellisch, J.P. Bertini intra-nuclear cascade implementation in Geant4. *arXiv* **2003**, arXiv:nucl-th/0306008v1.
15. Folger, G.; Wellisch, J.P. String Parton Models in Geant4. *arXiv* **2003**, arXiv:nucl-th/0306007.
16. The CALICE Collabration; Repond, J.; Yu, J.; Hawkes, C.M.; Mikami, Y.; Miller, O.; Watson, N.K.; Wilson, J.A.; Mavromanolakis, G.; Thomson, M.A.; et al. Design and electronics commissioning of the physics prototype of a Si-W electromagnetic calorimeter for the International Linear Collider. *J. Instrum.* **2008**, *3*, P08001.
17. Bilki, B.; Repond, J.; Schlereth, J.; Xia, L.; Deng, Z.; Li, Y.; Wang, Y.; Yue, Q.; Yang, Z.; Eigen, G.; et al. Testing hadronic interaction models using a highly granular silicon–tungsten calorimeter. *Nucl. Instrum. Methods Phys. Res. A* **2015**, *794*, 240–254. [CrossRef]

instruments

Article

The (Un)reasonable Effectiveness of Neural Network in Cherenkov Calorimetry [†]

Nural Akchurin, Christopher Cowden, Jordan Damgov, Adil Hussain and Shuichi Kunori *

Advanced Particle Detector Laboratory, Department of Physics and Astronomy, Texas Tech University, Lubbock, TX 79409, USA
* Correspondence: shuichi.kunori@ttu.edu
† Inspired by E.P Wigner, 1960.

Abstract: We report a greater than factor of two improvement in the hadronic energy resolution of a simulated Cherenkov calorimeter by estimating the energy with machine learning over traditional techniques. The prompt signal formation and energy threshold properties of Cherenkov radiation provide identifiable features that machine learning techniques can exploit to produce a superior model for energy reconstruction. We simulated a quartz-fiber calorimeter via the GEANT4 framework to study the reconstruction techniques in single events. We compared the machine learning-based reconstruction performance to the traditional simple sum of signal and dual-readout techniques that use both Cherenkov and scintillation signals. We describe why this game-changing approach to Cherenkov hadron calorimetry excels and our plans for a dedicated beam test to validate these findings with a fast, radiation-hard hadron calorimeter prototype.

Keywords: calorimetry; Cherenkov calorimeter; high-granularity; neural network; GNN; CNN

Citation: Akchurin, N.; Cowden, C.; Damgov, J.; Hussain, A.; Kunori, S. The (Un)reasonable Effectiveness of Neural Network in Cherenkov Calorimetry. *Instruments* **2022**, *6*, 43. https://doi.org/10.3390/instruments6040043

Academic Editors: Fabrizio Salvatore, Alessandro Cerri, Antonella De Santo and Iacopo Vivarelli

Received: 16 August 2022
Accepted: 9 September 2022
Published: 20 September 2022

Publisher's Note: MDPI stays neutral with regard to jurisdictional claims in published maps and institutional affiliations.

1. Introduction

The response of Cherenkov calorimeters to hadrons is vastly different compared to the response to electrons. It displays strong non-linearity and rather poor energy resolution for hadrons. For instance, the CMS Hadronic Forward (HF) calorimeter is comprised of fused-silica fibers embedded in a steel absorber [1]. It covers the forward region ($3.0 \leq \eta \leq 5.2$) where the radiation levels are measured in hundreds of Megarads. As expected, the fused-silica fibers have shown good radiation tolerance, and fast Cherenkov signals make the energy reconstruction free from signal pileup at the LHC. The HF has been working well for the measurement of tagging jets from pp $\rightarrow$ qqH, where typically $p_T^{\mathrm{jet}} \approx 50$ GeV, corresponding to $E \approx 500$ GeV at $\eta \approx 3.0$. On the other hand, lower typical jet energies in the barrel and end-cap regions, well below 100 GeV, render the use of Cherenkov calorimetry challenging outside the forward region. We explore ways of overcoming this limitation for future experiments in this paper.

We used a Monte Carlo (MC) simulation to study the performance of a neural network (NN) to reconstruct the energy of hadronic showers in a highly granular sampling ionization calorimeter [2]. In hadron-nucleus interactions, a large fraction of hadron energy goes into nuclear dissociation and becomes invisible. In the MC study, we viewed each hadronic shower as a 3D image with a $2 \times 2 \times 2$ cm^3 pixel resolution and let the Convolutional Neural Network (CNN) estimate the invisible energy based on the features of the "image". The CNN was trained with single charged pions of 0.5–150 GeV. The CNN trained in this way was able to estimate the invisible energy and restore the response linearity to single pions, electrons (0.5–150 GeV), and u-quark jets (20–1000 GeV), and it surpassed the hadronic resolution attainable with a comparable dual-readout (DR) calorimeter [3]. We used only the fast part of the signal ($\leq$5 ns) to create these images that are mostly produced by relativistic particles and the source of Cherenkov signals.

In the following, we discuss salient characteristics of hadronic showers based on a simple MC calorimeter setup (Section 2) and the performance of a NN in reconstructing the hadronic energy using Cherenkov signals in a highly granular fiber calorimeter (Sections 3 and 4).

2. Characteristics of Hadronic Showers

Using Geant4 [4], we simulated hadronic showers in both a solid copper absorber and a sampling calorimeter configuration with alternating plates of 17 mm copper absorber and 3 mm active material. We used the FTFP_BERT physics list to describe hadronic interactions. The calorimeter was 1.5 m deep (8.8 interaction lengths) and 1×1 m^2 in the transverse plane. In the sampling calorimeter configuration, two active materials were chosen: silicon plates to generate the ionization and quartz plates to generate the Cherenkov signals. Neither the details of charge collection in the silicon plates and light collection in the quartz plates nor the details of the readout electronics or photo-detectors were simulated.

2.1. Time Structure

The time structure of the ionization signals in silicon in the sampling configuration is shown in Figure 1. Here, the time is defined as the local calorimeter time, $t = t_{G4} - z/c$, which is corrected for the travel time of all particles along the z-axis. Three significant features emerge: (1) a very fast signal ($t \leq 1$ ns) due to pure electromagnetic shower and prompt pions, (2) a semi-fast signal ($t \leq 5$ ns) due to protons from the intra-nuclear cascade process [5], and (3) a slow component ($t \leq 100$ ns and beyond) due to electrons produced by photons from neutron capture in Cu/Si. As silicon plates do not contain free protons, no slow proton signal from the neutron-proton scattering is observed. We used very fast signals ($t \leq 1$ ns) in this part of the analysis.

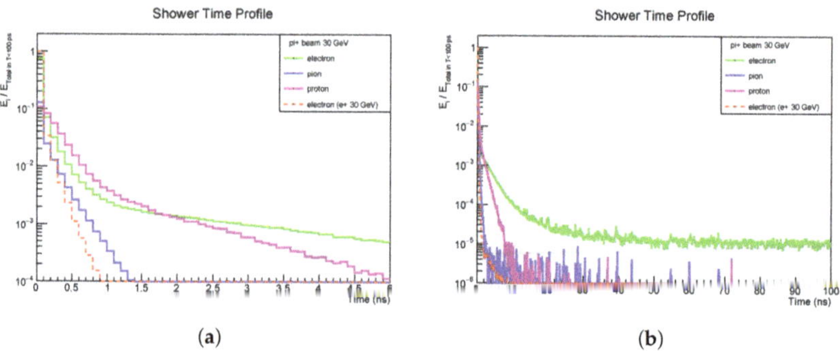

Figure 1. Time structure of hadronic showers in a sampling calorimeter composed of 17 mm Cu absorber with 3 mm silicon layer as active material. The time is the calorimeter local time, $t = t_{G4} - z/c$, which is corrected for the travel time along the z-axis for all particles. (**a**) ($t < 5$ ns) is a zoomed-in view of (**b**) ($t < 100$ ns).

2.2. Images of Hadronic Showers

The spatial distributions of particles in the solid copper absorber from a single 30 GeV π^+ are shown in Figure 2. Pions and protons display a clear vertex-track structure. Positrons are produced in e^+e^- pairs following the gamma emissions from $\pi^0 \to \gamma\gamma$ decays that are associated with the hadronic vertices. These pions, protons, and positrons make up the fast and semi-fast components of the shower. Neutrons and gammas are slow components and spread widely in the calorimeter. Electrons also spread widely. Some electrons are partners of positrons from π^0 decays, and many others are from the Compton process of the widespread gammas. As seen in these images, the fast components of hadronic showers provide a distinctive vertex-track structure for the network to utilize in improved energy reconstruction.

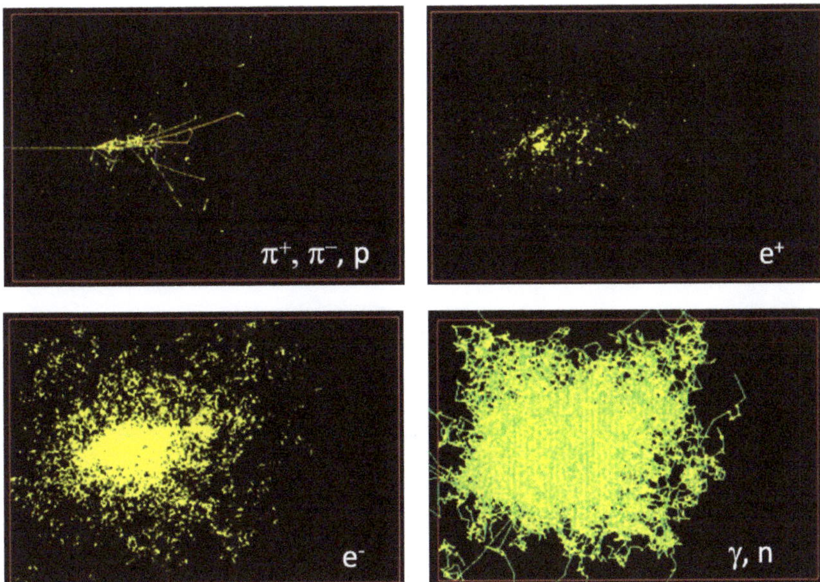

Figure 2. Images of shower particles for a single 30 GeV π^+ in a solid copper absorber ($100 \times 100 \times 150$ cm^3): Slow neutrons and γs from neutron capture by nuclei spread widely and form a fuzzy image (**bottom right**), while fast π^+, π^-, and protons display a clear view of hadronic vertices and tracks (**top left**). The e^+ (**top right**) and e^- (**bottom left**) are shown separately. The e^+s arise mainly from the fast component of the shower in e^+e^- pairs by gammas from $\pi^0 \rightarrow \gamma\gamma$ decays, whereas the e^-s also arise from the slow component due to the Compton scattering of widely spread γs in addition to counterparts of positions in the fast component.

2.3. Loss of Kinetic Energy in Hadron-Nucleus Interactions

Substantial kinetic energy is lost in hadron-nucleus interactions. As shown in Figure 3, the average energy loss is 2 or 4 GeV in π^+Cu interaction for 30 and 200 GeV incident pions, respectively. The multiplicity of secondary hadrons from hadron-nucleus interactions correlates with the lost kinetic energy (Figure 3c). The CNN likely recognizes this correlation at each hadronic vertex.

2.4. Invisible Energy in Hadronic Showers

The lost kinetic energy becomes invisible and fluctuates event by event. Figure 4 shows the strong correlation between invisible energy and the number of hadronic vertices produced in hadron-nucleus inelastic interactions in the form of fast ionization (<5 ns) and Cherenkov signal in silicon and quartz plates. Thus, the invisible energy can be estimated with good precision if we can count the number of vertices. The "image" of a vertex is easily distinguishable: a point with outgoing tracks. We use Dynamic Graph CNN (GNN) [6] to estimate the invisible energy in the energy reconstruction.

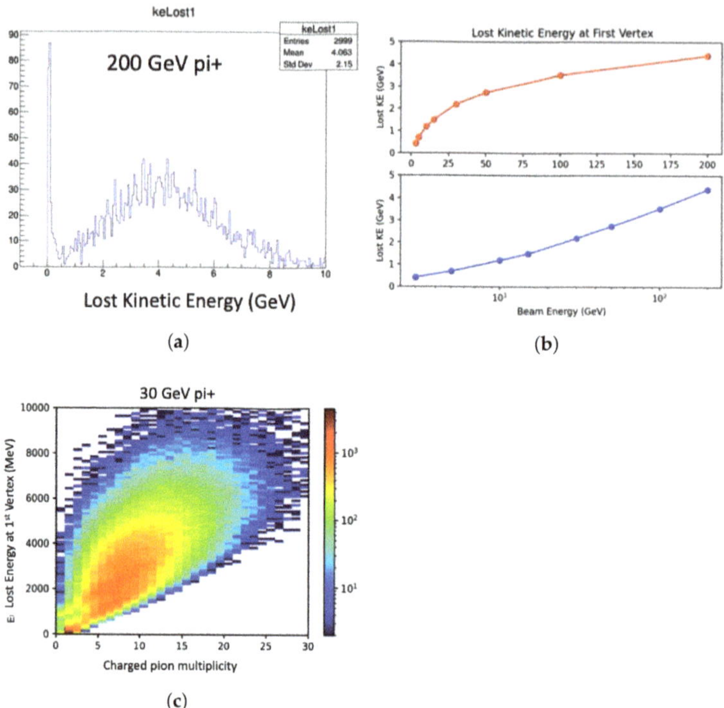

Figure 3. Loss of kinetic energy in hadron–nucleus interactions: (**a**) in the 200 GeV π^++Cu interaction, (**b**) average energy loss in π^++Cu interactions as a function of π^+ beam energy, shown in linear and log scales, and (**c**) correlation between the kinetic energy loss and the charged pion multiplicity in the 30 GeV π^++Cu interactions [2].

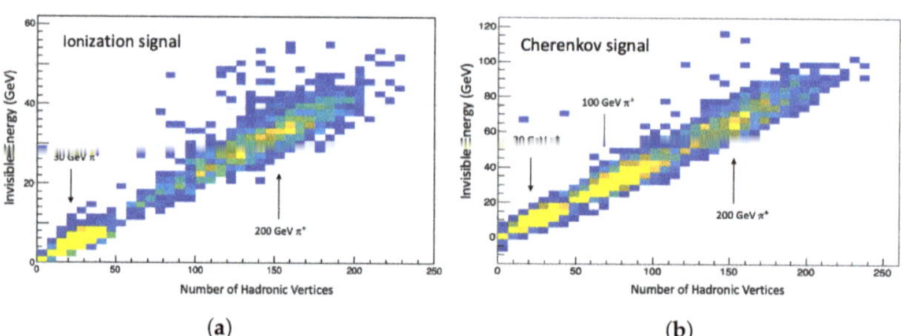

Figure 4. Invisible energy *vs* the number of hadronic vertices in a sampling calorimeter: (**a**) with ionization signal in Cu (17 mm) + Si (3 mm) and (**b**) with a Cherenkov signal in Cu (17 mm) + Quartz plate (3 mm) for 30 and 200 GeV π^+s. "Invisible energy" is defined as the difference between the beam energy and the simple sum of the ionization signal or Cherenkov signal. "Hadronic vertex" is defined as a vertex of hadron–nucleus inelastic interaction excluding neutron–nucleus interaction. The energy scale of the signals was calibrated with electrons.

3. Analysis and Results Using GNN

We simulated a dual-readout fiber calorimeter (using GEANT4 with the FTFP-BERT physics list). We reconstructed the energy of hadrons using scintillation and Cherenkov signals by a simple sum of signals and the application of a GNN. The DR technique [3] was applied and evaluated to form a benchmark as well. We used Dynamic Graph CNN

(GNN) [6], as in our previous analysis [2]. Its new neural network module (EdgeConv) incorporates local neighborhood information because it can be stacked or recurrently applied to learn global shape properties. The GNN was trained to reproduce the incident particle energy with a large set of simulated pions in the range of 0.5–150 GeV: training (700,000), validation (100,000), and testing (300,000).

Cherenkov and scintillation fibers (1-mm diameter) were placed in a copper absorber (2-m long) with center-to-center fiber spacing of 1.5 mm. The simulated light yields were set at 100 and 400 ph/GeV for the Cherenkov and scintillation photons, respectively. Signals from multiple fibers were summed to form a transversely segmented structure. In the case of 2D, the segmentation was 1×1 cm^2 in the transverse plane. In the case of 3D, we segmented the calorimeter into 3×3 cm^2 in the transverse plane and utilized the time of arrival of Cherenkov photons to measure the z-position along fibers. The binning was 50 ps (about 2.5 cm), and the signal was smeared by a Gaussian distribution to evaluate the effect of timing resolution. The response of the photo-detectors and readout electronics were not simulated.

As shown in Figure 5, the 2D segmented calorimeter response reconstructed by GNN is constant within 2% except for ~10% deviation at the lowest energy of 4 GeV.

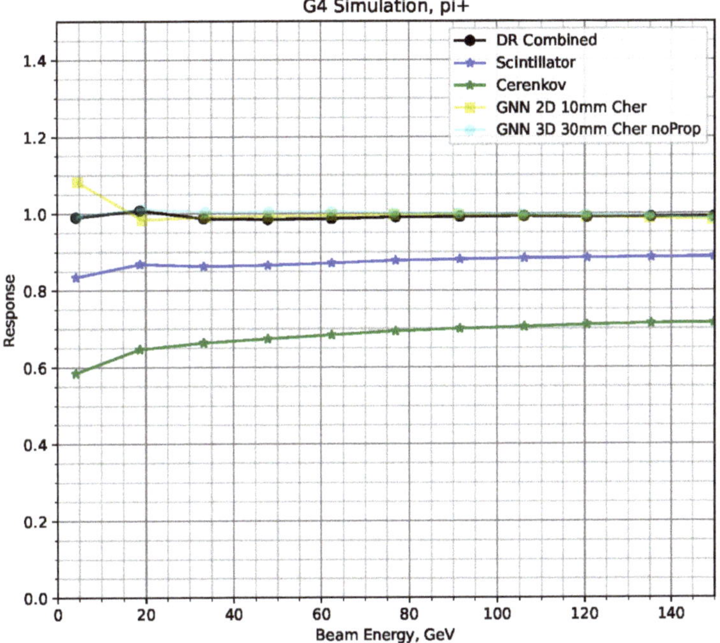

Figure 5. Response of the calorimeter to 4–150 GeV π^+: (green) the simple sum of the Cherenkov signal, (blue) simple sum of scintillation signal, (black) dual-readout method, (yellow) GNN 2D reconstruction, and (light blue) GNN 3D reconstruction. In all cases, the calorimeter is calibrated with electrons and the signal distributions are fitted by a Gaussian distribution.

The energy resolutions obtained by the simple sum, GNN, and the dual-readout (DR) technique are shown in Figure 6. The resolution by the DR method is shown as a reference in both plots in Figure 6. In the scintillation case, the 2D GNN improves the resolution from 5% to 3% and surpasses the resolution by the DR method. In the Cherenkov case, the GNN methods improve the resolution from 13% to 5% with 2D and to 4% with 3D. The resolution by the 3D reconstruction is comparable to the DR resolution and better in the energy range below 20 GeV.

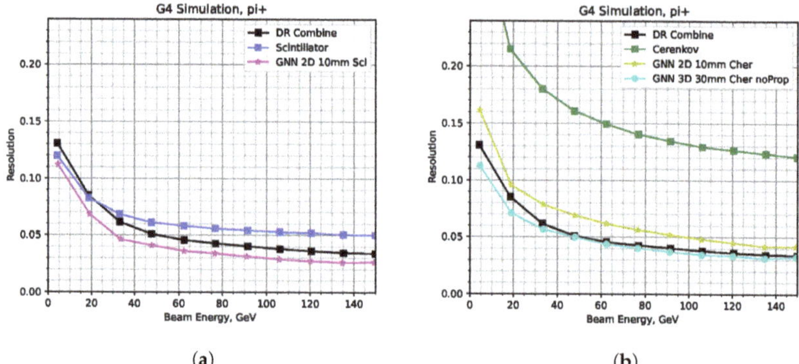

Figure 6. Energy resolution of various reconstruction methods for 4–150 GeV π^+ beams: (**a**) simple sum of signal and GNN reconstruction using only the scintillation and (**b**) only the Cherenkov signal. The resolution by the DR technique is shown in both plots as a reference.

4. Discussion

As with our previous work [2], the ideas presented here focused on the fast components in hadronic showers in a highly segmented calorimeter. In the approach outlined here, there is neither the traditional compensation ($e/h = 1$) mechanism with slow neutrons nor the event-by-event evaluation of f_{em} *a lá* DR-based. Although further studies are clearly needed, we posit that the network effectively takes advantage of the distinctive features of the shower images, such as hadronic vertices, that result in unusually good energy resolution. The strong correlation between the invisible energy and the number of hadronic vertices observed gives us a basis to form this conjecture.

In a hadron-nucleus interaction, the total momentum is conserved. If the interaction produces several protons and neutrons in an inter-cascade process, the initial momentum gets split among protons and neutrons in addition to the prompt pions. For example, when $p = 1$ GeV is imparted to a proton, it has a kinetic energy of 0.43 GeV, which is available to create ionization or a Cherenkov signal, while a large fraction of the initial momentum is taken away by the proton mass. The number of inelastic hadronic vertices reflects the number of protons and neutrons produced in the shower and may indicate the invisible energy due to the mass effect in the shower. The GNN is capable of recognizing the vertices in the 3D view of the shower.

4.1. A Simple Check of 2D Cherenkov Reconstruction

The 2D GNN reconstruction improved the energy resolution by more than a factor of two from the simple-sum method. To check if the 2D image of a Cherenkov signal has enough information to estimate the invisible energy and restore the beam energy, we scanned the 2D images of the Cherenkov signal and found that more activity in the 2D image implies more invisible energy. A couple of examples are shown in Figure 7. The trend is summarized in Figure 8. It shows a clear correlation between the number of hits in the 2D area and the invisible energy. This correlation does not depend on the incident beam energy. We believe that the GNN can easily utilize this information (simple hit counting) to estimate the invisible energy and restore the beam energy. It may use more complex information, such as the amplitude of each hit and the correlation of hits in the 2D space, to further improve the invisible energy estimation.

In the case of a 3D shower image, GNN may use its superior image recognition capability to improve the invisible energy estimation over the 2D reconstruction. The 2D images may be easily saturated by dense multi-particles in a jet, while 3D images are more tolerant to such saturation effects.

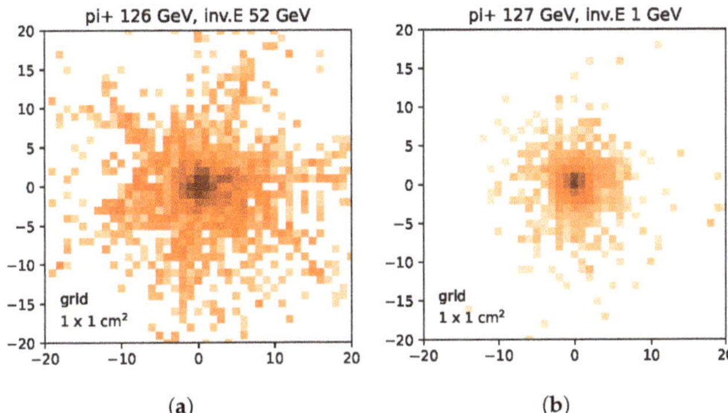

Figure 7. A 2D (xy) heat map of a Cherenkov signal in a 40×40 cm^2 area: the grid size (1×1 cm^2) corresponds to the transverse segmentation of the fiber calorimeter and the color scheme indicates the signal amplitude in $log(E)$. (**a**) An example of a large invisible energy event (beam energy 126 GeV and invisible energy 53 GeV) and (**b**) an example of a small invisible energy event (beam energy 127 GeV and invisible energy 1 GeV).

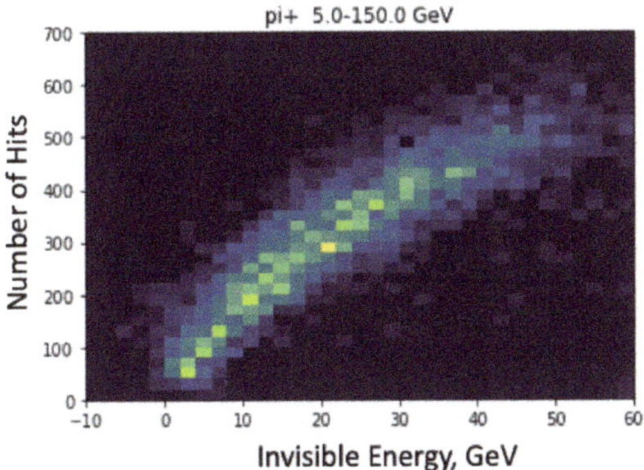

Figure 8. The correlation between the number of hits in the 2D map and the invisible energy is calculated with the energy sum of a Cerenkov signal.

4.2. Longitudinal Segmentation by Timing

In a fiber calorimeter, the shower particles and Cherenkov photons travel approximately in the direction of the incoming particle but at different speeds. The arrival time of the photons at the downstream end of the calorimeter can be expressed as $t = L1/c + L2/kc$, where c = speed of light and kc = speed of light in fiber ($k \sim 0.6$) and $L1$ and $L2$ as shown in Figure 9. Thus, a 2D fiber calorimeter can be turned into a 3D segmented one. The main advantages of such a device are (1) fewer channel counts than with a fully 3D segmented device, (2) protection of photo-detectors and readout electronics from radiation as they can be located behind the absorber, and (3) easier calibration without the need for depth calibration. Of course, the resolution of the timing measurement determines the effective longitudinal segmentation. The performance of the 3D GNN with various timing resolutions is summarized in Table 1. The energy resolution with $\Delta(t) = 150$ ps matches the resolution by the DR method, Figure 6.

Shower particles may hit the same fibers at different depths. A readout system (photo-detectors and readout electronics) requires the multi-hit capability for this kind of longitudinal segmentation. Silicon Photomultipliers (SiPMs) show excellent timing resolution for single hits, but there is currently no straightforward capability for analyzing the time structure of signals. This technology requires further R&D.

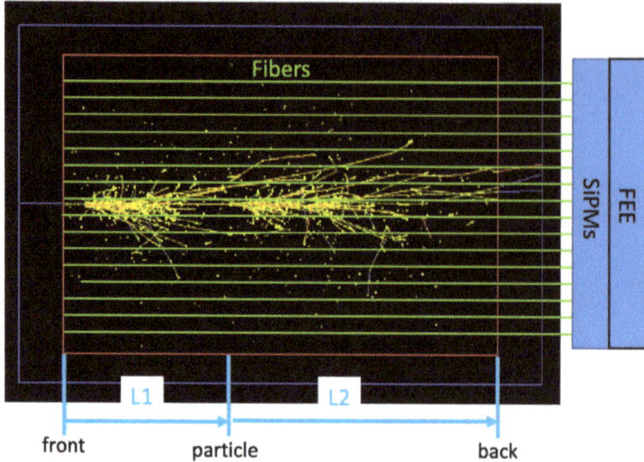

Figure 9. Schematic view of a Cherenkov fiber calorimeter using timing for the longitudinal segmentation.

Table 1. The energy resolution of the 3D GNN reconstruction with various timing resolutions for longitudinal segmentation.

Timing Resolution $\Delta(t)$, ps	Position Resolution $\Delta(z)$, cm	Energy Resolution σ/E, %
0	0.0	3.6
100	5.0	3.9
150	7.5	4.0
200	10.0	4.2

4.3. Verification of GNN Cherenkov Calorimetry

Our Monte Carlo study shows the effectiveness of neural networks in reconstructing the energy of incident hadrons from highly granular Cherenkov calorimeters. Although the result is impressive, it is based purely on a Monte Carlo simulation and needs to be verified with a prototype calorimeter in a real beam.

The effectiveness already appeared in the 2D reconstruction. We plan to modify the transverse segmentation of an existing fiber calorimeter prototype module to test the 2D reconstruction without timing information. Once we confirm the effectiveness, we move to 3D tests and more detailed simulations to optimize calorimeter designs and NN techniques for use in future experiments.

5. Conclusions

High granularity Cherenkov calorimeters combined with NN technology are poised to provide excellent performance in future high-energy experiments. The results presented here are unusually impressive and call for more detailed and systematic simulations complemented by data from a calorimeter prototype in a beam test. The combination of high granularity and powerful networks enables us to look into the mechanism of energy loss in hadron-nucleus interactions beyond the traditional views of hadronic showers. An improved understanding of the interplay between shower images and the commensurate interpretation of energy loss mechanisms will help us develop new detectors and algorithms for precision energy measurements.

Longitudinal segmentation by the timing of the Cherenkov photons will be a cost-efficient approach toward a highly segmented calorimeter and will naturally lend itself to energy reconstruction using NNs. It is clear that high-performance readout electronics and fast photo sensors are needed for this purpose.

Author Contributions: Conceptualization, methodology, N.A., C.C., J.D., A.H. and S.K.; Software, C.C., J.D. and S.K.; Validation, N.A. and C.C.; Formal Analysis, J.D., A.H. and S.K.; Data Curation, J.D. and A.H.; writing—original draft preparation, J.D., A.H. and S.K.; writing—review and editing, N.A. and C.C.; Visualization, A.H.; Project administration S.K.; Funding acquisition, N.A. All authors have read and agreed to the published version of the manuscript.

Funding: This work has been supported by the US Department of Energy, Office of Science (DE-SC0015592) and Texas Tech University, Office of the Vice President for Research and Innovation.

Data Availability Statement: Not applicable.

Acknowledgments: The authors acknowledge the High Performance Computing Center (HPCC) at Texas Tech University for providing computational resources that have contributed to the research results reported within this paper. http://www.hpcc.ttu.edu (accessed on 15 August 2022).

Conflicts of Interest: The authors declare no conflict of interest.

References

1. Abdullin, S.; Abramov, V.; Acharya, B.; Adams, M.; Akchurin, N.; Akgun, U.; Anderson, E.W.; Antchev, G.; Arcidy, M.; Ayan, S.; et al. Design, performance, and calibration of CMS forward calorimeter wedges. *Eur. Phys. J. C* **2008**, *53*, 139. [CrossRef]
2. Akchurin, N.; Cowden, C.; Damgov, J.; Hussain, A.; Kunori, S. On the Use of Neural Networks for Energy Reconstruction in High-granularity Calorimeters. *JINST* **2021**, *16*, P12036. [CrossRef]
3. Akchurin, N.; Carrell, K.; Hauptman, J.; Kim, H.; Paar, H.P.; Penzo, A.; Thomas, R.; Wigmans, R. Hadron and jet detection with a dual-readout calorimeter. *Nucl. Instr. Meth. A* **2005**, *537*, 537. [CrossRef]
4. GEANT4 Collaboration, GEANT4—A simulation toolkit. *Nucl. Instrum. Meth. A* **2003**, *506*, 250. [CrossRef]
5. Ribon, A. Hadronic Physics. Geant4 Tutorial. In Proceedings of the CERN, 22–23 January 2019. Available online: https://indico.cern.ch/event/781244/ (accessed on 15 August 2022). [CrossRef]
6. Wang, Y.; Sun, Y.; Liu, Z.; Sarma, S.E.; Bronstein, M.M.; Solomon, J.M. Dynamic Graph CNN for Learning on Point Clouds. *arXiv* **2018**, arXiv:1801.07829. [CrossRef]

Article

Energy Resolution Studies in Simulation for the IDEA Dual-Readout Calorimeter Prototype

Andreas Loeschcke Centeno on behalf of the IDEA Dual-Readout Collaboration

Department of Physics and Astronomy, University of Sussex, Brighton BN1 9RH, UK; a.loeschcke-centeno@sussex.ac.uk

Abstract: Precision measurements of Z, W, and H decays at the next generation of circular lepton colliders will require excellent energy resolution for both electromagnetic and hadronic showers. The resolution is limited by event-to-event fluctuations in the shower development, especially in the hadronic system. Compensating for this effect can greatly improve the achievable energy resolution. Furthermore, the resolution can benefit greatly from the use of particle-flow algorithms, which requires the calorimeters to have a high granularity. The approach of dual-readout calorimetry has emerged as a candidate to fulfil both of these requirements by allowing to reconstruct the fluctuations in the shower development event-by-event and offering a high transverse granularity. An important benchmark of such a calorimeter is the electromagnetic energy resolution; a prototype of the IDEA calorimeter has been built for use in testbeams. In parallel, a simulation of this prototype has been developed in Geant4 for a testbeam environment. Here, we outline how this simulation was used to study the electromagnetic energy resolution and conclude that a resolution of $14\%/\sqrt{E}$ is achievable.

Keywords: dual-readout calorimetry; electromagnetic showers; Cherenkov light; optical fibres

Citation: Loeschcke Centeno, A., on behalf of the IDEA Dual-Readout Collaboration. Energy Resolution Studies in Simulation for the IDEA Dual-Readout Calorimeter Prototype. *Instruments* **2022**, *6*, 44. https://doi.org/10.3390/instruments6040044

Academic Editor: Antonio Ereditato

Received: 18 August 2022
Accepted: 5 September 2022
Published: 20 September 2022

Publisher's Note: MDPI stays neutral with regard to jurisdictional claims in published maps and institutional affiliations.

1. Introduction

The main purpose of a dual-readout calorimeter is to significantly improve the resolution of the energy measurement for hadronic showers. This is obtained thanks to an event-by-event determination of the electromagnetic fraction of the shower.

The energy resolution of hadronic calorimeters is, in fact, in part, limited by fluctuations in the electromagnetic (EM) component from neutral pion decays in the induced particle shower [1]. Generally, the EM response (e) and non-EM response (h) of the calorimeter—where response is defined as the conversion efficiency from energy deposit to generated signal—are different ($e/h \neq 1$) mainly due to *invisible* energy (mostly energy that is lost to binding energy in nuclear reactions). In this case, the hadronic energy resolution is dominated by the fluctuation in the fraction of energy deposited in the EM component f_{EM} [2]. There are calorimeters, called *compensating* calorimeters, which counteract this effect by achieving $e/h = 1$, for instance the ZEUS barrel calorimeter [3]. However, in the past 25 years, a new approach has been developed to mitigate the effect of fluctuations in f_{EM}. In *dual-readout* calorimetry, two independent readout channels are used, namely a scintillation and a Cherenkov channel, which deliver complementary information about the EM and non-EM shower development. This allows us to measure f_{EM} event-by-event and correct for it when reconstructing the total shower energy, which can be calculated from the Cherenkov and scintillation signals with the dual-readout formula outlined in [2] for which the calorimeter specific value e/h needs to be known. The combination of the two channels yields a superior result in the energy resolution than an equivalent single-readout calorimeter.

For more information on dual-readout calorimetry, please see [1,2].

2. Prototype for Testbeams

A prototype of the IDEA dual-readout calorimeter [4], shown in Figure 1a, has been built for use in testbeams to study the EM energy resolution performance of this calorimeter. The dimensions of this prototype are $10 \times 10 \times 100\,\text{cm}^3$, sufficient to contain an EM shower with incoming energies up to 100 GeV better than 94%. This prototype also serves as a testing ground for the "Bucatini" layout structure, where the passive material is shaped in the form of capillary brass tubes housing the fibres, which are read out in the back. In this structure, the fibres are assembled in alternating rows, as shown in Figure 1b. The prototype is subdivided into 3×3 towers, each containing 320 fibres, half of which are Saint Gobain BC-10 scintillation fibres and the other half Mitsubishi SK-40 Cherenkov fibres. In the central tower, the fibres are connected to one Hamamatsu S14160-1315 PS Silicon Photomultiplier (SiPM) each. Those are then read out by five CAEN A5202 readout boards. For the surrounding eight towers, all fibres of a respective channel are bundled and connected to one Hamamatsu R-5900 Photomultiplier (PMT), resulting in a total of 16 PMTs. These connections can be seen on the backside of the prototype in Figure 1a.

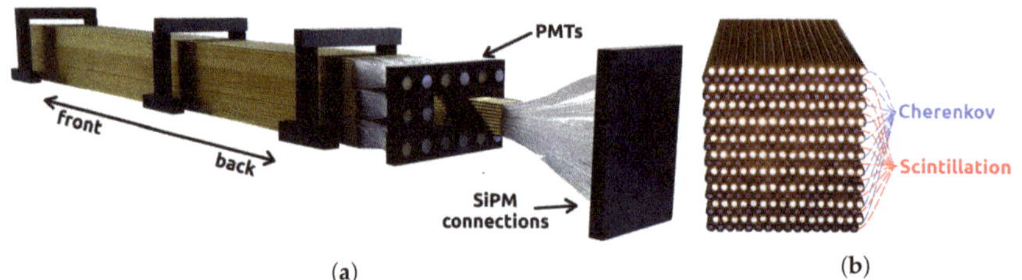

Figure 1. Dual-readout calorimeter prototype built for testbeams [5]. (**a**) The full prototype with the readout in the back, where the central tower has one SiPM per fibre, and the others have two PMTs each. (**b**) The layout of the fibres in one tower. The illuminated fibres are the Cherenkov channels.

A simulation of this prototype in a 'testbeam configuration' in Geant4 has been developed [6]. This configuration includes a rotation angle of 1° around the vertical axis, which was also present at the testbeam activities.

The simulation was validated using data from two testbeam campaigns using an e^+ beam, one at DESY in June 2021 with energies of 1–6 GeV and one at SPS in August 2021 with energies of 10–120 GeV. The validation was conducted by comparing shower profiles as a function of distance from the shower central axis [7].

3. Electromagnetic Energy Resolution

The first step in the simulation to study the EM energy resolution was to investigate the dependence on the beam impact point position; indeed, testbeam data showed varying calorimeter responses correlated with the alternating row layout of the prototype. In an actual experiment, knowing the behaviour in the calorimeter response as a function of the particle impact point allows us to correct for any observed effects once the impact point has been determined, e.g., by the use of particle flow algorithms [8].

3.1. Simulation Setup

We simulated a pencil-like beam with no radial extension at 19 different positions, as shown in Figure 2. The positions were chosen such that the beam centrally hits either one of the brass tubes, the gap in between the tubes, a scintillation fibre, or a Cherenkov fibre. Having this many different beam positions allows us to check for the periodicity of the response with three half-periods, which should exhibit the same behaviour.

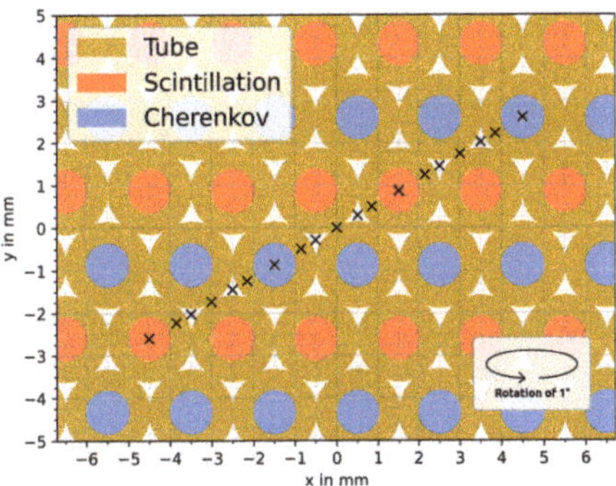

Figure 2. Simulated beam impact positions with a pencil-like beam of no radial extension. The beam travels in the (negative) z direction.

3.2. Impact Point Dependence

For each position, 50,000 e^+ were simulated at 20 GeV. As the output of the simulation is internally converted into the number of photoelectrons detected by the SiPMs and PMTs of all towers, the expected light yield per deposited GeV has to be calibrated back to energy in order to calculate the *combined* channel response. The calibration factors are calculated using the Monte Carlo truth information in such a way that both energy deposit distributions peak at 20 GeV averaged over all 19 simulated positions. Any leakage losses are neglected.

A Gaussian fit is performed to the energy deposit distribution at each position to determine the peak position μ and the width of the distribution σ at this particular position. An example of these fits for a single beam spot position can be seen in Figure 3 for both scintillation and Cherenkov channels.

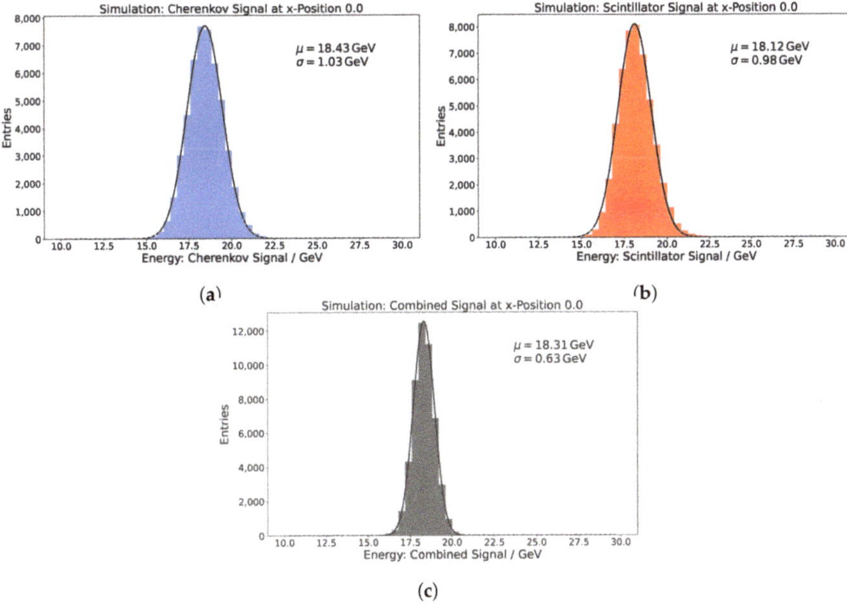

Figure 3. Energy deposit distributions for the (**a**) Cherenkov, (**b**) scintillation, and (**c**) combined channels. A Gaussian fit determines the peak position and width of the distribution, which later on is used to calculate the energy resolution.

The width at each position can be used to calculate the *combined* signal channel. The combined channel energy E_{comb} is estimated event-by-event by using a weighted average of the scintillation and Cherenkov channel deposits where the width of the deposit distribution serves as weight, according to Equation (1):

$$E_{\text{comb}} = \frac{\frac{E_C}{\sigma_C^2} + \frac{E_S}{\sigma_S^2}}{\frac{1}{\sigma_C^2} + \frac{1}{\sigma_S^2}}. \tag{1}$$

For each beam position, this yields a new distribution for which the peak and width can be determined via a Gaussian fit. An example is given in Figure 3c.

Figure 4 shows the fit result for the peak as a function of the beam positions for all three channels (scintillation, Cherenkov, and combined). The position is indicated with the y-coordinate due to the alternating row structure in the y-direction. A clear oscillating behaviour can be seen for all channels. It is not surprising to see this behaviour in the scintillation and Cherenkov channels, as the region of maximal energy deposit will develop closely around the respective signal fibres. From simulation studies, roughly 90% of the shower energy is deposited within a distance of 14 mm of the shower axis and even 50% within 5 mm. The oscillation for the scintillation channel is larger because the shower development for the scintillating component of the shower is narrower than the Cherenkov light-emitting component [9].

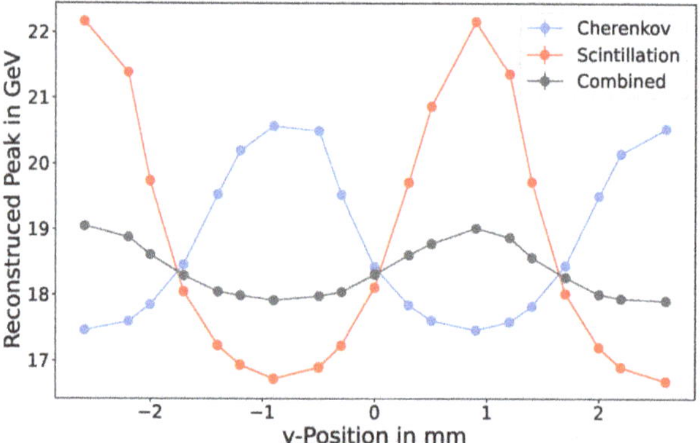

Figure 4. Reconstructed energy deposit as a function of the beam impact position for the Cherenkov, scintillation, and combined channels.

The oscillating behaviour carries through to the combined channel. This indicates that a position-dependent equalisation might be needed to accurately determine the energy in the combined channel.

We define the energy resolution as the width of the energy deposit σ over the reconstructed energy μ, where, due to consistent terminology, the resolution is labelled as σ/E.

Figure 5 shows σ/E for scintillation, Cherenkov, and the combined channels. The combined channel achieves a superior resolution to each single channel, as is expected [1]. However, a clear dependence on the impact position can be seen in all channels. The dependence is largest for the scintillation channel, where the resolution σ/E follows a similar trend to the reconstructed peak μ. One would naively expect the scintillation channel to show the best performance when hitting a scintillating fibre directly, i.e., the positions of the maxima in the energy deposit in Figure 4. However, this is not the case.

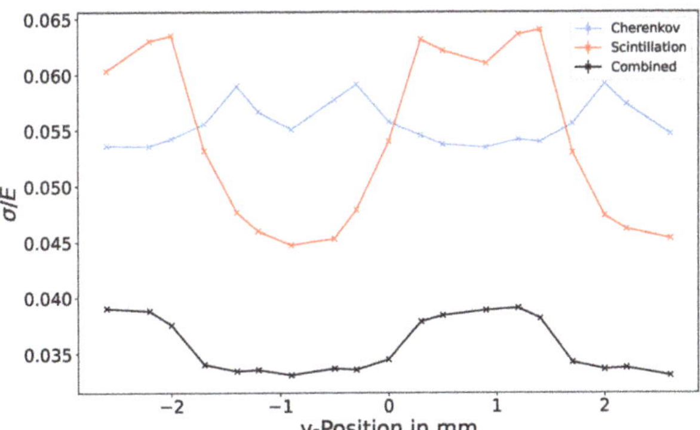

Figure 5. Energy resolution as a function of the beam impact position for the Cherenkov, scintillation, and combined channel.

Comparing the calorimeter signal distributions for positions $(-1.5, -0.87)$ (hitting a Cherenkov fibre) and $(+1.5, +0.87)$ (hitting a scintillating fibre) for the scintillation channel (Figure 6) shows that even though the reconstructed energy E increases when hitting the scintillating fibre, so does the width of the distribution σ, resulting in an overall worsened resolution σ/E. We conclude that the fluctuation in the energy deposit far from the shower central axis is much lower than close to it. This leads to a broad energy deposit distribution close to the shower axis and a narrow distribution further away. Very close to the shower central axis, where the fibre extends to a radial distance of 0.5 mm, the resolution also shows a dip, meaning that at this distance, the fluctuation is not as large as at medium distances between 0.5 mm and 1.4 mm.

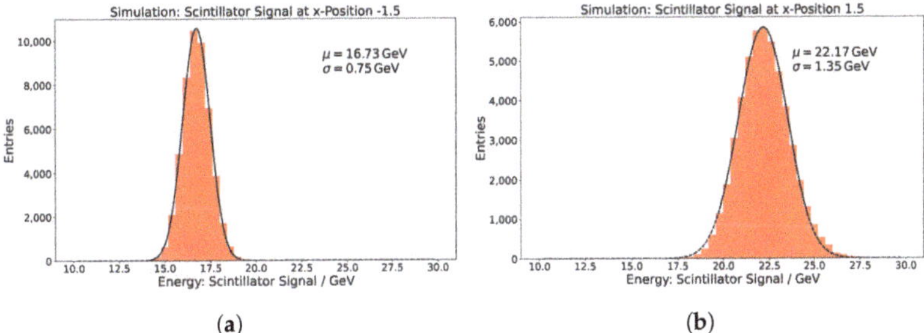

(**a**)   (**b**)

Figure 6. Energy deposit distributions for the scintillation channel when hitting (**a**) a Cherenkov and (**b**) a scintillating fibre (note the identical x scale). The narrower distribution for hitting a Cherenkov fibre is driving the energy resolution improvement in Figure 5.

The same is true for the Cherenkov component of the shower, though here, the effect is not as pronounced as in the scintillation channel. The combined channel is dominated (by definition) by the channel with a lower width at a given position, but through the combination is able to achieve two plateaus for the resolution around the position of the fibres. All this, again, indicates that knowing the exact position of the shower centre will be instrumental in extracting the best energy resolution.

3.3. Energy Dependence

We expect the energy resolution to evolve with the energy according to:

$$\frac{\sigma}{E} = \frac{a}{\sqrt{E}} + b,\qquad(2)$$

where a, named the *stochastic term*, represents the resolution due to fluctuations in the shower development and thus the main value of interest for this study. The constant term is in part influenced by shower leakage escaping the detector [10]. It is, therefore, a property of the prototype and not interesting for studying the performance of this particular layout for the dual-readout calorimeter. Additional contributions to the constant term, such as the position dependence of the signal response, are furthermore suppressed by keeping a slight rotation angle of the prototype with respect to the beam axis, as well as limiting this part of the study to a single beam impact point position [1].

To extract the value of a, we simulated 10,000 e^+ for energies of 5 GeV, 10 GeV, 20 GeV, 40 GeV, 60 GeV, and 80 GeV. The beam position $(-1.5, -0.87)$, which had the best overall performance in the energy resolution from Section 3.2, was chosen. Therefore, the value for the stochastic term to the energy resolution is determined under the assumption that a position-dependent calibration can take place.

The energy resolution as a function of $1/\sqrt{E}$ can be seen in Figure 7 for all three channels. Through a fit, we can extract the value for the slope, i.e., the stochastic term a, and the constant term.

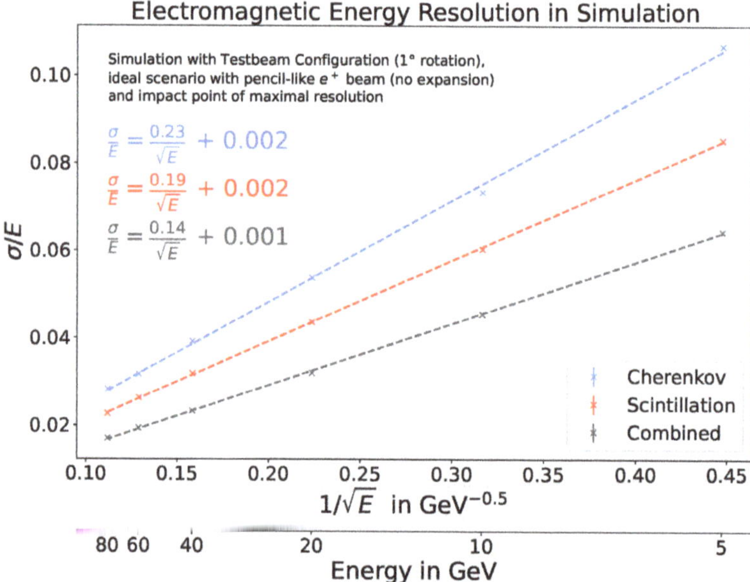

$$\frac{\sigma}{E} = \frac{0.23}{\sqrt{E}} + 0.002$$

$$\frac{\sigma}{E} = \frac{0.19}{\sqrt{E}} + 0.002$$

$$\frac{\sigma}{E} = \frac{0.14}{\sqrt{E}} + 0.001$$

Figure 7. Electromagnetic energy resolution as a function of $1/\sqrt{E}$ for Cherenkov, scintillation, and combined channels. The simulation was conducted under an ideal scenario with exact knowledge of beam impact point, no material in front of the calorimeter prototype, and a slight rotation angle of 1°.

For the Cherenkov and scintillation channel we extracted energy resolutions of $23\%/\sqrt{E}$ and $19\%/\sqrt{E}$, respectively, neglecting the constant term. This is in line with the expectation that the scintillation channel performs better than the Cherenkov one [1]. We can also see that in the combined channel, we achieve the best resolution of $14\%/\sqrt{E}$, assuming that the impact position of the impinging particle is known.

4. Conclusions

Using simulations, we have characterised the EM performance of a dual-readout calorimeter prototype and outlined a strong position dependence in the calorimeter response and energy resolution. A position-dependent calibration for the equalisation of this effect will be needed to extract the maximal performance of the calorimeter. Due to the

$\mathcal{O}$(mm) pitch of the readout fibres, it is likely possible to determine the shower axis to a sub-mm precision by analysing the shower profile.

Assuming a successful calibration, an EM energy resolution of $14\%/\sqrt{E}$ is achieved in a preliminary way to calculate the combined channel signal.

Funding: This project has received funding from the European Union's Horizon 2020 Research and Innovation programme under grant agreement No 101004761.

Data Availability Statement: Not applicable.

Conflicts of Interest: The authors declare no conflict of interest.

Abbreviations

The following abbreviations are used in this manuscript:

EM   Electromagnetic
SiPM   Silicon Photomultiplier
PMT   Photomultiplier

References

1. Akchurin, N.; Bedeschi, F.; Cardini, A.; Cascella, M.; Cei, F.; De Pedis, D.; Ferrari, R.; Fracchia, S.; Franchino, S.; Fraternali, M.; et al. The electromagnetic performance of the RD52 fiber calorimeter. *Nucl. Instrum. Methods Phys. Res. Sect. A Accel. Spectrom. Detect. Assoc. Equip.* **2014**, *735*, 130–144. [CrossRef]
2. Lee, S.; Livan, M.; Wigmans, R. Dual-readout calorimetry. *Rev. Mod. Phys.* **2018**, *90*, 025002. [CrossRef]
3. Derrick, M.; Gacek, D.; Hill, N.; Musgrave, B.; Noland, R.; Petereit, E.; Repond, J.; Stanek, R.; Sugano, K. Design and construction of the ZEUS barrel calorimeter. *Nucl. Instrum. Methods Phys. Res. Sect. A Accel. Spectrom. Detect. Assoc. Equip.* **1991**, *309*, 77–100. [CrossRef]
4. Antonello, M. IDEA: A detector concept for future leptonic colliders. *Nuovo Cim. C* **2020**, *43*, 27. [CrossRef]
5. First Beam Tests for Highly Granular Dual-Readout Calorimeter Prototype. Available online: https://aidainnova.web.cern.ch/first-beam-tests-highly-granular-dual-readout-calorimeter-prototype (accessed on 31 July 2022).
6. DREMTubes—GitHub Repository. Available online: https://github.com/lopezzot/DREMTubes (accessed on 31 July 2022).
7. Pezzotti, L. Including Calorimeter Test-Beams into Geant-Val. CALOR 2022. Available online: https://indi.to/rT6st (accessed on 31 July 2022).
8. Brient, J.C.; Videau, H. The Calorimetry at the Future $e^+ e^-$ Linear Collider. *arXiv* **2002**, arXiv:hep-ex/0202004. https://doi.org/10.48550/ARXIV.HEP-EX/0202004.
9. Antonello, M.; Caccia, M.; Cascella, M.; Dunser, M.; Ferrari, R.; Franchino, S.; Gaudio, G.; Hall, K.; Hauptman, J.; Jo, H.; et al. Tests of a dual-readout fiber calorimeter with SiPM light sensors. *Nucl. Instrum. Methods Phys. Res. Sect. A Accel. Spectrom. Detect. Assoc. Equip.* **2018**, *899*, 52–64. [CrossRef]
10. Kolanoski, H.; Wermes, N. *Particle Detectors: Fundamentals and Applications*; Oxford University Press: Oxford, UK, 2020; pp. 603–604.

instruments

Article

Development of an Argon Light Source as a Calibration and Quality Control Device for Liquid Argon Light Detectors

Mehmet Tosun [1,2,3,*], Burak Bilki [1,2,4], Fatma Boran [1,5], Furkan Dolek [1,5] and Kutlu Kagan Sahbaz [1,2,3]

1   Department of Mathematics, Beykent University, Istanbul 34500, Turkey
2   Turkish Accelerator and Radiation Laboratory, Ankara 06830, Turkey
3   Institute of Nuclear Sciences, Ankara University, Ankara 06100 , Turkey
4   Department of Physics and Astronomy, University of Iowa, Iowa City, IA 52242, USA
5   Department of Physics, Cukurova University, Adana 01330, Turkey
*   Correspondence: mehmet.tosun@cern.ch

Abstract: The majority of future large-scale neutrino and dark matter experiments are based on liquid argon detectors. Since liquid argon is also a very effective scintillator, these experiments also have light detection systems. The liquid argon scintillation wavelength of 127 nm is most commonly shifted to the visible range by special wavelength shifters or read out by the 127 nm sensitive photodetectors that are under development. The effective calibration and quality control of these active media is still a persisting problem. In order to respond to this need, we developed an argon light source which is based on plasma generation and light transfer across a MgF_2 window. The light source was designed as a small, portable and easy-to-operate device to enable the acquisition of performance characteristics of several square meters of light detectors. Here, we report on the development of the light source and its performance characteristics.

Keywords: liquid argon; plasma light source; scintillation light

Citation: Tosun, M.; Bilki, B.; Boran, F.; Dolek, F.; Sahbaz, K.K. Development of an Argon Light Source as a Calibration and Quality Control Device for Liquid Argon Light Detectors. *Instruments* **2022**, *6*, 45. https://doi.org/10.3390/instruments6040045

Academic Editors: Fabrizio Salvatore, Alessandro Cerri, Antonella De Santo and Iacopo Vivarelli

Received: 16 August 2022
Accepted: 5 September 2022
Published: 21 September 2022

Publisher's Note: MDPI stays neutral with regard to jurisdictional claims in published maps and institutional affiliations.

1. The Argon Light Source

Most future large-scale neutrino and dark matter experiments will rely on liquid argon detectors (see, e.g., [1–5]). For this reason, detectors to measure the scintillation light generated inside liquid argon detectors are needed. The number of photosensors to measure the 127 nm wavelength argon scintillation light is quite limited and usually a wavelength shifter such as tetraphenyl-butadiene (TPB) is employed (see, e.g., [6]). The calibration and quality control of these detectors is still an ongoing problem.

In order to meet this need, we made an argon plasma light source that produces light with a wavelength of 127 nm. The argon light was transferred to the outside of the light source body through a MgF_2 window. We made the body of the light source from polyoxymethylene and used titanium wires as the electrodes for the light source. The light source was put under a vacuum of 5×10^{-6} mbar and flushed a few times with high purity argon prior to be put in operation. The final filling was done to the target pressure and the chamber was sealed. The operating voltage and pressure were scanned in order to obtain the optimal operating conditions.

Figure 1 shows a picture of the light source in operation. The plasma light can be seen through the MgF_2 window. The final filling still contains ppm levels of contaminants which limit the fraction of the 127 nm light. In order to identify the optimal operating conditions, the argon pressure was scanned from 1000 mbar to 2000 mbar in steps of 100 mbar; the operating high voltage was scanned from 2600 V down to the point where the light is lost (usually around 1200 V) in steps of 100 V; and the average spectrum of the light was measured. Figure 2(left) shows a sample average spectrum which shows the argon and impurity peaks in the 200–1000 nm range. The peaks are identified, and the relevant intensity integrals are calculated. Figure 2(right) shows the intensity integrals due to argon

at the light source under vacuum. The data were recorded with self triggering on the light pulses 20 mV above baseline. The main pulse for the clean SiPM was mostly due to the impurities in the argon, and partly due to the red–infrared emission of argon. Compared to the clean SiPM-overlaid signals, the height of the triggering pulse for the TPB-coated SiPM was decreased and the readout window was populated with many additional pulses suggesting a nearly continuous emission of 127 nm light, which was not visible with the clean SiPM.

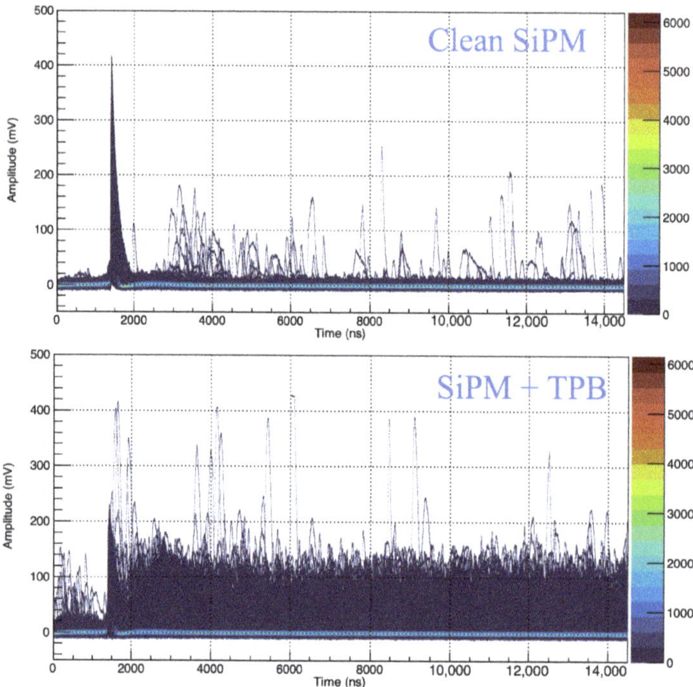

Figure 5. The overlaid signals measured with the clean (**top**) and TPB-coated (**bottom**) SiPM looking at the light source under vacuum. The SiPMs were placed right across the MgF$_2$ window of the light source in a vacuum assembly. The data taking was triggered with the SiPM signals themselves slightly above the single avalanche threshold. The trigger was timed to be around 1500 ns and all the waveforms were overlaid to make the plots.

Figure 6(left) shows the number of pulses with peak amplitudes above 30 mV in the 15 μs window per triggered event. The triggered events with the clean SiPM mostly contain single pulses with peaks above 30 mV and the number of two or more peaks is significantly reduced. For the case of TPB-coated SiPM, the number of pulses in the readout window with peaks larger than 30 mV is much higher. As the only difference was the introduction of the TPB on the SiPM window, which simply increased the sensitivity to 127 nm light, the operation of the light source was validated.

Figure 6(right) shows the full width at half-maximum (FWHM) of all the pulses in the 15 μs readout window for the clean and TPB-coated SiPMs. The majority of the pulses have less than 500 ns width. On the other hand, the TPB-coated SiPM pulses have an accumulation around 800 ns. Figure 7(left) shows an example of the pulse with an FWHM less than 500 ns , and Figure 7(right) shows an example with an FWHM larger than 500 ns for the TPB-coated SiPM. The wider pulses are attributed to the 127 nm light. The 127 nm light seems to be originating in bursts within which the individual pulses are a few nanoseconds apart. The time structure of the light is under further investigation.

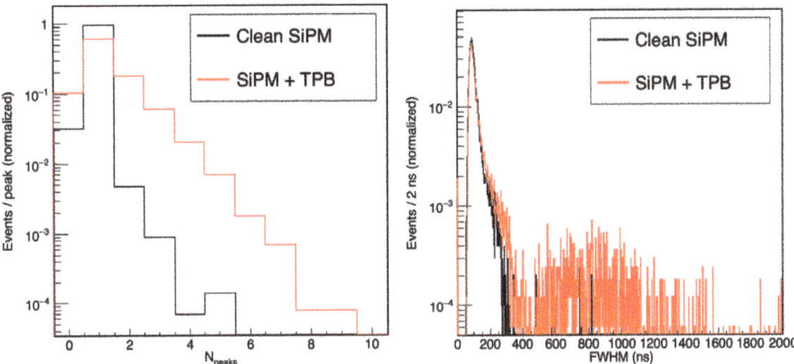

Figure 6. The number of pulses with peak amplitudes above 30 mV per triggered event (**left**) and the full width at half-maximum of all the pulses in the 15 μs readout window (**right**) for the clean and TPB-coated SiPMs.

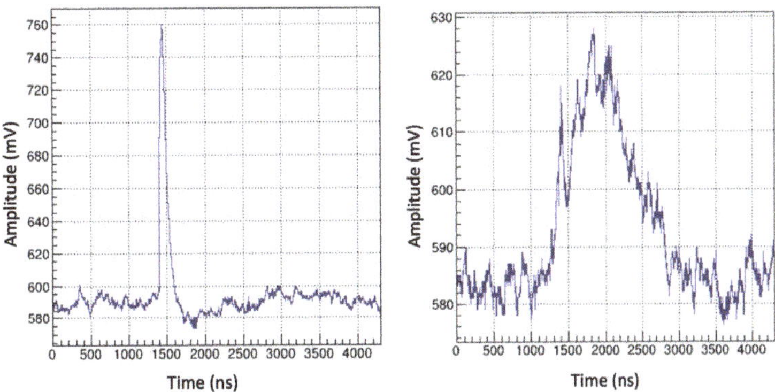

Figure 7. Example pulses with full width at half-maximum less (**left**) and larger (**right**) than 500 ns for the TPB-coated SiPM.

4. Conclusions

An argon light source envisaged to be a practical calibration and quality control device for liquid argon light detectors was developed. The preliminary characterization of the device indicated that 127 nm argon scintillation light was transferred through the MgF_2 window and could be identified with its specific waveform. The complete characterization of the light source is underway. The complete pulse shape discrimination, intensity stability and single filling lifetime are under investigation.

Author Contributions: Data curation, M.T.; Investigation, M.T., K.K.S., F.B., F.D.; Project administration, B.B. All authors have read and agreed to the published version of the manuscript.

Funding: This work is supported by Tübitak grant no 118C224.

Data Availability Statement: The data presented in this study are available on request from the corresponding author.

References

1. Abi, B.; Acciarri, R.; Acero, M.A.; Adamov, G.; Adams, D.; Adinolfi, M.; Ahmad, Z.; Ahmed, J.; Alion, T.; Monsalve, S.A.; et al. Deep Underground Neutrino Experiment (DUNE), Far Detector Technical Design Report. Volume I: Introduction to DUNE. *arXiv* **2020**, arXiv:2002.03005.
2. Acciarri, R.; Antonello, M.; Baibussinov, B.; Benetti, P.; Calaprice, F.; Calligarich, E.; Cambiaghi, M.; Canci, N.; Cao, C.; Carbonara, F.; et al. The WArP experiment. *J. Phys. Conf. Ser.* **2010**, *203*, 012006. [CrossRef]

3. Aalseth, C.E.; Acerbi, F.; Agnes, P.; Albuquerque, I.F.M.; Alexander, T.; Alici, A.; Alton, A.K.; Antonioli, P.; Arcelli, S.; Ardito, R.; et al. DarkSide-20k: A 20 tonne two-phase LAr TPC for direct dark matter detection at LNGS. *Eur. Phys. J. Plus* **2018**, *133*, 131. [CrossRef]

4. Acciarri, R.; Adams, C.; An, R.; Aparicio, A.; Aponte, S.; Asaadi, J.; Auger, M.; Ayoub, N.; Bagby, L.; Baller, B.; et al. Design and construction of the MicroBooNE detector. *J. Instrum.* **2017**, *12*, P02017. [CrossRef]

5. Pietropaolo, F. Review of Liquid-Argon Detectors Development at the CERN Neutrino Platform. *J. Phys. Conf. Ser.* **2017**, *888*, 012038. [CrossRef]

6. Kuźniak, M.; Szelc, A.M. Wavelength Shifters For Applications In Liquid Argon Detectors. *Instruments* **2021**, *5*, 4. [CrossRef]

7. Bilki, B. Study of Light Production with a Fifty Liter Liquid Argon TPC. Technology Furthermore, Instrumentation in Particle Physics 2021 (TIPP2021). Available online: https://indico.cern.ch/event/981823/contributions/4293603/ (accessed on 31 August 2022).

Article

Using Artificial Intelligence in the Reconstruction of Signals from the PADME Electromagnetic Calorimeter

Kalina Dimitrova * and on behalf of the PADME collaboration †

Faculty of Physics, Sofia University "St. Kliment Ohridski", 5 J. Bourchier Blvd., 1164 Sofia, Bulgaria
* Correspondence: kalina@phys.uni-sofia.bg
† The PADME collaboration: A.P. Caricato, M. Martino, I. Oceano, F. Oliva, S. Spagnolo (INFN Lecce and Salento Univ.), G. Chiodini (INFN Lecce), F. Bossi, R. De Sangro, C. Di Giulio, D. Domenici, G. Finocchiaro, L.G. Foggetta, M. Garattini, A. Ghigo, P. Gianotti, I. Sarra, T. Spadaro, E. Spiriti, C. Taruggi, E. Vilucchi (INFN Laboratori Nazionali di Frascati), V. Kozhuharov (Faculty of Physics, Sofia Univ. "St. Kl. Ohridski" and INFN Laboratori Nazionali di Frascati), S. Ivanov, Sv. Ivanov, R. Simeonov (Faculty of Physics, Sofia Univ. "St. Kl. Ohridski"), G. Georgiev (Sofia Univ. "St. Kl. Ohridski" and INRNE Bulgarian Academy of Science), F. Ferrarotto, E. Leonardi, P. Valente, A. Variola (INFN Roma1), E. Long, G.C. Organtini, G. Piperno, M. Raggi (INFN Roma1 and "Sapienza" Univ. Roma), S. Fiore (ENEA Frascati and INFN Roma1), V. Capirossi, F. Iazzi, F. Pinna (Politecnico of Torino and INFN Torino), A. Frankenthal (Princeton University).

Abstract: The PADME apparatus was built at the Frascati National Laboratory of INFN to search for a dark photon (A') produced via the process $e^+e^- \rightarrow A'\gamma$. The central component of the PADME detector is an electromagnetic calorimeter composed of 616 BGO crystals dedicated to the measurement of the energy and position of the final state photons. The high beam particle multiplicity over a short bunch duration requires reliable identification and measurement of overlapping signals. A regression machine-learning-based algorithm has been developed to disentangle with high efficiency close-in-time events and precisely reconstruct the amplitude of the hits and the time with sub-nanosecond resolution. The performance of the algorithm and the sequence of improvements leading to the achieved results are presented and discussed.

Keywords: dark photon; calorimetry; signal reconstruction; machine learning

Citation: Dimitrova, K.; on behalf of the PADME Collaboration. Using Artificial Intelligence in the Reconstruction of Signals from the PADME Electromagnetic Calorimeter. *Instruments* **2022**, 6, 46. https://doi.org/10.3390/instruments6040046

Academic Editors: Fabrizio Salvatore, Alessandro Cerri, Antonella De Santo and Iacopo Vivarelli

Received: 1 August 2022
Accepted: 13 September 2022
Published: 21 September 2022

Publisher's Note: MDPI stays neutral with regard to jurisdictional claims in published maps and institutional affiliations.

1. Introduction

In recent years, the search for an explanation of the Dark Matter phenomenon has led to the development of various hypotheses for an extension of the Standard Model, e.g., Weakly Interacting Massive Particles (WIMPs) [1]. However, the non-observation of new states with mass in the order of 100 GeV led scientists to explore other Dark Matter explanations. The main goal of PADME (Positron Annihilation into Dark Matter Experiment) [2] is to search for the dark photon A', a hypothetical gauge boson connecting the dark and the visible sector. In the case of non-vanishing interaction strength α' with the electrons, A' can be produced in the annihilation process of beam positrons with electrons from the target:

$$e^+e^- \rightarrow A'\gamma. \tag{1}$$

Knowing the four-momenta of the beam's positrons, the electrons at rest and the photon produced in the process, the missing mass of the dark photon can be calculated:

$$M^2_{miss} = (P_{e^+} + P_{e^-} - P_\gamma)^2. \tag{2}$$

The positron beam provided by the DAΦNE LINAC [3] can reach energies up to 550 MeV, providing a limit for the missing mass of 23.7 MeV, and is composed of bunches with a 50 Hz rate. Each bunch contains about 2×10^4 particles and its length can be varied with typical values of 200–300 ns.

The two main processes contributing to the background, are the annihilation $e^+e^- \rightarrow \gamma\gamma(\gamma)$ and Bremsstrahlung events $e^+N \rightarrow e^+N\gamma$.

To suppress the background from $e^+e^- \rightarrow \gamma\gamma(\gamma)$, the PADME experiment should have high photon detection efficiency, while for the rejection of $e^+N \rightarrow e^+N\gamma$, the radiating positron should be detected and a reliable matching in time between the positron and the emitted photon should be assured.

The initial studies based on a full Geant4 [4] simulation indicate that the PADME experiment can reach sensitivies in α' down to 10^{-8} [5] with high-efficiency detectors (greater than 99%) and time resolution better than 1 ns. In addition, due to the high bunch multiplicity, double-pulse separation capabilities are required for each of the chosen detectors.

2. The Padme Experiment

A sketch of the PADME experiment is shown in Figure 1. A short description of its major detector components [6] follows.

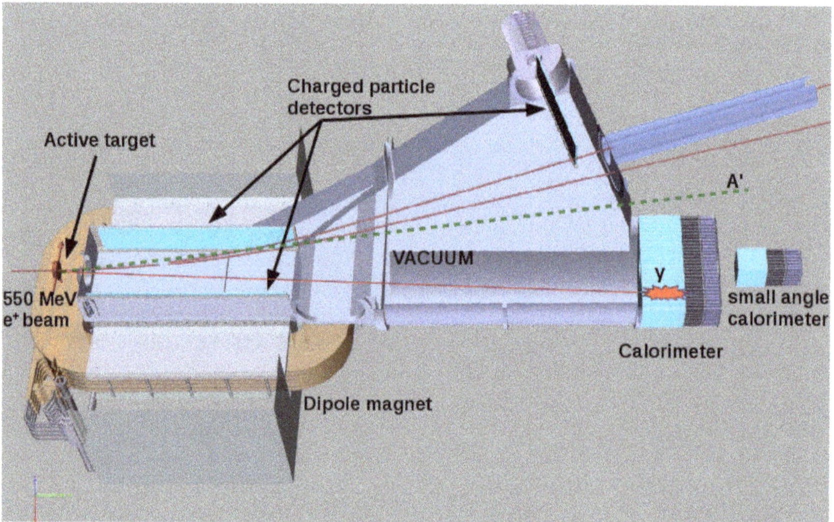

Figure 1. Outline of the PADME experiment.

2.1. Active Target

The target [7] is composed of polycrystalline diamond ($Z = 6$) since low Z is required to increase the annihilation to bremsstrahlung cross-section ratio. The target has a 100 μm thickness and 20 mm width and length. Apart from providing the target for the annihilation process, it also measures the beam's multiplicity and XY profile. For this reason 16 horizontal and 16 vertical graphite electrodes of 1 mm width are engraved onto the target using an excimer laser.

2.2. Charged Particle Detectors

Three sets of detectors register the charged particles. The beam positrons may lose energy in the target and produce Bremsstrahlung photons, detected by the electromagnetic calorimeter, which need to be rejected. This is achieved by coinciding these photons with the particles that produced them. These particles are detected by the positron and high energy positron vetoes. In case the A' decays into an e^+e^- pair, the electron will be registered by the electron veto. All three charged particle detectors are composed of $10 \times 10 \times 178$ mm^3 plastic scintillators with WLS fibers coupled to 3×3 mm^2 Hamamatsu S13360 silicon photomultipliers with 25 μm pixel size and are placed in 10^{-5} mbar vacuum. The positron and the electron vetoes are located inside the magnet and are composed of 90 and 96 scintillating bars, respectively. Both detect particles with momenta between

50 and 450 MeV. The high energy positron veto is located next to the beam exit window and is composed of 16 scintillating bars, with scintillation light read out on both sides. It allows the detection of positrons with momenta between 450 and 500 MeV.

The charged particle detectors segmentation provides measurement of the e^+/e^- momentum with a resolution of $\approx$5%. The time resolution is 700 ps [8].

2.3. Calorimeters

The PADME calorimetric system is composed of an Electromagnetic Calorimeter (ECal) and a small-angle calorimeter (SAC). The ECal (Figure 2) is composed of 616 BGO cystals measuring $2.1 \times 2.1 \times 23$ cm^3, connected to HZC 1912 photomultipliers. The optical isolation of the crystals is achieved by covering them with diffuse reflective TiO$_2$ paint and additionally with 50–100 μm thin black Tedlar foils. The ECal is placed 3.45 m away from the target and has a radius of 29 cm, thus achieving an angular coverage between 15 and 84 mrad. The lower limit is due to a square hole in its center, which is covered by the SAC. The scintillation light decay time is 300 ns. Calibration was performed both with a ^{22}Na source before constructing the calorimeter and then subsequently with cosmic rays. The energy resolution is $\sim$7% at $E_\gamma = 100$ MeV [9].

Figure 2. The PADME Electromagnetic Calorimeter.

The Small Angle Calorimeter (SAC) [10] is located downstream of the ECal and mainly detects photons produced by Bremsstrahlung events. To suppress this type of background, the data from the SAC is matched in time with the data from the charged particle detectors. In addition, the SAC detects photons from multiphoton annihilation events. The SAC is composed of 25 PbF$_2$ crystals measuring $3 \times 3 \times 14$ cm^3 and covers an angle between 0 and 15 mrad.

2.4. Readout System

The PADME Data Acquisition System consists of 29 CAEN V1742 ADC boards, each equipped with 32 analog and 2 trigger input channels. The V1742 switch capacitor digitizer employs the DRS4 chip, capable of sampling the input signal at 750 MS/s, 1 GS/s, 2.5 GS/s, and 5 GS/s. Complete waveforms of 1024 samples for each channel are recorded upon a beam-based trigger signal. In the case of the electromagnetic calorimeter which is sampled at 1 GS/s, this corresponds to a $\approx$1 μs recorded waveform.

3. Application of Neural Networks for Waveform Description

The high multiplicity of the positron beam in combination with the short bunch duration leads to many overlapping or close-in-time signals recorded in a single event. One way to solve this problem is to use neural networks (NNs) to count the signals from the electromagnetic calorimeter in each event, to identify and separate overlapping signals, and to extract signal parameters—the arrival times of the individual signals and their amplitudes. To train the NNs, an event simulation was developed. Each event represents a waveform of 1024 samples, for which the number of signals as well as the individual signal parameters can be varied. Various sets of waveforms $A(t)$ were generated, containing a random number of signals with shapes defined by the subtraction of two exponents

$$A(t) = A_0(e^{-(t-t_0)/\tau_1} - e^{-(t-t_0)/\tau_2}), \quad t \geq t_0, \tag{3}$$

where t_0 is the arrival time of the signal, τ_1 is the signal decay time, taken to be 300 ns, and τ_2 is related to the signal rise time, taken to be 10 ns. These values are typical for the PADME PMT + BGO crystal assembly. A_0 is the signal amplitude parameter, chosen to follow a Gaussian distribution. The arrival time follows a uniform distribution with a minimum value of $t_0 = 100$ ns, to account for the trigger specifications. For the training of the networks presented here, a mean value $A_0 = 200$ mV with $\sigma = 200$ mV was used and an additional lower limit $A_0 > 20$ mV was set. The signals from the ECal are digitised by an ADC with 1 V dynamic range, which should be sufficient for the maximal energy cell within an electromagnetic shower. Since the selected photon energy is between 50 MeV and 450 MeV, a 200 mV mean amplitude was chosen to increase the training statistics to signals corresponding to the 100 MeV range. All waveforms include a Gaussian noise with a mean value of 10 mV added in each bin. A predefined maximum number of four signals was used for all generated waveforms. Many of the events, recorded by the ECal contain only one or two signals, however, there are events with more recorded signals which requires the inclusion of such cases in the training. Different NNs were trained on the thus generated events and each network is trained on 100,000 events.

For the implementation of all neural networks presented in this study and for the output analysis were used the ROOT [11], TensorFlow [12] and Keras [13] frameworks. Three different convolutional neural networks (CNNs) [14] were developed, starting with the classification task of dividing the events into categories based on the number of signals in them and moving to regression tasks with the aim of signal parameter estimation.

The first CNN performs a classification task aimed at counting the number of signals in each waveform. It consists of a single convolutional layer, followed by three fully connected layers. This network was trained using labels containing only information about the number of signals in each event. The obtained model was then applied for the simulated set of events with two generated signals and the output was compared to the true labels. Figure 3 shows the efficiency of the signal counting as a function of $\Delta t = |t_2 - t_1|$ where t_2 and t_1 are the arrival times of the two signals. The efficiency is 50% for $\Delta t = 10$ ns and 100% for $\Delta t > 50$ ns.

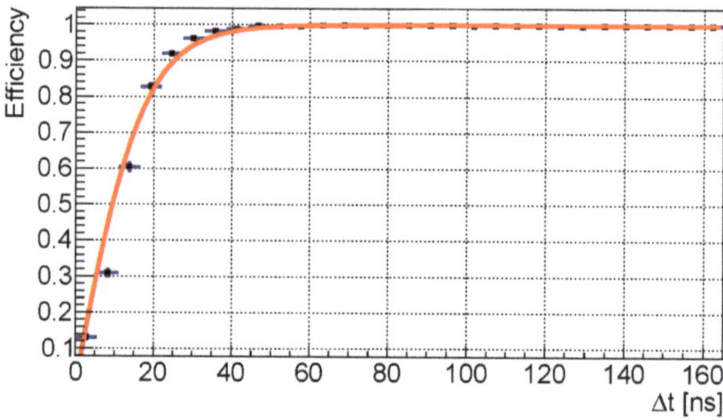

Figure 3. Efficiency of a signal-counting CNN as a function of the time difference between two signals. A sigmoid curve for the efficiency $\text{Eff} = 2 \cdot \left(\frac{1}{(1+exp(-\Delta t/70)} - 0.5 \right)$ is fitted (red). For signals with $\Delta t = 10$ ns, the efficiency is 50%. The efficiency reaches 100% for $\Delta t > 50$ ns.

The estimation of signal parameters requires the development of networks with more complex architectures. Convolutional autoencoders [15] can be used for extracting useful data from waveforms.

An autoencoder was developed with the targeted output replicating the original waveform. The neural network architecture consists of three convolutional layers followed by three deconvolutional layers with the same parameters and a single final deconvolutional layer for setting the output dimensions. Figure 4 (left) shows an example of the original and the output waveform for an event with two signals. It is observed that such networks can reproduce the signals, which represents the main motivation for applying this architecture for signal parameter estimation through supervised learning.

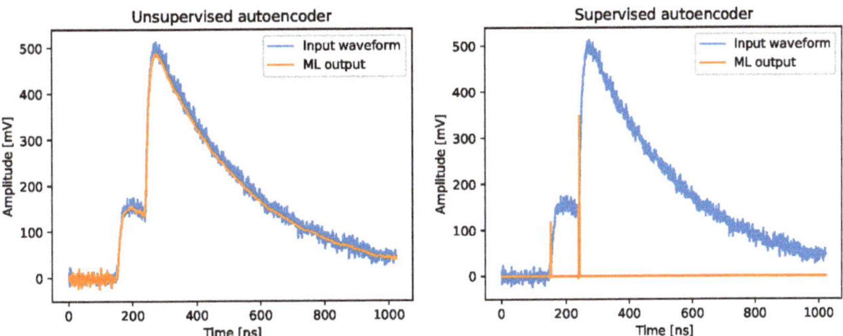

Figure 4. Autoencoder NN output and the original event with two signals. (**Left**) classical autoencoder. (**Right**) modified autoencoder.

A modified autoencoder trained on labeled data was developed with training based on output arrays of the same length as the input waveforms. On the positions of signal arrival, the value is set to the signal amplitude and all other values are set to 0. An example of the output of the modified autoencoder and its corresponding waveform is presented in Figure 4 (right).

To assess the modified autoencoder output and compare it with the original data labels, a reconstruction algorithm was developed. The data labels contain an array of 1024 numbers with a single non-zero amplitude value at the arrival time t_0 for each generated signal. The NN output gives multiple amplitude values over a number of time positions for each recognised signal, usually with a maximum on the most probable position and decreasing values on both of its sides. The reconstruction algorithm locates the maximum and adds the values of the three positions before it and the three positions after it to the maximum value. The result is taken to be the amplitude of the reconstructed signal and the time position of the maximum is taken to be the arrival time of this signal. The reconstruction is also applied to the original data labels and the results for the reconstructed output are then compared to them. This allows both for comparison of the amplitude value of the original and reconstructed signal, as well as evaluation of the arrival time and analysis of the neural network efficiency.

4. Signal Parameter Reconstruction

The probability for a signal to be discriminated and the accuracy of the reconstructed signal parameters were studied. The modified autoencoder model was trained on a set of 100,000 events with up to four signals each and was applied to a statistically independent test set, again with 100,000 events containing up to four signals.

4.1. Time Reconstruction

To study the double-pulse separation abilities of the machine learning algorithm, each simulated signal is associated with the closest-in-time one from the NN output. The left panel in Figure 5 represents the difference between the reconstructed and the original time of the signal arrival. The distribution of this difference (Figure 5, right panel) is symmetric, with $\sigma \sim 520$ ps and RMS ~ 3.2 ns if assumed Gaussian, however, non-Gaussian tails do exist.

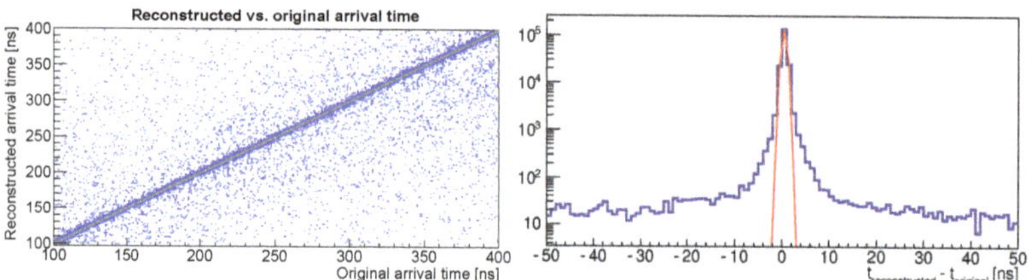

Figure 5. (**Left**) Reconstructed vs. original time of arrival for all events in the test set. (**Right**) Distribution of the signals according to the difference between the reconstructed and the original signal arrival time. The red curve represents a Gaussian distribution with $\sigma \approx 520$ ps and mean value of 0.

A signal is considered successfully identified if the difference between the original and reconstructed output is less than 2 ns.

4.2. Signal Recognition

Figure 6 shows all of the events in the test set, divided into bins based on the number of reconstructed signals and the originally generated ones. Ideally, these two numbers should be the same for all events. However, two major factors influence signal discrimination: a small difference in signal arrival time may cause two or more signals to be merged into one and signals with small amplitudes may not be identified above the noise.

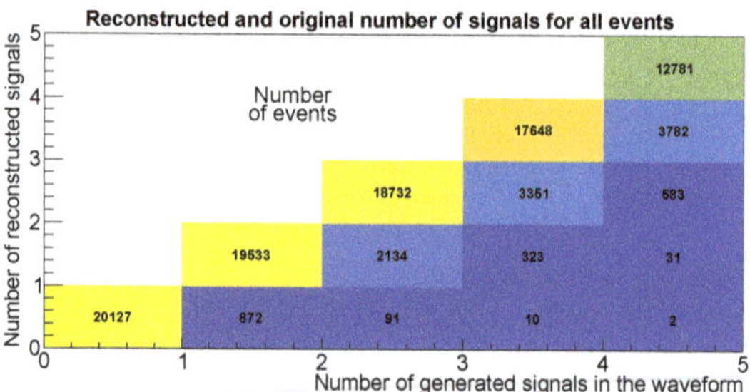

Figure 6. Number of reconstructed signals versus the original number of generated signals in the test sample.

Signals with time differences less than 10 ns are merged into a single hit and most events with amplitudes smaller than 50 mV are not likely to be identified, which results in decreased efficiency, as seen in Figure 7.

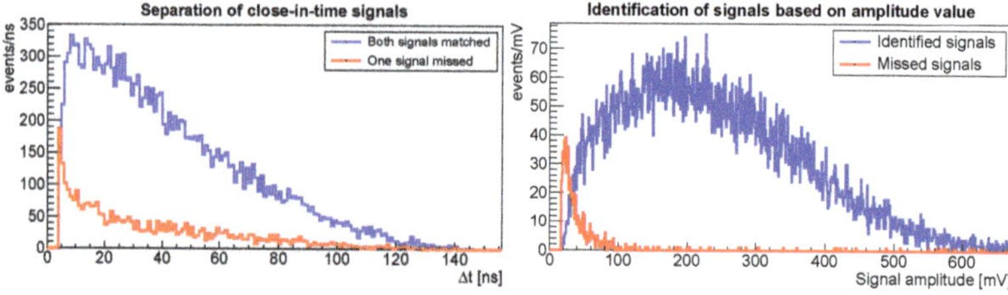

Figure 7. (**Left**) Matched (blue) and missed (red) events as a function of the arrival time difference Δt for events with two generated signals. (**Right**) Matched (blue) and missed (red) events as a function of the amplitude value for events with one generated signal. Missed events with high time differences are due to small amplitudes.

4.3. Amplitude Reconstruction

The developed CNN provides reconstruction of the signal amplitude values. The reconstructed versus the original amplitudes for the successfully identified signals are shown in the left panel of Figure 8. The correlation is well pronounced. To quantitatively compare the quality of signal identification, the average difference between the reconstructed and the true amplitude value for each 2 mV interval of generated amplitude is shown in the right panel of Figure 8.

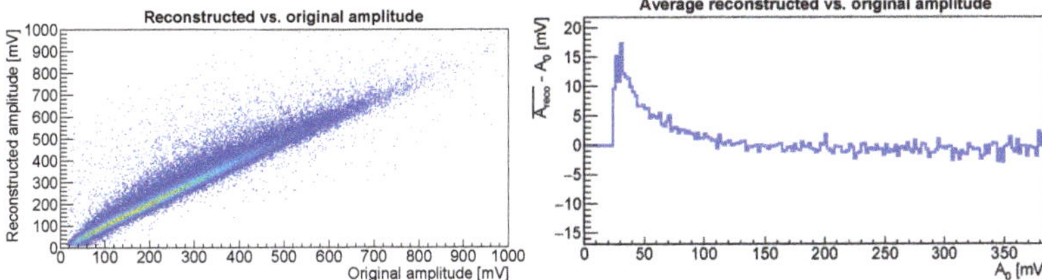

Figure 8. (**Left**) All events based on their reconstructed amplitude value and the original one (**Right**) Difference between the average value of the reconstructed amplitude $\overline{A_{reco}}$ and the original value A_0 for each original value A_0, divided into 2 mV bins. It can be observed that for small amplitudes the reconstructed value is higher than the original one.

The inaccuracy of the small amplitude values can be compensated for by a dedicated calibration of the NN output.

5. Conclusions

The high particle rate in the PADME calorimeter requires implementation of advanced reconstruction algorithms to achieve less than 1 ns time resolution of the reconstructed showers. Machine learning methods were applied for the successful identification of calorimeter signals and for the extraction of signal parameters. A CNN with a single convolutional layer was used to count the number of signals in an event. A modified CNN autoencoder was probed for the estimation of signal parameters—arrival time and amplitude. The performance of the networks was assessed through a specially designed algorithm, comparing the network output with the original data labels. The CNN provides time reconstruction with ~500 ps time resolution and the amplitude is reconstructed in the 30–700 mV range. There is an inaccuracy for amplitudes less than 100 mV, which can be solved by an additional calibration of the NN output or by developing networks with modified architectures specifically targeting small amplitudes.

Author Contributions: The presented material is a result of the joint work of the PADME collaboration. All authors have read and agreed to the published version of the manuscript.

Funding: This work was partially supported by BNSF: KP-06-D002_4/15.12.2020 within MUCCA, CHIST-ERA-19-XAI-009 and by LNF-SU MoU.

Data Availability Statement: Not applicable.

References

1. Alexander, J.; Battaglieri, M.; Echenard, B.; Essig, R.; Graham, M.; Izaguirre, E.; Jaros, J.; Krnjaic, G.; Mardon, J.; Morrissey, D.; et al. Dark Sectors 2016 Workshop: Community Report. *arXiv* **2016**, arXiv:hep-ph/1608.08632.
2. Raggi, M.; Kozhuharov, V. Proposal to Search for a Dark Photon in Positron on Target Collisions at DAΦNE Linac. *Adv. High Energy Phys.* **2014**, *2014*, 959802. [CrossRef]
3. Valente, P.; Belli, M.; Bolli, B.; Buonomo, B.; Cantarella, S.; Ceccarelli, R.; Cecchinelli, A.; Cerafogli, O.; Clementi, R.; Di Giulio, C.; et al. Linear Accelerator Test Facility at LNF Conceptual Design Report. *arXiv* **2016**, arXiv:physics.acc-ph/1603.05651.
4. Agostinelli, S.; Allison, J.; Amako, K.A.; Apostolakis, J.; Araujo, H.; Arce, P.; Asai, M.; Axen, D.; Banerjee, S.; Barrand, G.J.N.I.; et al. GEANT4—A simulation toolkit. *Nucl. Instrum. Methods Phys. Res. Sect. A Accel. Spectrometers Detect. Assoc. Equip.* **2003**, *506*, 250–303. [CrossRef]

5.	Raggi, M.; Kozhuharov, V.; Valente, P. The PADME experiment at LNF. *EPJ Web Conf.* **2015**, *96*, 01025. [CrossRef]
6.	Albicocco, P.; Assiro, R.; Bossi, F.; Branchini, P.; Buonomo, B.; Capirossi, V.; Capitolo, E.; Capoccia, C.; Caricato, A.P.; Ceravolo, S.; et al. Commissioning of the PADME experiment with a positron beam. *arXiv* **2022**, arXiv:physics.ins-det/2205.03430.
7.	Assiro, R.; Caricato, A.P.; Chiodini, G.; Corrado, M.; De Feudis, M.; Di Giulio, C.; Fiore, G.; Foggetta, L.; Leonardi, E.; Martino, M.; et al. Performance of the diamond active target prototype for the PADME experiment at the DAPHNE BTF. *Nucl. Instrum. Methods Phys. Res. Sect. A Accel. Spectrometers Detect. Assoc. Equip.* **2018**, *A898*, 105–110. [CrossRef]
8.	Ferrarotto, F.; Foggetta, L.; Georgiev, G.; Gianotti, P.; Kozhuharov, V.; Leonardi, E.; Piperno, G.; Raggi, M.; Taruggi, C.; Tsankov, L.; et al. Performance of the Prototype of the Charged-Particle Veto System of the PADME Experiment. *IEEE Trans. Nucl. Sci.* **2018**, *65*, 2029–2035. [CrossRef]
9.	Albicocco, P.; Alexander, J.; Bossi, F.; Branchini, P.; Buonomo, B.; Capoccia, C.; Capitolo, E.; Chiodini, G.; Caricato, A.P.; de Sangro, R.; et al. Characterisation and performance of the PADME electromagnetic calorimeter. *J. Instrum.* **2020**, *15*, T10003. [CrossRef]
10.	Frankenthal, A.; Alexander, J.; Buonomo, B.; Capitolo, E.; Capoccia, C.; Cesarotti, C.; De Sangro, R.; Di Giulio, C.; Ferrarotto, F.; Foggetta, L.; et al. Characterization and performance of PADME's Cherenkov-based small-angle calorimeter. *Nucl. Instrum. Methods Phys. Res. Sect. A Accel. Spectrometers Detect. Assoc. Equip.* **2019**, *919*, 89–97. [CrossRef]
11.	Brun, R.; Rademakers, F. ROOT: An object oriented data analysis framework. *Nucl. Instrum. Methods Phys. Res. Sect. A Accel. Spectrometers Detect. Assoc. Equip.* **1997**, *389*, 81–86. [CrossRef]
12.	Abadi, M.; Agarwal, A.; Barham, P.; Brevdo, E.; Chen, Z.; Citro, C.; Corrado, G.S.; Davis, A.; Dean, J.; Devin, M.; et al. TensorFlow: Large-Scale Machine Learning on Heterogeneous Systems. 2015. Available online: https://tensorflow.org (accessed on 11 July 2022).
13.	Chollet, F. Keras. 2015. Available online: https://keras.io (accessed on 11 July 2022).
14.	O'Shea, K.; Nash, R. An Introduction to Convolutional Neural Networks. *arXiv* **2015**, arXiv:cs.NE/1511.08458.
15.	Zhang, Y. A better autoencoder for image: Convolutional autoencoder. In Proceedings of the ICONIP17-DCEC, Guangzhou, China, 14–18 October 2017. Available online: http://users.cecs.anu.edu.au/~Tom.Gedeon/conf/ABCs2018/paper/ABCs2018_paper_58.pdf (accessed on 11 July 2022).

Project Report

Machine Learning Techniques for Calorimetry

Polina Simkina on behalf of the CMS Collaboration

IRFU, CEA, Université Paris-Saclay, 91190 Gif-sur-Yvette, France; polina.simkina@cern.ch

Abstract: The Compact Muon Solenoid (CMS) is one of the general purpose detectors at the CERN Large Hadron Collider (LHC), where the products of proton–proton collisions at the center of mass energy up to 13.6 TeV are reconstructed. The electromagnetic calorimeter (ECAL) is one of the crucial components of the CMS since it reconstructs the energies and positions of electrons and photons. Even though several Machine Learning (ML) algorithms have been already used for calorimetry, with the constant advancement of the field, more and more sophisticated techniques have become available, which can be beneficial for object reconstruction with calorimeters. In this paper, we present two novel ML algorithms for object reconstruction with the ECAL that are based on graph neural networks (GNNs). The new approaches show significant improvements compared to the current algorithms used in CMS.

Keywords: machine learning; graph neural network; high energy physics; calorimeter reconstruction

Citation: Simkina, P., on behalf of the CMS Collaboration. Machine Learning Techniques for Calorimetry. *Instruments* **2022**, *6*, 47. https://doi.org/10.3390/instruments6040047

Academic Editors: Fabrizio Salvatore, Alessandro Cerri, Antonella De Santo and Iacopo Vivarelli

Received: 31 July 2022
Accepted: 15 September 2022
Published: 21 September 2022

Publisher's Note: MDPI stays neutral with regard to jurisdictional claims in published maps and institutional affiliations.

1. Introduction

The Compact Muon Solenoid (CMS) experiment [1] is a general-purpose detector at the CERN Large Hadron Collider (LHC). The physics scope of the CMS is to probe the standard model of particle physics and search for the physics beyond the standard model with proton–proton collisions at a center of mass energy from 7 TeV (first collisions in 2010) to 13.6 TeV (collisions recorded since July 2022). In order to do so, it has to be able to efficiently reconstruct the particles coming from these collisions.

Along with the traditional algorithms, the Machine Learning (ML) approach is being broadly implemented, both for event reconstruction and data analysis. Algorithms such as boosted decision trees (BDT) and neural networks (NN) have already been successfully widely applied to the data from Run 2 (e.g., [2,3]). However, more sophisticated algorithms are becoming available, which may bring advantages to the reconstruction techniques in particle physics, using more and more low-level information (e.g., [4,5]).

Graph neural network (GNN) [6–8] is currently one of the most promising ML models. Its main distinguishing characteristics are:

1. GNNs can be applied on the data from complex detector geometries.
2. They are easily applied to sparse data with variable input sizes.
3. GNNs can be applied on non-Euclidean data (unlike convolutional neural networks).
4. In GNNs, the information can flow between close-by nodes of the graph.

In this paper, we will describe two models based on GNNs implemented for the reconstruction of electrons and photons in the CMS electromagnetic calorimeter (ECAL), along with the results achieved by these models and their comparison to the previously used algorithms.

2. e/γ Reconstruction

Photons and electrons play a crucial role in various physics analyses, including, for example, Higgs boson decays. The reconstruction of the energy and the position of these particles is done using mainly the ECAL. It is also necessary for the measurement of jets' momenta and missing transverse momentum.

2.1. ECAL

The ECAL is a homogenous calorimeter made of 75,848 lead tungstate (PbWO$_4$) crystals [9]. It is situated between the tracker and the hadronic calorimeter inside the solenoid, delivering a 3.8 T magnetic field, and is divided into two main parts:

- The barrel with crystal size: 2.2 × 2.2 × 23 cm, covering pseudorapidity $|\eta| < 1.479$.
- The endcaps with crystal size: 2.9 × 2.9 × 22 cm, covering pseudorapidity $1.479 < |\eta| < 3.0$.

2.2. Reconstruction in the ECAL

An electron or a photon is reconstructed from the electromagnetic shower in the ECAL. A cluster is built by collecting together the energy deposits (called "rechits") left by this shower in the detector. Each cluster represents a single particle or several overlapping particles. However, electrons and photons can interact with the material in front of the ECAL: electrons emit bremsstrahlung photons, and photons convert into electron–positron pairs, resulting in multiple nearby clusters in the ECAL. These clusters have to be combined to reconstruct the energy of the initial particle. The combination of the sub-clusters is called a SuperCluster [10].

Currently, a geometrical approach is used, called the "Mustache" algorithm. The idea is to combine all the clusters that fall into a specified window around the cluster with the highest energy ("seed") into a SuperCluster. This window has a shape resembling a mustache in the (η, ϕ) plane. This shape is chosen because the clusters are wider along the transverse ϕ-axis rather than the longitudinal η-axis, due to the CMS magnetic field (3.8 T). The size of the Mustache window depends on the η-position of the seed and the energy of the cluster.

This algorithm is very efficient; however, there are multiple effects that degrade its performance in terms of energy reconstruction:

- Energy lost before reaching the ECAL, and in detector gaps.
- Energy leakage out of the back of the ECAL.
- The use of finite energy thresholds to suppress noise in the detector electronics.
- Energy deposited by the multiple additional interactions, so-called pileup interactions.

Currently, to mitigate the effect of these issues, a multivariate regression technique (Boosted Decision Tree) trained on simulated photons is used to define an energy correction. The inputs to the BDT are ≈30 high-level variables that describe the shower.

Both for the SuperClustering and energy regression tasks, we propose new methods based on state-of-the-art ML tools.

3. SuperClustering

3.1. DeepSC Model

We developed a new model, called DeepSC, for the SuperClustering. The first step of this algorithm is similar to the Mustache: a window (rectangular shape) is opened around the seed. In the second step, the model predicts whether each cluster in this window belongs to the SuperCluster associated with the corresponding seed, instead of simply taking all of the clusters. Apart from the cluster classification, the DeepSC model also predicts energy correction for each identified SuperCluster [11]. This is the first ML method developed for cluster assignment to the SuperCluster in CMS.

The architecture of the new DeepSC model is presented on Figure 1. The main building blocks of the model are the following:

- Dense layers are used to extract the vectors of the latent features.
- Self-Attention Layers [12,13] that help the network to focus on the most important features.
- Graph Convolutional Network/Graph Highway Network (GHN) [14], where the information can be shared and aggregated between the close-by clusters. The two

algorithms are very similar to each other, with GHN being more robust to over-smoothing during the training.

The architecture of the new DeepSC model, based on GNNs and self-attention layers, is presented in Figure 1.

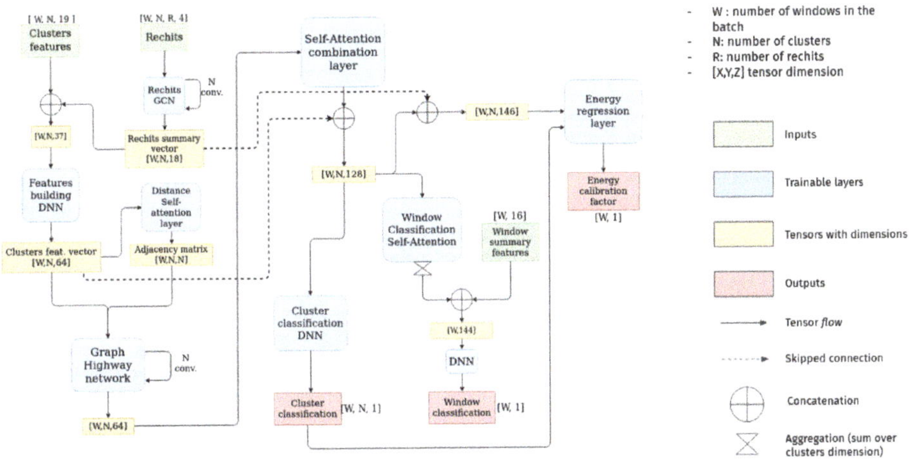

Figure 1. DeepSC model architecture. The input to the network consists of selected features and rechits of the clusters that fall into a predefined geometrical window. Using dense layers, the latent features are extracted from the initial input; they are processed and combined together using different types of graph architectures: Graph Convolutional Network (GCN) and Graph Highway Network (GHN). Self-attention layers are used as well, to help the network with focusing on the most important features/inputs. The final outputs are the following: information on whether each of the clusters belongs to the SuperCluster (cluster classification), the type of the particle from which the SuperCluster originated (window classification), and energy correction (energy calibration factor).

In addition to the SuperCluster reconstruction, the model can also predict the flavor of the particle from which the SuperCluster has emerged. The discrimination is done between three flavors: photon, electron, and jet. However, we do not aim at reconstructing the energy of the jets, since it is performed using a standard jet cone algorithm [15]. Therefore, to avoid performance degradation in terms of energy reconstruction for electrons/photons when adding jet discrimination, a ML technique called transfer learning [16] is used. Transfer learning consists of two steps. First, the model is trained only on an electron/photon sample to achieve the optimal performance for the energy reconstruction. Then, the model is re-trained, adding the jet sample, but "freezing" all the parts of the model that are not connected with particle identification. In this way, the reconstruction of the SuperCluster will not be affected by the jet sample.

The DeepSC algorithm is the first attempt to predict the particle flavor, cluster assignment, and energy correction at the same time with ML using raw detector level information.

3.2. Dataset Description

A dataset is generated to test the performance of the algorithm. Events are simulated using a full CMS Monte Carlo simulation at 14 TeV, with particles (electrons, photons, and partons) being generated uniformly in pseudorapidity and in a p_T range from 1 to 100 GeV. A pileup scenario with the number of true interactions uniformly distributed in the range of 55 to 75 is used. For the jet sample, every event is required to have at least one photon pair coming from a π_0.

One entry of the dataset is created in the following way: first, a rectangular window is opened around the seed (an energy threshold of 1 GeV). The size of the window depends on the position in η. All the clusters that fall in the specified window around the seed are passed to the network as an input. In more detail, the input for the network contains: cluster information (E, E_T, η, ϕ, z, number of crystals, and information relative to seed: $\Delta\eta$, $\Delta\phi$, ΔE, ΔE_T), list of rechits for each cluster, and summary window features (max, min, mean of E_T, E, $\Delta\eta$, $\Delta\phi$, ΔE, ΔE_T of all the clusters in the window).

3.3. Results

3.3.1. Energy Resolution

In the case of the DeepSC algorithm, the energy of the initial particle is reconstructed in two steps. First, the energy sum of all the clusters of the network assigned to a SuperCluster is calculated (E_{Raw}). Second, the energy correction coefficient is applied to E_{Raw} to achieve a better resolution. In this work, the impact of the first step on energy resolution is presented.

Both Mustache and DeepSC algorithms were applied to the same dataset to compare the performance. Figure 2 shows the resolution of the reconstructed uncorrected SuperCluster energy (E_{Raw}) divided by the true energy deposits in ECAL (E_{Sim}) versus the transverse energy of the generated particle E_T^{Gen} (left) and the number of simulated pileup (PU) interactions (right). The resolution is computed as being half the difference between the 84% quantile and the 16% quantile (one σ) of the E_{Raw}/E_{Sim} distribution in each bin. The lower panel shows the ratio of the resolution of the two algorithms: $\sigma_{DeepSC}/\sigma_{Mustache}$. The results are presented for photons; the performance for electrons is similar.

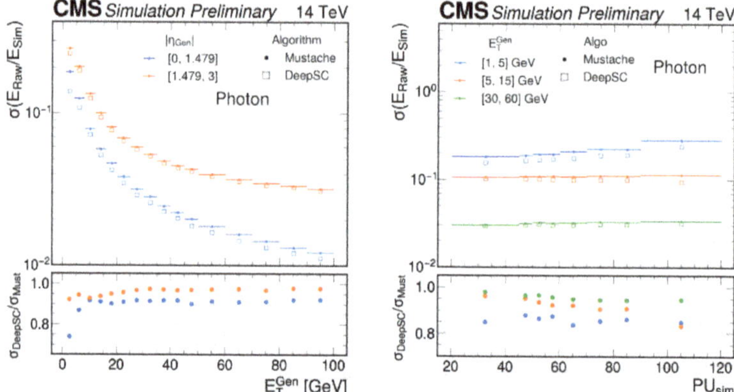

Figure 2. Energy resolution for DeepSC and Mustache algorithms. The resolution of the reconstructed uncorrected energy divided by the true energy deposits vs. generated transverse energy of the particle (**left**) and the number of simulated pileup interactions (**right**) is presented. The bottom panels show the ratio of the energy resolutions quantifying the improvement of the DeepSC model over the Mustache algorithm.

The DeepSC algorithm achieves improved performance, especially in the low-E_T and high-pileup regions, where the pileup and the noise significantly degrade the Mustache algorithm resolution. The performance of the DeepSC model in terms of the energy correction results are still under study. In Section 4 of this paper, we discuss another model for energy correction prediction with a similar approach using GNNs on low-level detector information.

3.3.2. Particle Identification

The output of the model for particle identification is the likelihood for the clusters in the window to originate from electron/photon/jet (score). In Figure 3, we show the results obtained for the jet scores in the jet and photon data samples (left), and the electron scores in the photon and electron samples (right).

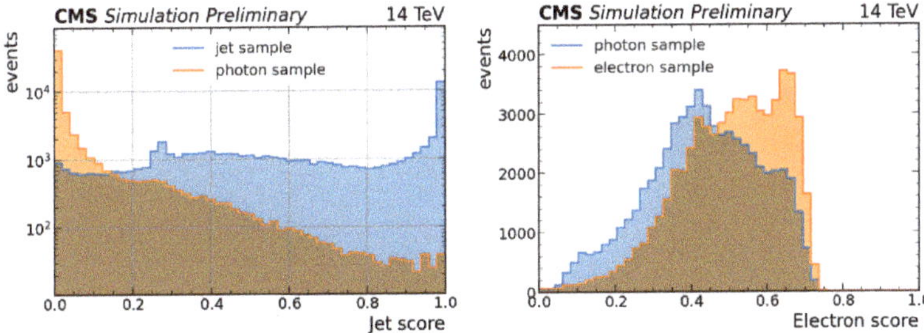

Figure 3. Model score distributions for particle identification. The jet score (**left**) represents the likelihood of the SuperCluster to originate from a photon. Clear discrimination between jet and photon samples is visible for this case. The electron score (**right**) represents the likelihood of the SuperCluster to originate from an electron. It demonstrates that despite the absence of the tracker information, some discrimination can be achieved between photon and electron samples.

We can see a clear discrimination between photon and jet samples achieved using the DeepSC model. This information could be used to improve the global CMS event reconstruction (Particle Flow reconstruction [17]), as well as provide extra input information for photon identification algorithms in offline analyses.

The efficient separation between electrons and photons can only be achieved by adding the tracker information to the ECAL. However, it is interesting to see the level of discrimination obtained by the model using only ECAL, and it is worth further investigation.

End-to-end comparison and complementarity with existing algorithms used in CMS are still under study.

4. Energy Regression

4.1. The Dynamic Reduction Network

For the energy regression task, we also propose to use a neural network. It is the first time raw detector information is used in the ML algorithm for the energy correction in CMS. Generally, neural networks perform best when low-level features are included, and in our case, we use rechits as an input. This will also mitigate the bias coming from human-engineered features, used in the current approach based on a multivariate regression with a BDT. Moreover, the rechits in the calorimeter are quite sparse and vary in number for each particle (from 1 to 100). In this case, it is natural to represent them as points of the graph. Therefore, the new architecture, the dynamic reduction network (DRN) [18] that we have developed for this task, is built on point cloud graph neural network techniques. The input to the model is a point cloud of rechits in the (position, energy) space, and graphs are formed by drawing edges between neighboring hits in a high-dimensional latent space.

The DRN is based on dynamic graph neural networks with the addition of a pooling step analogous to subsampling in CNNs. Our architecture [19] is summarized in Figure 4; the main steps are as follows:

1. The position and energy coordinates of each RecHit are mapped into a high-dimensional latent space using a fully-connected neural network.
2. The message-passing process is performed to aggregate the information between the neighbors and learn the global information.
3. Additional human-engineered features are added to the learned features that were not encoded in the initial hit collection. In particular, two additional features that describe the amount of energy leakage at the back of the ECAL and the energy density from pileup events, are concatenated to the learned features.
4. The resulting set of high-level features is passed through another fully connected neural network to produce the regression output.

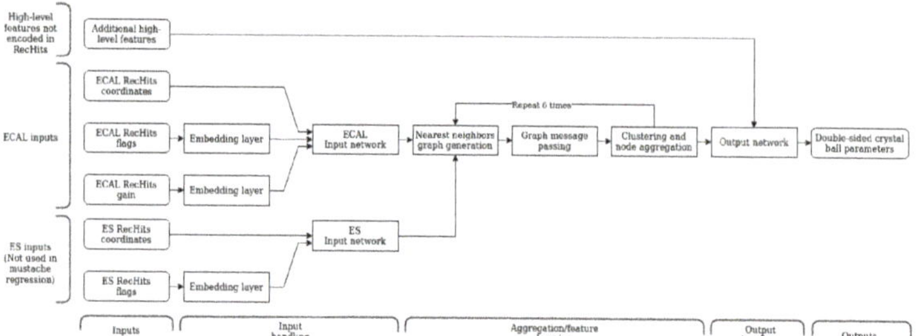

Figure 4. Flowchart of the operation of the Dynamic Reduction Network. A point cloud of rechits is mapped into a high-dimensional latent space using a fully-connected neural network, where it is then iteratively transformed and pooled using graph operations. This resulting high-level learned features are then concatenated with extra high-level information not available from the raw collection of rechits, and passed through another fully-connected neural network to obtain the regression output.

4.2. Training

The model was trained on realistic detector simulation data, which accurately models particle interactions and detector effects, including pileup. This gives us access to the truth energy values, allowing for supervised training. Our training sample consists of simulated photons with a flat p_T distribution in the range from 25 to 300 GeV fired directly into the detector. Our training data is generated under exactly the same conditions as that used to train the current BDT model.

4.3. Results

To compare the performance of the BDT that is currently used in the CMS reconstruction and the DRN model, we applied both of the algorithms to the same photon sample. To obtain the energy resolution, the histograms of E_{pred}/E_{true} were constructed for different transverse momentum ranges p_T and then fitted with the Cruijff function [20] to obtain the key metrics: mean response (μ) and relative resolution (σ/μ).

Figure 5 shows the obtained relative resolutions as a function of the particle's transverse momentum p_T.

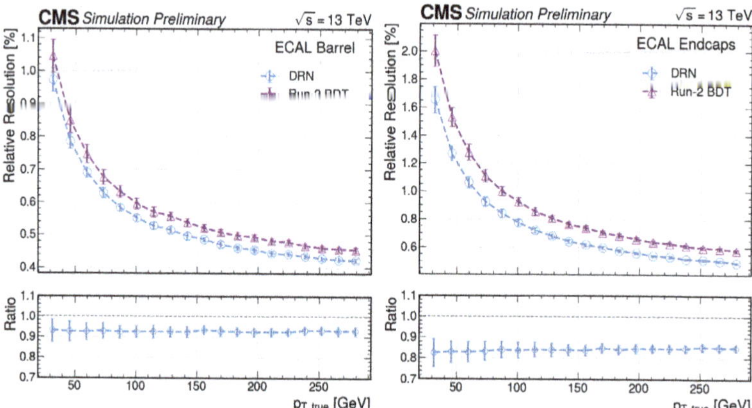

Figure 5. Dynamic Reduction Network (DRN) and Boosted Decision Tree performance in the ECAL barrel (**left**) and endcaps (**right**) as a function of generated transverse momentum. The DRN shows an improved resolution by >10%.

The DRN shows an improved resolution by a factor of >10% compared to the BDT for the whole momentum range.

To compare the performance in the actual analysis, the algorithms were also applied on the simulated data for the di-photon invariant mass distributions of H $\rightarrow \gamma\gamma$. The results are shown in Figure 6.

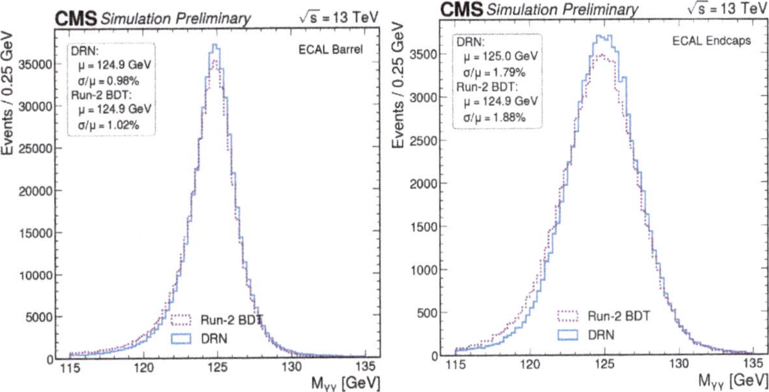

Figure 6. Di-photon invariant mass distributions of H $\rightarrow \gamma\gamma$ events for both the Dynamic Reduction Network (**DRN**) and Boosted Decision Tree architectures in the ECAL barrel (**left**) and endcaps (**right**). The DRN shows an improved resolution by >5% in both detector regions.

In this case, the DRN is able to obtain an improved resolution with respect to the BDT by a factor of >5%, both in the barrel and endcaps of the ECAL.

5. Conclusions

In this paper, we presented two novel ML approaches for the reconstruction in calorimetry. Particularly, two different GNN-based architectures were developed for the reconstruction of electromagnetic objects. The DeepSC model can be used for the clustering of energy deposits in the ECAL, as well as to bring extra information on particle identification. The DRN model predicts the energy corrections to be applied to electrons and photons. Both methods show significantly improved performance in energy resolution by about 10 % in comparison to the current reconstruction algorithms used for the ECAL.

Even though the two discussed models are currently developed independently from each other, it is possible to apply them consequently in order to achieve better performance. First, the DeepSC model can be used to retrieve the optimal cluster assignment, and afterwards, the energy correction can be calculated using the DRN. In the future, we plan to combine these two methods for energy reconstruction in the ECAL.

Funding: P.S. was supported by the CEA NUMERICS program, which has received funding from the European Union's Horizon 2020 research and innovation program under the Marie Sklodowska-Curie grant agreement No. 800945.

Data Availability Statement: Restrictions apply to the availability of these data. Data was obtained from the CMS collaboration and are available with its permission.

Conflicts of Interest: The author declares no conflict of interest. The funders had no role in the design of the study; in the collection, analyses, or interpretation of data; in the writing of the manuscript, or in the decision to publish the results.

References

1. The CMS Collaboration. The CMS experiment at the CERN LHC. *J. Instrum.* **2008**, *3*, S08004.
2. The CMS Collaboration. Measurements of Higgs boson production cross sections and couplings in the diphoton decay channel at √s= 13 TeV. *J. High Energy Physic.* **2021**, *2021*, 27. [CrossRef]
3. The CMS Collaboration. Search for nonresonant Higgs boson pair production in final state with two bottom quarks and two tau leptons in proton-proton collisions at √s = 13 TeV. *arXiv* **2021**, arXiv:2011.12373.

4. Stoye, M. and on Behalf of the CMS Collaboration. Deep learning in jet reconstruction at CMS. *J. Phys. Conf. Ser.* **2018**, *1085*, 042029. [CrossRef]
5. Pata, J.; Duarte, J.; Mokhtar, F.; Wulff, E.; Yoo, J.; Vlimant, J.R.; Pierini, M.; Girone, M. Machine Learning for Particle Flow Reconstruction at CMS. *arXiv* **2022**, arXiv:2203.00330. Available online: https://cds.cern.ch/record/2802826 (accessed on 20 July 2022).
6. Shlomi, J.; Battaglia, P.; Vlimant, J.R. Graph Neural Networks in Particle Physics. *Mach. Learn. Sci. Technol.* **2021**, *2*, 021001. [CrossRef]
7. Qasim, S.R.; Kieseler, J.; Iiyama, Y.; Pierini, M. Learning representations of irregular particle-detector geometry with distance-weighted graph networks. *Eur. Phys. J. C* **2019**, *79*, 608. [CrossRef]
8. Zhou, J.; Cui, G.; Hu, S.; Zhang, Z.; Yang, C.; Liu, Z.; Wang, L.; Li, C.; Sun, M. Graph Neural Networks: A Review of Methods and Applications. *arXiv* **2018**, arXiv:1812.08434.
9. The CMS Collaboration. The CMS Electromagnetic Calorimeter Project: Technical Design Report; 1997. Available online: http://cds.cern.ch/record/349375 (accessed on 20 July 2022).
10. The CMS Collaboration. Electron and photon reconstruction and identification with the CMS experiment at the CERN LHC. *J. Instrum.* **2021**, *16*, P05014. [CrossRef]
11. Valsecchi, D. Deep Learning Techniques for Energy Clustering in the CMS ECAL. 2022. Available online: https://cds.cern.ch/record/2803235 (accessed on 20 July 2022).
12. Vaswani, A.; Shazeer, N.; Parmar, N.; Uszkoreit, J.; Jones, L.; Gomez, A.N.; Kaiser, L.; Polosukhin, I. Attention Is All You Need. *arXiv* **2017**, arXiv:1706.03762.
13. Niu, Z.; Zhong, G.; Yu, H. A Review on the Attention Mechanism of Deep Learning. 2021. Available online: https://www.sciencedirect.com/science/article/pii/S092523122100477X (accessed on 6 September 2022).
14. Xin, X.; Karatzoglou, A.; Arapakis, I.; Jose, J.M. Graph Highway Networks. *arXiv* **2020**, arXiv:2004.04635.
15. Khachatryan, V.; Sirunyan, A.M.; Tumasyan, A.; Adam, W.; Asilar, E.; Bergauer, T.; Brandstetter, J.; Brondolin, E.; Dragicevic, M.; Ero, J.; et al. Jet energy scale and resolution in the CMS experiment in pp collisions at 8 TeV. *J. Instrum.* **2017**, *12*, P02014. [CrossRef]
16. Zhuang, F.; Qi, Z.; Duan, K.; Xi, D.; Zhu, Y.; Zhu, H.; Xiong, H.; He, Q. A Comprehensive Survey on Transfer Learning. *Proc. IEEE* **2021**, *109*, 43–76. [CrossRef]
17. The CMS Collaboration. Particle-flow reconstruction and global event description with the CMS detector. *J. Instrum.* **2017**, *12*, P10003. [CrossRef]
18. Gray, L.; Klijnsma, T.; Ghosh, S. A dynamic reduction network for point clouds. *arXiv* **2020**, arXiv:2003.08013.
19. Rothman, S. Calibrating Electrons and Photons in the CMS ECAL Using Graph Neural Networks. 2021. Available online: https://cds.cern.ch/record/2799575 (accessed on 20 July 2022).
20. del Amo Sanchez, P.; Lees, J.P.; Poireau, V.; Prencipe, E.; Tisser, V.; Tico, J.G.; Grauges, E.; Martinelli, M.; Palano, A.; Pappagallo, M.; et al. Study of $B \rightarrow X\gamma$ decays and determination of $|Vtd/Vts|$. *Phys. Rev. D* **2010**, *82*, 051101. [CrossRef]

Article

Secondary Emission Calorimetry

Burak Bilki [1,2,3,*], Kamuran Dilsiz [4], Hasan Ogul [5], Yasar Onel [2], David Southwick [2], Emrah Tiras [6], James Wetzel [2] and David Roberts Winn [7]

1 Department of Mathematics, Beykent University, Istanbul 34500, Turkey
2 Department of Physics and Astronomy, University of Iowa, Iowa City, IA 52242, USA
3 Turkish Accelerator and Radiation Laboratory, Ankara 06830, Turkey
4 Department of Physics, Bingöl University, Bingöl 12000, Turkey
5 Department of Nuclear Engineering, Sinop University, Sinop 57000, Turkey
6 Department of Physics, Erciyes University, Kayseri 38030, Turkey
7 Department of Physics, Fairfield University, Fairfield, CT 06824, USA
* Correspondence: burak.bilki@cern.ch

Abstract: Electromagnetic calorimetry in high-radiation environments, e.g., forward regions of lepton and hadron collider detectors, is quite challenging. Although total absorption crystal calorimeters have superior performance as electromagnetic calorimeters, the availability and the cost of the radiation-hard crystals are the limiting factors as radiation-tolerant implementations. Sampling calorimeters utilizing silicon sensors as the active media are also favorable in terms of performance but are challenged by high-radiation environments. In order to provide a solution for such implementations, we developed a radiation-hard, fast and cost-effective technique, secondary emission calorimetry, and tested prototype secondary emission sensors in test beams. In a secondary emission detector module, secondary emission electrons are generated from a cathode when charged hadron or electromagnetic shower particles penetrate the secondary emission sampling module placed between absorber materials. The generated secondary emission electrons are then multiplied in a similar way as the photoelectrons in photomultiplier tubes. Here, we report on the principles of secondary emission calorimetry and the results from the beam tests performed at Fermilab Test Beam Facility as well as the Monte Carlo simulations of projected, large-scale secondary emission electromagnetic calorimeters.

Keywords: secondary electron emission; radiation hardness; forward calorimetry; electromagnetic calorimetry

Citation: Bilki, B.; Dilsiz, K.; Ogul, H.; Onel, Y.; Southwick, D.; Tiras, E.; Wetzel, J.; Winn, D.R. Secondary Emission Calorimetry. *Instruments* **2022**, *6*, 48. https://doi.org/10.3390/instruments6040048

Academic Editors: Fabrizio Salvatore, Alessandro Cerri, Antonella De Santo and Iacopo Vivarelli

Received: 21 August 2022
Accepted: 8 September 2022
Published: 21 September 2022

Publisher's Note: MDPI stays neutral with regard to jurisdictional claims in published maps and institutional affiliations.

1. Introduction

The development of radiation-hard calorimeter systems is a long-standing problem. Despite the continuous need for this development, the amount of effort dedicated to R&D in this area is quite limited. In addition to a lack of novel developments, the currently operational detector systems suffer considerably from the lack of solid predictions of the effect of radiation on the active elements and the readout systems. Along this line, we attempted developing a novel, intrinsically radiation-hard calorimeter system based on the secondary emission (SE) principle. The detector modules envisaged will primarily utilize metal channel dynode chains, similar to that of the photomultiplier tubes, each coated with high secondary electron emission yield materials. The considered detector modules will be planar, of high granularity and tileable. The secondary emission technology is envisaged to be an asset for future implementations requiring radiation-hard, robust and cost-effective electromagnetic calorimeters [1,2].

Here we report on the principles of secondary emission calorimetry and the results from the beam tests of a dedicated secondary emission module constructed with basic principles. The Monte Carlo simulations of projected, large-scale secondary emission electromagnetic calorimeters are also presented.

2. Secondary Emission Detector Modules

In an SE detector module, SE electrons (SEe) are generated from an SE surface in the form of the cathode and the dynodes when charged hadronic or electromagnetic particles (shower particles) penetrate an SE sampling module either placed between absorber materials (Fe, Cu, Pb, W, etc.) in calorimeters or as a homogeneous calorimeter consisting entirely of dynode sheets as the absorbers. An SE cathode is a thin film, similar to the dynodes of photomultiplier tubes (PMTs). These films are typically simple metal oxides Al_2O_3, MgO, CuO/BeO, or other higher-yield materials. These materials are known to be very radiation-hard, as they are used in PMTs for up to 50 Grad dose and in accelerator beam monitors exposed to fluxes of higher than 10^{20} mip/cm^2 (see, e.g., [3] or [4]).

On the inner surface of a metal plate in vacuum, which serves as the entrance window to a compact vacuum vessel which is either metal or metal-ceramic, an SE film cathode is analogous to a photocathode, and the shower particles are similar to incident photons. The SEe produced from the top SE surface by the passage of shower particles, as well as the SEe produced from the passage of the shower particles through the dynodes, are similar to photoelectrons. The SEe are then amplified by sheets of dynodes, which could be metal meshes or other planar dynode structures. The SEe yield is a strong function of momentum, following dE/dx as in the Sternglass formula [5]. This variation with particle energy gives rise to quasi-compensation effects as the low-energy nuclear fragments of hadron showers have high yields, e.g., a 1 MeV alpha particle produces around 20 SEe. The comparison between SEe and photoelectrons should be emphasized: both are the result of dynode amplification. In a scintillation calorimeter, many photons are made per GeV, but typically only around 1–0.1% are collected and converted to photoelectrons; in an SE calorimeter, relatively few SEe from the shower particles are generated as the showers pass through the dynodes, but essentially all those SEe are amplified by the downstream dynodes. The result is that the statistics of photoelectrons and SEe are similar [6].

The construction requirements for an SE sensor module compared to the requirements for the construction of PMTs have several simplifications:

- The entire final assembly can be done in air. Dynodes used as particle detectors in mass spectrometers or in beam monitors cycle to air repeatedly.
- There are no critically controlled thin film vacuum depositions as in the case of photocathodes.
- Bake-out can be at refractory temperatures, unlike a photocathode, which degrades at temperatures higher than 300 °C.
- The SE module is sealed by normal vacuum techniques, and the necessary vacuum is 100 times worse compared to the PMTs.

The modules envisaged are compact, high gain, high speed, exceptionally radiation damage resistant, rugged, and cost effective, and can be fabricated in arbitrary tileable shapes.

3. Tests of the First SE Prototype Module

Due to the intrinsic similarities between the PMTs and the envisaged SE modules, the concept of an SE module can be validated by implementing relevant modifications to the PMTs. Since the photocathode functionality is not present in an SE module, the PMTs selected to construct the first SE module had excessive usage, and therefore had potentially degraded photocathodes. In addition, the photocathodes of the PMTs had the option of being disconnected from the multiplication chain so that the PMTs would not be responsive to any photons entering through or created at the window. Therefore, the entire dynode chain is utilized as SE surfaces. The largest signal is produced by an SEe produced at the first dynode (or the cathode).

The first SE prototype module was constructed with seven Hamamatsu single anode R7761 PMTs and was extensively tested at the Fermilab Test Beam Facility [7] with 4, 8 and 16 GeV electron beams. The characterization of the PMTs for the first SE sensor can be found in [8].

Figure 1 shows pictures of the first SE module. The module is designed to house the seven SE detectors in a closest-packed structure. A special electronics board was designed and produced for the first SE module. Figure 2 shows the circuit diagram of the electronics board for powering and readout of a single PMT.

Figure 1. Pictures of the first SE module. Each sensor was 39 mm in diameter with an active window diameter of 27 mm. The length of the sensors was 50 mm.

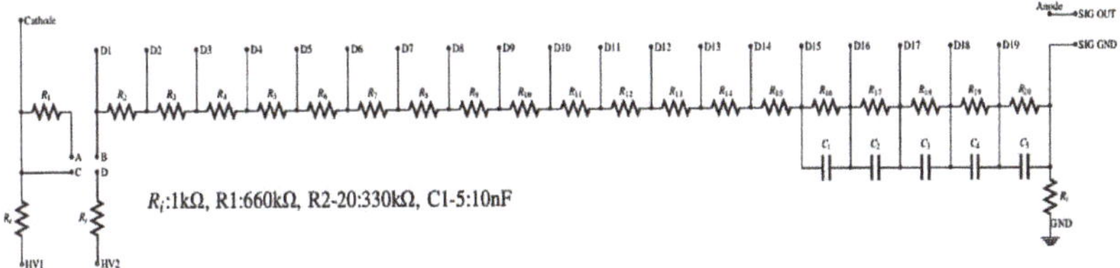

Figure 2. The circuit diagram of the baseboards for the powering and readout of a single PMT in the SE module.

Three different modes of operation exist on the baseboard for R7761 PMTs:

- Mode 1—normal divider mode: In this mode, the photomultiplier voltage divider chain is not modified and has equal potential differences across the dynodes, except the one across the cathode-first dynode (C–D1) gap, which is twice as large. This is the reference design from Hamamatsu.
- Mode 2—cathode-first dynode shorted: In this mode, jumpers on the board enable the bridging of R1, so that there is zero potential across the C–D1 gap (VC − VD1 = 0 V).
- Mode 3—cathode float mode: The design of the board allows the cathode to be separated from the remainder of the divider chain and be powered separately by another high voltage source. The potential across the C–D1 gap can also be adjusted such that it becomes positive with respect to the gap of D1–D2. If a second high voltage source is not used, the photocathode can still be charging up slightly. Dedicated tests resulted in no noticeable change in the response in particle beams when the photocathode was slightly reverse biased.

All of these modes can be examined in Figure 2, where the A-B bridge forms normal operation mode (Mode 1) with HV input on HV1, the B-C bridge forms Mode 2 with HV input on HV1, and the B-D bridge forms Mode 3 with HV input on HV2. Mode 2 was the default mode of operation for the beam tests.

Steel and tungsten absorbers were placed upstream of the SE module at increasing thicknesses to measure the shower development. With the 20 cm × 20 cm × 1.9 cm steel absorbers, all seven SE detectors were read out, and with the 3 cm × 3 cm × 0.35 cm tungsten absorbers, only the center module was read out. The lateral coverage of the SE module was not sufficient to produce a shower signal that scales with the shower depth with the steels absorbers. Therefore, the tests with the tungsten absorbers were taken as the baseline.

Figure 3 shows the response of the module to 8 (left) and 16 GeV (right) positrons with the tungsten absorbers. With a careful design of the trigger counters and the event selection based on the wire chambers, the electromagnetic shower profiles are accurately produced. The measurements (black) are also validated with Monte Carlo (MC) simulations (red). Figure 3 validates the concept of secondary emission sensors utilizing dynode chains similar to that of the photomultiplier tubes.

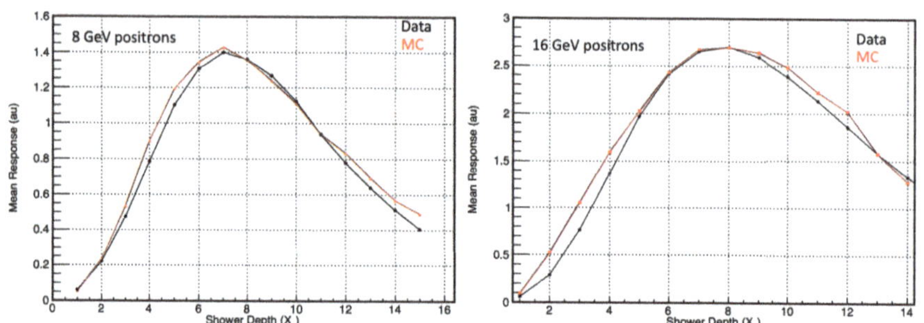

Figure 3. The response of the SE module to 8 (**left**) and 16 GeV (**right**) positrons with the tungsten absorbers.

4. Enhancement of Secondary Electron Emission

In order to enhance the production of secondary electrons in the SE modules, the cathode and the dynodes of the SE sensors can be made by coating the mesh copper foils with secondary emitters such as Al_2O_3, SnO_2, TiO_2 or ZrO_2. The coating can be done with vapor deposition techniques such as magnetron sputtering, for which Al_2O_3 and TiO_2 are very common targets.

Figure 4 (left) shows the simulated efficiencies for different thicknesses of Al_2O_3, SnO_2, TiO_2 and ZrO_2. The best performance is with a 100 nm thick Al_2O_3. The secondary electron emission efficiency is between 2 and 5% for all simulated secondary emitter coatings.

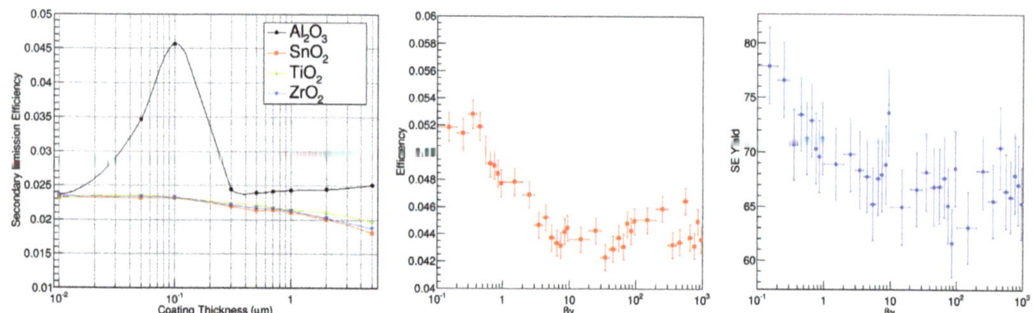

Figure 4. The simulated efficiencies for different thicknesses of Al_2O_3, SnO_2, TiO_2 and ZrO_2 (**left**), the secondary electron emission efficiency (**center**) and the SE yield (**right**) of the cathode once efficient as a function of the $\beta\gamma$ of the traversing particle for a 100-nm Al_2O_3-coated copper foil.

Figure 4 shows the secondary electron emission efficiency (center) and the secondary emission yield (right) of the cathode once efficient, i.e., there is at least one secondary electron produced at the cathode, as a function of the $\beta\gamma$ of the traversing particle for a 100-nm Al_2O_3-coated copper foil. The minimum ionization occurs around a $\beta\gamma$, of 40 which corresponds roughly to 4 GeV of muon energy. The average secondary electron yield is roughly around 68 with an increasing trend for lower $\beta\gamma$.

5. Projection of a Large-Scale SE Calorimeter Performance

In order to perform a simulation study for a large-scale SE calorimeter system, SE modules with 9-stage dynode chains were modeled. The number of dynodes was chosen so that the total number in one layer is minimum and the signal is still measurable. The dynodes are 150 μm apart and have 10–100 μm diameter holes, which are 50–100 μm apart. Figure 5 (left) shows the simulated charge spectrum for a 9-stage secondary emission device for a minimum ionizing particle that is efficient at the cathode. With an average of 300 fC, the signal can be recorded with commercial oscilloscopes.

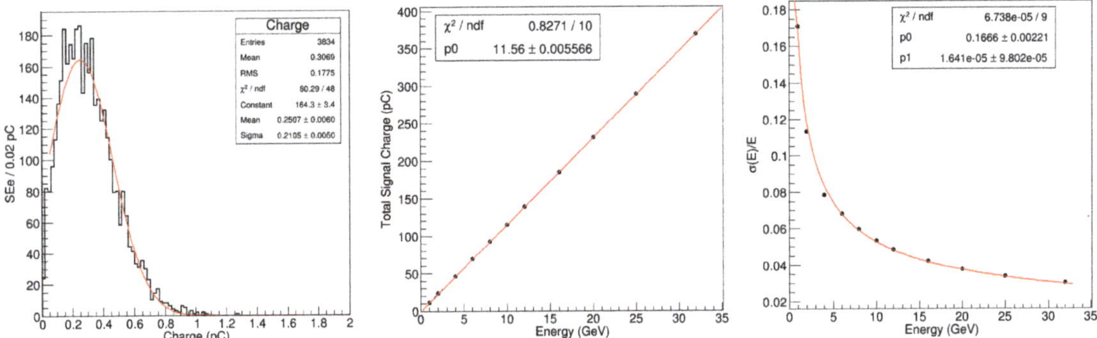

Figure 5. The simulated charge spectrum of a single layer (**left**), the MC predictions of the response linearity (**center**) and the energy resolution (**right**) of the 16-layer SE calorimeter prototype.

The electromagnetic response of an SE calorimeter prototype with 16 active layers interleaved with 1 X_0 tungsten absorbers was also simulated. The lateral size of the dynodes and the calorimeter layer was 1 m with no dead areas. The simulated electrons were normally incident on the front face of the calorimeter stack. Figure 5 (center and right) shows the MC predictions of the performance of the SE calorimeter prototype. The predictions are obtained for available Fermilab test beam energies of positrons/electrons for practical reference. The detector response is linear in the energy range of 1–32 GeV (center), and the electromagnetic energy resolution is obtained as $(16.7\%)/\sqrt{E}$ with a negligible constant term (right).

6. Conclusions

Secondary emission calorimetry is a feasible option particularly for electromagnetic calorimetry in high-radiation environments, as well as other implementations such as beam loss monitors and Compton polarimeters. The structure of the secondary emission sensors is quite similar to the dynode chain of photomultiplier tubes. The construction of the sensor modules have less strict vacuum requirements compared to photomultiplier tubes.

The first secondary emission sensor module was constructed with photomultiplier tubes with deactivated photocathodes. The preliminary tests validate the idea and suggest a full-scale secondary emission calorimeter prototype. The Monte Carlo simulations predict good response linearity and an energy resolution of $(16.7\%)/\sqrt{E}$ for a 16 layer calorimeter prototype up to 32 GeV. The secondary electron emission can also be enhanced by special surface coatings, such as Al_2O_3, applied on the dynodes.

Highly segmented readout for imaging calorimetry is possible with the envisaged secondary emission modules.

Author Contributions: Project administration, B.B., Y.O. and D.R.W.; Investigation, K.D., H.O., D.S., E.T. and J.W. All authors have read and agreed to the published version of the manuscript.

Funding: This research received no external funding.

Data Availability Statement: The data presented in this study are available on request from the corresponding author.

Acknowledgments: B.B. acknowledges support under Tübitak grant no. 118C224.

Conflicts of Interest: The authors declare no conflict of interest.

References

1. Winn, D.; Onel, Y. Secondary Emission Calorimeter Sensor Development. *J. Phys. Conf. Ser.* **2012**, *404*, 012021. [CrossRef]
2. Ozok, F.; Yetkin, T.; Yetkin, E.A.; Iren, E.; Erduran, M.N. Geant4 simulation of a conceptual calorimeter based on secondary electron emission. *J. Instrum.* **2017**, *12*, P07014. [CrossRef]
3. Ferioli, G.; Jung, R. Evolution of the Secondary Emission Efficiencies of various materials measured in the CERN SPS secondary beam lines. In Proceedings of the Third European Workshop on Beam Diagnostics and Instumentation for Particle Accelerators (DIPAC 97), Frascati, Italy, 12–14 October 1997; pp. 168–170.
4. Gusev, Y.; Lukianov, V.; Mamaeva, G.; Musienko, Y.; Reucroft, S.; Seliverstov, D.; Shusterman, L.; Swain, J. Radiation-hard photodetectors based on fine-mesh phototubes for calorimetry in very forward rapidity. *Nucl. Instrum. Ad Methods A* **2007**, *581*, 438–442. [CrossRef]
5. Sternglass, E.J. Theory of Secondary Electron Emission by High-Speed Ions. *Phys. Rev.* **1957**, *108*, 1. [CrossRef]
6. Ashida, Y.; Friend, M.; Ichikawa, A.K.; Ishida, T.; Kubo, H.; Nakamura, K.G.; Sakashita, K.; Uno, W. A new electron-multiplier-tube-based beam monitor for muon monitoring at the T2K experiment. *Prog. Theor. Exp. Phys.* **2018**, *2018*, 103H01. [CrossRef]
7. Fermilab Test Beam Facility. Available online: https://ftbf.fnal.gov/ (accessed on 1 August 2022).
8. Tiras, E.; Dilsiz, K.; Ogul, H.; Southwick, D.; Bilki, B.; Wetzel, J.; Nachtman, J.; Onel, Y.; Winn, D. Characterization of photomultiplier tubes in a novel operation mode for Secondary Emission Ionization Calorimetry. *J. Instrum.* **2016**, *11*, P10004. [CrossRef]

instruments

Article

Preliminary Results from ADRIANO2 Test Beams

Corrado Gatto [1,2,*,†], Gerald C. Blazey [2,†], Alexandre Dychkant [2,†], Jeffrey W. Elam [3,†], Michael Figora [2,†], Todd Fletcher [2,†], Kurt Francis [2,†], Ao Liu [4,†], Sergey Los [5,†], Cole Le Mahieu[6,†], Anil U. Mane [3,†], Juan Marquez [6,†], Michael J. Murray [6,†], Erik Ramberg [5,†], Christophe Royon [6,†], Michael J. Syphers [2,†], Robert W. Young [6,†] and Vishnu Zutshi [2,†]

[1] Istituto Nazionale di Fisica Nucleare, Sezione di Napoli, 80126 Naples, Italy
[2] Department of Physics, Northern Illinois University, DeKalb, IL 60115, USA
[3] Argonne National Laboratory, Lemont, IL 60493, USA
[4] Euclide Techlabs, Bolingbrook, IL 60440, USA
[5] Fermilab, Batavia, IL 60510, USA
[6] Department of Physics & Astronomy, University of Kansas, Laurence, KS 66045, USA
* Correspondence: corrado.gatto@na.infn.it
† These authors contributed equally to this work.

Abstract: A novel high-granularity, dual-readout calorimetric technique (ADRIANO2) is under development as part of the research program of T1604 Collaboration. (Talk Presented at the 19th International Conference on Calorimetry in Particle Physics (CALOR 2022), University of Sussex, Sussex, UK, 16–20 May 2022). The building block of such a calorimeter consists of a pair of optically isolated, small size tiles made of scintillating plastic and lead glass. The prompt Čerenkov light from the glass can be exploited to perform high resolution timing measurements, while the high granularity provides good resolution of the spatial components of the shower. Dual-readout compensation and particle flow techniques can be applied simultaneously to the scintillation and to the Čerenkov section, providing excellent energy resolution as well as PID particle identification. These characteristics make ADRIANO2 a 6-D detector, suited for High Energy as well as High Intensity experiments. A report on the status of the ADRIANO2 project, preliminary measurements of light yield, and current and future R&D plans by T1604 Collaboration are discussed.

Keywords: calorimetry; dual-readout; ADRIANO2

Citation: Gerald C. Blazey; Blazey, G.; Dychkant, A.; Elam, J.; Figora, M.; Fletcher, T.; Francis, K.; Liu, A.; Los, S.; Le Mathieu, C.; et al. Preliminary Results from ADRIANO2 Test Beams. *Instruments* 2022, 6, 49. https://doi.org/10.3390/instruments6040049

Academic Editors: Fabrizio Salvatore, Alessandro Cerri, Antonella De Santo and Iacopo Vivarelli

Received: 13 August 2022
Accepted: 5 September 2022
Published: 22 September 2022

1. Introduction

The physics program at future high intensity and high energy experiments encompasses a very large number of processes, involving final states, in many cases, with complicated topologies and overlapping showers. In such an environment, calorimeters will play an important role, especially at energies above 100 GeV, as their energy resolution scales, in most cases, is $1/\sqrt{E}$. An intensive detector $R\&D$ and Monte Carlo simulation activity is already in progress within the lepton and hadron colliders communities [1]. When used in a lower-energy environment, a dual-readout calorimeter has excellent Particle IDentification (PID) capabilities, since the two independent information, obtained from each readout, provide a much better separation between particle species than a conventional, single-readout calorimeter. Consequently, a dual-readout calorimeter also has several applications for high-intensity experiments, where the final states typically have simpler topologies. In those cases, the knowledge of the particle ID is, often, more desirable than an excellent energy resolution.

The general consensus within the lepton collider community is that the jet energy resolution needed to successfully distinguish the W from the Z signal at energy ($E > 500$ GeV, scales as $\sigma(E)/E \approx 30\%/\sqrt{E}$) or better. Such a resolution is unprecedented for conventional, single-readout hadronic calorimeters, and it has been reached in the past only by compensating calorimeters with "with a large fraction of active material [2]. We note, incidentally, that a similar compensation effect can also be obtained with a small sampling fraction[3], and that the best resolution achieved is close to 30%/sqrt(E) [4]. The large volume needed to contain the showers in that class of calorimeters would make them an impractical choice for experiments with colliding beams. Furthermore, the resolution of conventional, single-readout calorimeters is limited by the fluctuations in the electromagnetic (*EM*) content of the hadronic shower and by the unequal response of such devices to the *EM* and hadronic components of the shower itself ($e/h \neq 1$) [5].

In recent years, dual-readout calorimetry [6] has been introduced as an alternative technique to cope with those effects. The dual-readout technique relies on the concept of event-by-event energy compensation by measuring, independently, the *EM* and hadronic component of each shower. As already noted, the two measurements can also be exploited to identify the particle that initiated the shower.

Dual read-out calorimetry falls under two categories: sampling and integrally active. Sampling dual-readout techniques are currently investigated by several collaborations (cfr., for example, [6–8]). While advantageous from the costing point of view, the sampling approach introduces in the energy measurement process two extra sources of energy fluctuations: (a) Poisson fluctuations in the Čerenkov signal, induced by the low photo-electron statistics, and (b) sampling fluctuations, associated to the use of a totally passive absorber. Such fluctuations not only degrade the energy measurement, but they also have detrimental consequences on particle identification. While available space and cost constraints would justify the adoption of a sampling dual-readout calorimeter in High Energy experiments, a preferred choice for High Intensity experiments would be an integrally active dual-readout technique. In such cases, in fact, the experiments are typically performed at a lower energy, therefore requiring smaller volumes to contain the particles. The showers produced have lower occupancy than those generated in High Energy experiments, and jets are rarely observed, justifying, in that case, the adoption of integrally active calorimeters, where the absorber is also active and it participates in the compensation mechanism by producing a Čerenkov signal.

The precursor of the integrally active dual readout calorimetry is the ADRIANO technique [9,10]. The central idea of ADRIANO was to mix layers of scintillator and Čerenkov radiators to independently measure the hadronic and the electromagnetic components of the energy deposited in the calorimeter. Several ADRIANO prototypes have been built and tested over the years in order to determine the relevant detector parameters and to optimize the performance in either High Energy or High Intensity applications. The baseline structure of a Čerenkov module in ADRIANO consists of long lead glass plates read out with wavelength-shifting (WLS) fibers, the latter optically coupled to the plates. The number of such fibers can be varied, depending on the application and on the desired performance. Different scintillating modules were also built using different techniques, consisting, for example, of long plates of scintillating plastics or sparsified scintillating fibers, embedded in the volume of the Čerenkov radiator, and optically decoupled from the latter.

A picture of three ADRIANO prototypes during the assembly phase is shown in Figure 1.

The R&D on ADRIANO spanned almost a decade, and several test beams were performed to characterize the technique and optimize the performance of the detector. The results [10] have indicated that the light yield (LY) of several prototypes met or exceeded the requirements set for several High Energy and High Intensity experiments. The development of ADRIANO has set the stage for the new generation of integrally active, dual-readout techniques: ADRIANO2, where the advantages of dual-readout compensation and a highly granular layout are integrated. The ADRIANO2 technique will be discussed in detail in the rest of this article.

Figure 1. *Three 105 cm long ADRIANO prototypes during the assembly phase at Fermilab's Thin Film Facility.*

2. Description of *ADRIANO*2 Techniques

All ADRIANO prototypes built within the T1015 project demonstrated consistently high light yield and good uniformity. While the non-segmented, log-style ADRIANO modules offer a low-cost solution for certain calorimetric applications, where a small number of readout channels is desired, they lack the high granularity and fast timing characteristics that are becoming increasingly important in today's experiments. This limitation is intrinsic to the chosen layout and to the inherent slowness of WLS's fibers, a characteristic that spoils the prompt aspect of Cerenkov light. The ADRIANO2 technique aims at resolving the above limitations by choosing small tiles of a plastic scintillator and of a Čerenkov radiator as building blocks of the calorimeter. The light generated in each tile would be individually read out with one or more silicon photomultipliers (SiPM) directly coupled to the tile (on-tile-SiPM). This relatively novel approach maintains the benefits of dual readout calorimetry, while opening up the possibility of applying Particle Flow Analysis (PFA) algorithms [11] to track the showers as they develop in the calorimeter, and to associate them with tracks upstream and muons downstream. Furthermore, since the Čerenkov signal is prompt, it can be exploited to accurately determine the time of passage of a charged particle in each tile for Time-of-Flight (ToF) measurements or for fast-triggering the data acquisition. Thus, three distinct measurements of the energy deposition in every scintillator-radiator tile pair can be made: the amplitude of the charge deposited, the component of the charge that is from electrons, and its precise time of arrival. The key ingredient of ADRIANO2 is the collection of the Čerenkov light produced inside small lead glass tiles, using fast SiPMs directly coupled to the glass and a fast electronics readout. Multiple tiles are sandwiched to build a calorimeter tower or a module. Typical dimensions of a tile are several cm for the side and about 1 cm for the thickness. The SiPM's and the front end electronics (FEE) are mounted on a printed circuit board (PCB) facing the tile. The latter might eventually have one or more dimples to accommodate the SiPM. A similar technique has been extensively developed at NIU [12,13] for the plastic tiles employed for the HGCAL of CMS. If several SiPM are used for reading each tile, a weighted mean algorithm can be applied to determine the position of the impinging particle. This is possible since the lead glass is a highly dispersive optical medium and the, mostly-blue Čerenkov light has a typical light path inside the tile of about 1 cm. Therefore, the amount of light reaching each sensor is a direct function of the distance it travels before being collected. This curbs the time jitter of the detector due to the small size of the tile, since only the photons traveling directly toward the sensor have a good probability of being collected, while those that follow a path with multiple bounces are, typically, absorbed. This effect portends to a very good timing resolution for ADRIANO2.

The R&D ongoing in T1604 Collaboration aims at studying the performance (light yield and timing resolution) of the glass tiles with regards to several fabrication parameters. The latter are: dimensions, surface finish, type of coating, and the eventual presence of a dimple to accommodate the SiPM. All tiles are considered have a footprint of 30×30 mm^2, matching the size of the scintillating tiles employed for the HGCAL of CMS. Thicknesses of 10 mm, 20 mm, and 30 mm were used for the prototypes. The tiles were cut to size from larger blocks of Schott SF57HTUltra, and either left ground or polished with a commercial procedure (Cat-i-glass, Elgin, IL 60177, USA). A cylindrical dimple is imparted to some of the tiles with a small-grit, diamond grinder. The dimple was subsequently polished with a water-based diamond paste. A picture of several dimpled tiles is shown in Figure 2 with blue masking fluid protecting the dimple before the coating process.

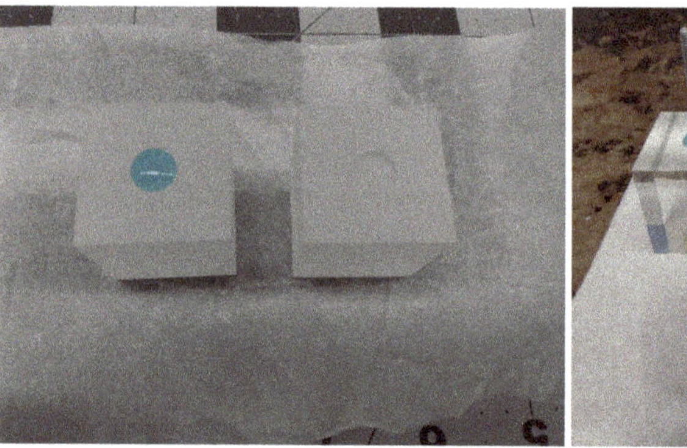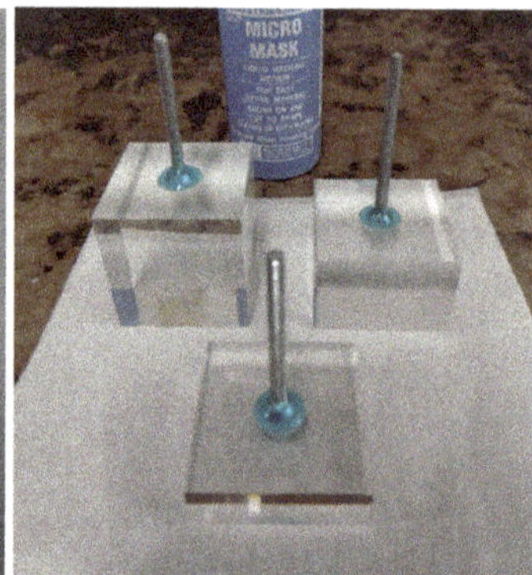

Figure 2. Ground (**left**) and polished (**right**) SF57HHT glass tiles of various dimensions.

Finally, the tiles were either wrapped or coated to suppress any leakage of the light generated internally by the particles above the Cerenkov threshold. Two diffuse wrapping (Teflon and Tyvek), two reflective wrappings (Esr2000 and aluminized Mylar), one diffuse coating (AvianB, BaSO4-based paint), and six reflective coatings (Al sputtering, Al paint, Ag sputtering, Ag paint, Mo-ALD, W-ALD) were considered. The Al sputtering was performed independently in the Chemistry Dept. of NIU and at Euclide Techlabs [14]. All atomic layer deposition (ALD) coating was conducted at the Argonne National Lab (IL).

A picture of several coated and wrapped tiles is shown in Figure 3. The area facing the SiPM is masked with Kevlar tape or masking fluid to allow the light to reach the photo-sensor.

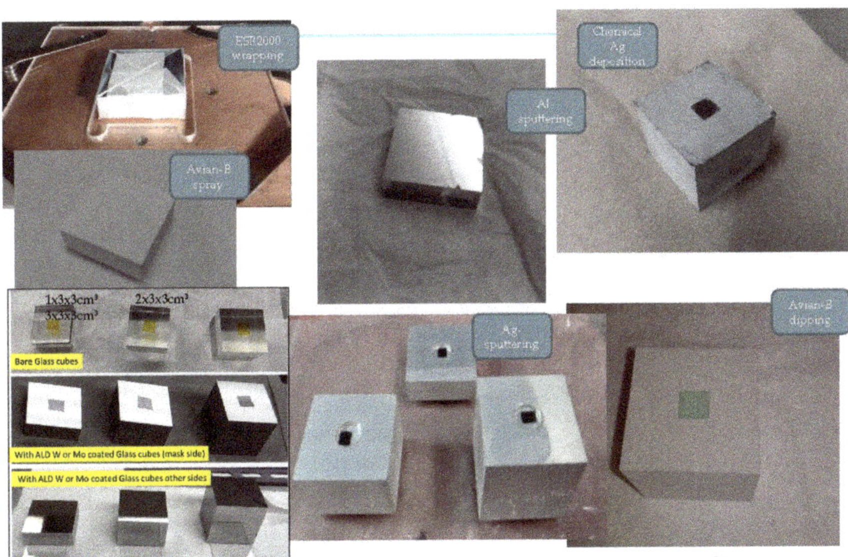

Figure 3. Sample of coated and wrapped ADRIANO2 glass tiles.

3. ADRIANO2 Readout System

The Front End Electronic (FEE) board, holding the photon sensor and the readout electronics, consists of a small board originally designed for the ORKA project, and subsequently modified for applications where fast timing is required [15]. The active component is a GALI-S66+ amplifier with a 12× gain and a bandwidth of 0.05–1500 MHz. An optional Peltier element could be accommodated. The board has multiple pads for hosting a variety of SiPM's, spanning several Hamamatsu families and sensor dimensions. One special version was designed at Fermilab, in which the central sensor was replaced by four peripheral sensors, actively ganged into one amplifier. A picture of a single-sensor and of a quadruple-sensor FEE boards are shown in Figure 4. Two species of Hamamatsu SiPM's were used for the measurements: the S14160-6050 and the older S13360-6050.

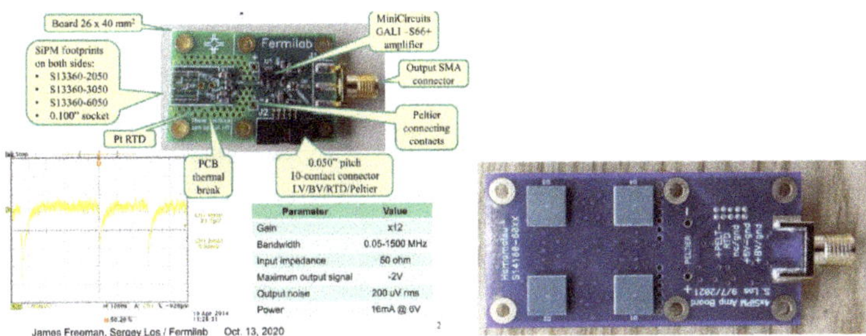

Figure 4. Single-sensor (**left**) and quadruple-sensor (**right**) FEE boards for ADRIANO2 light capture.

The FEE board is complemented by a 4-channel control board [15], to which up to four SIPM Amp boards can be connected. The control board supplies a common low voltage power for the amplifiers. All channels feature individually regulated bias and Peltier power voltages, along with a Pt10K RTD readout.

Two different DAQ systems were employed to acquire the signals from the FEE boards. A 32-ch Sampic Time Digitizer [16] was used for timing measurements, while a 16-ch Wavecatcher [17] system was used to digitize the waveforms and extract light yield information. The calibration of each board was performed by self triggering the acquisition and by fitting the distribution with multiple Gaussian curves.

4. Preliminary Results and Discussion

The current R&D focuses mainly on the lead-glass tiles, since the response of plastic tiles has been extensively studied in the past. The ADRIANO2 tiles with the same coating and surface finish were assembled into triplets consisting of 10 mm, 20 mm, and 30 mm tiles. Up to seven such triplets were positioned in a dark box and exposed to a beam of 120 GeV protons at the Fermilab test Beam facility (FTBF). A picture of the test beam setup is shown in Figure 5. Several test beams were necessary to test all combinations of surface finish and coating/wrapping. The layout chosen has the advantage that up to three tiles of the same species (belonging to the same triplet) can be used in coincidence to trigger the DAQ. Therefore, one could perform, at the same time, measurements of light yield, timing, and efficiency. A small rod of quartz with dimensions $3 \times 3 \times 6$ mm^3 readout by a Hamamatsu 4160-4050 SiPM was also used as an external trigger and as a beam position monitor. The rod was mounted on a remotely controlled x-y stage and used to scan the tile response with regards to the position of the impinging proton. All triplets were equipped with single-sensor FEE boards. One triplet, consisting of ground-surface tiles coated with Avian-B paint, was equipped with a set of 4-sensor boards.

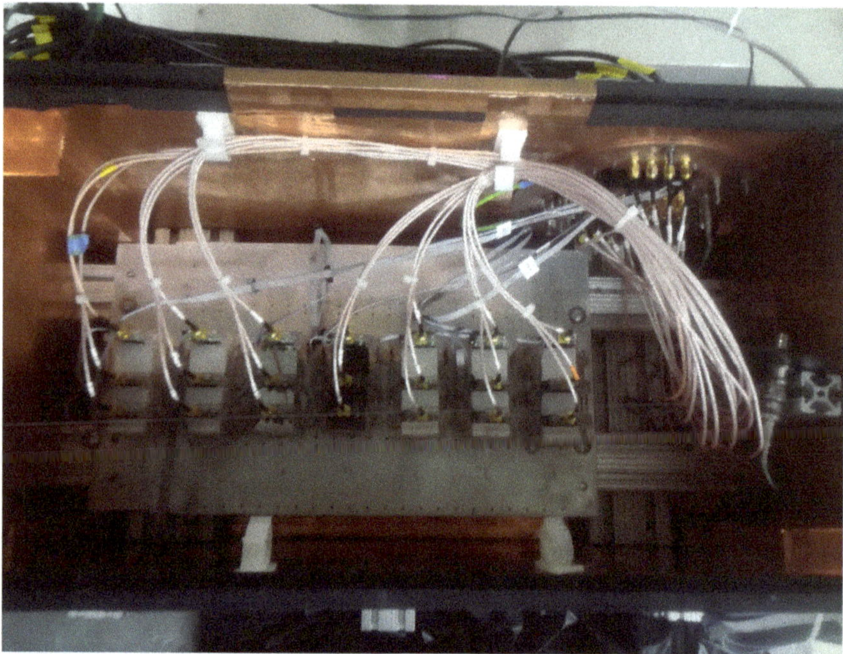

Figure 5. Setup of a test beam at Fermilab's *FTBF* of seven triplets of *ADRIANO2* tiles.

The average light yield (photoelectron/mip), measured at the center of the tile for twenty-one triplets is summarized in Figure 6. The x-axis indicates the thickness of the tile.

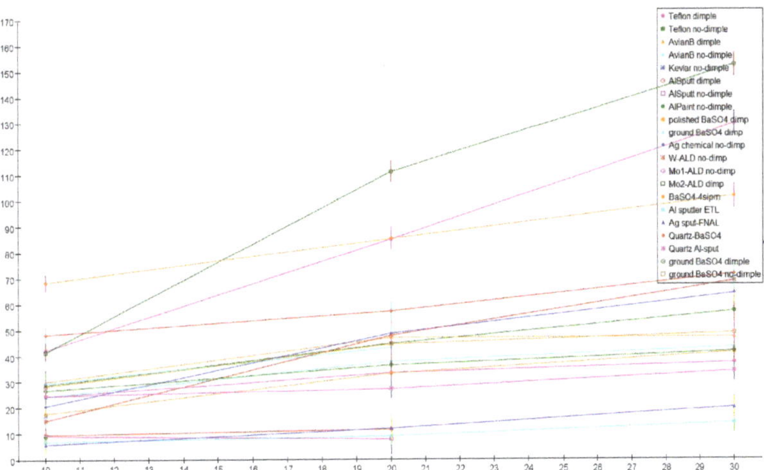

Figure 6. Light yield for 21 groups of tiles corresponding to several surface finish and coatings/wrappings.

4.1. Light Yield Measurements

The average energy deposited by a MIP particle in SF57HHT glass is ~9 MeV. The LY shows a linear dependence from the tile thickness, although the rate of change is lower than the unity. The same behavior was also reported for the plastic tiles of CMS's HGCAL [18], so it was not unexpected. The tiles instrumented with the 4-sensor FEE board consistently have a larger LY compared to all others, thanks to a $4\times$ larger sensitive area. However, the improvement in the LY is only a factor ~2.3$\times$ compared to tiles with analogous surface finish and coatings, but instrumented with single-sensor FEE boards. We also observe that the presence of a dimple does not appreciably change the LY. All tiles with mirror coating have a LY consistently lower than the tiles with diffuse coating. The only exception was observed in the two triplets coated with an ALD thin-film of Mo (differing by the thickness of the Mo film: 50 nm vs. 80 nm), which shows a LY unusually large, and a rate of increase in LY vs. tile thickness approximately two times as large as all other tiles. We are further investigating this effect, to make sure that it does not have instrumental origins. The plot in Figure 6 also shows the light yield for two tile triplets made of JGS1 glass. The measurements for those will be discussed in an upcoming article.

The Esr2000 and Al-Mylar wrapping exhibit a strong fluorescence component concomitant with the Čerenkov signal. This can be observed in the left plot of Figure 7, showing the emission spectrum of both films at 440 nm, measured with a TI QuantaMaster4/2006SE spectrofluorimeter. The right picture in Figure 7 shows a typical waveform obtained from a tile wrapped in Esr2000 when exposed to a 120 GeV proton beam, along with the waveform obtained from mirror coated tiles. For these measurements, the sampling rate of the Sampic was set at 6.4 Gsa/s (a S14160-6050 sensor was used for all waveforms). The risetime of the Esr2000 is ~6 ns, about five time larger than that measured for the other tiles, confirming that the fluorescence component is a non-negligible fraction of the total light collected by the sensor. A similar behavior was already observed by other experimenters [19]. The longer risetime makes these kind of films unsuitable for fast timing measurements. Therefore, the corresponding tiles were dropped from further measurements.

Analysis of the efficiency measurements with regards to the position of the beam are still in progress. Results will appear in an upcoming publication.

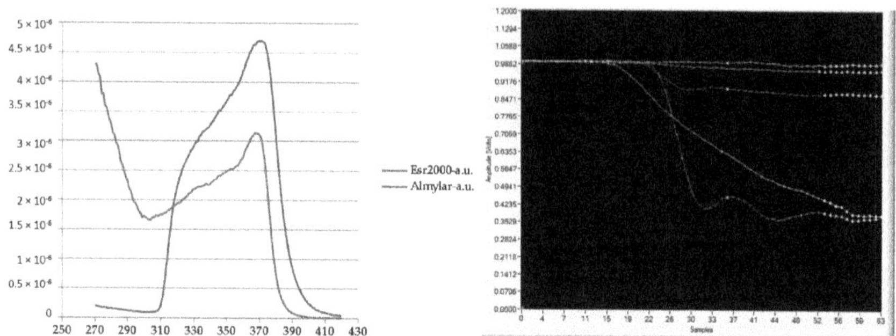

Figure 7. *A 440 nm emission spectrum (**left**) and digitized waveform (**right**) of Esr2000 and Al-Mylar films.* The sampling rate of the Sampic was set to 6.4 Gsa/s.

4.2. Timing Measurements

The analysis of the timing measurements obtained with the Sampic are still in progress and will be published in an upcoming article. Nonetheless, the behavior across the tiles families appears to be quite consistent. All mirror coated tiles have a consistently fast risetime ($\sim$1 ns with the S14160-6050 sensor and $\sim$3 ns with the S13660-6050 sensor). Timing resolution in the range $\sim$50–100 ps have been estimated when a constant fraction discrimination is applied (via software). On the other hand, all tiles coated or wrapped with diffuse materials exhibit a long risetime ($\sim$8 ns) and, consequently, a much worse timing resolution in the range of $\sim$150–250 ps. The effect is till being investigated, although it suggests that the photons are bouncing several times off the diffuse coating before eventually reaching the light sensor. The tiles instrumented with four sensors have a time resolution of $\sim$80 ps, regardless of the fact that they are coated with the Avain-B white paint. Further tests should help in clarifying the above behavior.

5. Conclusions

Several ADRIANO2 tiles, with thicknesses ranging from 10 mm to 30 mm, have been fabricated, using different surface finishes and coatings. Preliminary results from several test beams at FTBF have been reported. The light yield for nineteen groups of three tiles with the same fabrication parameters has been measured using a 120 GeV proton beam. Studies for the determination of the timing resolution of ADRIANO2 tiles are still in progress. Our goal is to identify a fabrication technique such that the timing measurement for each tile has a resolution of 80 ps (or better) when traversed by a minimum ionizing particle. Furthermore, the light yield must be consistent with an EM energy resolution of $\sigma(E)/E \approx 2\%/\sqrt{E}$ or better (stochastic fluctuations only). Five of the groups tested exhibit a light yield consistent with that goal. Preliminary analysis of the tiles' timing response suggests that three groups also possess the desired timing properties. T1604 collaboration will address these issues in the future and it will eventually exploit new coatings with improved performance for the Čerenkov light.

Author Contributions: These authors contributed equally to this work. All authors have read and agreed to the published version of the manuscript.

Funding: This research was funded by Department of Energy and Environment grant number G1A62765.

Data Availability Statement: Not applicable.

Acknowledgments: We would like to thank Evgueni Nesterov (Chem. Dept.—NIU) for their help with the optical measurements and for providing the infrastructure for the Al sputtering of the tiles. We also thank Euclide Techlabs for assisting with the Al sputtering of the tiles. The work was supported by the grant #G1A62765 funded by the U.S. Department of Energy, Office of Science and by the Fermilab's Detector R&D program.

Conflicts of Interest: The authors declare no conflict of interest.

References

1. Available online: http://www.linearcollider.org/clic-study.web.cern.ch/clic-study/ (accessed on 4 September 2022).
2. Buontempo, S.; Capone, A.; Cocco, A.G.; De Pedis, D.; Di Capua, E.; Dore, U.; Ereditato, A.; Ferroni, M.; Fiorillo, G.; Loverre, P.F.; et al. Construction and test of calorimeter modules for the CHORUS experiment. *Nucl. Instr. and Meth. A* **1994**, *349*, 250. [CrossRef]
3. Section on Calorimetry. Available online: https://journals.aps.org/rmp/abstract/10.1103/RevModPhys.90.025002 (accessed on 4 September 2022).
4. Available online: https://www.sciencedirect.com/science/article/abs/pii/016890029190062U?via%3Dihub (accessed on 4 September 2022).
5. Wigmans, R. Calorimetry energy measurement in particle physics. In *International Series of Monographs on Physics*; Oxford University Press: Oxford, UK, 2000; Volume 107.
6. Ziman, J.M. *International Series of Monographs on Physics*; Oxford University Press: Oxford, UK, 2000; Volume 107.
7. Antonello, M.; Caccia, M.; Ferrari, R.; Gaudio, G.; Pezzotti, L.; Polesello, G.; Proserpio, E.; Santoro, R. Expected performance of the IDEA dual-readout fully projective fiber calorimeter. *J. Instrum.* **2020**, *15*, C06015. [CrossRef]
8. Pezzotti, I.; Newman, H.; Freeman, J.; Hirschauer, J.; Ferrari, R.; Gaudio, G.; Polesello, G.; Santoro, R.; Lucchini, M.; Giagu, S.; et al. Dual-Readout Calorimetry for Future Experiments Probing Fundamental Physics. *arXiv* **2022**, arXiv:2203.04312. https://doi.org/10.48550/arXiv.2203.04312.
9. Gatto, C.; Di Benedetto, V.; Hahn, E.; Mazzacane, A. Status of Dual-readout R&D for a linear collider in T1015 Collaboration. *arXiv* **2016**, arXiv:1603.00909.
10. Gatto, C.; Di Benedetto, V.; Mazzacane, A.; T1015 Collaboration. Status of ADRIANO R&D in T1015 collaboration. *J. Phys. Conf. Ser.* **2015**, *587*, 012060
11. Thomson, M.A. Particle flow calorimetry and the PandoraPFA algorithm. *Nucl. Instrum. Meth.* **2009**, *A611*, 25. [CrossRef]
12. Abu-Ajamieh, F.; Blazey, G.; Cole, S.; Dyshkant, A.; Hedin, D.; Johnson, E.; Viti, I.; Zutshi, V.; Ramberg, E.; Rivera, R.; et al. Beam tests of directly coupled scintillator tiles with MPPC readout. *Nucl. Instrum. Meth.* **2011**, *A659*, 348–354 [CrossRef]
13. Blazey, G.; Chakraborty, D.; Dyshkant, A.; Francis, K.; Hedin, D.; Hill, J.; Lima, G.; Powell, J.; Salcido, P.; Zutshi, V.; Demarteau, M. Directly coupled tiles as elements of a scintillator calorimeter with MPPC readout. *Nucl. Instrum. Meth.* **2009**, *A605*, 277–281. [CrossRef]
14. Available online: https://www.euclidtechlabs.com/ (accessed on 4 September 2022).
15. Available online: https://inspirehep.net/files/1957eb3a77c24d631f51040ad2f58fd7 (accessed on 4 September 2022).
16. Available online: https://www.sciencedirect.com/science/article/pii/S0168900216308373?via%3Dihub (accessed on 4 September 2022).
17. Breton, D.; Delagnes, E.; Maalmi, J.; Rusquart, P. The WaveCatcher family of SCA-based 12-bit 3.2-GS/s fast digitizers. In Proceedings of the 19th IEEE-NPSS Real Time Conference, Nara, Japan, 26–30 May 2014; pp. 1–8. [CrossRef]
18. Belloni, A.; Chen, Y.M.; Dyshkant, A.; Edberg, T.K.; Eno, S.; Zutshi, V.; Freeman, J.; Krohn, M.; Lai, Y.; Lincoln, D.; et al. Test Beam Study of SiPM-on-Tile Configurations. *arXiv* **2021**, arXiv:2102.08499. https://arxiv.org/abs/2102.08499
19. Janecek, M. Reflectivity Spectra for Commonly Used Reflectors. *IEEE Trans. Nucl. Sci.* **2012**, *59*, 490–497. [CrossRef]

*instruments*

Project Report

TAO—The Taishan Antineutrino Observatory

Hans Theodor Josef Steiger [1,2] on behalf of the JUNO Collaboration

[1] Cluster of Excellence PRISMA+, Staudingerweg 9, 55128 Mainz, Germany; hsteiger@uni-mainz.de
[2] Institute of Physics, Johannes Gutenberg University Mainz, Staudingerweg 7, 55128 Mainz, Germany

Abstract: The Taishan Antineutrino Observatory (TAO or JUNO-TAO) is a satellite detector for the Jiangmen Underground Neutrino Observatory (JUNO). JUNO aims at simultaneously probing the two main frequencies of three-flavor neutrino oscillations, as well as their interference related to the mass ordering, at a distance of ~53 km from two powerful nuclear reactor complexes in China. Located near the Taishan-1 reactor, TAO independently measures the antineutrino energy spectrum of the reactor with unprecedented energy resolution. The TAO experiment will realize a neutrino detection rate of about 2000 per day. In order to achieve its goals, TAO is relying on cutting-edge technology, both in photosensor and liquid scintillator (LS) development which is expected to have an impact on future neutrino and Dark Matter detectors. In this paper, the design of the TAO detector with a special focus on calorimetry is discussed. In addition, an overview of the progress currently being made in the R&D for a photosensor and LS technology in the frame of the TAO project will be presented.

Keywords: JUNO; JUNO-TAO; neutrino detectors; reactor neutrinos

1. Science Case and Motivation

JUNO aims at simultaneously probing the two main frequencies of three-flavor neutrino oscillations, as well as their interference related to the mass ordering, at a distance of ~53 km from two powerful nuclear reactor complexes in China [1]. The present information on the reactor spectra is not meeting the requirements of an experiment like JUNO, with a design resolution of 3% at 1 MeV. Unknown fine structures in the reactor spectrum might cause severe uncertainties, which could even make the interpretation of JUNO's reactor neutrino data impossible. TAO is aiming for a measurement of the reactor neutrino spectrum at very low distances (<30 m) to the 4.6 GW$_{th}$ strong core with a groundbreaking resolution better than 2% at 1 MeV [2]. Furthermore, TAO will make a major contribution to the investigation of the so-called reactor anomaly [3]. Present calculations of the reactor neutrino spectrum indicate a deficit of approx. 3% in the measured reactor fluxes. Currently, these anomalies can be interpreted as indications of the existence of right-handed sterile neutrinos. Nevertheless, it should be mentioned that recent refinements of the reactor flux models have been able to reduce this tension [4]. Beyond that, the reactor neutrino spectra recorded by Double Chooz [5], Reno [6] and Daya Bay [7] show an excess in the neutrino flux from 5 MeV to 6 MeV of unknown origin. This can be considered one of the most puzzling questions in the physics of reactor neutrinos today. Beyond these studies, TAO will search for signatures of sterile neutrinos in the mass range of 1 eV, which have just regained importance in light of the recently reconfirmed gallium anomaly by the BEST Experiment [8]. An additional goal of the TAO experiment is the verification of the detector technology for reactor monitoring and safeguard applications for the future effective fight against the proliferation of nuclear weapons material.

2. The JUNO—TAO Detector Design

The TAO experiment (see Figure 1) will realize a neutrino detection rate via the inverse beta decay (IBD) of about 2000 per day, which is approximately 30 times the rate in the

Citation: Steiger, H.T.J. TAO—The Taishan Antineutrino Observatory. *Instruments* **2022**, *6*, 50. https://doi.org/10.3390/instruments6040050

Academic Editors: Fabrizio Salvatore, Alessandro Cerri, Antonella De Santo and Iacopo Vivarelli

Received: 31 August 2022
Accepted: 20 September 2022
Published: 22 September 2022

Publisher's Note: MDPI stays neutral with regard to jurisdictional claims in published maps and institutional affiliations.

JUNO main detector [9]. In order to achieve its goals, TAO is relying on cutting-edge technology, both in photosensor and liquid scintillator (LS) development which is expected to have an impact on future neutrino and Dark Matter detectors. The experiment will realize an optical coverage of the 2.8 tons of Gd-loaded LS close to 95% with novel silicon photomultipliers (SiPMs), with a photon detection efficiency (PDE) above 50%. To efficiently reduce the dark count of these light sensors, the entire detector will be cooled down to $-50\,^\circ$C. The combination of SiPMs with cold LS will lead to an increase in the photoelectron yield by a factor of 4.5 compared to the JUNO central detector [9].

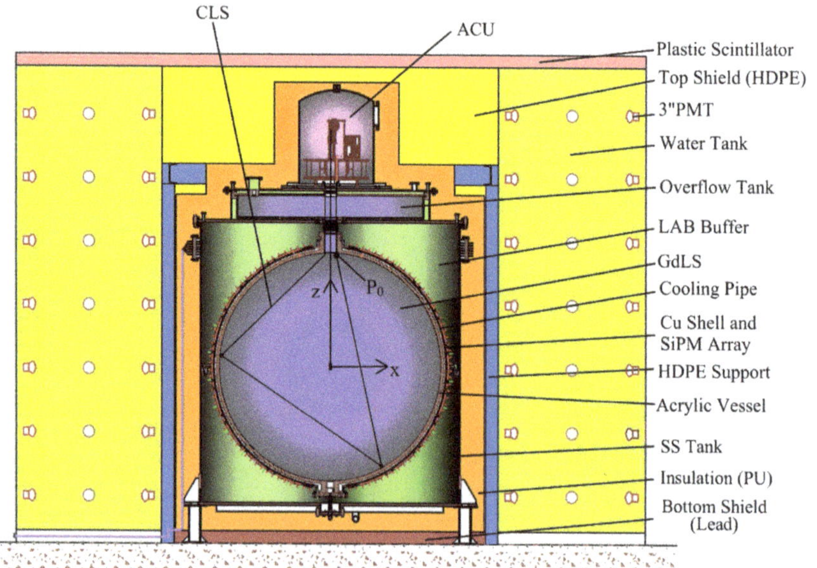

Figure 1. Conceptual design of the TAO detector, which consists of a central detector with the LS neutrino target and a buffer liquid, a calibration system, an outer shielding, and a veto system. The ambient and cosmogenic background will be reduced by the veto system based on plastic scintillator detectors. Layers of high-density polyethylene (HDPE) and polyurethane (PU) will act as passive shielding. A water tank equipped with PMTs acts as a surrounding water Cherenkov detector. The buffer vessel containing LAB and the inner detector (1.8 m acrylic vessel) filled with 2.8 tons of novel LAB based Gd-loaded LS as well as a 10 m² array of ~4000 SiPMs [8,9] mounted on a copper shell will be cooled down to $-50\,^\circ$C. The complete inner detector is housed in a stainless steel (SS) tank. Insulation is realized by means of PU layers. The calibration system consists of the Automated Calibration Unit (ACU) and a Cable Loop System (CLS). A segment of the CLS cable is located in the GdLS where P0 marks its starting point. A coordinate system (see x and z) is defined for the detector with its origin in the center of the target vessel. Figure taken from [10].

3. Cold Gd-loaded Liquid Scintillator

The TAO detector will use a Gadolinium-loaded Liquid Scintillator (GdLS) as the target material for the electron antineutrinos undergoing the IBD reaction

$$\overline{\nu_e} + p \rightarrow e^+ + n$$

on a proton of the scintillator. While the prompt positron signal is exploited mainly for event energy and vertex reconstruction, the neutron capture on Gd will be used as a clean well distinguishable delayed signal to reduce the accidental background [9]. While the neutron capture on hydrogen in the LS has a (n, γ) cross-section of ~0.332 barns and the energy of the emitted gamma is 2.2 MeV the advantage of adding natural Gd with

~49,000 barns and a gamma cascade of a total energy ~8 MeV is obvious. Furthermore, the neutron-capture time in the LS is significantly shortened to ~28 μs for loading with 0.1% Gd (by mass), as compared to ~200 μs in the unloaded scintillator [11].

As a consequence of the detector design and the necessity to lower the dark noise of the SiPMs, the GdLS and the buffer liquid will be cooled down to −50 °C or lower. Linear Alkylbenzene (LAB) similar to the one used in Daya Bay [11] and JUNO [12] will be the solvent of the liquid scintillator. Especially the high flash point > 130 °C and low volatility make LAB very suitable for using it in close proximity to a nuclear reactor. Commercially available LAB is mainly a mixture of molecules with 9 to 14 carbon atoms in the linear chain. It has a freezing point below −60 °C. Nonetheless, the LAB's water content may precipitate at low temperatures greatly reducing the LS's transparency. Therefore, extensive drying of the solvent is required. Furthermore, the solubility of fluors and wavelength shifters is greatly reduced at low temperatures. By adding Dipropylenglykol-n-butylether (DPnB) in sub-percent quantities as a freezing inhibitor and antioxidant, the latter problem is cured. Currently, a fluor (2,5-Diphenyloxazole, PPO) concentration of 3 g/L and 2 mg/L of the secondary wavelength shifter BisMSB (1,4-Bis (2-methylstyryl) benzene) are considered as the baseline option for TAO. New R&D for JUNO using one Daya Bay detector [12,13] shows that 1 to 3 mg/L of BisMSB will result in the highest light output for the JUNO main detector. Slightly dependent on the solvent's purity, for most detectors of various sizes, 3 mg/L BisMSB is sufficient to reach the optimal photoelectron yield. For the TAO scintillation cocktail, the emission spectrum is dominated by BisMSB (shown in Figure 2) matches well the spectral detection efficiency maximum (>50%) at ~430 nm. For TAO the solvent is loaded via the procedure developed by Daya Bay [11] using the Gd-complex with the ligand 3,5,5-trimethylhexanoic acid (TMHA). A final mass concentration of 0.1% Gd in the liquid scintillator is foreseen.

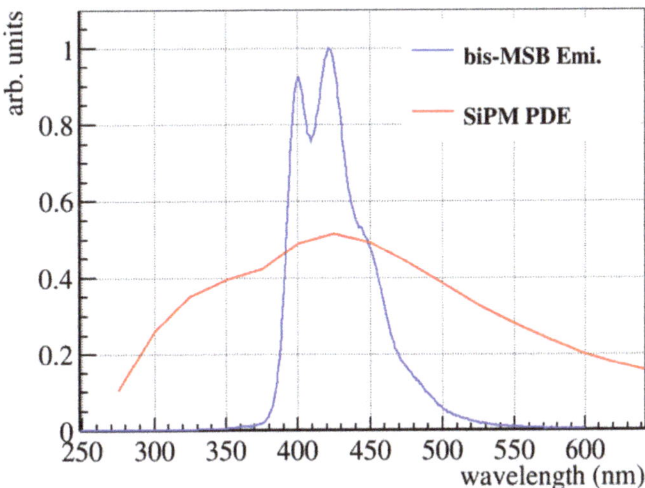

Figure 2. The effective emission spectrum of BisMSB (blue) dissolved in cyclohexane under excitation with UV light in a 10 mm × 10 mm × 40 mm quartz glass cuvette. The red graph represents a typical spectral response spectrum of a TAO-like SiPM at −50 °C matching well the BisMSB emission.

4. Calibration System

To achieve its goals, the absolute energy scale, nonlinearities, position dependencies, and detector resolution have to be understood precisely. Therefore, meticulous calibration is crucial. The calibration system contains the Automated Calibration Unit (ACU) which is reused and modified from the Daya Bay experiment [14] and a Cable Loop System (CLS), as shown in Figure 1. While the ACU (see Figure 3) can be used to calibrate TAO's energy response precisely on the Z-axis, the CLS will allow off-axis calibrations. Moreover, the ACU contains a system based on a pulsed UV LED light source enabling timing calibration.

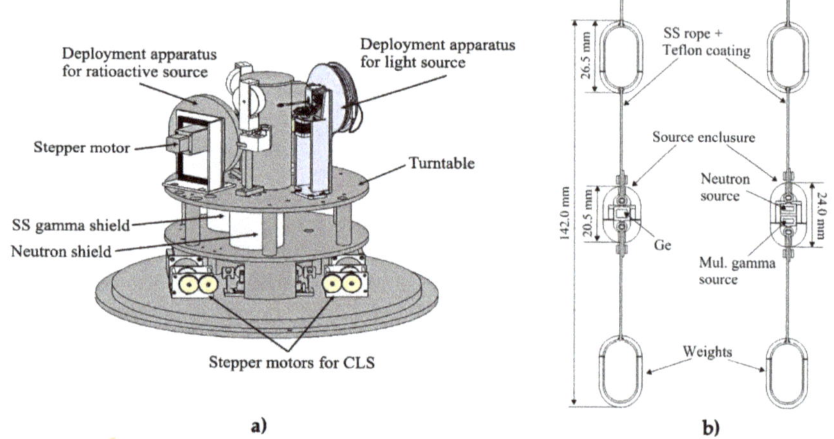

Figure 3. (**a**) Automated calibration system as used before in the Daya Bay Experiment [14]. The steel dome housing the ACU on top of the detector is not shown here. (**b**) Schematic drawing of the source holding structure. The ^{68}Ge source (left) and a combined radioactive source (right). The combined source contains multiple γ emitting isotopes (^{137}Cs, ^{54}Mn, ^{40}K, ^{60}Co) and one neutron source (^{241}Am–^{13}C). Figures taken from [10] and modified.

4.1. ACU

The entire ACU (see drawing in Figure 3a) is placed below a gas and liquid tight stainless steel dome, which provides a fully enclosed system. The ACU is equipped with a turntable with three possible positions for the source deployment wheels. By that, different calibration sources (see Figure 3b) can be deployed into the detector without opening the sealed steel dome. For the TAO two of this deployment, systems will be used for radioactive sources mounted in dedicated capsules and held by PTFE-coated stainless-steel wires. The third wheel is foreseen to be used for the UV light injection system. All three deployment systems feature a wire load monitoring device, which is able to stop the motors of the ACU automatically before a source holding cable breaks in case of sticking somewhere in the detector [10,14].

4.2. Ultraviolet LED Calibration System

The UV light source of the ACU is equipped with a diffuser improving the isotropy of the emission. The wavelength of the UV light is (265 ± 5) nm by default but is changeable to 420 nm or any other values if necessary. The light source can be used to monitor the performance and fundamental properties of the TAO detector. This includes monitoring the state of each channel and calibrating its timing, SiPM gain, and quantum efficiency. Moreover, the UV source can be used to test the data acquisition and offline analysis pipeline. A detailed study of the central detector's pileup is feasible and foreseen.

Such a wide use of the UV LED calibration subsystem requires a highly specific design of all components. The driver of the nanosecond flasher (LED driver) applies Kapustinsky's concept [15]. Two consecutive signals are generated by two LED channels in the flasher. Increasing the amplitude of a single output signal by merging the first and second signals with each other is also possible. Nonetheless, both LEDs work independently and the respective light pulses can be set to different intensities. This is achieved by programming the steering board (see Figure 4) with its dedicated microcontroller. There also the repetition rate can be adjusted. The resulting output is monitored pulse to pulse with a control line. Therefore, the two signals are merged into a single time sequence and subsequently copied by using an X-shape combiner-splitter. One of its outputs is guided to the detector while the other one is furnished into the control line. The latter signal is measured with a Photomultiplier Tube (Hamamatsu H7732P-01) which is connected to TAO's DAQ. The design outlined here is the development of the concept proposed in references [16,17] and realized with some changes in the JUNO Laser Calibration System [18]. Full documentation of the system is published in [10].

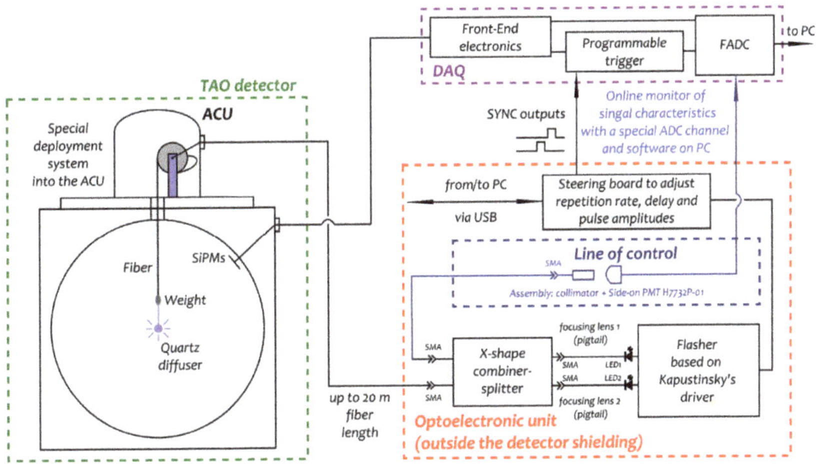

Figure 4. Simplified schematics of the UV LED calibration system. The dashed boxes show structural parts of the experimental setup namely the detector (green), the DAQ electronics (purple), and the optoelectronic unit (red) of the UV LED subsystem. The latter generates light pulses which are transferred to the detector target via fiber optics. Simultaneously, the DAQ is triggered from the steering board of the LED subsystem. Subsequently, the ADC digitizes also the signals from the control line. Figure taken from [10].

4.3. CLS

The Cable Loop System (CLS) uses one radioactive source (^{137}Cs), that can be deployed to an off-axis position in the target of TAO. The system development was based on experiences from a similar but drastically larger system in JUNO [19]. The radioisotope is positioned in a small area of the stainless steel cable, which is coated with PTFE along its entire length. This should prevent contamination of the Gadolinium-loaded scintillator and makes cleaning of the cable simpler. Glued anchors on the inner surface of the acrylic vessel guide the cable within the detector. Two stepper motors are used for pulling it in either direction to position the radioactive source in the target with high accuracy.

5. Conclusions

JUNO aims at simultaneously probing the two main frequencies of three-flavor neutrino oscillations, as well as their interference related to the mass ordering. The present information on the reactor spectra is not meeting the requirements of a high-resolution experiment like JUNO. Therefore, the TAO experiment aims for a measurement of the reactor neutrino spectrum with the unprecedented resolution of 2% at 1 MeV to identify unknown fine structures. Furthermore, TAO will make a major contribution to the investigation of the so-called reactor anomaly. Beyond that, the reactor neutrino spectra recorded by Double Chooz, Reno, and Daya Bay show an excess in the neutrino flux from 5 MeV to 6 MeV of unknown origin. By its excellent statistics and resolution, the TAO experiment can be expected to shed light on this excess and clarify if it is caused by non-standard neutrino interactions.

TAO will be built directly outside the containment of the new reactor core Taishan 1, which is one of the strongest nuclear reactors in the world. In order to realize its goals, TAO relies on a completely new concept for liquid scintillator detectors. The experiment will realize an optical coverage of the 2.6 tons of Gd-loaded LS close to 95% with novel SiPMs with a photon detection efficiency above 50%. Cooling down the detector to $-50\ ^\circ$C will allow a drastic reduction in the SiPM's dark noise. The combination of SiPMs with cold LS will lead to an increase in the photoelectron yield by a factor of 4.5 compared to the JUNO central detector.

The chemical development of a gadolinium-loaded and cold-resistant scintillator is of essential interest for TAO. Extensive R&D projects aiming toward the optimization and full characterization of the scintillation performance at low temperatures are ongoing. However, the use of cold scintillators is not limited to TAO: Due to the high light yield, an interesting future application might be the detection of coherent neutrino-nucleon scattering on carbon in the scintillator. While intrinsic decays of ^{14}C define a lower energy threshold of ~150 keV, neutrinos from nuclear reactors, stopped-pion sources and supernovae would produce nuclear recoil signals above this threshold. Naturally, a cold scintillator might be interesting as well for other experiments where excellent energy resolution or sub-nanosecond timing is required.

To achieve TAO's goals, the absolute energy scale, nonlinearities, position dependencies and detector resolution have to be determined with high accuracy. Therefore, meticulous calibration is crucial. Thanks to the ACU and CLS it will be possible to deploy different radioactive sources on and off the central axis of the detector mapping its response. The degradation of the energy resolution and energy scale uncertainty will be controlled by 0.05% and 0.3%, respectively, making TAO able to measure the electron antineutrino spectrum of a nuclear reactor with unprecedented precision.

Funding: This work is supported by the National Natural Science Foundation of China under Grant Number 12022505, 11875282 and 11775247, the joint RSF-NSFC project under Grant Number 12061131008, the Youth Innovation Promotion Association of Chinese Academy of Sciences, and by the Program of the Ministry of Education and Science of the Russian Federation for higher education establishments with project Number FZWG-2020-0032 (2019-1569). Beyond that, this work is supported by the Cluster of Excellence PRISMA+ at the Johannes Gutenberg-University Mainz.

Data Availability Statement: Not applicable.

Conflicts of Interest: The authors declare no conflict of interest.

References

1. An, F.; An, G.; An, Q.; Antonelli, V.; Baussan, E.; Beacom, J.; Bezrukov, L.; Blyth, S.; Brugnera, R.; Avanzini, M.B.; et al. Neutrino Physics with JUNO. *J. Phys. G Nucl. Part. Phys.* **2016**, *43*, 030401. [CrossRef]
2. Abusleme, A.; Adam, T.; Ahmad, O.; Aiello, S.; Akram, M.; Ali, N.; An, F.; An, G.; An, Q.; Andronico, G.; et al. TAO Conceptual Design Report: A Precision Measurement of the Reactor Antineutrino Spectrum with Sub- percent Energy Resolution. *arXiv* **2020**, arXiv:2005.08745.
3. Mention, G.; Fechner, M.; Lasserre, T.; Mueller, T.A.; Lhuillier, D.; Cribier, M.; Letourneau, A. The Reactor Antineutrino Anomaly. *Phys. Rev. D* **2011**, *83*, 073006. [CrossRef]
4. Giunti, C.; Li, Y.F.; Ternes, C.A.; Xin, Z. Reactor antineutrino anomaly in light of recent flux model refinements. *Phys. Lett. B* **2022**, *829*, 137054. [CrossRef]
5. de Kerret, H.; Abrahao, T.; Almazan, H.; dos Anjos, J.C.; Appel, S.; Barriere, J.C.; Bekman, I.; Bezerra, T.J.C.; Bezrukov, L.; Blucher, E.; et al. First Double Chooz θ_{13} Measurement via Total Neutron Capture Detection. *arXiv* **2019**, arXiv:1901.09445.
6. Bak, G.; Choi, J.H.; Jang, H.I.; Jang, J.S.; Jeon, S.H.; Joo, K.K.; Ju, K.; Jung, D.E.; Kim, J.G.; Kim, J.H.; et al. Measurement of Reactor Antineutrino Oscillation Amplitude and Frequency at RENO. *Phys. Rev. Lett.* **2018**, *121*, 201801. [CrossRef] [PubMed]
7. An, F.; Balantekin, A.B.; Band, H.R.; Bishai, M.; Blyth, S.; Butorov, I.; Cao, D.; Cao, G.F.; Cao, J.; Cen, W.R.; et al. Measurement of the Reactor Antineutrino Flux and Spectrum at Daya Bay. *Phys. Rev. Lett.* **2016**, *116*, 061801, Erratum in *Phys. Rev. Lett.* **2017**, *118*, 099902. [CrossRef] [PubMed]
8. Barniov, V.V.; Cleveland, B.T.; Danshin, S.N.; Ejiri, H.; Elliott, S.R.; Frekers, D.; Gavrin, V.N.; Gorbachev, V.V.; Gorbunov, D.S.; Haxton, W.C.; et al. Results from the Baksan Experiment on Sterile Transitions (BEST). *Phys. Rev. Lett.* **2022**, *128*, 232501. [CrossRef]
9. Abusleme, A.; Adam, T.; Ahmad, S.; Ahmed, R.; Aiello, S.; Akram, M.; An, F.; An, G.; An, Q.; Andronico, G.; et al. JUNO physics and detector. *Prog. Part. Nucl. Phys.* **2022**, *123*, 103927. [CrossRef]

10. Xu, H.; Abusleme, A.; Anfimov, N.V.; Callier, S.; Campeny, A.; Cao, G.; Cao, J.; Cerna, C.; Chen, Y.; Chepurnov, A.; et al. Calibration strategy of the JUNO-TAO experiment. *arXiv* **2022**, arXiv:2204.03256.
11. Beriguete, W.; Cao, J.; Ding, Y.; Hans, S.; Heeger, K.M.; Hu, L.; Huang, A.; Luk, K.; Nemchenok, I.; Qi, M.; et al. Production of a gadolinium-loaded liquid scintillator for the Daya Bay reactor neutrino experiment. *Nucl. Instrum. Meth.* **2014**, *763*, 82–88. [CrossRef]
12. Abusleme, A.; Adam, T.; Ahmad, S.; Aiello, S.; Akram, M.; Ali, N.; An, F.P.; An, G.P.; An, Q.; Andronico, G.; et al. Optimization of the JUNO liquid scintillator composition using a Daya Bay antineutrino detector. *Nucl. Instr. Meth. A* **2021**, *988*, 164823. [CrossRef]
13. Zhang, Y.; Yu, Z.-Y.; Li, X.-Y.; Deng, Z.-Y.; Wen, L.-J. A complete optical model for liquid-scintillator detectors. *Nucl. Instrum. Meth. A* **2020**, *967*, 163860. [CrossRef]
14. Liu, J.; Cai, B.; Carr, R.; Dwyer, D.A.; Gu, W.Q.; Li, G.S.; Qian, X.; McKeown, R.D.; Tsang, R.H.M.; Wang, W.; et al. Automated Calibration System for a High-Precision Measurement of Neutrino Mixing Angle θ_{13} with the Daya Bay Antineutrino Detectors. *Nucl. Instrum. Meth. A* **2014**, *750*, 19–37. [CrossRef]
15. Kapustinsky, S.; DeVries, R.M.; DiGiacomo, N.J.; Sondheim, W.E.; Sunier, J.W.; Coombes, H. A Fast Timing Light Pulser for Scintillation Detectors. *Nucl. Instrum. Meth. A* **1985**, *241*, 612–613. [CrossRef]
16. Chepurnov, A.S.; Gromov, M.B.; Shamarin, A.F. Online Calibration of Neutrino Liquid Scintillator Detectors above 10 MeV. *J. Phys. Conf. Ser.* **2016**, *675*, 012008. [CrossRef]
17. Chepurnov, A.S.; Gromov, M.B.; Litvinovich, E.A.; Machulin, I.N.; Skorokhvatov, M.D.; Shamarin, A.F. The Calibration System Based On the Controllable UV/Visible LED Flasher for the Veto System of the DarkSide Detector. *J. Phys. Conf. Ser.* **2017**, *798*, 012118. [CrossRef]
18. Zhang, Y.; Liu, J.; Xiao, M.; Zhang, F.; Zhang, T. Laser Calibration System in JUNO. *J. Instrum.* **2019**, *14*, P01009. [CrossRef]
19. Abusleme, A.; Adam, T.; Ahmad, S.; Ahmed, R.; Aiello, S.; Akram, M.; An, F.; An, G.; An, Q.; Andronico, G.; et al. Calibration strategy of the juno experiment. *J. High Energy Phys.* **2021**, *2021*, 4. [CrossRef]

Article

The SiD Digital ECal Based on Monolithic Active Pixel Sensors

James E. Brau [1,*], **Martin Breidenbach** [2], **Alexandre Habib** [2], **Lorenzo Rota** [2] and **Caterina Vernieri** [2]

[1] Department of Physics, University of Oregon, Eugene, OR 97403, USA
[2] SLAC National Accelerator Laboratory, 2575 Sand Hill Road, Menlo Park, CA 94025, USA
* Correspondence: jimbrau@uoregon.edu

Abstract: The SiD detector concept capitalizes on high granularity in its tracker and calorimeter to achieve the momentum resolution and particle flow calorimetry physics goals in a compact design. The collaboration has had a long interest in the potential for improved granularity in both the tracker and ECal with an application of monolithic active pixel sensors (MAPS) and a study of MAPS in the SiD ECal was described in the ILC TDR. Work is progressing on the MAPS application in an upgraded SiD design with a prototyping design effort for a common SiD tracker/ECal design based on stitched reticles to achieve 10×10 cm^2 sensors with 25×100 micron^2 pixels. Application of large area MAPS in these systems would limit delicate and expensive bump-bonding, provide possibilities for better timing, and should be significantly cheaper than the TDR concept due to being a more conventional CMOS foundry process. The small pixels significantly improve shower separation. Recent simulation studies confirm previous performance projections, indicating electromagnetic energy resolution based on digital hit cluster counting provides better performance than the SiD TDR analog design based on 13 mm^2 pixels. Furthermore, the two shower separation is excellent down to the millimeter scale. Geant4 simulation results demonstrate these expectations.

Keywords: calorimetry; MAPS; silicon detectors; linear collider

Citation: Brau, J.E.; Breidenbach, M.; Habib, A.; Rota, L.; Vernieri, C. The SiD Digital ECal Based on Monolithic Active Pixel Sensors. *Instruments* **2022**, *6*, 51. https://doi.org/10.3390/instruments6040051

Academic Editors: Fabrizio Salvatore, Alessandro Cerri, Antonella De Santo and Iacopo Vivarelli

Received: 11 August 2022
Accepted: 10 September 2022
Published: 23 September 2022

Publisher's Note: MDPI stays neutral with regard to jurisdictional claims in published maps and institutional affiliations.

1. Introduction

The physics goals of future electron–positron colliders motivate measurements of unprecedented precision. These will improve understanding of the Higgs boson and top quark, and enable searches for new phenomena that might discover beyond the standard model electroweak particles. With the consequential challenging demands for tracking, heavy-quark tagging, jet energy measurement, and beam polarization measurement, for example, advances in silicon detectors bear significant potential impact. Hadronic jet measurements by particle flow analysis (PFA), based on the combination of a precision tracker and a highly granular calorimeter, will be improved with new low-mass and highly granular sensors. To achieve these goals, several hundred square meters of silicon sensors are needed for low-mass trackers and sampling calorimetry, as envisioned in the SiD design of the ILC Technical Design Report (TDR) [1]. Trackers will require nearly one hundred square meters from the multiple layers at large radii, with micron-scale resolution. Sampling calorimeters need an order of magnitude larger areas of thinned overall packages for reduced Moliere radius, providing excellent shower containment. Silicon monolithic active pixel sensors (MAPS) present an attractive approach to this application.

2. Large-Area MAPS for Future Linear e^+e^- Collider

MAPS devices, with silicon diodes and readout circuitry contained in the same die, can be fabricated in standard CMOS processes, and were identified as a promising technology for high-granularity, low-material budget detectors many years ago [2–5]. They have several advantages over traditional hybrid pixel detector technologies. They can be built inexpensively, thinned to needs, offer individual pixel readout, are reasonably radiation-hard, and low in mass, while operating at high speeds and low power consumption from a single low-voltage supply. The close connection between a sensor and front-end amplifier, without externally added interconnections, significantly reduces the input capacitance, reducing the noise floor. Operation with smaller signals from thinner sensors is possible with satisfactorily high S/N.

Interconnects, routinely realized with some type of metal spheres or pillars bump-bonding, are on one hand comparative in size with the most recently desired pixel sizes O (10 μm). On the other hand, they are responsible for the dominating contribution to the capacitance seen at the input node of an in-pixel amplifier, effectively dominating otherwise very small sensor capacitance. A reduction in capacitance, owing to low-temperature direct bonding technology, was shown in comparative studies [6] to significantly improve pixel detector performance.

Magnitude, speed and efficiency of collecting charge carriers liberated in interactions of radiation with a sensor can be completely controlled in hybrid pixel detectors. MAPS are dependent on the active sensor volume and possible depletion of the substrate in the CMOS process. To retain the advantage of using standard CMOS processes for building MAPS, the older-generation MAPS, such as those used in the Heavy Flavor Tracker (HFT) at STAR [7], were built in the commonly used low-resistivity (about 10 Ω·cm) epitaxial wafer fabrication processes. At that time, thermal diffusion charge collection allowed signal magnitudes on the order of 1 k e$^-$, shared among the neighboring pixels and collected in times on the order of 100 ns [8]. The use of only one type of transistor in pixel circuitry was a significant limitation of the early MAPS. Radiation hardness was on the order 10^{13} of 1 MeV neutrons equivalent per cm^2 and 10 Mrads of the Total Ionizing Dose [9]. These parameters were suitable for the HFT at STAR, the first practical use of MAPS. MAPS simultaneously developed using processes suitable for high-voltage applications have not been used in any experiment yet.

It was shown many years ago that full depletion of the Active Sensitive Volume (ASV) of a device is required for charge to be collected by drift, enabling fast response and radiation hardness. Additionally, both N- and P-type transistors are required in a pixel area to develop efficient processing blocks. As a result of these requirements, new inner trackers, such as the ITS1/2 for ALICE, have employed the increased resistivity of ASV. Figure 1 shows cross-sections of structures built in the TowerSemi aka Tower-Jazz 180 nm. The implantations and distributions of resulting electric fields were optimized [10,11]. A reasonably thick, higher resistivity silicon film, translates to the maximally depleted ASV. Shielding of wells, visible in Figure 1, allows the pixel electronics to employ more than one type of transistor. The key benefit of this in tracking or vertexing, is the increased spatial resolution. While the TowerSemi 180 nm process has been available for the MAPS design for a while, the TowerSemi 65 nm process recently opened up, allowing its exploration to address needs for future applications. The TowerSemi 65 nm process allows a more than four-fold increase in the number of transistors per pixel compared to the 180 nm process. A new generation of MAPS, thinned and on an increased-resistivity substrate (>1 k Ω·cm) has been enabled by the developments for ITS3 of the ALICE experiment [12–14].

Another key feature of MAPS is the limited chip dimensions of a reticle. The standard 65 nm process limits single chips to about 2.6×3.2 cm^2 in process nodes, and even smaller dimensions in earlier processes. This limitation can be overcome by stitching of reticles, or partially by butting [15,16]. While stitching has been available for sometime in commercial CMOS applications, it is less common in particle detector applications; a demonstration has been achieved for X-ray detection [17]. An example of butting is the STAR Heavy Flavor Tracker at BNL RHIC [18], where chips were butted one next to another on the staves and the readout areas of the individual MAPS units were limited to one edge. On the other hand, a few years after the HFT at RHIC, the upgrade of the ALICE Inner Tracker System 3 (ITS3) at the LHC is going to adopt the stitched arrangement of the MAPS devices [12]. Both developments are aimed at targeting inner tracking or vertexing layers for Nuclear Physics experiments, where requirements on readout speed and radiation hardness are less stringent than for many high-energy physics experiments.

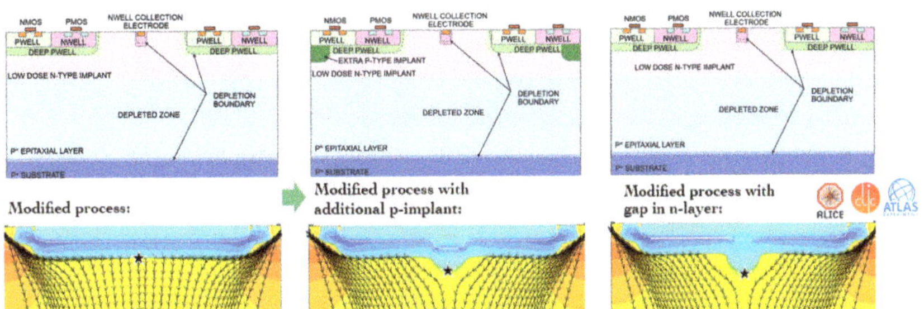

Figure 1. Cross-sections of MAPS structures and electric field lines in the improved process.

At CERN, the ALICE ITS3 upgrade based on MAPS is driving the CERN WP1.2 collaboration investigating wafer-scale MAPS devices on the TowerSemi 65 nm process. Within this collaboration, SLAC is investigating the challenges of wafer-scale designs optimized for detectors at linear machines, focusing in particular on tracking and calorimetry. This effort will help identify the risks that wafer-scale MAPS pose at the system level, such as yield, power distribution and fill factor, as well as evaluate essential aspects of the integration of such devices into a detector system: i.e., cooling, assembly procedures, wafer thinning and handling, and power delivery.

SLAC is developing a readout circuitry optimized for low duty cycle linear collider with bunch spacing as low as a few ns. The beam time structure of these machines (with bunch trains separated by tens-to-hundreds of milliseconds) makes this possible. Two techniques are being implemented. The readout electronics will adopt a power pulsing scheme: the analog front-end circuitry will be powered off during the dead-time between bunch trains. With low duty cycle machines such as ILC and C^3 [19], this technique enables a power reduction by more than two orders of magnitude. Power pulsing techniques were previously developed and characterised with the KPiX ASIC [20]. Second, the pixel front-end circuitry will be based on a synchronous readout architecture, where the operation of the circuitry is timed with the accelerator bunch train. In this way, the noise and timing performance of the circuitry can be optimized while maintaining low-power consumption. SLAC will leverage a decade of expertise via synchronous readout architectures operating with fast integration times [21], which have been implemented in all ASICs developed for the Linear Coherent Light Source (LCLS).

By combining all these techniques, the goal of the current R&D at SLAC is to achieve the specifications described in Table 1. We have derived the initial specifications from the C^3 configuration, as it bears the most challenging requirements for timing resolution and is compatible with the current limits of MAPS technology. Moreover, a sparse read-out mechanism, based on asynchronous read-out logic, will minimise the digital power as well as render the circuitry more robust to local variations of transistor performance. A first, small-scale prototype of such a device is expected in late 2022.

The development of wafer-scale MAPS will allow designers to investigate the following challenges:

- **Power pulsing:** Current drawn from the supply needs to reach the peak value in the shortest time possible to take full advantage of the power pulsing technique. This minimizes the duty cycle and thus decreases the average power consumption. However, the instantaneous current consumption of the pixel matrix can reach several Amperes over a few microseconds.
- **Power distribution:** Distribution of the power supply over a large area is challenging because of the non-negligible voltage drop over long metal distribution lines.
- **Yield:** Since the probability of fabrication defects scales with the area of the device, it is essential to develop new techniques to mitigate the effects of fabrication defects, such as shorts between supply and ground lines. A defect on one reticle-size MAPS

would result in a lower number of usable dies per wafer and a defect on a wafer-scale device is almost inevitable, possibly resulting in the loss of a full wafer.

- **Stitching techniques:** Design of stitching MAPS introduces additional layout design rules and methodologies, with the goal to increase the fabrication yield. This additional set of rules is not traditionally encountered by ASIC designers. Exposing ASIC designers to such design rules is an essential first step towards the development of wafer-scale devices.
- **Assembly and power delivery:** Preliminary mechanical and assembly tests need to be conducted to evaluate sensor-power delivery techniques, while minimizing detector dead material.

Table 1. Target specifications for 65 nm prototype.

Parameter	Value
Min. Threshold	$140\,e^-$
Spatial resolution	$7\,\mu m$
Pixel size	$25 \times 100\,\mu m^2$
Chip size	$10 \times 10\,cm^2$
Chip thickness	$300\,\mu m$
Timing resolution (pixel)	$\sim$ns
Total Ionizing Dose	100 kRads
Hit density/train	$1000\,hits/cm^2$
Hits spatial distribution	Clusters
Power density	$20\,mW/cm^2$

With the successful resolution of the challenges described here and the development of the MAPS with the target specifications shown in Table 1, they will be applied to the upgrade of the SiD design.

3. SiD

SiD (Figure 2) has been designed to make precision measurements of the Higgs boson, *W*- and *Z*-boson, the top quark and other particles at the linear collider. With relatively benign linear collider experimental conditions, the detector can be optimized for these precision measurements. There are lower collision rates, lower complexity, and less background than experienced in a hadron collider.

The requirements in calorimetry performance are driven by excellent jet reconstruction and measurement. Detection and separation of *W* and *Z* bosons in their hadronic decay modes is essential, and this motivates application of a PFA and a goal of 3–4% jet mass resolution at energies above 100 GeV, about twice as good as achieved in hadron collisions at the LHC.

In addition, tracking requirements include precise reconstruction of the *Z*-boson mass in the Higgs recoil analysis, as well as separation of jet flavors for Higgs couplings measurement. The asymptotic momentum resolution requirement for high-momentum tracks is nearly an order of magnitude better than achieved at the LHC, with minimization of detector material to maintain excellent momentum resolution for lower-momentum tracks.

The combination of excellent jet mass resolution, extremely precise tracking, and triggerless running gives SiD at the ILC superb potential for discovery. Quantitatively, the requirements are the following:

- **Impact parameter resolution:** $5\,\mu m \oplus \frac{10\,\mu m\,GeV/c}{p\,\sin^{3/2}\theta}$, where p is particle momentum and θ is the angle between the particle and the beamline.
- **Momentum resolution:** $\Delta(1/p) < 5 \times 10^{-5}\,(GeV/c)^{-1}$ asymptotically at high momenta, maintaining excellent tracking efficiency and very good momentum resolution at lower momenta with an aggressive minimization of detector material budget.

- **Jet energy resolution:** $\Delta E/E = 3\text{–}4\%$ for light flavour jets with $E \gtrsim 100$ GeV based on a PFA; this requires good longitudinal and transverse segmentation, with a minimal Moliere radius.
- **Readout:** Triggerless.
- **Powering:** Power of major systems cycled between bunch trains to minimizing cooling requirements and level of inactive material within detector.

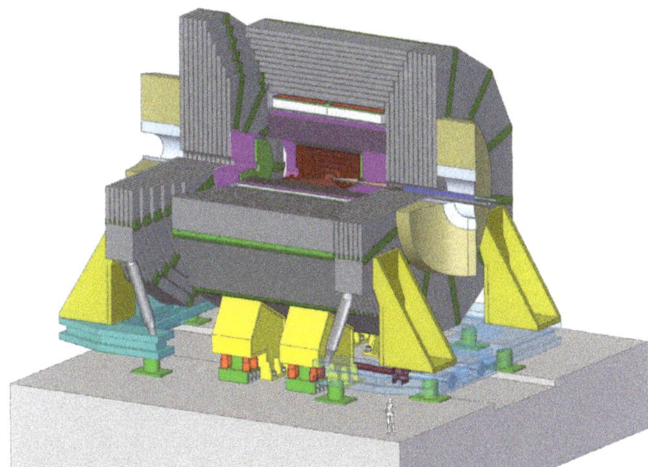

Figure 2. SiD on its platform, showing tracking (**red**), ECAL (**light green**), HCAL (**violet**) and reconfigured dodecagonal iron yoke.

The upgrade to MAPS for SiD's ECal and tracking sensors requires 1364 m^2 of silicon, including 54 m^2 in the barrel tracker, 20 m^2 in the end cap tracker, 1000 m^2 in the barrel ECal, and 290 m^2 in the end cap ECal. The readout for each bunch train will include a very small fraction of the 540 billion 25 $\times$ 100 µm^2 pixels of the system.

It has been shown that requiring a hit-time resolution below 5 ns for the vertex and tracking detectors, and 1 ns resolution for calorimeter hits, all the particle flow objects have sub-ns time resolution [22]. This is achieved from a truncated mean in the energy-weighted hit times of a cluster; therefore, we expect the $\sim$ns timing of a large number of MAPS hits to result in a particle flow object time resolution substantially below a nanosecond.

4. MAPS Performance for ECal

The finely granular, digital readout of the SiD ECal offered by application of MAPS sensors provides the potential for significantly enhanced performance over that envisioned in the ILC TDR [1]. One advantage of this digital approach over the TDR analog approach is the reduction in effects due to variations in energy deposition, such as Landau fluctuations. Fluctuations in the development of the shower remain as the main contribution to resolution. The fine granularity also reduces the likelihood of overlapping particles per pixel and improves the separation of nearby distinct showers, such as from high-energy π^0s or jets, and contributes to improved particle flow pattern recognition. Quantifying the nature of these effects has been investigated with GEANT4 simulations.

The longitudinal structure of the SiD ECal defined in the ILC TDR remains unchanged in this digital approach. The ECal has thirty total layers. The first twenty layers each have 2.5 mm tungsten thickness and 1.25 mm readout gap. The last ten layers each have 5 mm tungsten plus the same 1.25 mm readout gap. The total depth is 26 radiation lengths, providing reasonable containment for high-energy showers.

The 25 µm $\times$ 100 µm pixel geometry of the 2500 µm^2 area is chosen for the tracking precision from the 25 µm size in the bend plane. Excellent performance with a purely digital ECal based on this fine granularity is expected. A 25 µm $\times$ 100 µm pixel geometry is found to achieve ECal performance equivalent to 50 μm × 50 μm. Previous studies [23–25] have even indicated potential energy resolution advantages for a digital ECal solution and were cited in the ILC TDR. A recent prototype beam test of a similar concept has been published [26].

New simulation studies, based on this fine digital configuration, have confirmed the previous studies referred to in the ILC TDR and demonstrated additional details on the performance. These studies indicate the electromagnetic energy resolution based on counting clusters of hits in the MAPS sensors, and weighting them based on longitudinal and transverse position in shower, should provide better performance than the SiD original design based on 13 mm² analog pixels, as shown in Figure 3.

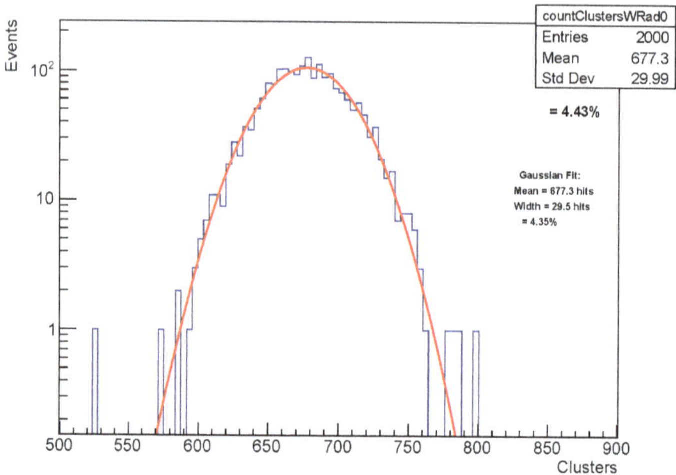

Figure 3. The distribution of weighted (by longitudinal and transverse position in shower) cluster counts for a 10 GeV gamma shower in the new SiD digital MAPs design based on a GEANT4 simulation.

GEANT4 simulations of the digital MAPS performance applied in the electromagnetic calorimeter have been under development and study since early 2021, and are continuing. These studies are aimed at understanding the ultimate performance and limitations, and to inform the ASIC designers on the requirements for the sensor chips. The expected performance has been found to exceed the requirements and performance of the SiD TDR ECal design. The 5 Tesla magnetic field was found to have a minor effect, degrading the resolution by a few per cent due to the field's impact on the lower energy electrons and positrons in a shower.

Separation of the two showers is excellent, as shown in Figure 4 for two 10 GeV electron showers separated by one cm and Figure 5 of two 20 GeV gamma showers from a 40 GeV π^0 decay. The fine granularity of pixels provides excellent separation. The performance for two electron showers versus their separation is summarized in Figure 6. The fine granularity allows for identification of two showers down to the mm scale of separation, and the energy resolution of each of the showers does not degrade significantly for the mm scale of shower separation.

While easily matching the energy measurement and consequential resolution of the larger, analog structure in the TDR pixel design (13 mm²), the measurement in the finely granular MAPS can be optimized. To understand this, we have begun by comparing the number of pixels with energy deposition above the threshold of about 1/4 MIP (MIP ~4 keV, including integration over angles of incidence), or 1 keV, and the number of particles with kinetic energies over 0.1 MeV passing through the sensors. These two counts are referred to as Hits and MIPs. For example, 20 GeV gammas in the SiD 5 Tesla field produce mean values of 2482 Hits and 1260 MIPs. The increase in Hits over MIPs results from electrons and positrons below the 0.1 MeV MIP threshold, delta ray production, gamma absorption, and pixel charge sharing for MIPs. The MIP count represents an ideal, potential signal, while the Hit count is the simplest measurement, without optimization considerations of the shower and hit properties. For the 20 GeV gammas, the resolutions are 4.2% and 2.4%. However, the properties of the digital Hit distributions can be used to improve the experimental resolution. The noise from electronics is ignored, but is not expected to significantly affect results.

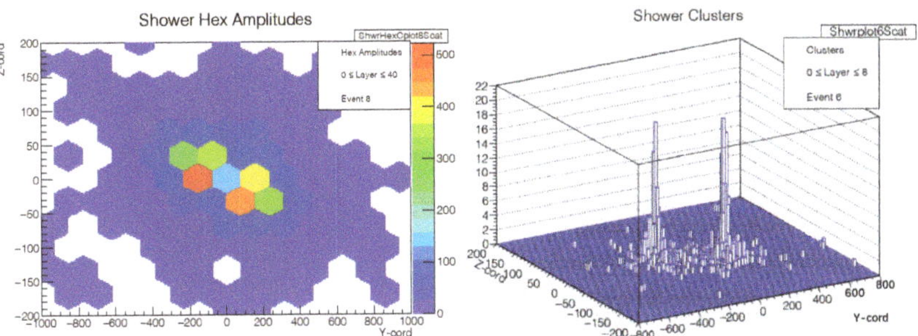

Figure 4. Transverse distribution of two 10 GeV showers separated by one cm. LEFT: Pixel amplitudes in the ILC 13 mm^2 TDR pixel design. RIGHT: Clusters in the first 5.4 radiation lengths in the new SiD digital MAPS design based on a GEANT4 simulation.

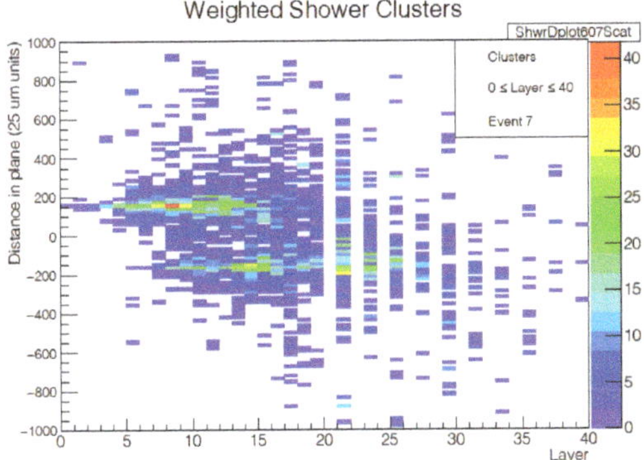

Figure 5. Projection in z-layer plane of the pixel clusters in two 20 GeV gamma showers emerging from a 40 GeV π^0 decay. The z-direction is the 100-μm pixel direction and the layers shown are 20 thin (0.64 X_0) followed by 10 thick tungsten layers. Each vertical bin is 400 μm wide. The two showers are separated by less than one cm.

The goal in the search for an improved experimental resolution is to find an algorithm of the Hits which reaches as closely as possible to the MIPs performance, the assumed ideal performance. An initial improvement is achieved by combining Hits into clusters and counting clusters. A cluster is constructed by combining all Hits that touch each other on any of the eight connections, boundaries or corners. Figure 7 shows the distribution of cluster counts for 20 GeV gamma showers, and the number of MIPs contained within specific cluster counts. An additional improvement comes from noting the probability that multiple MIPs appear in clusters as a function of the cluster size and the cluster position in the shower, longitudinally and transversely. By applying both of these corrections, the 20 GeV gamma resolution is brought to 3.3%. The resulting energy-dependent resolution between 1 and 50 GeV is well characterized by 12.2%/ $\sqrt{E} \oplus 1.4\%$, compared to the MIP resolution of 9.8%/ $\sqrt{E} \oplus 1.1\%$.

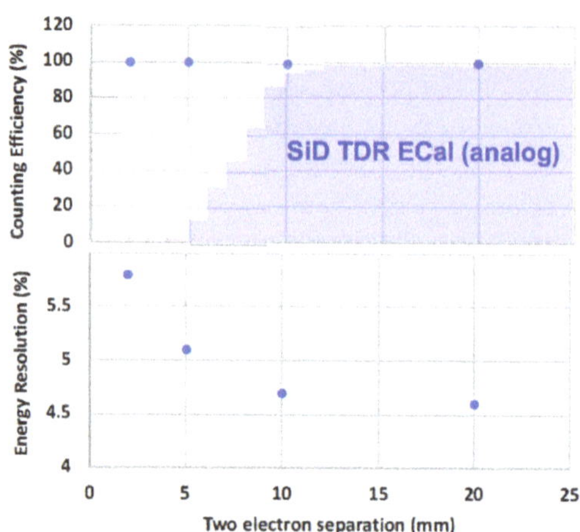

Figure 6. Efficiency for distinguishing two 10 GeV electron showers as a function of shower separation (upper curve) and the degradation of energy resolution as a function of separation due to overlap of cluster hits (lower curve) in the new SiD digital MAPS design based on a GEANT4 simulation.

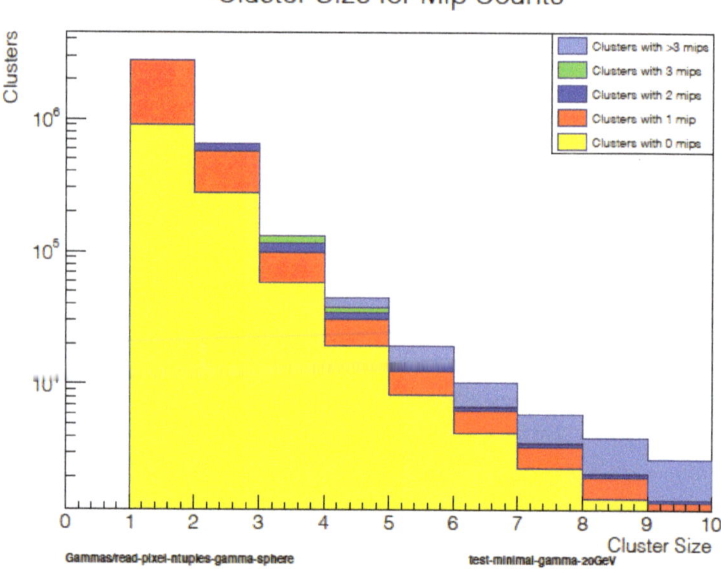

Figure 7. Distribution of cluster counts for 2000 showers from 20 GeV gammas, and the number of MIPs contained within specific cluster counts.

Figure 8 shows the gamma energy resolution performance for the range of measurements from the basic MIP counting (light blue, an idealized, best possible resolution) to that achieved by analyzing Hits in clusters (dark blue). These simulations are now mature and are well positioned to guide the design and production of the sensors. Future planned simulation studies include optimized reconstruction of showers and π^0s within jets, separation of electromagnetic showers and other depositions in the ECal, and the impact of these on jet energy resolution, particularly in the measurements of the Higgs branching ratios.

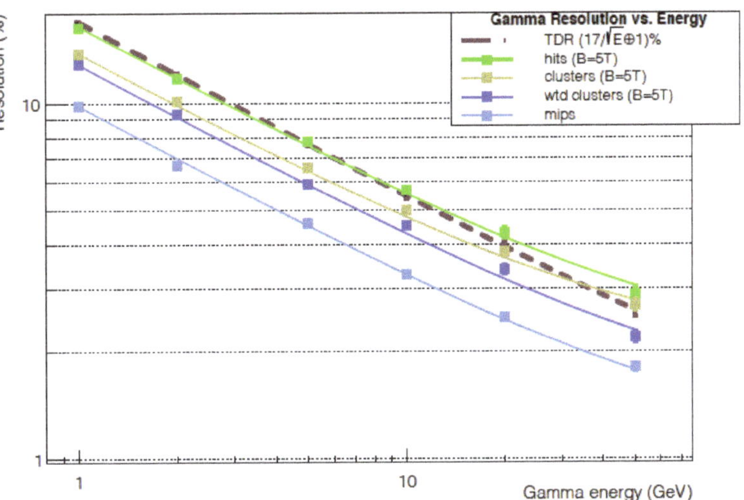

Figure 8. Energy resolution for gamma showers as a function of energy. The curves show (from lower up) the resolution based on counting minimum ionizing particles (light blue, MIPs), modified cluster counting (dark blue), pure simple cluster counting (light brown), active pixels (green, Hits), and the required performance from the ILC TDR (brown dash–dot).

More generally, future work will include:

- Potential of multi-bit digital operation;
- Jet reconstruction;
- Optimization of the overall ECal design, including consideration of manufacturability, possible with robots.

The large volume of data provided by a MAPS-based ECal reveals details of particle showers. The extraction of the most pertinent information, e.g., particle energy, particle type, and the separation of nearby and overlapping showers, provides an opportunity to apply Machine Learning techniques. We plan to apply such deep learning methods to particle and jet reconstruction in the SiD collider detector ECal based on MAPS technology.

5. MAPS Performance for Tracker Detectors

A MAPS-based tracker for SiD would feature a sensor similar in size to that described in the ILC TDR, 10×10 cm^2 devices. It would be constructed by stitching 2 cm $\times$ 2 cm reticles. However, such a device would provide exceptional granularity of 25 μm by 100 μm pixels, with the alignment placing the 25 μm pixel dimension in the bend direction, providing a resolution of 25 μm/$\sqrt{12} \approx 7$ μm without charge sharing. The 25 μm pixel size matches the KPiX-readout, silicon-strip width of the SiD TDR design which was recently assembled, tested, and shown to achieve 7 μm resolution [27]. The depleted 10 μm thick epi-layer charge collection of the MAPS allows a minimum threshold of 1/4 MIP, ensuring high efficiency. The pixel nature provides vastly improved pattern recognition for track finding over the strip devices. For the endcaps, such a sensor would eliminate the need for two sensors in a small-angle-stereo configuration, reducing both the material budget and cost.

6. Summary

The application of MAPS in SiD tracking and ECal systems offers the potential for significantly improved performance from that envisioned in the ILC TDR [1]. Motivated by this, the SiD MAPS development targets improvements to speed and resolution performance, and the system approaches needed for large-scale use of MAPS at a reasonable cost. Beyond the linear collider, future colliders in general will need large areas of silicon sensors, several hundred m^2, for low-mass trackers and sampling calorimetry. These developments, therefore, would have impacts beyond SiD. The requirements for trackers and calorimeters, particularly very thin, large areas with micron-scale resolution, are driving this work. An on-going effort focuses on developing readout electronics compatible with a power pulsing scheme: the analog front-end circuitry will be powered off during the dead-time between bunch trains. With low duty cycle machines such as ILC and C^3 , this technique reduces power by more than two orders of magnitude. The pixel front-end circuitry, as well, will be based on a synchronous readout architecture, where operation of the circuitry is timed with the accelerator bunch train. In this way, the noise and timing performance of the circuitry can be optimized while maintaining low-power consumption. The development of wafer-scale MAPS will allow designers to investigate the power pulsing, power distribution, yield, stitching techniques, assembly and power delivery.

Author Contributions: Conceptualization, M.B.; Investigation, A.H. and L.R.; Methodology, L.R.; Project administration, C.V.; Software, J.E.B.; Writing—original draft, J.E.B.; Writing—review & editing, J.E.B., M.B., C.V. and L.R. All authors have read and agree to the published version of the manuscript.

Funding: The authors were supported by the U.S. Department of Energy contract DE-AC02-76SF00515 and grant DE-SC0017996 for this research.

Data Availability Statement: The simulated data presented in the figures of this study are available on request from the corresponding author.

Conflicts of Interest: The authors declare no conflict of interest.

References

1. Behnke, T.; Brau, J.E.; Philip N. Burrows, P.N.; Fuster, J.; Peskin, M.; Stanitzki, M.; Sugimoto, Y.; Yamada, S.; Yamamoto, H. The International Linear Collider Technical Design Report—Volume 4: Detectors. *arXiv* **2013**, arXiv:1306.6329 .
2. Turchetta, R.; Berst, J.D.; Casadei, B.; Claus, G.; Colledani, C.; Dulinski, W.; Hu, Y.; Husson, D.; Le Normand, J.P.; Riester, J.L.; et al. A monolithic active pixel sensor for charged particle tracking and imaging using standard VLSI CMOS technology. *Nucl. Instrum. Methods Phys. Res. Sect. A Accel. Spectrom. Detect. Assoc. Equip.* **2001**, *458*, 677–689. [CrossRef]
3. Deptuch, G.; Berst, J.D.; Claus, G.; Colledani, C.; Dulinski, W.; Gornoushkin, Y.; Husson, D.; Riester,J.-L.; Winter, M.. Design and Testing of Monolithic Active Pixel Sensors for Charged Particle Tracking. *IEEE Trans. Nucl. Sci.* **2002**, *49*, 601–610. [CrossRef]
4. Perić, I. A novel monolithic pixel detector implemented in high-voltage CMOS technology. In Proceedings of the 2007 IEEE Nuclear Science Symposium and Medical Imaging Conference, Honolulu, HI, USA, 26 October–3 November 2007; Volume 2, pp. 1033–1039. [CrossRef]
5. Turchetta, R. Design and Process Development of CMOS Image Sensors with TowerJazz, Workshop on CMOS Active Pixel Sensors for Particle Tracking. 2014. Available online: https://indico.cern.ch/event/309449/contributions/1680002/attachments/591507/814229/cpix2014Turchetta.pdf (accessed on 7 September 2022).
6. Deptuch, G.W.; Carini, G.; Enquist, P.; Grybos, P.; Holm, S.; Lipton, R.; Maj, P.; Patti, R.; Siddons, D.P.; Szczygiel, R.; et al. Fully 3D-integrated Pixel Detectors for X-Rays. *IEEE Trans. Electron Devices* **2016**, *63*, 205–214. [CrossRef]
7. Greiner, L.; Anderssen, E.; Matis, H.S.; Ritter, H.G.; Schambach, J.; Silber, J.; Stezelberger, T.; Sun, X.; Szelezniak, M.; Thomas, J.; et al. A MAPS based vertex detector for the STAR experiment at RHIC. *Nucl. Instrum. Methods Phys. Res. Sect. A Accel. Spectrom. Detect. Assoc. Equip.* **2011**, *650*, 68–72. [CrossRef]
8. Deptuch, G.; Winter, M.; Dulinski, W.; Husson, D.; Turchetta, R.; Riester, J.L. Simulation and Measurements of Charge Collection in Monolithic Active Pixel Sensors. *Nucl. Instrum. Methods Phys. Res. Sect. A Accel. Spectrom. Detect. Assoc. Equip.* **2001**, *465*, 92–100. [CrossRef]
9. Deveaux, M.; Amar, S.; Besson, A.; Baudot, J.; Claus, G.; Colledani, C.; Deptuch, G.; Dorokhov, A.; Dulinski, W.; Goffe, M.; et al. Charge Collection Properties of Monolithic Active Pixel Sensors (MAPS) Irradiated with Non-Ionising Radiations. *Nucl. Instrum. Methods Phys. Res. Sect. A Accel. Spectrom. Detect. Assoc. Equip.* **2007**, *583*, 134–138. [CrossRef]
10. Snoeys, W.; Rinella, G.A.; Hillemanns, H.; Kugathasan, T.; Mager, M.; Musa, L.; Riedler, P.; Reidt, F.; Van Hoorne, J.; Fenigstein, A.; et al. A process modification for CMOS monolithic active pixel sensors for enhanced depletion, timing performance and radiation tolerance. *Nucl. Instrum. Methods Phys. Res. Sect. A Accel. Spectrom. Detect. Assoc. Equip.* **2017**, *871*, 90–96. [CrossRef]
11. Schioppa, E.J.; Tortajada, I.A.; Berdalovic, I.; Bortoletto, D.; Cardella, R.; Dachs, F.; Dao, V.; De Acedo, L.F.S.; Freeman, P.M.; Hemperek, T.; et al. Measurement results of the MALTA monolithic pixel detector. *Nucl. Instrum. Methods Phys. Res. Sect. A Accel. Spectrom. Detect. Assoc. Equip.* **2020**, *958*, 162404. [CrossRef]

12.	Contin, G. The MAPS-based ITS Upgrade for ALICE. *arXiv* **2020**, arXiv:2001.03042.

13.	Mager, M. The LS3 upgrade of the ALICE Inner Tracking System based on ultra-thin, wafer-scale, bent Monolithic Active Pixel Sensors. In Proceedings of the 15th Trento Workshop on Advanced Silicon Radiation Detectors, Vienna, Austria, 17–19 February 2020.

14.	Prabket, J.; Poonsawat, W.; Kobdaj, C.; Naeosuphap, S.; Yan, Y.; Jeamsaksiri, W.; Yamwong, W.; Chaowicharat, E.; Hruanun, C.; Poyai, A. Resistivity profile of epitaxial layer for the new ALICE ITS sensor. *J. Instrum.* **2019**, *14*, T05006. [CrossRef]

15.	Theuvissen, A. BUTTING versus STITCHING (1). Available online: https://harvestimaging.com/blog/?p=1568 (accessed on 7 September 2022).

16.	Theuvissen, A. BUTTING versus STITCHING (2). Available online: https://harvestimaging.com/blog/?p=1599 (accessed on 7 September 2022).

17.	Wunderer, C.B.; Marras, A.; Bayer, M.; Glaser, L.; Gottlicher, P.; Lange, S.; Pithan, F.; Scholz, F.; Seltmann, J.; Shevyakov, I.; et al. The PERCIVAL soft X-ray imager. *J. Instrum.* **2014**, *9*, C03056. [CrossRef]

18.	Contin, G. The MAPS-based vertex detector for the STAR experiment: Lessons learned and performance. *Nucl. Instrum. Methods Phys. Res. Sect. A Accel. Spectrom. Detect. Assoc. Equip.* **2016**, *831*, 7–11. [CrossRef]

19.	Bai, M.; Barklow, T.; Bartoldus, R.; Breidenbach, M.; Grenier, P.; Huang, Z.; Kagan, M.; Lewellen, J.; Li, Z.; Markiewicz, T.W.; et al. C^3: A "Cool" Route to the Higgs Boson and Beyond. *arXiv* **2021**, arXiv:2110.15800.

20.	Brau, J.; Breidenbach, M.; Dragone, A.; Fields, G.; Frey, R.; Freytag, D.; Freytag, M.; Gallagher, C.; Haller, G.; Herbst, R.; et al. KPiX—A 1024 channel readout ASIC for the ILC. In Proceedings of the 2012 IEEE Nuclear Science Symposium and Medical Imaging Conference Record (2012 NSS/MIC), Anaheim, CA, USA, 29 October–3 November 2012; pp. 1857–1860.

21.	Dragone, A.; Caragiulo, P.; Markovic, B.; Herbst, R.; Nishimura, K.; Reese, B.; Herrmann, S.; Hart, P.; Blaj, G.; Segal, J.; et al. ePix: A class of front-end ASICs for second generation LCLS integrating hybrid pixel detectors. In Proceedings of the 2013 IEEE Nuclear Science Symposium and Medical Imaging Conference (2013 NSS/MIC), Seoul, Korea, 27 October–2 November 2013; pp. 1–5.

22.	Linssen, L.; Miyamoto, A.; Stanitzki, M.; Weerts, H. Physics and Detectors at CLIC: CLIC Conceptual Design Report. *arXiv* **2012**, arXiv:1202.5940.

23.	Ballin, J.A.; Dauncey, P.D.; Magnan, A.M.; Noy, M.; Mikami, Y.; Miller, O.; Rajovic, V.; Watson, N.K.; Wilson, J.A.; Crooks, J.P.; et al. A Digital ECAL based on MAPS. *arXiv* **2009**, arXiv:0901.4457.

24.	Stanitzki, M.; SPiDeR Collaboration. Advanced monolithic active pixel sensors for tracking, vertexing and calorimetry with full CMOS capability. *Nucl. Instrum. Methods Phys. Res. Sect. A Accel. Spectrom. Detect. Assoc. Equip.* **2011**, *650*, 178–183. [CrossRef]

25.	Dauncey P.; SPiDeR Collaboration. Performance of CMOS sensors for a digital electromagnetic calorimeter. *PoS* **2010**, *502*, ICHEP2010.

26.	De Haas, A.P.; Nooren, G.; Peitzmann, T.; Reicher, M.; Rocco, E.; Rohrich, D.; Ullaland, K.; van den Brink, A.; van Leeuwen, M.; Wang, H.; et al. The FoCal prototype—An extremely fine-grained electromagnetic calorimeter using CMOS pixel sensors. *J. Instrum.* **2018**, *13*, P01014. [CrossRef]

27.	Brau, J.; Breidenbach, M.; Freytag, D.R.; Kleinwort, C.; Kraemer, U.; Reese, B.A.; Roelofs, S.; Stanitzki, M.; Steinhebel, A.; Tsionou, D.; et al. Lycoris—A large-area, high resolution beam telescope. *J. Instrum.* **2021**, *16*, P10023. [CrossRef]

instruments

Article

Tracker-in-Calorimeter (TIC) Project: A Calorimetric New Solution for Space Experiments [†]

Gabriele Bigongiari [1,2,*], Oscar Adriani [3,4], Giovanni Ambrosi [5], Philipp Azzarello [6], Andrea Basti [1], Eugenio Berti [3,4], Bruna Bertucci [5,7], Lorenzo Bonechi [4], Massimo Bongi [3,4], Sergio Bottai [4], Mirko Brianzi [4], Paolo Brogi [1,2], Guido Castellini [4,8], Enrico Catanzani [5,7], Caterina Checchia [1,2,], Raffaello D'Alessandro [3,4], Sebastiano Detti [4], Matteo Duranti [5], Noemi Finetti [4,9], Valerio Formato [10], Maria Ionica [5], Paolo Maestro [1,2], Fernando Maletta [4], Pier Simone Marrocchesi [1,2], Nicola Mori [4], Lorenzo Pacini [4,8], Paolo Papini [4], Sergio Bruno Ricciarini [4,8], Gianluigi Silvestre [5,7], Piero Spillantini [4], Oleksandr Starodubtsev [4], Francesco Stolzi [1,2], Jung Eun Suh [1,2], Arta Sulaj [1,2], Alessio Tiberio [3,4] and Elena Vannuccini [4]

1   INFN Pisa, Largo Bruno Pontecorvo 3, I-56127 Pisa, Italy
2   Dipartimento di Scienze Fisiche, Della Terra e dell'Ambiente, Università di Siena, Strada Laterina 8, I-53100 Siena, Italy
3   Dipartimento di Fisica e Astronomia, Università di Firenze, Via G. Sansone 1, Sesto Fiorentino, I-50019 Firenze, Italy
4   INFN Firenze, Via B. Rossi 1, Sesto Fiorentino, I-50019 Firenze, Italy
5   INFN Perugia, Via A. Pascoli, I-06100 Perugia, Italy
6   Département de Physique Nucléaire et Corpusculaire, University of Geneva, CH-1211 Geneva, Switzerland
7   Dipartimento di Fisica e Geologia, Università di Perugia, Via A. Pascoli, I-06100 Perugia, Italy
8   IFAC (CNR), Via Madonna del Piano 10, Sesto Fiorentino, I-50019 Firenze, Italy
9   Dipartimento di Scienze Fisiche e Chimiche, Università dell'Aquila, Via Vetoio, Coppito, I-67100 L'Aquila, Italy
10   INFN Roma Tor Vergata, I-00133 Rome, Italy
*   Correspondence: gabriele.bigongiari@pi.infn.it; Tel.: +39-050-2214-349
†   This paper is based on the talk at the 19th International Conference on Calorimetry in Particle Physics (CALOR 2022), University of Sussex, Brighton, UK, 16–20 May 2022.

Citation: Bigongiari, G.; Adriani, O.; Ambrosi, G.; Azzarello, P.; Basti, A.; Berti, E.; Bertucci, B.; Bonechi, L.; Bongi, M.; Bottai, S.; et al. Tracker-in-Calorimeter (TIC) Project: A Calorimetric New Solution for Space Experiments. *Instruments* **2022**, 6, 52. https://doi.org/10.3390/instruments6040052

Academic Editors: Fabrizio Salvatore, Alessandro Cerri, Antonella De Santo and Iacopo Vivarelli

Received: 30 August 2022
Accepted: 20 September 2022
Published: 26 September 2022

Publisher's Note: MDPI stays neutral with regard to jurisdictional claims in published maps and institutional affiliations.

Abstract: A space-based detector dedicated to measurements of γ-rays and charged particles has to achieve a balance between different instrumental requirements. A good angular resolution is necessary for the γ-rays, whereas an excellent geometric factor is needed for the charged particles. The tracking reference technique of γ-ray physics is based on a pair-conversion telescope made of passive material (e.g., tungsten) coupled with sensitive layers (e.g., silicon microstrip). However, this kind of detector has a limited acceptance because of the large lever arm between the active layers, needed to improve the track reconstruction capability. Moreover, the passive material can induce fragmentation of nuclei, thus worsening charge reconstruction performances. The Tracker-In-Calorimeter (TIC) project aims to solve all these drawbacks. In the TIC proposal, the silicon sensors are moved inside a highly-segmented isotropic calorimeter with a couple of external scintillators dedicated to charge reconstruction. In principle, this configuration has a good geometrical factor, and the angle of the γ-rays can be precisely reconstructed from the lateral profile of the electromagnetic shower sampled, at different depths in the calorimeter, by silicon strips. The effectiveness of this approach has been studied with Monte Carlo simulations and validated with beam test data of a small prototype.

Keywords: cosmic rays; astroparticles; γ-ray astronomy

1. Introduction

A calorimeter coupled to a charge detector is the typical instrument configuration for direct measurement experiments of the Cosmic Ray (CR) elemental spectra. This setup can easily be scaled to cover the desired energy range of the high-energy cosmic radiation. Recently, the ongoing CALET [1] and DAMPE [2,3] experiments, based on this concept, reported results on the H and He spectra over a wide energy range and revealed interesting spectral characteristics. However, current missions are limited by the acceptance, which prevents them from going beyond 100 TeV due to the decreasing flux at higher energies. In fact, to scan the "knee" regions of the individual H, He, and Nuclei spectra, an acceptance of at least 2.5 m^{2}sr $\times$ 5 years and an energy resolution better than 40% are needed. On the other hand, calorimetric CR measurements conducted in space also have the ability to study the inclusive electronic component (e.g., CALET and DAMPE experiments were specifically designed to study the high-energy electron+positron spectrum). To improve the quality of the electron + positron flux measurements, the electromagnetic showers must have an energy resolution of at least 2% and an electron/hadron rejection power greater than 10^6, as well as an acceptance larger than the hadron acceptance (of at least 3.6 m^{2}sr $\times$ 5 years). All these requirements place further constraints on the instrument. A full treatment of problems connected to the design and construction of an apparatus for high-energy cosmic rays can be found in [4], containing an extensive description of the future High-Energy cosmic-Radiation Detection (HERD) facility; the requirements and the expectations of such kinds of experiments are described as well.

Moreover, in a modern multi-messenger space experiment, γ-rays detection also plays an important role. A good angular resolution, which allows for the precise identification of astrophysical sources, is a fundamental requirement for γ-ray astronomy. The direction of the incoming photon can be reconstructed by a pair-conversion telescope. A pair-conversion instrument detects high-energy γ-rays by exploiting a passive material, generally thin foils of dense metal, commonly tungsten, in which electron–positron pairs are generated and then using standard particle-physics techniques, such as a silicon microstrip detectors, to detect these particles. This kind of detector is the standard tracker used for γ-ray physics in the energy region above 100 MeV, and it has been installed in recent experiments, such as Fermi [5] and AGILE [6] (see Figure 1 left panel). The minimum specifications for the Large-Area Telescope of Fermi experiment, in the energy range 20 MeV–300 GeV, are an angular resolution less than 0.15° above 10 GeV (less than 3.5° above 100 MeV) with an energy resolution better than 10% and a field of view larger than 2 sr. Such a geometry suffers some disadvantages. First of all, the acceptance is limited by the large lever arm between the active sensors needed to improve the tracking performance. In addition, the presence of dense passive material reduces the mass budget available for the calorimeter. Moreover, tungsten layers can induce fragmentation of nuclei, thus worsening the charge reconstruction performance of the apparatus. The main purpose of the Tracker-In-Calorimeter (TIC) project, approved and financed by INFN (Italy) in 2017, is the development of a detector with good angular resolution, needed for γ rays, and a good geometric factor, needed for charged particles. This work summarizes the results of the TIC collaboration, extensively described in [7].

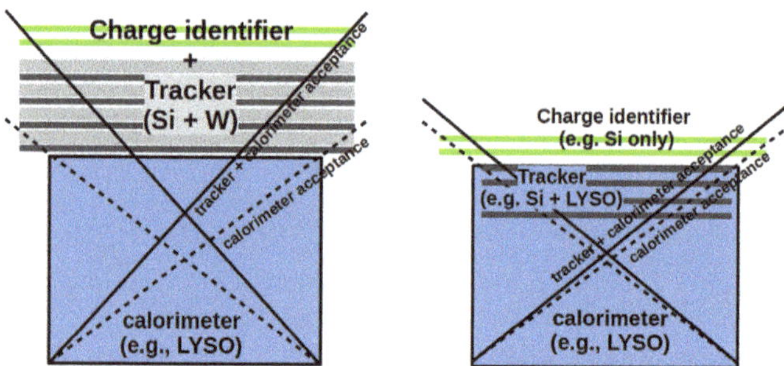

Figure 1. Conceptual designs of the two different detector geometries for γ-ray physics: on the **left**, the standard approach with a external tracker on the top of a calorimeter, and on the **right**, the TIC approach with a silicon tracker integrated inside a calorimeter.

2. Problems and Proposed Solution

The major constraint in achieving the above-mentioned performance is the weight limitation for these types of detectors (a few tons), which has a significant impact on both geometrical factors and energy resolution. A starting point for a possible solution could be the proposal for a homogeneous and isotropic calorimeter made by the CaloCube collaboration [8]. The CaloCube concept design was adopted by the future HERD experiment [4], which was proposed as one of several space astronomy payloads aboard the future Chinese Space Station. The suggested solution is a big cube composed of cubic scintillating crystals read out by photodiodes (PDs) (see Figure 2). The "Rubik" geometry and homogeneity allow particles to be collected from either the top or lateral faces, optimizing geometrical acceptance for a fixed mass budget. The active absorber gives good energy resolution, and the high granularity enables shower imaging, which provides criteria for both leakage correction and electron/hadron separation.

The CaloCube architecture was tuned for the detection of nuclei in the TeV–PeV energy region. A total weight of 2 tons, including both active and passive materials, was assumed. A comparative study of different scintillating materials with a wide variety of densities and calorimetric and optical characteristics was carried out. The results favor materials with superior shower confinement (e.g., Lutetium Yttrium Orthosilicate, LYSO), which compensate for the reduced volume due to these crystals' higher density and shorter interaction length. An effective geometric factor of up to 4 m^2sr can be obtained, with an energy resolution greater than 40%. The projected performance for electrons and γ-rays is a geometric factor of up to 3.4 m^2sr and an energy resolution greater than 2%. A detailed description of the Calocube project, the optimization of the detector design (also in terms of mass budget), and the results of beam tests of the prototype can be found in [9].

The TIC project aims to optimize a homogeneous calorimeter, such as CaloCube, to track γ-rays. A photon hitting the calorimeter starts an electromagnetic shower developing inside it. Such a segmented detector can sample the shower profile at different points. In principle, from this information, the shower axis can be reconstructed, and then the incoming direction and the impact point of the photon. Using a sampling calorimeter [10], it is possible to obtain a resolution better than 100 μm on the impact point of electrons above 100 GeV. The same results are expected for photons. In practice, in the TIC approach, the main tracker is removed and integrated inside a calorimeter, CaloCube-like, thus removing the problems previously described. The TIC conceptual design is shown in Figure 1 (right panel). In this scheme, one side of the calorimeter is instrumented with silicon microstrips (the sensitive layer of the tracker) interleaved with some layers of thin scintillating crystals (the γ-ray converter layers of the tracker). The proposed geometry is shown in Figure 3.

On the upper face, a stack of three layers of thin crystals interleaved by four silicon microstrip detectors has been added. A standard silicon sensor consists of an array of narrow strips (a few hundred um width, many cm long) aligned along a direction (X or Y), and it can sample the signals only along the perpendicular direction (Y or X). To measure both coordinates, coupling two sensors, with strips orthogonal to each other, is required. The gap (7 cm) between the first two silicon layers is necessary to mitigate the effect of the multiple scattering of the generated electron–positron pair on the tracking power. The key point is the fine sampling (hundreds of μm) of the initial part of the electromagnetic shower by the silicon strips (with respect to the crystals, a few cm thick) that allows increasing the tracking accuracy of the whole apparatus.

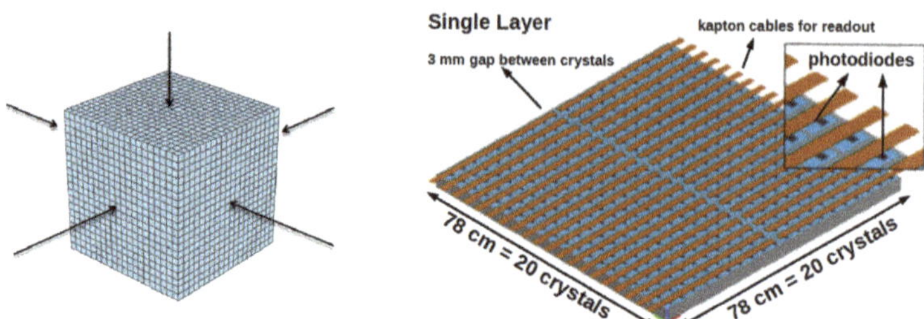

Figure 2. Conceptual design of the CaloCube 3D highly segmented calorimeter: on the **left**, the complete cubic detector; on the **right**, 1 of the 20 layers (from [8]).

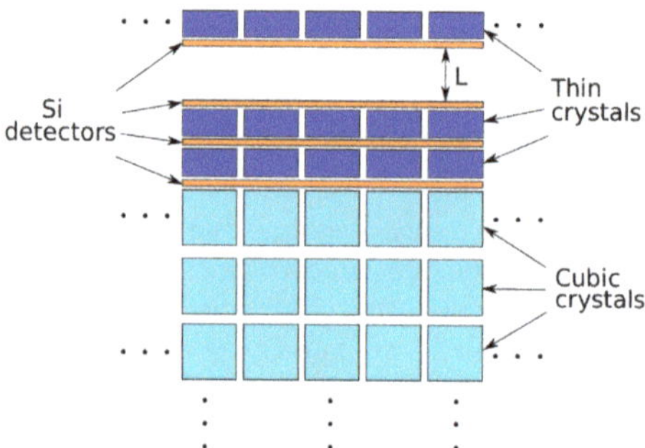

Figure 3. Schematic side view of the upper face of the simulated TIC: on the top, 3 scintillator layers made of thin crystals (blue) plus 4 silicon microstrip layers (orange) interleaved in; on the bottom, some layers made of cubic crystals (cyan) from CaloCube (from [7]).

3. Monte Carlo Simulations

A Monte Carlo simulation, based on the FLUKA package [11], has been set up in order to verify the validity of the proposal and quantify the performance of the detector. The simulated geometry derives from the design of the future HERD calorimeter [4], composed of cubic crystals, made up of LYSO (see Figure 3). The basic calorimeter is an ensemble of $21 \times 21 \times 21$ small cubic bricks, 3.5 cm thick, interleaved by 5 mm carbon fibers, simulating the support structure. The total depth is $\sim$3.2 interaction lengths (λ_I) and $\sim$58 radiation lengths (X_0). On the top of the designated upper face, the fine tracker, i.e., a stack of silicon microstrips and thin crystals, is installed. Monochromatic beams of γ-rays at different energies (1, 10, and 100 GeV), uniformly illuminating the entire upper face, with an isotropic angular distribution, have been simulated.

The track of a photon impinging the calorimeter can be reconstructed by applying a multi-step procedure to the crystal and Si-strip signals, described in the Section 4. The angular resolution obtained with the simulated data as a function of the γ-ray energy is shown in Figure 4, compared to those of Fermi and DAMPE. The performance of the tracking algorithm also depends on the efficiency of event selection used for the analysis. In Figure 4, the results of the analysis with two different selections are presented. A complete description of this analysis can be found in [7]. For Fermi and DAMPE efficiency calculations, see [12,13].

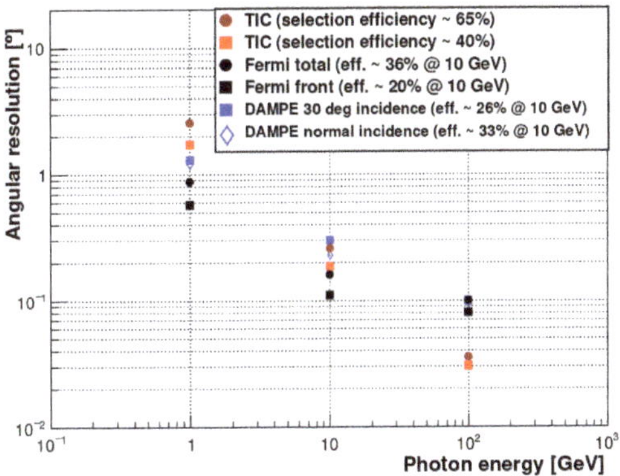

Figure 4. Angular resolution vs. Photon Energy: TIC compared to Fermi and DAMPE (from [7]).

The performance improves as energy increases for all the experiments. This behavior depends on two effects. At lower energies, the multiple scattering in the converting materials limits the tracking accuracy. The TIC crystals are thicker than the pair-conversion telescopes of FERMI and DAMPE, and this means a worse resolution. At higher energies, FERMI and DAMPE are constrained by the position resolution of their tracking systems, which then improves to an asymptote. Instead, in the TIC calorimeter, the statistical fluctuations of the electromagnetic shower decrease, leading to a more precise sampling of the signals and then improved tracking performance. At around 100 GeV, the resolution of TIC is better than that of FERMI and DAMPE.

4. Track Reconstruction Method

A particle (photon) traversing the detector releases energy in both crystals and silicon strips. Starting from this information (signals), it is possible in principle to reconstruct the particle track and then the original direction of the incoming particle. This is the goal of an apparatus dedicated to γ-rays physics. Therefore, the track reconstruction algorithm, used for the data analysis, is fundamental to verifying the potential of the detector design. A multi-step procedure has been developed based on the Principal Component Analysis method (PCA). The starting point is the selection, event by event, of the crystals with a signal above a given threshold (i.e., a few sigmas above the electronic noise). Let N be the number of the selected crystals. From the coordinates $c_i^{(n)}$ (with i = X, Y, Z and $n = 1, \ldots, N$) of these crystals, the corresponding covariance matrix M_{ij}^{CAL} can be calculated:

$$M_{ij}^{CAL} = \frac{1}{\sum_{n=1}^{N} S_{CAL}^{(n)}} \sum_{n=1}^{N} S_{CAL}^{(n)}(c_i^{(n)} - C_i)(c_j^{(n)} - C_j) \quad \text{where} \quad C_{i/j} = \frac{\sum_{n=1}^{N} S_{CAL}^{(n)} c_{i/j}^{(n)}}{\sum_{n=1}^{N} S_{CAL}^{(n)}}.$$

In these formulas, $C_{i/j}$ (with i/j = X,Y,Z) is the center of gravity (c.o.g.) of the coordinates of the selected crystals, whereas $S_{CAL}^{(n)}$ (with n from 1 to N) are the crystal signals, used as weights for both calculations (covariance matrix and c.o.g.). From the M_{ij}^{CAL} matrix, the eigenvectors can be extracted. The eigenvector with the largest eigenvalue represents a first estimation of the particle track, based on the calorimeter signals only.

The next step is to introduce the silicon detector. As explained before, a silicon strip sensor can measure only one transverse coordinate (X or Y in TIC geometry). So, the algorithm described below has to be repeated twice, for X–Z and Y–Z views. Let J be the strips in a view with a signal exceeding a threshold based on the system noise. From the coordinates $\mathbf{d}_i^{(k)}$ (with i = X, Z or Y, Z and k = 1, ..., J) of the selected strips, the covariance matrix $\mathbf{M}_{ij}^{\text{SIL}}$ is calculated by the formula:

$$\mathbf{M}_{ij}^{\text{SIL}} = \frac{1}{\sum_{k=1}^{J} \mathbf{W}_{\text{SIL}}^{(k)}} \sum_{k=1}^{J} \mathbf{W}_{\text{SIL}}^{(k)}(\mathbf{d}_i^{(k)} - \mathbf{D}_i)(\mathbf{d}_j^{(k)} - \mathbf{D}_j) \quad \text{where} \quad \mathbf{D}_{i/j} = \frac{\sum_{k=1}^{J} \mathbf{W}_{\text{SIL}}^{(k)} \mathbf{d}_{i/j}^{(k)}}{\sum_{k=1}^{J} \mathbf{W}_{\text{SIL}}^{(k)}}.$$

This time, $\mathbf{D}_{i/j}$ (with i/j = X, Z or Y, Z) is the c.o.g. of the coordinates of the selected strips. The core of the whole algorithm is the definition of the weights $\mathbf{W}_{\text{SIL}}^{(k)}$ (with k = 1, ..., J) used in the matrix calculation. A simple estimation can be the ratio between the signal $\mathbf{S}^{(k)}$ from the k-th strip and the total energy $\mathbf{S_P}$ measured by the plane containing the k-th strip. However, the impact point of the particle on each silicon plane can be estimated using the calorimeter track. The extrapolated transverse coordinate $\mathbf{X_P}$ of this point is distributed according to a Gaussian function. The combination of signal information and coordinate spatial dispersion has been found as the best estimation for the weighting factor of the k-th strip, according to the formula:

$$\mathbf{W}_{\text{SIL}}^{(k)} = \frac{\mathbf{S}^{(k)}}{\mathbf{S_P}} \cdot \frac{1}{\sqrt{2\pi}\sigma_P} \exp\left[-\frac{1}{2}\left(\frac{\mathbf{d}_{X/Y}^{(k)} - \mathbf{X_P}}{\sigma_P} \right) \right]$$

where σ_P is the standard deviation of Gaussian dispersion, estimated by simulation. This definition connects the calorimeter and the silicon strip tracking. The eigenvector with the largest eigenvalue of the matrix $\mathbf{M}_{ij}^{\text{SIL}}$ represents a reconstructed track improved with respect to the previous one. Since the weights can be recalculated using the new estimate of the impact point from this new track, the whole procedure can be repeated until convergence. A detailed description of the algorithm can be found in [7].

5. Prototype and Beam Test Results

The simulation results have been further validated by building a small prototype. Its configuration is shown in Figure 5. The main part has been obtained from the refurbished CaloCube prototype [9]. In front of the calorimeter (represented by the horizontal blue modules in Figure 5), a module based on the TIC design has been added. This part was in turn composed of two sub-parts. The first one was made of two layers of thin crystals with, in the middle, two silicon layers separated by a gap, as in the schematic of Figure 3. The second one was a kind of "sandwich" made of alternate layers of silicon strips and crystals. The thin crystals of the first two layers were CaloCube spare cubes sawn in half. The silicon detectors were spare sensors from the DAMPE experiment [14]. The material, CsI(Tl), used for CaloCube is different from the LYSO of the simulations of the full-scale detector. Then, a dedicated simulation has been developed.

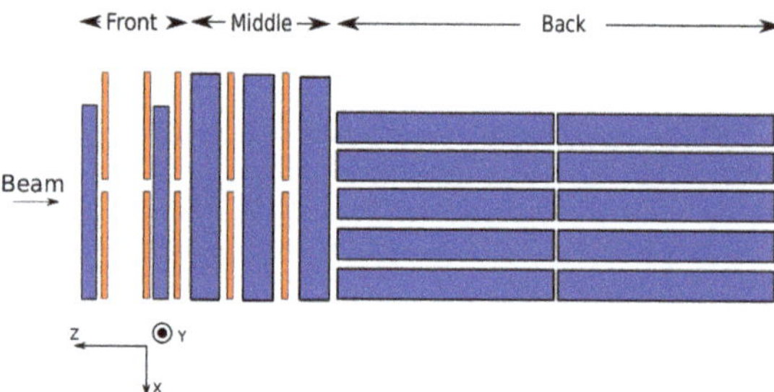

Figure 5. Top view of the TIC prototype: the blue modules represent the trays containing the crystals; the silicon sensors are in orange (from [7]).

The limited number of available silicon modules was sufficient only to instrument one view, and then the tracker was able to reconstruct only the XZ projection of the particle trajectory. The prototype has been tested at the CERN accelerators with electron beams. The choice of electrons instead of photons has been made only for practical reasons. Nevertheless, since the tracking method is based on the sampling of electromagnetic showers, the performance of the detector with electrons is expected to be virtually identical to those with photons. In the beam test area, the support structure allowed the whole apparatus to move up and down and rotate and then modify the angle and the impact point of the particles. During the test, the energy of the beam was varied from 1 GeV to 100 GeV and the incidence angle from 0 to 10 degrees. The response calibration and the alignment of the instrument have been realized with a muon beam. The analysis of the beam test data is based on the same reconstruction procedure applied to simulations described in Section 4.

The algorithm reconstructs the tracks of particles impinging the detector. The dispersion of the reconstructed incoming directions around the true incidence angles is a good estimator of the TIC performance. The true incidence is unknown by definition, but it can be evaluated from the mean of the distribution of reconstructed angles with the detector in a fixed position. This evaluation is naturally affected by a bias that was considered in the simulation. The dedicated simulation, based on FLUKA, included an accurate description of the experimental environments. The particle generator has been set also to reproduce the real beam line, according to specifications provided by CERN. Moreover, some instrumental effects, such as the capacitive coupling of microstrips and the electronic noise, were taken into account as well.

In Figure 6. the distributions of dispersion around the mean angle, in the case of 1 GeV and 5 GeV electrons, are shown: the agreement between the real and simulated data is very good. In this energy region, below 50 GeV, the angular resolution is intrinsically worse, and the tuning of the simulation seems sufficient to correctly model the detector behavior. However, the accuracy of simulations becomes progressively more important at higher energies. Above 50 GeV, it is necessary to introduce an additional spread (of about 0.04°) to the nominal one of the simulated beam to obtain a quite good agreement with real data. The effect of this extra "fine-tuning" to accurately reproduce the real angular volatility is evident in Figure 7, showing the dispersion distributions of test beam data (black curve) in comparison with simulations (red and blue curves).

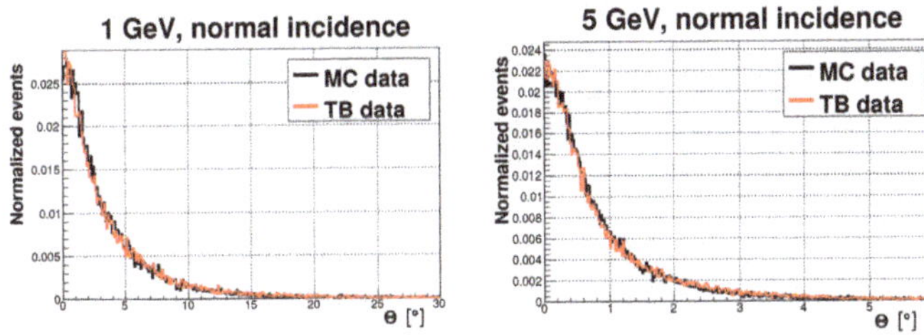

Figure 6. Angular resolution of the reconstructed tracks for 1 and 5 GeV electrons (from [7]).

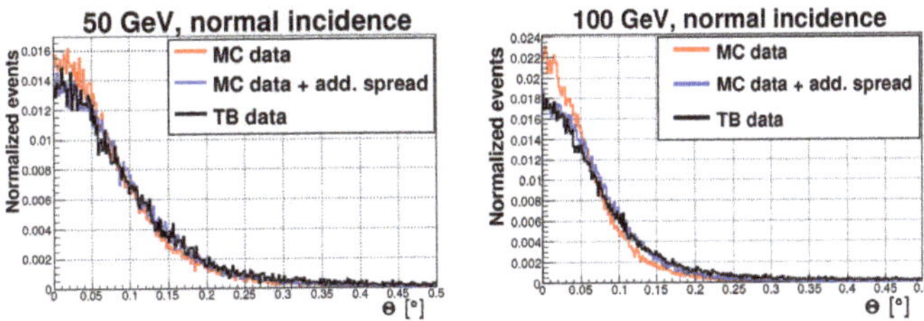

Figure 7. Angular resolution of the reconstructed tracks for 50 and 100 GeV electrons (from [7]).

This correction accounts for unknown systematic effects affecting the apparatus in the whole energy range. The residual test beam–Monte Carlo resolution difference gives an estimation of systematic error. A detailed description of Monte Carlo tuning and the study of systematic errors can be found in [7]. The angular distributions of reconstructed photons are well described by a Point Spread Function (PSF). The width of these dispersion distributions measures the precision of the tracking performance of TIC at different energies. A description of the determination method of PSF for a γ-ray space experiment (FERMI) can be found in [15]. In Table 1, the results for 68% PSF containment radius are summarized: the agreement between simulated and real data is quite satisfactory.

Table 1. Summary of results: XZ angular resolution from test beam (TB) and Monte Carlo (MC) data for 68% PSF (derived from [7]).

Energy (GeV)	Angle (Deg)	Res. 68% (Deg)—TB	Res. 68% (Deg)—MC
1	0	3.72 ± 0.11	3.87 ± 0.12
5	0	0.985 ± 0.016	0.955 ± 0.014
10	0	0.4095 ± 0.0087	0.3971 ± 0.0050
50	0	0.1205 ± 0.0023	0.0995 ± 0.0050
100	0	0.0897 ± 0.0010	0.0680 ± 0.0008
100	10	0.0884 ± 0.0013	0.0646 ± 0.0004

Instead, the agreement at 95% PSF is not as good as it looks in Table 2. The worsening is probably due to residual instrumental effects not reproduced by Monte Carlo simulation. This results in a different shape of tails of the distributions. On the other hand, the effect on the angular resolution of the particle impact angle is negligible. In Tables 1 and 2 (last two lines), the results for 100 GeV electrons with 0° and 10° incidence angles are shown: the resolutions are perfectly compatible.

Table 2. Summary of results: XZ angular resolution from test beam (TB) and Monte Carlo (MC) data for 95% PSF (derived from [7]).

Energy (GeV)	Angle (Deg)	Res. 95% (Deg)—TB	Res. 95% (Deg)—MC
1	0	10.63 ± 0.15	11.44 ± 0.33
5	0	3.300 ± 0.063	3.282 ± 0.054
10	0	1.946 ± 0.062	1.790 ± 0.068
50	0	0.921 ± 0.029	0.710 ± 0.058
100	0	0.678 ± 0.033	0.308 ± 0.010
100	10	0.662 ± 0.036	0.259 ± 0.019

6. Conclusions

The TIC collaboration developed a new detector for Cosmic Ray experiments, starting from the experience of the CaloCube project, which developed a compact cubic calorimeter. This design maximizes instrument acceptance, respecting the limit on the mass budget of space-based missions. This is crucial to extend the energy measure of charged Cosmic Rays until the PeV region. The TIC idea was the integration of a silicon tracker inside this high-segmented calorimeter to optimize its design also for γ-rays physics: the installation of silicon microstrips permits the tracking accuracy needed for high-energy photons. Different geometries have been investigated using simulations: the expected performances are comparable with the main γ-ray experiments (Fermi, DAMPE). A small-scale prototype has been built and tested at CERN laboratories with electron beams. The analysis of the collected data confirmed the expectations in the energy range from 1 GeV to 100 GeV; above 50 GeV, the measured angular resolution is even better than Fermi and DAMPE. At 68% PSF, the test data are in good agreement with Monte Carlo studies. This validates the simulation predictions of the full-scale detector and provides a robust validation of the measurement principle. These promising results represent an important contribution to the development of new instruments for astroparticle experiments. The TIC design has been considered during the planning of the HERD mission mentioned above. In addition, the TIC geometry can be further improved. In the original scheme, the deep homogeneous calorimeter can provide a very precise measure of photon energy, but the γ-ray acceptance is constrained by the fact that only one face of a cubic calorimeter is instrumented with silicon sensors. A possible development could be instrumenting the other faces and then extending the energy range beyond 300 GeV (the limit of the FERMI experiment). Interesting perspectives could open up in the coming years for γ-ray physics.

Author Contributions: Data curation, E.B., L.P. and E.V.; Formal analysis, S.B. and P.P.; Resources, S.D. and O.S.; Supervision, O.A., G.A., B.B., G.C., M.D., P.S.M., P.S. and R.D.; Visualization, A.B., L.B., M.B. (Massimo Bongi), M.B. (Mirko Brianzi), P.B., E.C., C.C., N.F., V.F., M.I., P.M., F.M., S.B.R., G.S., F.S., J.E.S., A.S., A.T. and P.A.; Writing—original draft, G.B.; Writing—review & editing, N.M. All authors have read and agreed to the published version of the manuscript.

Funding: This research received no external funding.

Data Availability Statement: Not applicable.

Conflicts of Interest: The authors declare no conflict of interest. The funders had no role in the design of the study; in the collection, analyses, or interpretation of data; in the writing of the manuscript, and in the decision to publish the results.

References

1. Adriani, O.; Akaike, Y.; Asano, K.; Asaoka, Y.; Bagliesi, M.G.; Berti, E.; Bigongiari, G.; Binns, W.R.; Bonechi, S.; Bongi, M.; et al. Direct measurement of the cosmic-ray proton spectrum from 50 GeV to 10 TeV with the Calorimetric Electron Telescope on the International Space Station. *Phys. Rev. Lett.* **2019**, *122*, 181102. [CrossRef] [PubMed]
2. An, Q.; Asfandiyarov, R.; Azzarello, P.; Bernardini, P.; Bi, X.J.; Cai, M.S.; Chang, J.; Chen, D.Y.; Chen, H.F.; Chen, J.L.; et al. Measurement of the cosmic-ray proton spectrum from 40 GeV to 100 TeV with the DAMPE satellite. *Sci. Adv.* **2019**, *5*, eaax3793. [PubMed]
3. Alemanno, F.; An, Q.; Azzarello, P.; Barbato, F.C.T.; Bernardini, P.; Bi, X.J.; Cai, M.S.; Catanzani, E.; Chang, J.; Chen, D.Y.; et al. Measurement of the cosmic ray helium energy spectrum from 70 GeV to 80 TeV with the DAMPE space mission. *Phys. Rev. Lett.* **2021**, *126*, 201102. [CrossRef] [PubMed]
4. Zhang, S.N.; Adriani, O.; Consortium, H.; Albergo, S.; Ambrosi, G.; An, Q.; Azzarello, P.; Bai, Y.; Bao, T.; Bernardini, P.; et al. Introduction to the High Energy cosmic-Radiation Detection (HERD) Facility onboard China's FutureSpace Station. *Proc. Sci.* **2017**, *301*, 1077.
5. Atwood, W.B.; Bagagli, R.; Baldini, L.; Bellazzini, R.; Barbiellini, G.; Belli, F.; Borden, T.; Brez, A.; Brigida, M.; Caliandro, G.A.; et al. Design and initial tests of the Tracker-converter of the Gamma-ray Large Area Space Telescope. *Astropart. Phys.* **2007**, *28*, 422–434. [CrossRef]
6. Bulgarelli, A.; Argan, A.; Barbiellini, G.; Basset, M.; Chen, A.; Di Cocco, G.; Foggetta, L.; Gianotti, F.; Giuliani, A.; Longo, F.; et al. The AGILE silicon tracker: Pre-launch and in-flight configuration. *Nucl. Instruments Methods Phys. Res. Sect.* **2010**, *95*, 213–226. [CrossRef]
7. Adriani, O.; Ambrosi, G.; Azzarello, P.; Basti, A.; Berti, E.; Bertucci, B.; Bigongiari, G.; Bonechi, L.; Bongi, M.; Bottai, S.; et al. Tracker-In-Calorimeter (TIC): A calorimetric approach to tracking gamma rays in space experiments. *J. Instrum.* **2020**, *15*, P09034. [CrossRef]
8. Starodubtsev, O.; Adriani, O.; Bongi, M.; Bottai, S.; D'Alessandro, R.; Detti, S.; Lenzi, P.; Mori, N.; Papini, P.; Vannuccini, E.; et al. Development of a homogeneous, isotropic and high dynamic range calorimeter for the study of primary cosmic rays in space experiments. *Proc. Sci.* **2015**, *189*, 1–9. [CrossRef]
9. Adriani, O.; Albergo, S.; Auditore, L.; Basti, A.; Berti, E.; Bigongiari, G.; Bonechi, L.; Bongi, M.; Bonvicini, V.; Bottai, S.; et al. The CaloCube project for a space based cosmic ray experiment: Design, construction, and first performance of a high granularity calorimeter prototype. *J. Instrum.* **2019**, *14*, P11004.
10. Adriani, O.; Bonechi, L.; Bongi, M.; Castellini, G.; Ciaranfi, R.; D'Alessandro, R.; Grandi, M.; Papini, P.; Ricciarini, S.; Tricomi, A.; et al. The construction and testing of the silicon position sensitive modules for the LHCf experiment at CERN. *J. Instrum.* **2010**, *5*, P01012.
11. Battistoni, G.; Boehlen, T.; Cerutti, F.; Chin, P.W.; Esposito, L.S.; Fassò, A.; Ferrari, A.; Lechner, A.; Empl, A.; Mairani, A.; et al. Overview of the FLUKA code. *Ann. Nucl. Energy* **2015**, *82*, 10–18. [CrossRef]
12. Fermi Collaboration, Fermi LAT Performance. Available online: https://www.slac.stanford.edu/exp/glast/groups/canda/lat_Performance.htm (accessed on 19 September 2022). [CrossRef]
13. Chang, J.; Ambrosi, G.; An, Q.; Asfandiyarov, R.; Azzarello, P.; Bernardini, P.; Bertucci, B.; Cai, M.S.; Caragiulo, M.; Chen, D.Y.; et al. The DArk Matter Particle Explorer mission. *Astropart. Phys.* **2017**, *95*, 6–24. [CrossRef]

14. Azzarello, P.; Ambrosi, G.; Asfandiyarov, R.; Bernardini, P.; Bertucci, B.; Bolognini, A.; Cadoux, F.; Caprai, M.; De Mitri, I.; Domenjoz, M.; et al. The DAMPE silicon-tungsten tracker. *Nucl. Instruments Methods Phys. Res. Sect.* **2016**, *831*, 378–384.
15. Ackermann, M.; Ajello, M.; Allafort, A.; Asano, K.; Atwood, W.B.; Baldini, L.; Ballet, J.; Barbiellini, G.; Bastieri, D.; Bechtol, K.; et al. Determination of the Point-Spread Function for the Fermi Large Area Telescope from On-Orbit Data and Limits on Pair Halos of Active Galactic Nuclei. *Astrophys. J.* **2013**, *54*, 765. [CrossRef]

Article

The Impact of Crystal Light Yield Non-Proportionality on a Typical Calorimetric Space Experiment: Beam Test Measurements and Monte Carlo Simulations

Lorenzo Pacini [1,*], Oscar Adriani [1,2], Eugenio Berti [1,2], Pietro Betti [2], Gabriele Bigongiari [3,4], Lorenzo Bonechi [1], Massimo Bongi [1,2], Sergio Bottai [1], Paolo Brogi [3,4], Guido Castellini [1,5], Caterina Checchia [3,4], Raffaello D'Alessandro [1,2], Sebastiano Detti [1], Noemi Finetti [1,6], Paolo Maestro [3,4], Pier Simone Marrocchesi [3,4], Nicola Mori [1], Miriam Olmi [1,2], Paolo Papini [1], Claudia Poggiali [1,2], Sergio Ricciarini [1,5], Piero Spillantini [1,2], Oleksandr Starodubtsev [1], Francesco Stolzi [3,4], Alessio Tiberio [1] and Elena Vannuccini [1]

1   INFN Firenze, Via B. Rossi 1, I-50019 Firenze, Italy
2   Department of Physics and Astronomy, University of Florence, Via G. Sansone 1, I-50019 Firenze, Italy
3   Department of Physical Sciences, Earth and Environment, University of Siena, I-53100 Siena, Italy
4   INFN Pisa, Largo B. Pontecorvo, I-56127 Pisa, Italy
5   IFAC CNR, Via Madonna del Piano 10, I-50019 Firenze, Italy
6   Department of Physical and Chemical Sciences, University of L'Aquila, Via Vetoio, Coppito, I-67100 L'Aquila, Italy
*   Correspondence: lorenzo.pacini@fi.infn.it

Citation: Pacini, L.; Adriani, O.; Berti, E.; Betti, P.; Bigongiari, G.; Bonechi, L.; Bongi, M.; Bottai, S.; Brogi, P.; Castellini, G.; et al. The Impact of Crystal Light Yield Non-Proportionality on a Typical Calorimetric Space Experiment: Beam Test Measurements and Monte Carlo Simulations. *Instruments* **2022**, *6*, 53. https://doi.org/10.3390/instruments6040053

Academic Editors: Fabrizio Salvatore, Alessandro Cerri, Antonella De Santo and Iacopo Vivarelli

Received: 30 August 2022
Accepted: 22 September 2022
Published: 27 September 2022

Publisher's Note: MDPI stays neutral with regard to jurisdictional claims in published maps and institutional affiliations.

Abstract: Calorimetric space experiments were employed for the direct measurements of cosmic-ray spectra above the TeV region. According to several theoretical models and recent measurements, relevant features in both electron and nucleus fluxes are expected. Unfortunately, sizable disagreements among the current results of different space calorimeters exist. In order to improve the accuracy of future experiments, it is fundamental to understand the reasons of these discrepancies, especially since they are not compatible with the quoted experimental errors. A few articles of different collaborations suggest that a systematic error of a few percentage points related to the energy-scale calibration could explain these differences. In this work, we analyze the impact of the nonproportionality of the light yield of scintillating crystals on the energy scale of typical calorimeters. Space calorimeters are usually calibrated by employing minimal ionizing particles (MIPs), e.g., nonshowering proton or helium nuclei, which feature different ionization density distributions with respect to particles included in showers. By using the experimental data obtained by the CaloCube collaboration and a minimalist model of the light yield as a function of the ionization density, several scintillating crystals (BGO, CsI(Tl), LYSO, YAP, YAG and BaF2) are characterized . Then, the response of a few crystals is implemented inside the Monte Carlo simulation of a space calorimeter to check the energy deposited by electromagnetic and hadronic showers. The results of this work show that the energy scale obtained by MIP calibration could be affected by sizable systematic errors if the nonproportionality of scintillation light is not properly taken into account.

Keywords: cosmic rays; calorimetry; scintillation; light yield

1. Introduction

Several relevant open questions regarding astroparticles and dark-matter physics require accurate measurements of cosmic rays (CRs). For instance, the direct observation of electron and positron spectra above a few TeV provides unique information regarding high-energy CR sources near Earth and dark-matter models, while CR acceleration and propagation models benefit from the accurate measurement of proton and nuclei spectra [1]. Space spectrometers such as PAMELA [2] and AMS-02 [3] are capable of separating matter and antimatter, but they cannot detect particles above a few TeV due to the limited acceptance and maximal detectable rigidity (MDR). Thus, the calorimetric technique is employed to explore higher energies: a few examples of running space calorimeters are CALET [4] and DAMPE [5], while a future detector is HERD [6]. Even if the precision of recent direct CR measurements is strongly increased with respect to previous instruments, evidence of the disagreement among different experiments exists.

Figure 1 [7] shows the recent measurements of the electron and positron flux. Here, two groups of experiments are clearly present: from ~ 100 GeV to ~ 1 TeV DAMPE [8] and Fermi-LAT [9] feature higher fluxes with respect to those of CALET [10] and AMS-02 [11]. The differences are larger than the errors quoted by the experiments, which could be explained if unaccounted systematic errors are present. Besides electrons and positrons, other examples of tensions in CR measurements are the carbon, oxygen and iron spectra: CALET [12,13] results feature higher overall normalization with respect to the result of AMS-02 [14,15], even if the spectral shape is similar.

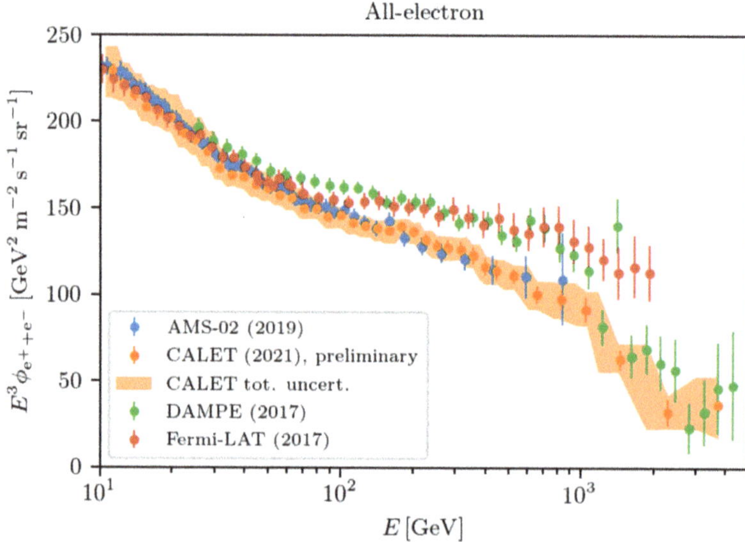

Figure 1. Recent direct measurements of the electron flux [7].

Since space calorimeters are usually composed of inorganic scintillators, the main goal of this work is to study the effect of the nonproportionality of the scintillation light yield for the measurement of high-energy showers, which could translate into a sizable systematic error on the energy scale of space experiments. A list of experiments employing inorganic crystals is shown in Table 1. The work presented in this contribution is also discussed in detail in [16].

Table 1. List of space cosmic-ray experiments based on calorimetric instruments produced with inorganic scintillators, and the main characteristics of the calorimeters.

Experiment	Material	Electromagnetic Depth (X_0)	Hadronic Depth (λ_I)	Launch Year
CALET [4]	PWO	27	1.2	2015
DAMPE [5]	BGO	32	1.6	2015
FERMI [17]	CsI(Tl)	8.6	0.4	2008
HERD [18]	LYSO	55	3.0	2027 (expected)

2. Nonproportional Light Response: Minimalist Approach

The light response of both organic and inorganic scintillators depends on the ionization density [19], which is proportional to the energy deposit inside the crystals per unit length

(dE/dx). In this paper, the Minimalist Approach model [20] was employed to approximate the nonproportionality of scintillators since it fit the data used to characterize the materials well (see Section 4). This model takes into account two effects. The first, the quenching [21] (or Birks) effect, dominates at high excitation density. A slightly modified version of this model was employed here: as proposed in [22], assuming a division of the energy deposition into cylindrical core and halo regions surrounding the particle trajectory, only the charge carriers inside the core are affected by the quenching effect. The relative light-emission efficiency formula for the modified Birk effect is then:

$$L_B = \frac{1 - \eta_H}{1 + B(1 - \eta_H) \times \frac{dE}{dx}} + \eta_H \tag{1}$$

where B is the Birks parameter, and η_H is the fraction of carriers in the halo region.

The second, called the Onsager effect, dominates at low excitation density and is described in [23]. Electrons and holes that initially do not form excitons can be recombined afterwards only if they are closer than the Onsager radius. This effect improves the efficiency of light emission, which can be written as follows:

$$L_O = 1 - \eta_{e/h} \exp\left(-\frac{(dE/dx)}{(dE/dx)_O}\right) \tag{2}$$

where $(dE/dx)_O$ is the strength of the Onsager term, and $\eta_{e/h}$ is the fraction of initial carriers that do not form excitons.

By combining Equations (1) and (2), the relative light-emission efficiency can be expressed as follows:

$$L = \left[1 - \eta_{e/h} \exp\left(-\frac{(dE/dx)}{(dE/dx)_O}\right)\right] \times \left[\frac{1 - \eta_H}{1 + B(1 - \eta_H) \times \frac{dE}{dx}} + \eta_H\right]. \tag{3}$$

3. Monte Carlo Simulation of the Ionization Density

To study the dE/dx in different materials, a simulation code based on FLUKA [24] was employed. In order to improve the simulation accuracy, the minimal energy thresholds for particle tracking were set to be 1 keV for electrons and 100 eV for photons. Furthermore, all physical processes that contribute to ionization were activated. The output of the simulation is the amount of energy released for every bin of ionization density. Since the light signal depends on energy loss and light-emission efficiency, it can be computed as follows.

$$S_L = \sum_i \Delta E_i \times L_i. \tag{4}$$

where S_L is the light signal in arbitrary units, and ΔE_i and L_i are the energy loss and light-emission efficiency in a given bin of ionization density, respectively.

Two examples of the simulation output are shown in Figure 2. The mean energy deposit due to nonshowering protons (helium nuclei) crossing 2 cm of LYSO is shown in the black (red) histogram. These histograms also include the energy deposited by secondary particles (e.g., δ rays). As expected, the helium ionization density profile is different with respect to the one of protons, and it features the main peak at $\sim$24 MeV/cm, and a secondary peak at $\sim$6 MeV/cm, which is due to δ-ray emission. Figure 2 also shows the typical light-emission efficiency [25] for alkali (green) and silicate (blue) scintillators (e.g., CsI(Tl) and LYSO). Even if the ratio between the mean energy deposit of helium nuclei and protons is 4, the ratio of the light signals is less (greater) than 4 for silicate (alkali) scintillators due to the different luminous efficiencies.

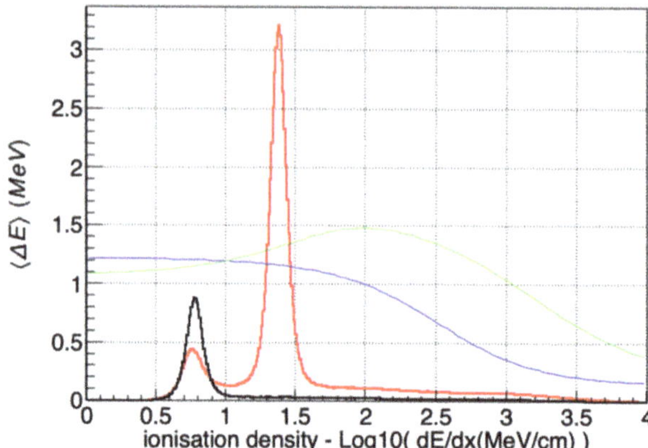

Figure 2. Black and red histograms represent the mean energy deposit for each bin of ionization density due to 100 GeV nonshowering protons and helium in 2 cm of LYSO, respectively. The green and blue curves are the typical luminous efficiencies for alkali and silicate scintillators, reported here in arbitrary units [16].

4. Characterization of Scintillators with CaloCube Beam Test Data

Typical methods employed to study the scintillator nonproportionality are based on Compton electrons and photon response, but in this work, the ionization produced by high energy nuclei is used; this technique was exploited by GLAST/Fermi-LAT [26] and DAMPE [27]. Different nonshowering nuclei feature different mean ionization densities that can be used to measure light signals corresponding to different ionization densities. By employing nuclei from protons ($Z = 1$) to argon ($Z = 18$), the light signals corresponding to ionization densities between 5 MeV/cm and 2 GeV/cm can be measured.

These measurements were performed with the CaloCube collaboration [28], which was an R & D project that exploited new concepts for the design of a space calorimeter. The main design is a 3D-segmented, homogeneous, isotropic, cubic calorimeter produced with cubic scintillating crystals. The acceptance of this instrument is larger than that of typical space telescopes since it measures particles coming from each side and not only from the zenith. The read-out system of the scintillating light consists of a pair of photodiodes with different active areas and double-gain custom front-end electronics. In order to optimize the design, different scintillating materials were tested with the Monte Carlo simulations of the calorimeter [29]. Furthermore, few CaloCube prototypes were built [30]. For instance, the performance of the large-scale prototype with high-energy electrons was described in [31]. In this work, data acquired with the prototype tested at the CERN SPS accelerator with high-energy nuclei in 2015 are discussed. This prototype consisted of several trays equipped with CsI(Tl) crystals, while the last tray allocated different cubic scintillators (test crystal), as shown in Figure 3.

The properties of the test crystals are summarized in Table 2.

The scintillating light was read out with a photodiode (VTH2090) coupled with custom front-end electronics, which mainly consisted of CASIS [32] chips.

During the test, the beam consisted of nucleus fragments with $30 \cdot A \, GeV$ kinetic energy, an A/Z ratio equal to 2, and charge ranging from 1 to 18. Specific runs were employed to acquire events in which the beam directly hit each test crystal. During the data analysis, nonshowering nuclei were selected with the information of the nearby CsI crystals, while the impact position and the charge of the particle were reconstructed using a silicon tracker placed upstream of the CaloCube prototype.

The main results of this test are summarized in Figure 4, where different markers show the mean value of different crystal signals divided by the square of the nucleus charge. Considering an ideal scintillator featuring constant light-emission efficiency, the points would be displaced on a horizontal line; for the tested scintillators, a clear deviation from this ideal condition was shown. Different trends of the point series are related to the different material nonproportionality of the scintillation light.

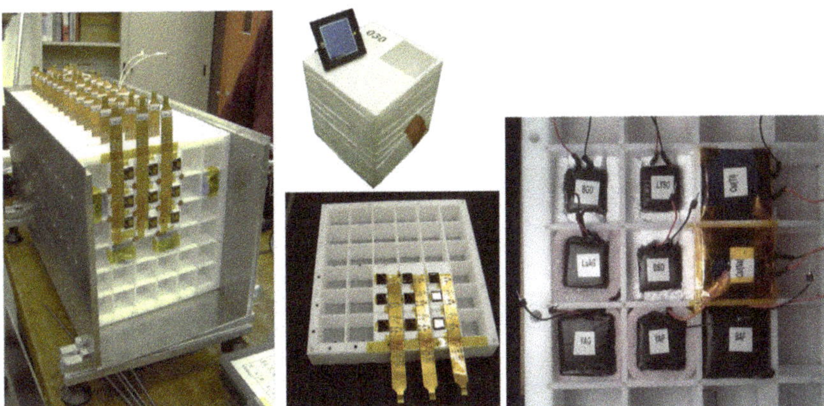

Figure 3. Left panel: image of the CaloCube prototype made of CsI crystals. Central panels: image of a prototype layer with CsI crystals (bottom panel) and a crystal with a VTH2090 PD (top panel). Right panel: image of the last tray, which includes different scintillators.

Table 2. Main properties of the tested materials, where ρ is the density of the crystal, λ_i and X_0 are the interaction and radiation lengths, respectively, λ_{max} is the wavelength of the scintillation light at the emission maximum, and τ_{decay} is the decay time.

Material	Size (cm)	ρ (g/cm^3)	λ_I (cm)	X_0 (cm)	λ_{max} (nm)	τ_{decay} (ns)
BGO	2.0	7.1	23	1.1	480	300
CsI(Tl)	3.6	4.5	40	1.9	550	1220
LYSO	2.0	7.4	21	1.1	420	40
YAP	2.2	5.5	22	2.7	370	27
YAG	2.5	4.6	25	3.5	550	70
BaF$_2$	3.1	4.9	31	2.0	300	650

By exploiting the data shown in Figure 4, the relative light yield of each scintillator as a function of the nuclei charge was computed. For instance, the black points of Figure 5 show the relative light yield for CsI(Tl) and LYSO normalized to $Z = 18$. The minimalist approach was then used to fit the data by exploiting the dE/dx profile obtained with the FLUKA simulation, as explained in the previous section; the red points of Figure 5 show the fit results. The simple model was able to reproduce the experimental trends of each crystal, reducing χ^2_{red} from 0.64 to 1.64. The fit results regarding the parameters of Equation (3) are summarized in Table 3.

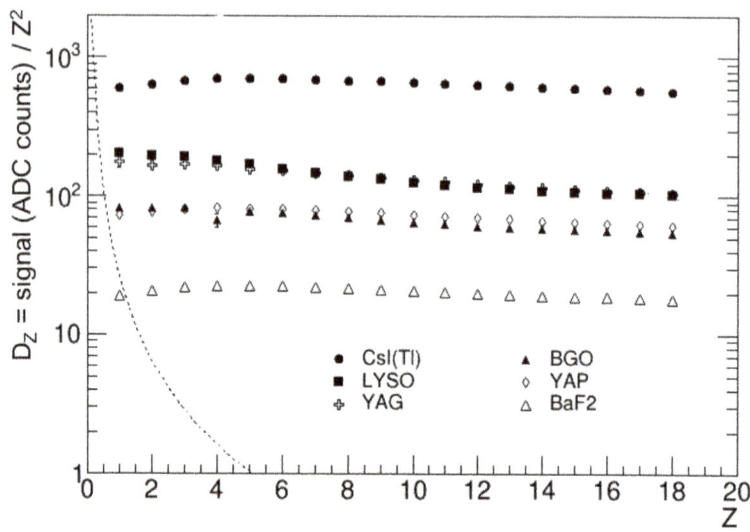

Figure 4. Different markers show the mean of the light signals obtained with different crystals crossed by different high-energy nuclei, divided by Z^2, and plotted as a function of Z. The dashed line represents the noise level [16].

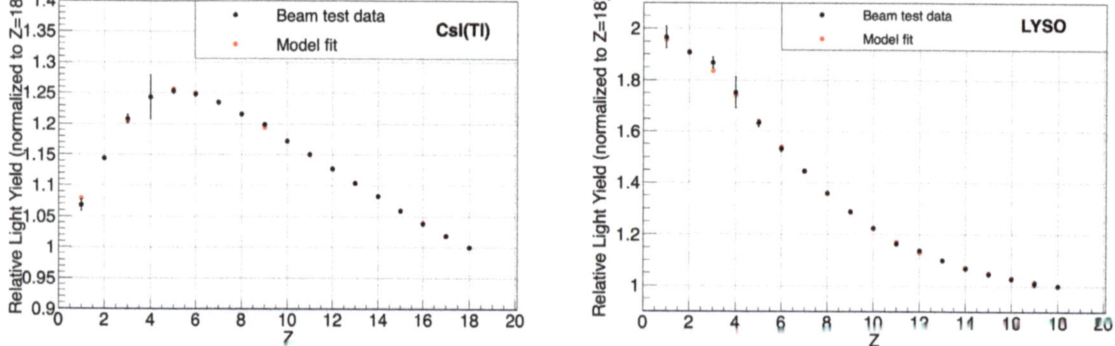

Figure 5. Relative light yield normalized to the argon one of (left panel) CsI(Tl) and (right panel) LYSO. Black points are the result of beam test data analysis, while red points are the result of the fit with the minimalist model; Equation (3) [16].

Table 3. Parameters of Equation (3) obtained with the fit of CaloCube data [16].

Material	$\eta_{e/h}$	$(dE/dx)_O$ MeV/cm	η_H	$(1/B)$ MeV/cm	χ^2_{red}
BGO	0.159 ± 0.033	98 ± 45	0.1884 ± 0.0039	364 ± 42	1.64
CsI(Tl)	0.326 ± 0.010	34.1 ± 2.8	0.121 ± 0.012	1338 ± 64	0.81
LYSO	0.758 ± 0.045	164.7 ± 8.4	0.0274 ± 0.0048	45.1 ± 9.1	0.64
YAP	0.2212 ± 0.0085	90 ± 11	0.174 ± 0.012	873 ± 70	1.24
YAG	0.0912 ± 0.015	73 ± 29	0.1052 ± 0.0055	462 ± 31	1.23
BaF$_2$	0.322 ± 0.024	35.8 ± 6.2	0.3440 ± 0.0071	546 ± 36	1.11

5. Simulation of a Space Calorimeter

The final section of this work is the study of a possible systematic error on the energy measurement obtained with a space calorimeters. A very simple geometric configuration, i.e. a homogeneous cube of 1 m^3, is simulated with the FLUKA package: the active materials alternatively employed in the simulation are LYSO, BGO and CsI. In a real experiment, the effect of the nonproportionality of the light yield depends on several parameters e.g., the crystal manufacturer, the specific detector geometry and calibration, the front-end electronics and acquisition system. With the approach described in this contribution, the possible existence of systematic effects is discussed while quantitative results for specific running experiments can not be determined.

The typical calibration of space calorimeters involves several steps, e.g., the CALET calibration is described in [33]. Usually, one of the first steps makes use of the energy deposited by Minimum Ionizing Particles (MIP), i.e., nonshowering protons and helium nuclei. Thus, the calibration of the energy scale strongly relies on MIP measurements. Unfortunately, the ionization profile of the energy deposited by MIPs is different with respect to the one of an electromagnetic or hadronic shower. The simulation shows that even if the ionization profile of electron showers is different with respect one the MIP one, this is almost constant with the energy form 10 GeV to 1 TeV, while the profile of hadronic showers due to proton interactions depends on the energy. Figure 6 shows the ionization profile of electrons and protons obtained with the simulation and the light-emission efficiency of LYSO obtained with the fit described in previous section.

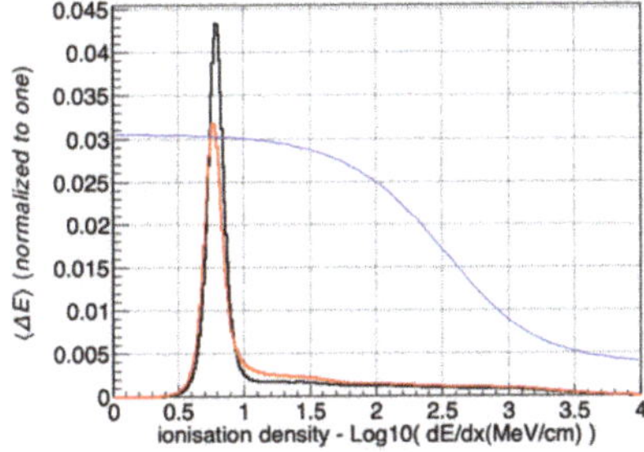
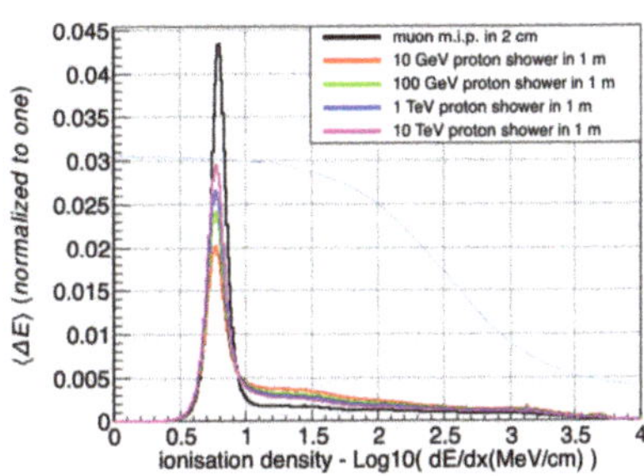

Figure 6. (**Left panel**) Ionization density profile of 10 GeV electrons (red line), MIP (black line), and the light-emission efficiency of LYSO (blue line) in an arbitrary unit. (**right panel**) Ionization density profile of protons at different energies (colored tick lines), MIP (black line), and the light-emission efficiency of LYSO (thin line) in an arbitrary unit [16].

As explained in Section 3, for a given energy deposit, a different ionization density profile translates into a different light output due to the nonproportional light-emission efficiency of the scintillator. For instance, a systematic shift of about 2.3% of the measured total energy deposit in LYSO crystals for electrons from 10 GeV to 1 TeV is obtained. Due to the steepness of cosmic-ray spectra, this energy shift translates into a sizable error on the normalization of the flux: assuming a spectral index equal to γ, and a systematic energy shift of Δ, the shift of the flux normalization was $(\gamma - 1) \cdot \Delta$. The electron spectral index was ~3; thus, the normalization shift obtained with the LYSO crystal was about ~5%. Furthermore, the energy shifts obtained for protons depend on energy; thus, these affect both the shape and the normalization of reconstructed spectra.

The main results of this work, i.e., the systematic energy shift obtained with a 1 m^3 homogeneous calorimeter produced with different crystals, are summarized in Table 4.

Table 4. Energy shift due to the nonproportionality of light yield for a 1 m^3 homogeneous calorimeter produced with different scintillating materials [16].

Scintillator	Electrons $\geq$ 10 GeV	Protons 10 GeV	Protons 100 GeV	Protons 1 TeV	Protons 10 TeV
LYSO	−2.3%	−7.1%	−5.6%	−4.6%	−3.4%
BGO	−1.1%	−4.3%	−3.0%	−2.3%	−1.8%
CsI(Tl)	+0.82%	+2.9%	+2.0%	+1.5%	+1.2%

6. Conclusions

In this contribution, CaloCube data and the minimalist approach were employed to characterize the nonproportionality of scintillator light outputs. These results were then used to study the impact of this effect on homogeneous space calorimeters. Assuming that the energy scale of this kind of detector is mainly calibrated with MIP, an effect on the energy measurement of few percentage points exists. This leads to systematic shifts of the reconstructed spectral normalization of up to 5%. Regarding the results published with the running experiments, it is not clear if this effect had already been considered for the energy scale estimation since it was not mentioned in the main papers regarding calibration and energy reconstruction.

In order to accurately take into account the nonproportionality of scintillators, future calorimetric experiments might apply a similar procedure to the one presented in this contribution. The first step is the characterization of the material by employing a read-out system that is used during inflight operation, e.g., by using the high-energy nucleus technique described here. The second step could be to estimate the impact of nonproportionality on the energy scale and on other important parameters related to shower reconstruction, e.g., electron or proton discrimination variables. Eventually, if the effect is sizable, as suggested in this work, the last step could be the implementation of this effect inside the simulated data reconstruction software by using the ionization density profile and the material characterization.

Author Contributions: Conceptualization, P.P., E.V., O.A. and R.D.; software, N.M.; investigation, E.B., P.B. (Pietro Betti), L.P., O.S. and S.D.; data curation, E.V., M.O., C.P. and E.B.; writing—original draft preparation, L.P.; writing—review and editing, P.P. and E.V.; supervision, G.B., L.B., M.B., S.B., P.B. (Paolo Brogi), G.C., C.C., N.F., P.M., P.S.M., S.R., P.S., F.S. and A.T. All authors have read and agreed to the published version of the manuscript.

Funding: This research received no external funding.

Data Availability Statement: Not applicable.

Conflicts of Interest: The authors declare no conflict of interest.

References

1. Evoli, C.; Amato, E.; Blasi, P.; Aloisio, R. Stochastic nature of Galactic cosmic-ray sources. *Phys. Rev. D* **2021**, *104*, 123029. [CrossRef]
2. Munini, R.; Boezio, M.; Bruno, A.; Christian, E.C.; de Nolfo, G.A.; Felice, V.D.; Martucci, M.; Merge', M.; Richardson, I.G.; Ryan, J.M.; et al. Evidence of Energy and Charge Sign Dependence of the Recovery Time for the 2006 December Forbush Event Measured by the PAMELA Experiment. *Astrophys. J.* **2018**, *853*, 76. [CrossRef]
3. Aguilar, M.; Ali Cavasonza, L.; Ambrosi, G.; Arruda, L.; Attig, N.; Barao, F.; Barrin, L.; Bartoloni, A.; Başeğmez-du Pree, S.; Bates, J.; et al. The Alpha Magnetic Spectrometer (AMS) on the international space station: Part II—Results from the first seven years. *Phys. Rep.* **2021**, *894*, 1–116. [CrossRef]
4. Adriani, O.; Akaike, Y.; Asano, K.; Asaoka, Y.; Bagliesi, M.G.; Berti, E.; Bigongiari, G.; Binns, W.R.; Bonechi, S.; Bongi, M.; et al. Direct Measurement of the Cosmic-Ray Proton Spectrum from 50 GeV to 10 TeV with the Calorimetric Electron Telescope on the International Space Station. *Phys. Rev. Lett.* **2019**, *122*, 181102. [CrossRef] [PubMed]
5. An, Q.; Asfandiyarov, R.; Azzarello, P.; Bernardini, P.; Bi, X.J.; Cai, M.S.; Chang, J.; Chen, D.Y.; Chen, H.F.; Chen, J.L.; et al. Measurement of the cosmic ray proton spectrum from 40 GeV to 100 TeV with the DAMPE satellite. *Sci. Adv.* **2019**, *5*, eaax3793. [CrossRef] [PubMed]
6. Gargano, F. The High Energy cosmic-Radiation Detection (HERD) facility on board the Chinese Space Station: Hunting for high-energy cosmic rays. In Proceedings of the 37th International Cosmic Ray Conference—PoS (ICRC2021), Berlin, Germany, 15–22 July 2021; Volume 395, p. 26. [CrossRef]
7. Mertsch, P. Rapporteur Talk: Cosmic Ray Direct. In Proceedings of the 37th International Cosmic Ray Conference—PoS(ICRC2021), Berlin, Germany, 15–22 July 2021; Volume 395, p. 44.
8. Ambrosi, G.; An, Q.; Asfandiyarov, R.; Azzarello, P.; Bernardini, P.; Bertucci, B.; Cai, M.S.; Chang, J.; Chen, D.Y.; Chen, H.F.; et al. Direct detection of a break in the teraelectronvolt cosmic-ray spectrum of electrons and positrons. *Nature* **2017**, *552*, 63–66. [CrossRef]

9. Abdollahi, S.; Ackermann, M.; Ajello, M.; Atwood, W.B.; Baldini, L.; Barbiellini, G.; Bastieri, D.; Bellazzini, R.; Bloom, E.D.; Bonino, R.; et al. Cosmic-ray electron-positron spectrum from 7 GeV to 2 TeV with the Fermi Large Area Telescope. *Phys. Rev. D* **2017**, *95*, 082007. [CrossRef]

10. Adriani, O.; Akaike, Y.; Asano, K.; Asaoka, Y.; Bagliesi, M.G.; Berti, E.; Bigongiari, G.; Binns, W.R.; Bonechi, S.; Bongi, M.; et al. Extended Measurement of the Cosmic-Ray Electron and Positron Spectrum from 11 GeV to 4.8 TeV with the Calorimetric Electron Telescope on the International Space Station. *Phys. Rev. Lett.* **2018**, *120*, 261102. [CrossRef]

11. Aguilar, M.; Ali Cavasonza, L.; Alpat, B.; Ambrosi, G.; Arruda, L.; Attig, N.; Azzarello, P.; Bachlechner, A.; Barao, F.; Barrau, A.; et al. Towards Understanding the Origin of Cosmic-Ray Electrons. *Phys. Rev. Lett.* **2019**, *122*, 101101. [CrossRef]

12. Adriani, O.; Akaike, Y.; Asano, K.; Asaoka, Y.; Berti, E.; Bigongiari, G.; Binns, W.R.; Bongi, M.; Brogi, P.; Bruno, A.; et al. Measurement of the Iron Spectrum in Cosmic Rays from 10 GeV/n to 2.0 TeV/n with the Calorimetric Electron Telescope on the International Space Station. *Phys. Rev. Lett.* **2021**, *126*, 241101. [CrossRef]

13. Adriani, O.; Akaike, Y.; Asano, K.; Asaoka, Y.; Bagliesi, M.G.; Berti, E.; Bigongiari, G.; Binns, W.R.; Bongi, M.; Brogi, P.; et al. Direct Measurement of the Cosmic-Ray Carbon and Oxygen Spectra from 10 GeV/n to 2.2 TeV/n with the Calorimetric Electron Telescope on the International Space Station. *Phys. Rev. Lett.* **2020**, *125*, 251102. [CrossRef] [PubMed]

14. Aguilar, M.; Cavasonza, L.A.; Allen, M.S.; Alpat, B.; Ambrosi, G.; Arruda, L.; Attig, N.; Barao, F.; Barrin, L.; Bartoloni, A.; et al. Properties of Iron Primary Cosmic Rays: Results from the Alpha Magnetic Spectrometer. *Phys. Rev. Lett.* **2021**, *126*, 041104. [CrossRef] [PubMed]

15. Aguilar, M.; Ali Cavasonza, L.; Alpat, B.; Ambrosi, G.; Arruda, L.; Attig, N.; Aupetit, S.; Azzarello, P.; Bachlechner, A.; Barao, F.; et al. Observation of the Identical Rigidity Dependence of He, C, and O Cosmic Rays at High Rigidities by the Alpha Magnetic Spectrometer on the International Space Station. *Phys. Rev. Lett.* **2017**, *119*, 251101. [CrossRef] [PubMed]

16. Adriani, O.; Berti, E.; Betti, P.; Bigongiari, G.; Bonechi, L.; Bongi, M.; Bottai, S.; Brogi, P.; Castellini, G.; Checchia, C.; et al. Light yield non-proportionality of inorganic crystals and its effect on cosmic-ray measurements. *J. Instrum.* **2022**, *17*, P08014. [CrossRef]

17. Atwood, W.B.; Abdo, A.A.; Ackermann, M.; Althouse, W.; Anderson, B.; Axelsson, M.; Baldini, L.; Ballet, J.; Band, D.L.; Barbiellini, G.; et al. The Large Area Telescope on the Fermi Gamma-Ray Space Telescope Mission. *Astrophys. J.* **2009**, *697*, 1071–1102. [CrossRef]

18. Dong, Y.; Zhang, S.; Ambrosi, G. Overall Status of the High Energy Cosmic Radiation Detection Facility Onboard the Future China's Space Station. *Proc. Sci.* **2019**, *358*, 62.

19. Murray, R.B.; Meyer, A. Scintillation Response of Activated Inorganic Crystals to Various Charged Particles. *Phys. Rev.* **1961**, *122*, 815–826. [CrossRef]

20. Moses, W.; Bizarri, G.; Williams, R.; Payne, S.; Vasil'ev, A.; Singh, J.; Li, Q.; Grim, J.; Choong, W.S. The Origins of Scintillator Non-Proportionality. *IEEE Trans. Nucl. Sci.* **2012**, *59*, 2038. [CrossRef]

21. Birks, J. (Ed.) Preface. In *The Theory and Practice of Scintillation Counting*; International Series of Monographs in Electronics and Instrumentation; Pergamon: London, UK, 1964; pp. xvii–xviii. [CrossRef]

22. Tarle, G.; Ahlen, S.P.; Cartwright, B.G. Cosmic ray isotope abundances from chromium to nickel. *APJ* **1979**, *230*, 607–620. [CrossRef]

23. Payne, S.A.; Cherepy, N.J.; Hull, G.; Valentine, J.D.; Moses, W.W.; Choong, W.S. Nonproportionality of Scintillator Detectors: Theory and Experiment. *IEEE Trans. Nucl. Sci.* **2009**, *56*, 2506–2512. [CrossRef]

24. Ferrari, A.; Sala, P.R.; Fasso, A.; Ranft, J. *FLUKA: A Multi-Particle Transport Code*; Stanford University: Stanford, CA, USA, 2005. [CrossRef]

25. Payne, S.; Moses, W.; Sheets, S.; Ahle, L.; Cherepy, N.; Sturm, B.; Dazeley, S.; Bizarri, G.; Choong, W.S. Nonproportionality of Scintillator Detectors: Theory and Experiment. II. *IEEE Trans. Nucl. Sci.* **2011**, *58*, 3392–3402. [CrossRef]

26. Lott, B.; Piron, F.; Blank, B.; Bogaert, G.; Bregeon, J.; Canchel, G.; Chekhtman, A.; d'Avezac, P.; Dumora, D.; Giovinazzo, J.; et al. Response of the GLAST LAT calorimeter to relativistic heavy ions. *Nucl. Instrum. Methods Phys. Res. Sect. Accel. Spectrom. Detect. Assoc. Equip.* **2006**, *560*, 395–404. [CrossRef]

27. Stanford University; Wei, Y.; Zhang, Y.; Zhang, Z.; Wu, L.; Dai, H.; Liu, C.; Zhao, C.; Wang, Y.; Zhao, Y.; et al. The Quenching Effect of BGO Crystals on Relativistic Heavy Ions in the DAMPE Experiment. *IEEE Trans. Nucl. Sci.* **2020**, *67*, 939–945. [CrossRef]

28. Adriani, O.; Albergo, S.; Auditore, L.; Basti, A.; Berti, E.; Bigongiari, G.; Bonechi, L.; Bongi, M.; Bonvicini, V.; Bottai, S.; et al. The CALOCUBE project for a space based cosmic ray experiment: Design, construction, and first performance of a high granularity calorimeter prototype. *J. Instrum.* **2019**, *14*, P11004. [CrossRef]

29. Adriani, O.; Albergo, S.; Auditore, L.; Basti, A.; Berti, E.; Bigongiari, G.; Bonechi, L.; Bonechi, S.; Bongi, M.; Bonvicini, V.; et al. CaloCube: An isotropic spaceborne calorimeter for high-energy cosmic rays. Optimization of the detector performance for protons and nuclei. *Astropart. Phys.* **2017**, *96*, 11–17. [CrossRef]

30. Pacini, L.; Adriani, O.; Agnesi, A.; Albergo, S.; Auditore, L.; Basti, A.; Berti, E.; Bigongiari, G.; Bonechi, L.; Bonechi, S.; et al. CaloCube: An innovative homogeneous calorimeter for the next-generation space experiments. *J. Phys. Conf. Ser.* **2017**, *928*, 012013. [CrossRef]

31. Adriani, O.; Agnesi, A.; Albergo, S.; Antonelli, M.; Auditore, L.; Basti, A.; Berti, E.; Bigongiari, G.; Bonechi, L.; Bongi, M.; et al. The CaloCube calorimeter for high-energy cosmic-ray measurements in space: Performance of a large-scale prototype. *J. Instrum.* **2021**, *16*, P10024. [CrossRef]

32. Zampa, G.; Bonvicini, V.; Orzan, G.; Zampa, N. CASIS: a Very High Dynamic Range Front-End Electronics with Integrated Cyclic ADC for Calorimetry Applications. In Proceedings of the 2006 IEEE Nuclear Science Symposium Conference Record, Dresden, Germany, 19–25 October 2006; Volume 2, pp. 845–849. [CrossRef]
33. Asaoka, Y.; Akaike, Y.; Komiya, Y.; Miyata, R.; Torii, S.; Adriani, O.; Asano, K.; Bagliesi, M.; Bigongiari, G.; Binns, W.; et al. Energy calibration of CALET onboard the International Space Station. *Astropart. Phys.* **2017**, *91*, 1–10. [CrossRef]

Article

Upgrade of ATLAS Hadronic Tile Calorimeter for the High-Luminosity LHC

Pavel Starovoitov on behalf of the Tile Calorimeter System

Kirchhoff Institute for Physics, Im Neuenheimer Feid 227, 69120 Heidelberg, Germany;
pavel.starovoitov@kip.uni-heidelberg.de

Abstract: The Tile Calorimeter (TileCal) is a sampling hadronic calorimeter covering the central region of the ATLAS experiment, with steel as the absorber and plastic scintillators as the active medium. The High-Luminosity phase of the LHC, delivering five times the LHC's nominal instantaneous luminosity, is expected to begin in 2029. TileCal will require new electronics to meet the requirements of a 1 MHz trigger, higher ambient radiation, and to ensure better performance under high pile-up conditions. Both the on- and off-detector TileCal electronics will be replaced during the shut-down of 2026–2028. The photomultiplier tube (PMT) signals from every TileCal cell will be digitized and sent directly to the back-end electronics, where the signals are reconstructed, stored, and sent to the first level of the trigger at a rate of 40 MHz. This will provide better precision in the calorimeter signals used by the trigger system and will allow the development of more complex trigger algorithms. The modular front-end electronics feature radiation-tolerant, commercial, off-the-shelf components and a redundant design to maintain system performance in case of single points of failure. The timing, control, and communication interface with the off-detector electronics is implemented with modern Field-Programmable Gate Arrays (FPGAs) and high-speed fiber optic links running up to 9.6 Gb/s. The TileCal upgrade program has included extensive R&D and test beam studies. A Demonstrator module with reverse compatibility with respect to the existing system was inserted in ATLAS in August 2019 for testing in actual detector conditions. The ongoing developments for on- and off-detector systems, together with expected performance characteristics and results of test-beam campaigns with the electronics prototypes, will be discussed.

Keywords: calorimetry; hadron calorimetry; ATLAS; tile calorimeter; large hadron collider; upgrade

Citation: Starovoitov, P., on behalf of the Tile Calorimeter System. Upgrade of ATLAS Hadronic Tile Calorimeter for the High-Luminosity LHC. *Instruments* **2022**, *6*, 54. https://doi.org/10.3390/instruments6040054

Academic Editors: Fabrizio Salvatore, Alessandro Cerri, Antonella De Santo and Iacopo Vivarelli

Received: 16 August 2022
Accepted: 20 September 2022
Published: 27 September 2022

Publisher's Note: MDPI stays neutral with regard to jurisdictional claims in published maps and institutional affiliations.

1. Introduction

The Tile Calorimeter (TileCal) [1,2] is the central hadronic calorimeter of the ATLAS experiment [3] in the Large Hadron Collider (LHC) [4] at CERN. TileCal contributes to the measurement and reconstruction of hadrons, jets, τ-leptons hadronic decays, and missing transverse momentum. It also assists in muon identification and provides inputs to the Level 1 calorimeter trigger system. TileCal is a sampling calorimeter consisting of staggered steel plates and plastic scintillating tiles that are oriented perpendicular to the beam. TileCal is divided into a long barrel, consisting of two central barrels and two extended barrels (cf Figure 1a). It covers the pseudo-rapidity range $-1.7 < \eta < 1.7$. Each barrel is segmented into 64 wedges (modules) in φ corresponding to 0.1 granularity in $\Delta\varphi$. Each module is further segmented in the radial direction into three layers. The granularity in $\Delta\eta$ in the two innermost layers is 0.1, and it is 0.2 in the outermost layer. The segmentation in the η, ϕ, and radial directions define the cell structure of the TileCal. In total, there are 5182 cells in 256 TileCal modules. A schematic with the TileCal module is depicted in Figure 1b. Charged particles produce light in scintillators, which is collected by wavelength shifting (WLS) fibers from two sides of each plastic tile and that then transport it to the photomultiplier tubes (PMTs). Each TileCal cell is read-out by two PMTs to provide signal redundancy and to improve the energy resolution.

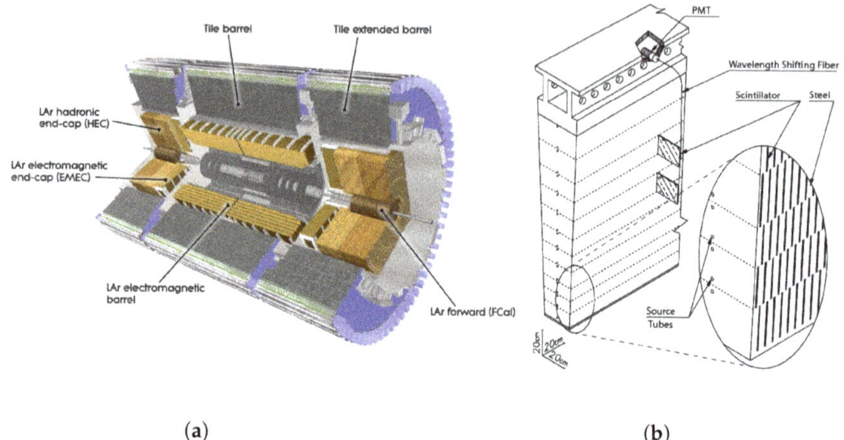

(**a**)   (**b**)

Figure 1. (**a**) Cut-away view of the ATLAS calorimeter system. The Tile Calorimeter, consisting of one barrel and two extended barrels, is shown with gray color in the outermost part of the picture. (**b**) The mechanical structure and optical readout of a single Tile Calorimeter module. The scintillating tiles are oriented normally to the beam-line and read-out by fibers in the radial direction.

In the present system, the PMT signals are digitized at a frequency of 40 MHz. The digital samples are stored in on-detector pipeline memories. At the same time, the trigger analog boards sum the PMT pulses into pseudo-projective 0.1×0.1 in η–φ space trigger towers and send them to the Level-1 calorimeter trigger system. Once the event is accepted by the hardware-based Level-1 trigger system, it is extracted from pipelines and sent for further processing. The Level-1 trigger system processes events at the maximal rate of 100 kHz. The software-based High-Level Trigger system (HLT) refines the events selected at the Level 1 trigger and transfers them to local storage at about a 1 kHz rate.

The datasets collected by the LHC experiments during the Run-1 and Run-2 operations allowed for testing the Standard Model (SM) of particle physics with remarkable precision up to unprecedented energy scales in the sectors of strong and electroweak interactions, heavy flavor physics, and Higgs boson production. The High-Luminosity upgrade of the LHC (HL-LHC) [5] will provide an instantaneous luminosity that is five times larger than the nominal LHC one, with a goal to collect 4000 fb^{-1} integrated luminosity by the end of the HL-LHC data taking. This dataset manifests a new precision frontier in the Higgs sector, opens up unique opportunities to study rare production processes and offers a chance to probe for Beyond the Standard Model physics with unprecedented sensitivity. Among most important are the measurements of the Higgs boson couplings to a percent accuracy and its self-coupling, studies of the di-Higgs production, and decays in Higgs boson couplings to invisible states, as well as the production of three massive vector bosons, searches for lepton flavor violation, and measurements of four top-quarks production mechanisms.

Such an ambitious physics program implies a high event rate and requires excellent read-out capabilities of the sub-detectors, supreme selectivity and functionality of the trigger system, and ultimate performance of the data-acquisition system (DAQ). The high-luminosity conditions impose significant challenges for the detector, trigger, and data-acquisition systems. Approximately 200 simultaneous proton–proton collisions will be produced in every bunch crossing on average, leading to a significant increase in the particle flux in the detector. TileCal on-detector electronics will receive up to about 160 Gy of total ionizing dose (TID) during the full HL-LHC data taking, which is an order of magnitude larger than one in Run-2. The HL-LHC Tile Calorimeter must satisfy the broad physics program goals and withstand extremely challenging conditions during a decade-long data taking.

In the HL-LHC system, the digitized PMT signals are sent to the off-detector electronics, where the energy reconstruction is performed for every TileCal cell. Trigger objects of different granularity, i.e., 0.1×0.1 and 0.2×0.2 in η–φ space trigger towers and individual cells with energy depositions above a certain threshold (e.g., cells compatible with minimum ionizing particle energy losses for muon triggering), are built and sent to the L0/L1 Calo, L0Muon, and Global trigger systems for further processing [6]. Therefore, the full readout system has to be replaced in order to handle the increase in the data rate, to enhance radiation tolerance of the on-detector electronics, and to be compatible with the fully digital ATLAS trigger and data-acquisition system for the HL-LHC.

In addition, 10% of the most exposed PMTs will be replaced by new PMTs, while the remaining optics will be retained. The higher radiation levels also require the redesign of the low-voltage (LV) and high-voltage (HV) power distribution and regulation systems.

The HL-LHC will start the proton–proton collisions in 2029. The ATLAS experiment has started the design and construction of the upgraded detector [7,8] to fully exploit the physics potential of the HL-LHC dataset.

This contribution presents the Tile Calorimeter upgrade program [9] and discusses its current status. It is organized as follows. In Section 2, the new design of the TileCal mechanical structure is described, and the problem of optical components aging is addressed in Section 3. The upgrades of both the on-detector and off-detector electronics are discussed in Section 4. Sections 5 and 6 describe the low-voltage and high-voltage power supply systems, respectively. Finally, the measurements of the hadron response performed in test-beam studies using the upgraded electronics are presented in Section 7. The results are summarized in Section 8.

2. Mechanical Structure of TileCal

TileCal is organized into three cylindrical volumes (barrels), one Long Barrel (LB), consisting of two central barrels, and two Extended Barrels (EB), centered along the beam axis. Each barrel consists of 64 identical modules.

An EB module is read-out by 32 PMTs housed in a super-drawer (SD) together with the associated electronics. Super-drawers are located at the outer radius of the calorimeter. An LB module is read out by 2×45 PMTs that are placed into two super-drawers together with related on-detector electronics. The total number of PMTs is 9852, and almost all calorimeter cells are read-out by two PMTs for redundancy, while only a few cells are read-out using the single-PMT scheme.

The SDs are constructed of mechanically linked Mini-Drawers (MD). They hold the PMTs and on-detector electronics. Each MD can hold up to 12 PMTs and is electrically independent from the adjacent MDs. Four MDs are linked to create one LB super-drawer, while only three MDs are required to build one EB super-drawer. In EB super-drawers, two additional micro-drawers hold PMTs in the right position. The micro-drawers do not hold digital electronics. This modular design was chosen to facilitate handling during the installation and maintenance. The MDs are produced from aluminum and high-density, fire-retardant polyethylene. Cooling of the electronics is ensured via a cooling bridge and a water channel inside the aluminum frame. Following extensive prototyping and tests in beam at CERN, the design has been finalized, and the production will be completed by the end of 2022.

3. Aging of the Optics and Long-Term Robustness Tests of New PMTs

Most of the optics elements in the TileCal cells cannot be replaced without dismounting the entire ATLAS detector and a complete disassembling of the TileCal. Dedicated analysis of the TileCal performance and aging during the Run-1 and Run-2 data-taking periods [9] indicate that the aging of these elements will not significantly impact the TileCal's performance throughout the full HL-LHC data taking. Therefore, neither scintillating tiles nor the WLS fibers will be replaced towards the HL-LHC upgrade.

The PMT response varies over time during data taking because it is affected by variations in the photo-cathode quantum efficiency and dynode multiplication gain due to the integrated anode charge over the running period. To ensure excellent detector performance during the physics collisions, the PMT response is monitored every 2–3 days with the Laser calibration system, sending laser light pulses to all the PMTs. Using the Laser calibration system, the PMT response variation during Run-1 and Run-2 was studied in great detail [10,11]. The PMT response is characterized by down-drifts and up-drifts (response recovery). The down-drifts coincide with the collision periods, while the recoveries occur during the technical stops. The observed down-drifts mostly affect the PMTs that are reading out the TileCal's innermost layer [12]. The estimated maximum integrated anode change at the end of the HL-LHC data taking is around 600 C for most exposed TileCal cells.

Since the degradation in the PMT response depends on the amount of integrated anode charge, the study of the PMT response variation as a function of the integrated anode charge is performed to understand the PMT's performance in the HL-LHC's conditions. In this study, PMTs are excited using the laser light, while a light-emitting diode (LED) is used to generate a charge that is integrated by the PMTs. All PMTs are equipped with high-voltage active dividers to ensure the PMTs' linearity over a wide range of anode currents. The results of the study are shown in Figure 2. Red triangular points represent the average response of PMTs model Hamamatsu R7877 dismounted from TileCal detector in February 2017. After being dismounted, these PMTs already integrated up to 40 C in previous laboratory tests. Black circular points represent the average response of PMTs model Hamamatsu R11187, an evolution of model Hamamatsu R7877. Over time, the number of PMTs in a given data bin can vary, due to the different charge integrated by a given PMT. Missing data points correspond to a period of time in Fall 2021, when many interruptions in data taking had taken place with interrupting the charge integration.

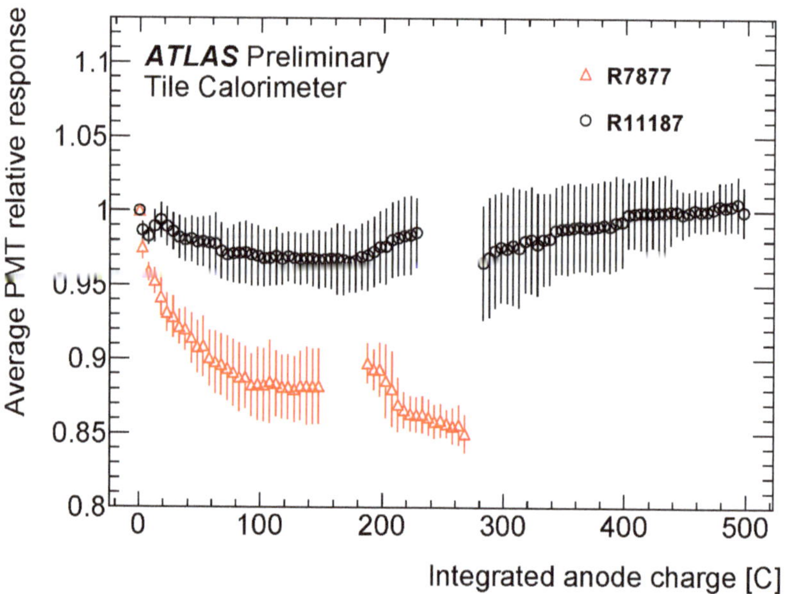

Figure 2. The PMT response as a function of the integrated anode charge. The response is plotted in the units of PMT response measured at zero integrated anode charge. Two different PMT models are shown. Each point is evaluated as the average of the response of the set of PMTs with the same integrated charge. The error bar corresponds to the RMS of the PMT response over the set.

With more than 400 C (several PMTs are already at 500 C) of integrated anode charge, all but a single PMT show excellent performance, retaining full response [12,13]. Both tested PMT models match the Tile Calorimeter performance requirements through the HL-LHC's data taking. Based on the detailed studies, approximately 10% of all PMTs, corresponding to the most exposed cells, will be replaced within the HL-LHC upgrade program.

4. Read-Out Electronics

The TileCal read-out electronics are divided into two main domains: on-detector electronics that are housed inside the SDs and must pass stringent requirements on the radiation hardness and off-detector electronics that are located in underground counting rooms about 100 m away from the ATLAS detector. There are no requirements on the radiation hardness for the off-detector electronics. The TileCal read-out scheme for the HL-LHC upgrade is presented in Figure 3.

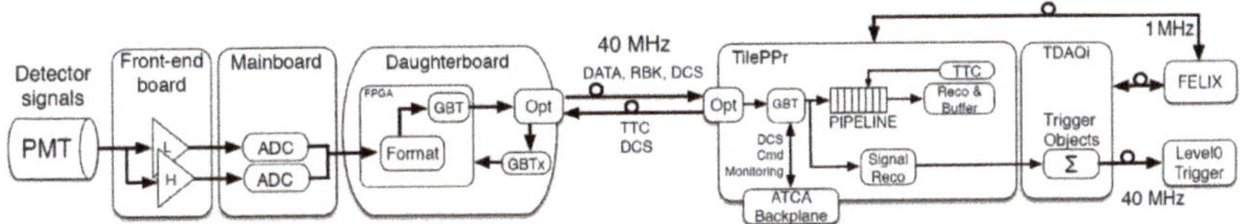

Figure 3. The TileCal Phase-2 upgrade read-out scheme. The on-detector electronics section (left diagram side) with the PMTs, FENICS, Main Board and Daughter Board is connected by the long fibers to the off-detector electronics section (right diagram side) with the Tile PreProcessor. The TDAQi module is optically interfaced to the trigger and DAQ systems.

4.1. On-Detector Electronics

The TileCal HL-LHC on-detector electronics chain is housed inside the mini-drawers. Each mini-drawer contains up to 12 PMT blocks, each with a light mixer, a PMT, its active high-voltage divider base, and an amplifier/shaper card (FENICS). The PMT block is connected to a Main Board that provides a low voltage and controls to FENICS cards, digitizes their signals and routes the data to the link Daughter Board that handles all high-speed communication with the back-end electronics. Four major constituents of the on-detector electronics are discussed in this section.

4.1.1. PMT Blocks

The PMTs are located inside metal cylinders that include active HV-dividers and analog read-out boards, FENICS. The FENICS boards shape and amplify the PMTs pulses and are based on the commercially available off-the-shelf (COTS) components [14]. Two types of signal processing are implemented. The "fast readout" for physics data taking operates in two different gains in order to cover the dynamic range from 200 fC up to 1000 pC to measure energies of a few hundreds MeV in a single particle or a multi-TeV hadronic jet. The "integrator read-out" integrates the PMT current for the calibration of the calorimeter with a ^{137}Cs source. It also provides a relative measurement of the accelerator luminosity at the ATLAS interaction point. The integrator read-out uses six different gains to precisely cover eight orders of magnitude in luminosity measurements, ranging from a dedicated van der Meer scan to physics collisions data. The FENICS board is also able to inject a precise charge and to measure the conversion from pC to ADC counts. The radiation qualification of FENICS, along with validation of the design, has been completed, and pre-production has been accomplished in Summer 2022.

4.1.2. High-Voltage Active Dividers

In order to ensure a 1% precision on the energy scale in TileCal, the high voltage fed to the eight-dynode PMTs must be stable within 0.5 V in the range 600–900 V. During the HL-LHC data taking, the PMT current can reach 40 µA in the most exposed TileCal cells. Therefore, active dividers are necessary to maintain a better than 1% linearity. The linearity of the active–divider–PMT system was successfully tested in a dedicated set up that emulates pulses from physics collisions on top of a continuous soft activity from pile-up. The active dividers were fully validated in radiation tests, and production has started.

4.1.3. Main Board

One Main Board installed in each MD digitizes the data from up to 12 PMTs. It is based on the COTS components. The fast-read out uses 24 12-bit ADCs operated at 40 MHz, while for the integrator readout, it uses 12 16-bit SAR ADCs. The Main Board routes the data to the Daughter Board and provides digital control of the FENICS to configure it for either a calibration run or for the physics data taking. For robustness and redundancy, both the Main Board and the Daughter Board are designed to have two electrically independent sides. As each calorimeter cell is read out by two PMTs, these two PMTs are connected to two different sides of the Main Board [14]. The Main Board was fully qualified for the expected radiation environment in the HL-LHC and has already entered the production stage.

4.1.4. Daughter Board

There is one link Daughter Board (DB) per MD. It is responsible for the high-speed communication (4.8/9.6 Gbps) with the off-detector electronics. The Daughter Board sends precision data as well as the slow control and monitoring data from the on-detector electronics. The Daughter Board receives the LHC clock and distributes it to the on-detector electronics and exchanges configuration and sends control commands to the front-end boards [15]. Each Daughter Board uses 2 Kintex Ultrascale FPGAs. The Daughter Board design has been extensively tested and is validated in several test-beam campaigns in the H8 line in the North Area of CERN. The active components of the DB, such as optical transceivers, FPGA, and isolation amplifiers, are having their radiation hardness evaluated. The design will be finalized in 2023.

4.2. Off-Detector Electronics

During HL-LHC operation, the PMT digital samples for every bunch crossing will be transferred from the Daughter Board to the off-detector electronics, where the data will be reconstructed at 40 MHz frequency.

The reconstructed information includes calibrated energy and time per calorimeter cell or group of cells depending on the trigger system. The energy and time of the PMT pulses are calculated using the digital filters. The trigger primitives are sent to the Level-1 trigger system, while data are stored in pipeline buffers waiting for a trigger decision. The TileCal off-detector electronics are hosted in four ATCA crates, and eight PreProcessor (PPr) blades are installed per crate. In order to have more flexibility and reduce the complexity of the designs, the PPr functionality is split between three different boards. The basic interface with the ATCA backplane and the power distribution is located on the custom ATCA Carrier Base Board (ACBB). The interface with the front-end electronics, pipeline buffering and signal reconstruction are implemented in the Compact Processing Module (CPM). Each ACBB hosts four slots to install CPM boards [16]. The construction of the trigger primitives and the interfaces with the L0/L1 trigger system and Front End LInk eXchange [17] (FELIX) systems are performed in the Trigger and DAQ interface (TDAQi) system [18]. The diagram of a single PPr blade is presented in Figure 4.

The PPr system provides multiple copies of the data for those parts of the detector where the area used in the trigger algorithms overlap. After the trigger decision, the selected data events are transferred to the FELIX system, which is the core element of the central ATLAS DAQ system.

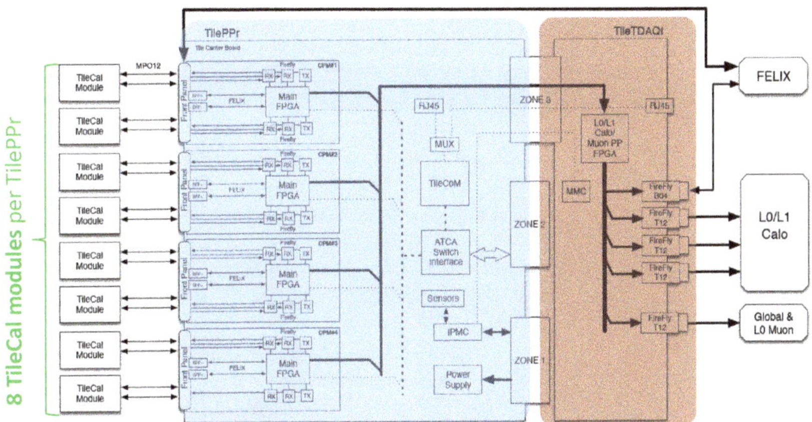

Figure 4. Block diagram of the TileCal Preprocessor. A complete PPr system is able to process the data from eight TileCal super-drawers.

5. Low1

A three-stage, low-voltage system is presented in Figure 5. Low-voltage power supplies (LVPS) are placed on-detector to convert 200 VDC to 10 VDC as required for the on-detector electronics. The LVPS design is based on the COTS components. However, due to their position in a MD, the LVPS are the TileCal components most exposed to radiation. Following a series of extensive irradiation test campaigns, radiation hard components were identified.

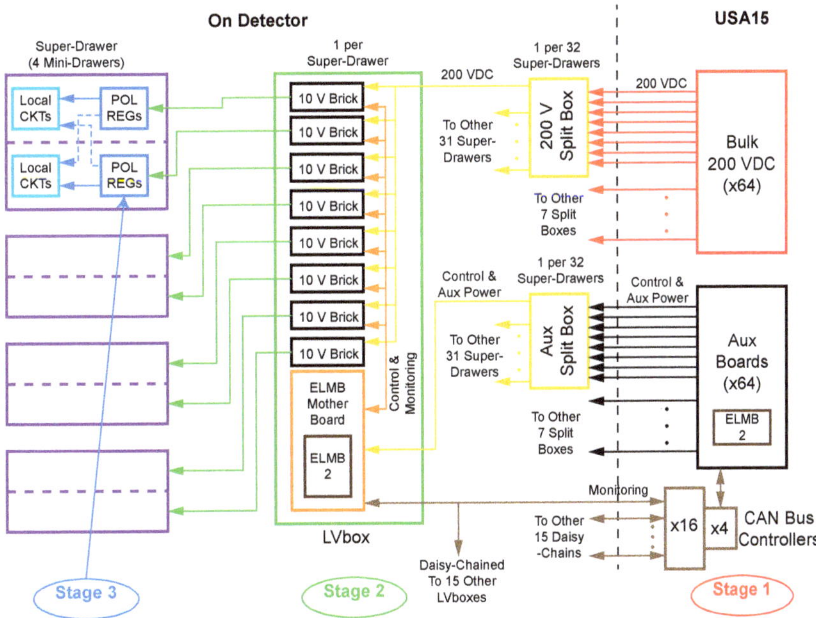

Figure 5. The block diagram of the TileCal low-voltage distribution system for the HL-LHC upgrade. The "10V bricks" stand for 200 VDC to 10 VDC converters. Stage 3 is implemented directly on the front-end electronics boards.

The monitoring and control functions of the LVPS were separated to improve robustness. The monitoring functions are ensured by the radiation hard version of the Embedded Local Monitor Board (ELMB). The new ELMB2 board is functionally equivalent to the legacy ELMB.

The on-detector part of the ON/OFF/ENABLE control of each low-voltage converter is built using a passive system implemented on a so-called ELMB-motherboard. It uses a tri-state voltage levels to minimize the necessary cabling. The control system allows for individual control of a single MD LVPS. In addition, each Main Board and Daughter Board half-board is powered by a separate power converter to minimize the impact of a single component failure on the system's performance. The ELMB-motherboard also routes power to the ELMB2. The production of the ELMB-motherboard will be completed in Summer 2022.

Point-of-load regulators located directly on the Main Board and the Daughter Board make up the third stage of the LV system. A full vertical slice of the LV system was built and tested at CERN in the period from 2019 to 2022. It demonstrated the required functionalities and robustness of all three stages of the LV system [19]. The LVPS boards will go into pre-production in Fall 2022.

6. High-Voltage Distribution and Regulation

The High-Voltage (HV) TileCal HL-LHC upgrade system consists of HV-remote boards and HV-supplies boards, located far from the detector in custom-designed HV-crates, connected to the on-detector components by $\sim$100 m long HV-cables [20]. Inside the detector, the passive HV-bus boards are used to bring the HV individually to each PMT located inside a mini-drawer. The HV-remote boards separately regulate each HV channel using a dedicated regulation loop. This regulation loop is a simplified version of the legacy one [21]. As in the HL-LHC, this board is located off-detector and is not affected by the radiation damage. The regulation scheme controls the high voltage in each channel to within 0.5 V. The total number of HV-remote boards is 256.

The main functional improvement in the HL-LHC HV system compared to the legacy one is an addition of ON/OFF control for each group of four channels complemented by a jumper for each individual channel.

A single input HV, either -830 V or -950 V, is provided for every group of 24 channels, while the voltage of each channel is regulated individually in a range of 360 V. The primary HV is provided by the Hamamatsu C12446-12 modules mounted on the HV-supply boards, and two primary HV inputs are used to provide HV for 48 (32) channels in a Long (Extended) Barrel module.

The HV-bus board is the only component of the HV distribution system that is installed on the detector. These boards are fully passive and have four layers to ensure the protection of the tracks with high-voltage in the inner layers of the board. The TileCal HL-LHC upgrade HV distribution system is shown in Figure 6.

The latest prototypes of the full HV system components were successfully used in the TileCal test-beam campaign of June 2022.

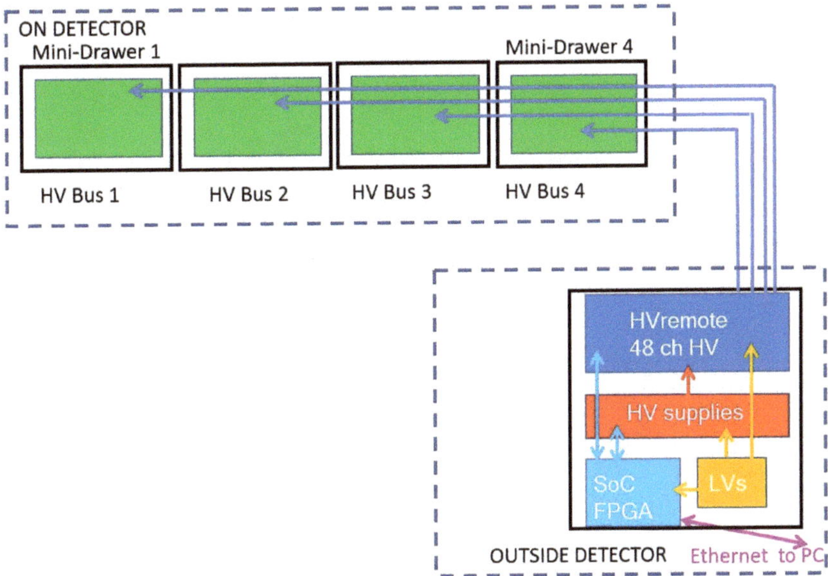

Figure 6. The high-voltage power supply system scheme for the Phase-2 upgrade. The regulation system is remote, located far from the detector, and a large number of 100 m long cables bring the HV to the detector modules. Inside the detector, the HV is distributed to 4 (3) mini-drawers in the Long Barrel (Extended Barrel) modules.

7. Test-Beam

Several test-beam campaigns were carried out at CERN with the HL-LHC TileCal upgrade electronics. Three Tile Calorimeter modules, two LBs and one EB, were exposed to hadron (pion, kaon and proton) beams with energies, E_{beam}, ranging from 16 to 30 GeV [22].

The MDs were equipped with FENICS cards, Main-Boards and Daughter Boards. The on-detector electronics were powered with the pre-production version of the LVPS, and the latest prototypes of the HV system were used to operate the PMTs. The on-detector electronics were configured through the upgrade version of the TileCal PreProcessor, which was also used to take both physics and calibration data. In addition, one module was equipped with a combination of the upgrade and legacy electronics. The aim was to study the performance of the different versions of upgrade electronics and to obtain a direct comparison with the legacy system. The good performance of the new electronics was demonstrated during these test beam campaigns.

In addition to the tests of the latest electronics prototypes, the test-beam campaigns are used to study the TileCal response to hadrons with different energies. The hadron identification is performed using a system of Cerenkov counters [22]. The energy deposited by an incident particle, E_c^{raw}, is reconstructed as a sum of energy depositions in all calorimeter cells with energies twice as large as the noise threshold, σ_{noise}. The noise is determined in events collected between beam bursts using a random trigger, typically $\sigma_{\text{noise}} \sim 30$ MeV. The energy response is defined as a ratio of the mean deposited energy to the beam energy, $\mathcal{R} = \langle E_c^{\text{raw}} \rangle / E_{\text{beam}}$. The measurements of the Tile Calorimeter response to positive pions, kaons and protons as a function of beam energy are presented in Figure 7. The measurements are also compared to the predictions obtained with the Geant4-based [23,24] ATLAS simulation toolkit [25] of the TileCal test-beam experimental setup.

The energy response results for pions, kaons and protons determined in data and in simulated events agree within the uncertainties. On average, the differences between the data and Monte Carlo simulations of the energy response was found to be 1.1% with an average total uncertainty in the energy response determination of 1.4%.

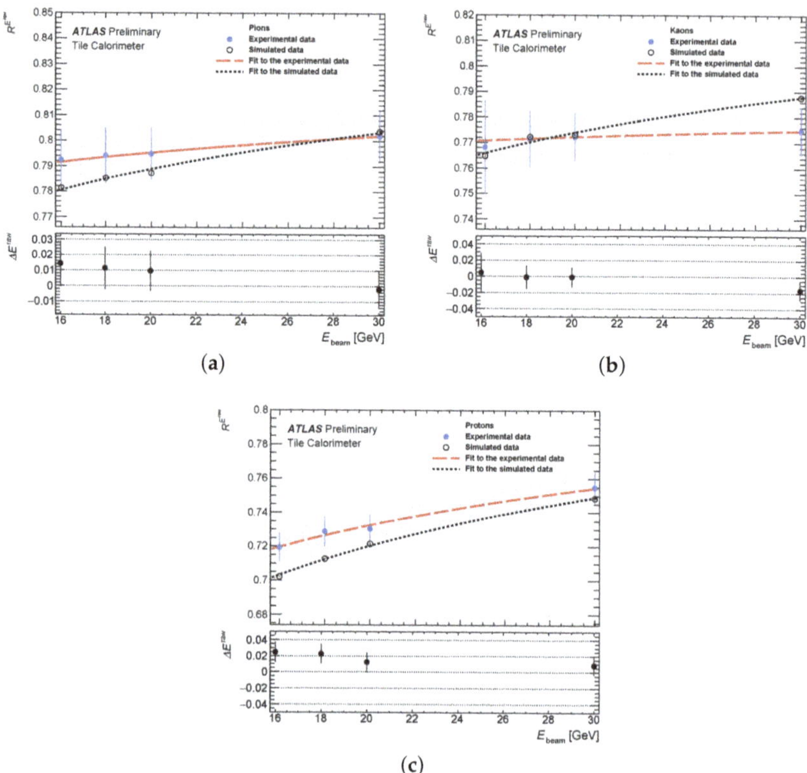

Figure 7. Energy response normalized to incident beam energy measured (blue dots) and predicted by MC simulation (black circles) as a function of beam energy obtained in the case of (**a**) pion, (**b**) kaon and (**c**) proton beams. The details of the fits to data (red line) and MC simulation (black line) are explained in Ref. [22]. The experimental uncertainties include statistical and systematic effects combined in quadrature. Simulated results show only statistical uncertainty.

8. Conclusions

The HL-LHC will open unprecedented possibilities to test the Standard Model of particle physics and take a glimpse on a possible manifestation of new phenomena or production of unknown particles. In order to withstand the tough radiation environment and exigent particle flux conditions and to provide superlative detector performance, the upgrade program was launched by the ATLAS experiment. It requires a full replacement of TileCal on-detector and off-detector electronics and the development of new approaches to power the detector electronics. The design of the ATLAS Tile Calorimeter upgrade for HL-LHC is essentially complete; all parts of the system have been prototyped and validated in standalone test-benches, as well as in integration tests together with other parts of the TileCal upgrade project. The radiation tests of the active components used for the pre-production and production series have been proven to be sufficiently radiation hard to resist estimated particle fluxes and guarantee the design performance. Several parts of the on-detector electronics are already in the final production stage, while the rest are in the pre-production phase. System-level tests for the low-voltage distribution system and the high-voltage power supply system have been successfully completed. Some elements of the LVPS system are already being produced, while the rest of LVPS and the high-voltage power supply system are entering the pre-production stage. Several test-beam campaigns were organized throughout the last decade, where the latest prototypes of various TileCal subsystems were simultaneously validated in the real data-taking environment. The accumulated data are used to constrain the modeling of a hadron interaction in matter in detector simulation frameworks.

The TileCal upgrade project is on schedule for the system on-surface integration in 2024 and installation in ATLAS by the end of 2026.

Funding: This research received no external funding.

Data Availability Statement: Data sharing is not applicable to this article.

Conflicts of Interest: The author declares no conflict of interest.

References

1. ATLAS Collaboration. *ATLAS Tile Calorimeter: Technical Design Report*; CERN-LHCC-96-42; ATLAS Collaboration: Geneva, Switzerland, 1996.
2. ATLAS Collaboration. Readiness of the ATLAS Tile Calorimeter for LHC collisions. *Eur. Phys. J. C* **2010**, *70*, 1193–1236. [CrossRef]
3. ATLAS Collaboration. The ATLAS Experiment at the CERN Large Hadron Collider. *JINST* **2008**, *3*, S08003. [CrossRef]
4. Evans, L.; Bryant, P. LHC Machine. *JINST* **2008**, *3*, S08001. [CrossRef]
5. Apollinari, G.; Brüning, O.; Nakamoto, T.; Rossi, L. High Luminosity Large Hadron Collider HL-LHC. *CERN Yellow Rep.* **2015**, 1–19. [CrossRef]
6. ATLAS Collaboration. *Technical Design Report for the Phase-II Upgrade of the ATLAS TDAQ System*; CERN-LHCC-2017-020; ATLAS Collaboration: Geneva, Switzerland, 2017.
7. ATLAS Collaboration. *Letter of Intent for the Phase-II Upgrade of the ATLAS Experiment*; CERN-LHCC-2012-022, LHCC-I-023; ATLAS Collaboration: Geneva, Switzerland, 2012.
8. ATLAS Collaboration. *ATLAS Phase-II Upgrade Scoping Document*; CERN-LHCC-2015-020, LHCC-G-166; ATLAS Collaboration: Geneva, Switzerland, 2015.
9. ATLAS Collaboration. *Technical Design Report for the Phase-II Upgrade of the ATLAS Tile Calorimeter*; CERN-LHCC-2017-019; ATLAS Collaboration: Geneva, Switzerland, 2017.
10. ATLAS Collaboration. Operation and performance of the ATLAS Tile Calorimeter in Run 1. *Eur. Phys. J. C* **2018**, *78*, 987. [CrossRef] [PubMed]
11. Di Gregorio, G. *Long Term Aging Test of the New PMTs for the HL-LHC ATLAS Hadron Calorimeter Upgrade*; ATL-TILECAL-PROC-2021-018; ATLAS Collaboration: Geneva, Switzerland, 2021.
12. Chiarelli, G.; Di Gregorio, G.; Leone, S.; Scuri, F. Long term aging test of the new PMTs for the HL-LHC ATLAS hadron calorimeter upgrade. In Proceedings of the 15th Pisa Meeting on Advanced Detectors, La Biodola-Isola D'elba, Italy, 22–28 May 2022; ATL-TILECAL-SLIDE-2022-213.
13. Chiarelli, G.; Di Gregorio, G.; Leone, S.; Scuri, F. Long Term Aging Test of the New PMTs for the HL-LHC ATLAS Hadron Calorimeter Upgrade. In Proceedings of the 15th Pisa Meeting on Advanced Detectors, La Biodola-Isola D'elba, Italy, 22–28 May 2022. ATL-TILECAL-PROC-2022-007.
14. Tang, F.; Akerstedt, H.; Anderson, K.; Bohm, C.; Hildebrand, K.; Muschter, S.; Oreglia, M. Upgrade Analog Readout and Digitizing System for ATLAS TileCal Demonstrator. *IEEE Trans. Nucl. Sci.* **2015**, *62*, 1045–1049. [CrossRef]
15. Santurio, E.V.; Silverstein, S.; Bohm, C. Readiness of the ATLAS Tile Calorimeter link daughterboard for the High Luminosity LHC era. *Top. Workshop Electron. Part. Phys. TWEPP* **2020**, *370*, 087. [CrossRef]
16. Duato, A.C.; Biot, A.V.; Argos, F.C.; Pais, J.T.; Medel, J.S.; Olcina, R.G. A new data transfer scheme for the HL-LHC upgrade of the ATLAS Tile Hadronic Calorimeter. In Proceedings of the International Conference on Technology and Instrumentation in Particle Physics, Online, 24–29 May 2021. ATL-TILECAL-PROC-2021-004.
17. Orellana, G.E. Projected ATLAS Electron and Photon Trigger Performance in Run 3. In Proceedings of the 8th Conference of Large Hadron Collider Physics (LHCP), Online, 25–30 May 2020; p. 244. [CrossRef]
18. Yue, X. Tile TDAQ interface module for the Phase-II Upgrade of the ATLAS Tile Calorimeter. In Proceedings of the 2019 IEEE Nuclear Science Symposium and Medical Imaging Conference (NSS/MIC), Manchester, UK, 26 October–2 November 2019. doi:10.1109/NSS/MIC42101.2019.9059861 [CrossRef]
19. Hibbard, M.; Moayedi, S.; Hadavand, H.; Davoudi, A. ATLAS TileCal low voltage power supply upgrade hardware and testing. *Nucl. Instrum. Methods Phys. Res. Sect. A Accel. Spectrom. Detect. Assoc. Equip.* **2019**, *936*, 112–114. [CrossRef]
20. Gomes, A.; Augusto, J.; Cuim, F.; Evans, G.; Fernandez, R.; Gurriana, L.; Marques, R.; Martins, F.; Pereira, C. Upgrade of the ATLAS Tile Calorimeter high voltage system. *JINST* **2022**, *17*, C01061. [CrossRef]
21. Gomes, A.; Soares Augusto, J.; Cuim, F.; Evans, G.; Fernandez, R.; Gurriana, L.; Martins, F. Upgrade of the ATLAS Tile Calorimeter High Voltage System. *PoS* **2020**, *370*, 062. [CrossRef]
22. Abdallah, J.; Angelidakis, S.; Arabidze, G.; Atanov, N.; Bernhard, J.; Bonnefoy, R.; Bossio, J.; Bouabid, R.; Carrio, F.; Davidek, T.; et al. Study of energy response and resolution of the ATLAS Tile Calorimeter to hadrons of energies from 16 to 30 GeV. *Eur. Phys. J. C* **2021**, *81*, 549. [CrossRef]
23. Agostinelli, S.; Allison, J.; Amako, K.A.; Apostolakis, J.; Araujo, H.; Arce, P.; Asai, M.; Axen, D.; Banerjee, S.; Barrand, G.; et al. GEANT4—A simulation toolkit. *Nucl. Instrum. Meth. A* **2003**, *506*, 250–303. [CrossRef]

24. Allison, J.; Amako, K.; Apostolakis, J.; Araujo, H.; Dubois, P.A.; Asai, M.; Barrand, G.; Capra, R.; Chauvie, S.; Chytracek, R.; et al. Geant4 developments and applications. *IEEE Trans. Nucl. Sci.* **2006**, *53*, 270. [CrossRef]
25. ATLAS Collaboration. The ATLAS Simulation Infrastructure. *Eur. Phys. J. C* **2010**, *70*, 823–874. [CrossRef]

 *instruments* 

Article

Noble Liquid Calorimetry for FCC-ee

Nicolas Morange on behalf of the Noble Liquid Calorimeters Study Group

IJCLab, CNRS/IN2P3, Université Paris-Saclay, 91405 Orsay, France; nicolas.morange@cern.ch;
Tel.: +33-1-64-46-83-24

Abstract: Noble liquid calorimeters have been successfully used in particle physics experiments for decades. The project presented in this article is that of a new noble liquid calorimeter concept, where a novel design allows us to fulfil the stringent requirements on calorimetry of the physics programme of the electron-positron Future Circular Collider at CERN. High granularity is achieved through the design of specific readout electrodes and high-density cryostat feedthroughs. Excellent performance can be reached through new very light cryostat design and low electronics noise. Preliminary promising performance is achieved in simulations, and ideas for further R&D opportunities are discussed.

Keywords: electromagnetic calorimeter; FCC; liquified noble gas

Citation: Morange, N., on behalf of the Noble Liquid Calorimeters Study Group. Noble Liquid Calorimetry for FCC-ee. *Instruments* **2022**, *6*, 55. https://doi.org/10.3390/instruments6040055

Academic Editors: Fabrizio Salvatore, Alessandro Cerri, Antonella De Santo and Iacopo Vivarelli

Received: 28 July 2022
Accepted: 19 September 2022
Published: 27 September 2022

Publisher's Note: MDPI stays neutral with regard to jurisdictional claims in published maps and institutional affiliations.

1. Introduction

Noble liquid calorimeters have been successfully used in particle physics experiments at colliders for decades, owing to the very good properties of this technology and reasonable construction prices. The latest example is the liquid argon (LAr) sampling calorimeter of the ATLAS experiment at the LHC [1], which has a central role in the whole ATLAS physics programme. The LAr technology allows us to achieve a very good stochastic term of 10% for the resolution of electromagnetic objects and a linearity at the per-mille level over four orders of magnitude, and it has been shown to have gained stability at the 10^{-4} level over years of operation.

Following the outcome of the latest European Strategy for Particle Physics that the next high-priority collider should be a Higgs boson factory, CERN is bringing forward the electron-positron Future Circular Collider (FCC-ee) project [2] with its 91 km long ring. The very broad physics programme of the FCC [3], encompassing much more than Higgs boson physics, has strong implications on the design of calorimeters [4]. Indeed, the specifications include excellent hadronic energy resolution (around 4% at 50 GeV for Higgs physics) that in turn points towards highly granular calorimeters optimised for particle flow reconstruction algorithms, very good energy resolution on low-energy photons (for b and τ physics), excellent shower shape discrimination for e/γ against hadrons (in particular for τ physics), and systematic uncertainties which should be reduced to very low levels. Noble liquid calorimeters have the potential to fulfil all these stringent requirements. As the FCC-ee will feature between two and four interaction points, different detector technologies can be used for future experiments, and noble liquid calorimetry can therefore be the basis of an FCC-ee detector concept.

However, unlike the mature highly granular (CALICE) and dual-readout (DREAM) concepts which have been heavily developed for a decade, the R&D on noble liquid calorimeters stopped twenty-five years ago, and a vigorous effort is therefore needed to prove the feasibility of such a concept. Initial studies towards a high-granular LAr calorimeter were started in the context of the FCC-hh project [5], where noble liquids were seen as the only viable candidates for the electromagnetic calorimeters because of the high radiation levels [6]. The much simpler environment of the FCC-ee (negligible radiation levels and low data rates) allows us to build upon this design and optimise it towards the ultimate performance required.

2. R&D Towards a Granular Noble Liquid Calorimeter at FCC-ee

Achieving the concept of a viable noble liquid calorimeter for FCC-ee requires several specific detector developments, which are in progress. These go in parallel with simulation studies that guide the design and provide the expected performance of the concept.

2.1. High-Granular Readout Electrodes

A sampling noble liquid calorimeter is made of absorber plates immersed in a liquid bath. A readout electrode also providing high voltage is inserted in the middle of two consecutive absorbers' plates. Ionization electrons created in the liquid of the two gaps by the showering of the particles therefore drift in the electric field towards the high-voltage pads of the readout electrode. The electrical signal created capacitively during the drift on the readout plane for a given calorimeter cell is routed to the back or front of the electrode, to be amplified and shaped by dedicated readout electronics. The ATLAS LAr calorimeter uses a simple copper-kapton electrode, where signals are routed on the same plane as where they are read out capacitively. To achieve a ten-fold increase in the granularity of the calorimeter compared to ATLAS (i.e., reaching a few million cells) while minimising gaps in the angular coverage, the signals should be routed in a separate layer from the capacitive readout.

This calorimeter concept for FCC-ee therefore uses multilayer printed circuit boards (PCB) to achieve this goal. A schematic side view of the readout is given in Figure 1. The signals are routed to the back of the electrode by traces located in the middle of the PCB, which are connected to the readout plane through vias. As the signal from a given cell travels below the readout planes of all cells located behind it, cross-talk should be prevented by the addition of a shielding (ground) layer between the traces and the readout. The outermost layers provide the high voltage and are capacitively coupled to the readout layer.

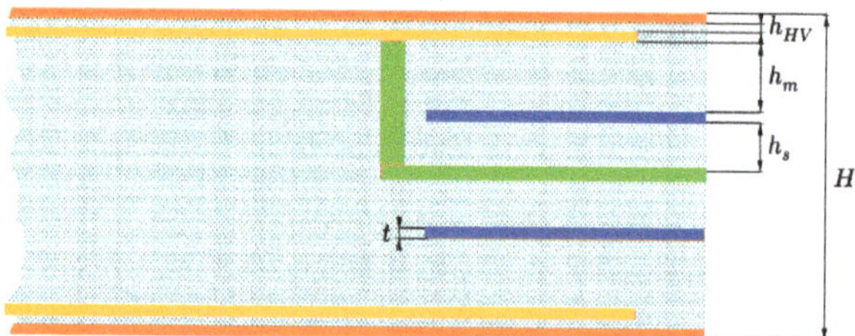

Figure 1. Side view of a readout electrode. The innermost layer contains the readout traces. Above and below it are shields to reduce cross-talk with other cells. The next layer is the readout layer, connected to the trace through vias. The outermost layer provides the high voltage.

The use of PCBs as readout electrodes drives the design of the concept of the electromagnetic calorimeter barrel shown in Figure 2. In order to have a full coverage in ϕ with good uniformity and no cracks, the absorber and readout planes are inclined by an angle of 50 degrees around the barrel. The overall dimensions (inner radius of 210 cm and outer radius of 270 cm including the cryostat) are chosen so that the detector can fit in a modified version of the IDEA detector concept [2]. This baseline design uses 2 mm lead plates as absorber and LAr as active material, with gaps ranging from about 2.5 mm at the inner radius to about 4.5 mm at the outer radius. The use of PCBs gives complete freedom in the drawing of the calorimeter cells and us allows therefore to create projective cells along θ and ϕ and to optimise the granularity for physics performance (similarly to the very fine η "strips" of the first layer of the ATLAS LAr calorimeter). This baseline design features 12 longitudinal layers and cells of about $2 \times 2\,\text{cm}^2$, with a finer segmentation in the second layer. The LAr gap widening effect between the inner and outer radius, which could significantly increase the constant term of the resolution, is compensated by the presence of the 12 longitudinal layers [6]. The preliminary design of the endcap calorimeters use readout and absorber planes perpendicular to the beam axis.

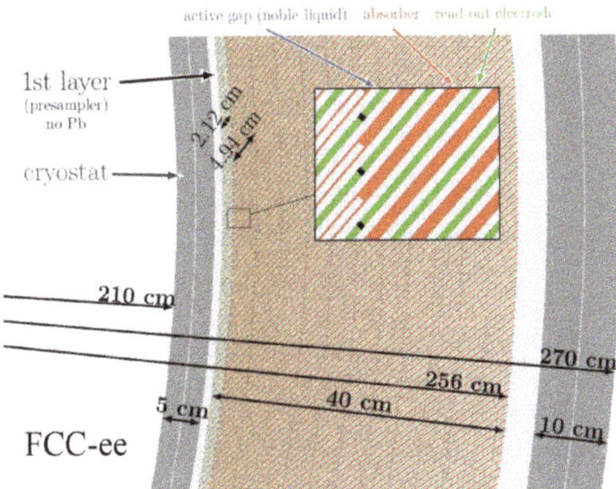

Figure 2. Sketch of the transverse view of the design of the electromagnetic calorimeter barrel. Only a slice of the ϕ coverage is shown.

Studies are ongoing on the design of the readout electrodes, with simulations and first prototypes, to optimise the performance. Indeed, the heights of the insulation layers between the copper layers and the widths of the traces and the shields, have a strong impact on the cell impedance, on the cell capacitance (and therefore the noise), and on the cross-talk. Simulations are performed both with the Sigrity tool of Cadence and with ANSYS HFSS. The first results show promising performance, with cross-talk limited to below the two per-mille level when using the integration time of a few hundred ns. The cell capacitances are dominated by the capacitance between the readout plane and the shields, indicating there is a trade-off between noise and cross-talk. The first validation of the accuracy of the simulations has been performed by measuring simple test structures (single transmission lines in a PCB). Reasonable agreement between the measurements and the simulation is achieved over the frequency range of interest: about 5 to 10% on the impedance between 1 MHz and 500 MHz. A second series of prototypes is being fabricated to study the impact of the size of the shields and of the depths of the layers and to conduct performance measurements of cells in a full-scale design.

2.2. High-Density Feedthroughs

A calorimeter with a few million cells implies that a few million signals have to be extracted from the cryostat. High-density feedthroughs therefore have to be developed, aiming for a density five times greater than that of ATLAS feedthroughs. The new concept being studied avoids the use of connectors. Instead, kapton cables can be slid into 3D-printed epoxy resin structures with slits. They are then glued in place, and the structure is fixed on the stainless steel flange with the use of a bolted compression plate. Leak and pressure tests have been performed at 300 K and 77 K with several designs and glues. Suitable materials showing negligible leaks after thermal cycles have been identified. The design of a complete flange is also making progress. Simulations of stress and deformations at 300 K and 77 K have been performed and allowed us to narrow down adequate solutions.

2.3. Cryostat

The cryostat is a crucial element of a noble liquid calorimeter, as it must sustain mechanically the body of the calorimeter, while its front wall should be as thin as possible to allow the measurement of low-energy particles in the calorimeter. Ongoing directions for the R&D involve new materials, in particular carbon fiber, and sandwiches of materials, benefiting from recent progress in the aerospace industry. These developments are expected to be used in all future detectors, as such cryostats will be built around the future solenoids. In particular, a wall made of a honeycomb aluminium structure sandwiched between two layers of carbon fiber can be robust enough while keeping a material budget of 0.04 radiation lengths (X_0) [7]. Ongoing developments at CERN aim to address issues specific to our field. One is that of sealing methods, as the cryostats have to be closed after inserting large structures such as solenoid magnets or a calorimeter. Leak and pressure tests, as well as microcrack resistance tests, are being performed on test structures using Belleville washers for bolting carbon fiber walls together. A second ongoing development addresses the interface between metal and carbon fiber, which is crucial for mounting feedthroughs on top of a carbon fiber cryostat.

2.4. Readout Electronics

The energy resolution of a calorimeter is the sum of a constant term, a sampling term, and a noise term. At FCC-ee, where there is no pile-up noise as in the LHC, the latter one is dominated by the electronics noise. The goal is therefore to minimise the noise of the readout chain to achieve a good energy resolution even on low-energy photons (around 200 MeV) and to measure the energy of an MIP with a good signal-to-noise ratio. The dominant noise term goes as $C\sqrt{4kT/(g_m \tau_p)}$, where C is the capacitance that depends on the cell capacitance and on the transmission line, g_m is a characteristic of the transistors in a given technology, and τ_p is the peaking time of the signal after shaping. Compared to the ATLAS LAr calorimeter case, this design features smaller cells (and therefore smaller C), and shaping times can be much longer because of the low number of interactions per bunch crossing (typically 200 ns instead of 50 ns).

Analytical simulations show that a readout chain similar to the one used in ATLAS, where all signals are routed outside of the calorimeter before being amplified and processed by warm electronics, could provide adequate performance, with a noise of a few MeV per cell, and a signal-to-noise ratio of about 3 for the energy of a MIP traversing a cell. An attractive alternative is to use cold electronics located in the cryostat at the back of the calorimeter for at least the amplification and shaping of the signals. This would reduce C, as transmission lines are much shorter, T is by definition lower, and the g_m of the transistors is larger. Simulations show that the noise could be reduced by a factor of 5 or more, making it negligible for all measurements and achieving a very high signal-to-noise ratio for MIP deposits. In addition, processing signals in the cold would simplify the requirements on the cryostats, at least on cross-talk, and possibly on signal density if multiplexing can be used. With very low radiation levels expected at FCC-ee, the ageing effects and failures of components in the cold are not expected to be an issue, however, power consumption and heat dissipation should be investigated to make this option possible.

2.5. Towards a First Prototype

As developments are ongoing on many technical aspects simultaneously, a natural goal is to build a small prototype of around 40×40 cm within a few years to prove in testbeam the feasibility of the whole concept. Mechanics studies are expected to start in autumn 2022, with the design of absorbers and spacers. The readout electronics of a first prototype will have to reuse existing components and avoid designing a whole new chain. Fortunately, the requirements of this concept are not far from those of other projects. The readout electronics developed for the DUNE experiment [8] can be used in the cold and could therefore be used for a first prototype, with the only caveat that their dynamic range cannot cover the charge expected from the highest-energy deposits. Conversely, the SKIROC ASIC developed for CALICE Si-W calorimeters [9] matches the physics requirements of this design but was not designed to work in an LAr cryogenic environment. It could, however, be used quite readily for a warm readout electronics option. For a first prototype, it is expected to reuse an existing aluminium cryostat or possibly a first prototype of a carbon fiber cryostat built at CERN, which may be ready on a similar timescale.

3. Expected Performance

In parallel with the R&D studies on the important items for the future realization of this concept, the expected performance of the design was evaluated in simulation. The baseline geometry of the barrel calorimeter was implemented in the FCC software suite using DD4HEP [10] and features 12 layers amounting to about 22 X^0. The layers are composed of 2 mm lead absorber plates, 1.2 mm for the readout PCBs, and a double LAr gap of 2×1.2 mm at the inner radius. The plates were inclined by about 50° around the barrel. The resulting typical calorimeter cell size was about $2 \times 2 \times 3$ cm^3. Simple fixed-size clusters' reconstruction and cluster-level corrections enabled the first performance studies.

The main goal of the performance studies was to optimise the design and guide important design decisions: the choice of the absorber (lead vs. tungsten), choice of active material (LAr vs. liquid krypton), and optimisation of the granularity of the cells. The electromagnetic energy resolution was evaluated on single unconverted photon simulations. The relative resolution of the baseline design was $8\%/\sqrt{E} \oplus 0.7\%$, as shown on Figure 3. Work is ongoing to evaluate the expected resolution of alternative choices for absorbers and active material. The use of PCBs as readout electrodes allows large flexibility in the size of the calorimeter cells, possibly using different sizes per layer. The first studies on π^0 identification efficiency and on classification efficiency of τ decay modes show promising performance and point towards using small cells in the first calorimeter layers and towards the use of liquid krypton to take advantage of its smaller Moliere radius compared to LAr.

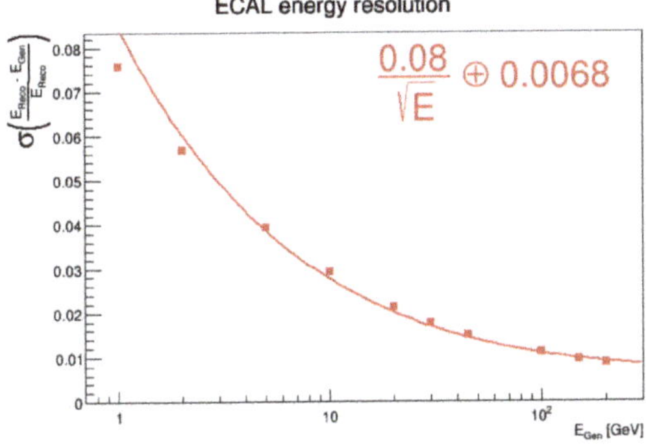

Figure 3. Unconverted photon energy resolution in the baseline design using LAr as active material and lead absorbers. No noise is included in the simulation, but it is expected to be negligible compared to the sampling term even for low-energy photons. The constant term is in particular due to an incomplete containment of the showers in the fixed size clusters, and it is expected to be reduced with future improvements in the reconstruction.

The next main milestone of the simulation studies is the study of jet physics, as a jet energy resolution of about 4% at 50 GeV is one of the main requirements for Higgs physics at FCC-ee. Jet reconstruction at future e^+e^- colliders will, however, rely on particle flow algorithms to fully exploit the performance of all detectors. To that end, the Pandora PFA reconstruction [11] is being integrated in the FCC software chain. Once a hadronic calorimeter is added to the detector concept (possibly using one of the CALICE designs), the jet performance of this electromagnetic calorimeter design can be studied and end-to-end detector optimisation can be performed, and in particular the optimal size of cells can be further studied.

4. Conclusions and Perspectives

Noble liquids have proved to be an excellent technology for electromagnetic calorimeters, featuring very good energy resolution, granularity, linearity, uniformity, and stability over time. They are therefore appealing candidates for detectors at FCC-ee, with several ongoing R&D efforts recently begun to show the feasibility of future detectors. Evidence is being gained that a high-granularity noble liquid calorimeter could be a feasible and versatile solution, fulfilling the stringent FCC-ee requirements. Good progress is being made on the design of high-granularity readout electrodes and on high-density feedthroughs. The concept takes advantage of the R&D on thin cryostats that will benefit all future experiments. The first simulations show that adequate performance on the electromagnetic objects can be achieved.

Time and resources permitting, other fields of R&D could be explored around this concept to possibly improve its performance. The doping of noble liquid, which increases the signal yield by enhancing the drift velocity, could be studied if the noise needs to be further reduced. The cell-level timing capabilities of the design could be also explored. The readout electronics can easily be optimised, but the performance will be limited by the Landau fluctuations in the sampling, and the usefulness of timing in the reconstruction will depend on the overall detector design and the presence (or the lack thereof) of particle identification detectors using dE/dx, timing, or Cerenkov light. Finally, the collection and use of noble liquid scintillation light could be studied. This fast signal is used in dark matter noble liquid detectors, but its use in a calorimeter at a collider would be a novelty and provide a complementary time and energy measurement. There are, however, significant challenges to overcome in order to collect and measure the scintillation light.

Funding: This research was funded through the AIDAInnova programme by the European Union's Horizon 2020 Research and Innovation programme under Grant Agreement No. 101004761.

Data Availability Statement: Not applicable.

Conflicts of Interest: The author declares no conflict of interest.

References

1. ATLAS Collaboration. *ATLAS Liquid Argon Calorimeter: Technical Design Report*; CERN-LHCC-96-41; 1996. Available online: https://inspirehep.net/files/291fdf1fd0dd27a8ee7e64ab8bc9d97f (accessed on 24 July 2022).
2. FCC Collaboration. FCC-ee: The Lepton Collider: Future Circular Collider Conceptual Design Report Volume 2. *Eur. Phys. J. Special Top.* **2019**, *228*, 261–623. [CrossRef]
3. FCC Collaboration. FCC Physics Opportunities: Future Circular Collider Conceptual Design Report Volume 1. *Eur. Phys. J. C* **2019**, *79*, 474. [CrossRef]
4. Aleksa, M.; Bedeschi, F.; Ferrari, R.; Sefkow, F.; Tully, C.G. Calorimetry at FCC-ee. *Eur. Phys. J. Plus* **2021**, *136*, 1066. [CrossRef]
5. FCC Collaboration. FCC-hh: The Hadron Collider: Future Circular Collider Conceptual Design Report Volume 3. *Eur. Phys. J. Special Top.* **2019**, *228*, 755–1107. [CrossRef]
6. Aleksa, M.; Allport, P.; Bosley, R.; Faltova, J.; Gentil, J.; Goncalo, R.; Helsens, C.; Henriques, A.; Karyukhin, A.; Viossler, J., et al. Calorimeters for the FCC-hh, arXiv 2019, arXiv:1912.09962.
7. Barba, M. R&D on Light-Weight Cryostats and on Highdensity Signal Feedthroughs. Presentation at 4th FCC Physics and Experiments Workshop. 2020 Available online: https://indico.cern.ch/event/932973/contributions/4062114/ (accessed on 24 July 2022).
8. Lopriore, E.; Braga, D.; Chen, H.; Christian, D.; Dabrowski, M.; Deptuch, G.; Fried, J.; Furic, I.; Gao, S.; Grace, C.R.; et al. Characterization and QC practice of 16-channel ADC ASIC at cryogenic temperature for Liquid Argon TPC front-end readout electronics system in DUNE experiment. *J. Instrum.* **2021**, *16*, T06005. [CrossRef]
9. Callier, S.; Dulucq, F.; de La Taille, C.; Martin-Chassard, G.; Seguin-Moreau, N. SKIROC2, front end chip designed to readout the Electromagnetic CALorimeter at the ILC. *J. Instrum.* **2011**, *6*, C12040. [CrossRef]
10. Frank, M.; Gaede, F.; Petric, M.; Sailer, A. AIDASoft/DD4HEP. 2018. Available online: https://zenodo.org/record/7002833#.YykiD0xBxPY (accessed on 24 July 2022). [CrossRef]
11. Thomson, M.A. Particle Flow Calorimetry and the PandoraPFA Algorithm. *Nucl. Instrum. Methods A* **2009**, *611*, 25–40. [CrossRef]

Article

Hadron-Induced Radiation Damage in Fast Heavy Inorganic Scintillators

Chen Hu [1], Fan Yang [1,2], Liyuan Zhang [1], Ren-Yuan Zhu [1,*], Jon Kapustinsky [3], Xuan Li [3], Michael Mocko [3], Ron Nelson [3], Steve Wender [3] and Zhehui Wang [3]

1   High Energy Physics, California Institute of Technology, Pasadena, CA 91125, USA
2   School of Physics, Nankai University, Tianjin 300071, China
3   Los Alamos National Laboratory, Los Alamos, NM 87545, USA
*   Correspondence: zhu@caltech.edu

Abstract: Fast and heavy inorganic scintillators with suitable radiation tolerance are required to face the challenges presented at future hadron colliders of high energy and intensity. Up to 5 GGy and 5×10^{18} n_{eq}/cm^2 of one-MeV-equivalent neutron fluence is expected by the forward calorimeter at the Future Hadron Circular Collider. This paper reports the results of an investigation of proton- and neutron-induced radiation damage in various fast and heavy inorganic scintillators, such as LYSO:Ce crystals, LuAG:Ce ceramics, and BaF_2 crystals. The experiments were carried out at the Blue Room with 800 MeV proton fluence up to 3.0×10^{15} p/cm^2 and at the East Port with one MeV equivalent neutron fluence up to 9.2×10^{15} n_{eq}/cm^2, respectively, at the Los Alamos Neutron Science Center. Experiments were also carried out at the CERN PS-IRRAD proton facility with 24 GeV proton fluence up to 8.2×10^{15} p/cm^2. Research and development will continue to develop LuAG:Ce ceramics and BaF_2:Y crystals with improved optical quality, F/T ratio, and radiation hardness.

Keywords: radiation hardness; irradiation; crystals; scintillators; protons; neutrons

Citation: Hu, C.; Yang, F.; Zhang, L.; Zhu, R.-Y.; Kapustinsky, J.; Li, X.; Mocko, M.; Nelson, R.; Wender, S.; Wang, Z. Hadron-Induced Radiation Damage in Fast Heavy Inorganic Scintillators. *Instruments* **2022**, *6*, 57. https://doi.org/10.3390/instruments6040057

Academic Editors: Fabrizio Salvatore, Alessandro Cerri, Antonella De Santo and Iacopo Vivarelli

Received: 10 August 2022
Accepted: 15 September 2022
Published: 5 October 2022

Publisher's Note: MDPI stays neutral with regard to jurisdictional claims in published maps and institutional affiliations.

1. Introduction

Future high-energy physics (HEP) experiments face an unprecedented challenge of a harsh radiation environment against ionization dose charged hadrons as well as neutral hadrons. The radiation environment expected by the forward calorimeter in the proposed Future Hadron Circular Collider (FCC-*hh*) reaches an ionization dose up to 5 GGy and a neutron fluence up to 5×10^{18} one-MeV-equivalent n_{eq}/cm^2 [1]. Bright and fast cerium-doped lutetium yttrium oxyorthosilicate ($Lu_{2(1-x)}Y_{2x}SiO_5$:Ce or LYSO:Ce) features high stopping power and suitable radiation tolerance against both ionization dose and hadrons. Under construction is the barrel timing layer (BTL) [2] using LYSO crystals for the compact muon solenoid (CMS) upgrade at the high-luminosity large hadron collider (HL-LHC). LYSO crystals were also proposed for the Mu2e experiment at Fermilab [3]. In addition, a total absorption LYSO calorimeter for the coherent muon to electron transition (COMET) experiment at the High-Energy Accelerator Research Organization (KEK) [4] and a 3D imaging calorimeter for the high-energy cosmic-radiation detection (HERD) experiment in space [5] are being built. Under investigation are cost-effective cerium-doped lutetium aluminum garnet ($Lu_3Al_5O_{12}$:Ce or LuAG:Ce) ceramics for the precision-timing, ultracompact, radiation-hard electromagnetic calorimetry (RADiCAL) concept for future HEP experiments [6]. BaF_2 crystals also attract attention due to their ultrafast light with 0.5 ns decay time, which is considered for the Mu2e-II upgrade [7]. While radiation damage in fast and heavy inorganic scintillators against ionization dose is well understood [8–14], investigations are still ongoing to understand radiation damage against hadrons, including both charged hadrons [8,10,13–28] and neutrons [29–33].

Starting in 2014, a series of experiments was performed to study radiation damage in fast and heavy inorganic scintillators against hadrons by using 800 MeV protons at the Blue Room and neutrons at the East Port, respectively, at the Los Alamos Neutron Science Center (LANSCE). Inorganic scintillators were irradiated up to 3.0×10^{15} p/cm^2 and 9.2×10^{15} one MeV equivalent n$_{eq}$/cm^2. LYSO:Ce crystals and LuAG:Ce ceramics were also irradiated at the European Organization for Nuclear Research (CERN) PS-IRRAD proton facility by 24 GeV protons up to 8.2×10^{15} p/cm^2. In this paper, we summarize our results obtained for LYSO:Ce crystals, LuAG:Ce ceramics, and BaF$_2$ crystals. The result of this work is of great importance for the CMS phase-II upgrade of the BTL detector [2] at the HL-LHC, as well as an ultra-radiation-hard RADiCAL calorimetry [6] for the proposed FCC-hh.

2. Materials and Methods

The samples were produced at the Beijing Glass Research Institute (BGRI), the Beijing Opto-Electronics Technology Company Ltd. (BOET or OET), the Shanghai Institute of Ceramics (SIC), the Shanghai Institute of Optics and Fine Mechanics (SIOM), the Chongqing Shengpu Electronics Co. LTD (SIPAT), and the Sichuan Tianle Photonics Company (Tianle). Four proton irradiation experiments 6501 [25–27], 6990 [25–27], 7324 [28,33] and 8051 [33] were conducted in 2014, 2015, 2016, and 2018, respectively. Three neutron irradiation experiments 6991 [31,32], 7332 [31,32] and 7638 [31–33] were conducted in 2015, 2016, and 2017, respectively. Some LYSO:Ce and LuAG:Ce samples were also irradiated at the CERN PS-IRRAD facility. The details of these experiments and samples are listed in Table 1.

Table 1. Samples used in various proton and neutron irradiation experiments.

Experiment	Samples	Dimension (mm^3)	Fluence (cm^{-2})
CERN PS-IRRAD	$4 \times$ SIC LYSO	$14 \times 14 \times 1.5$	7.4×10^{13}–6.9×10^{15}
	$10 \times$ BOET LFS	$14 \times 14 \times 1.5$	1.0×10^{14}–8.2×10^{15}
	$2 \times$ SIC LuAG	$\Phi 14.4 \times 1$	7.1×10^{13}–1.2×10^{15}
LANSCE-p-6990	OET LFS	$25 \times 25 \times 180$	1.8×10^{14}–2.9×10^{15}
LANSCE-p-7324	SIC LYSO	$25 \times 25 \times 200$	5.0×10^{13}–3.0×10^{15}
	$9 \times$ SIC LYSO	$10 \times 10 \times 3$	2.7×10^{13}–9.7×10^{14}
	$6 \times$ SIC BaF$_2$	$25 \times 25 \times 5$	2.7×10^{13}–9.7×10^{14}
	$6 \times$ SIC PWO	$25 \times 25 \times 5$	2.7×10^{13}–9.7×10^{14}
LANSCE-p-8051	SIPAT LYSO	$25 \times 25 \times 200$	3.8×10^{13}–1.6×10^{15}
	Tianle LYSO	$25 \times 25 \times 200$	2.2×10^{13}–1.8×10^{15}
LANSCE-n-6991	$18 \times$ OET LFS	$14 \times 14 \times 1.5$	9.4×10^{14}–9.2×10^{15}
LANSCE-n-7332	$12 \times$ SIC LYSO	$10 \times 10 \times 5$	1.7×10^{15}–8.3×10^{15}
	$12 \times$ SIC BaF$_2$	$15 \times 15 \times 5$	1.7×10^{15}–8.3×10^{15}
	$12 \times$ SIC PWO	$15 \times 15 \times 5$	1.7×10^{15}–8.3×10^{15}
LANSCE-n-7638	$6 \times$ SIC LYSO	$10 \times 10 \times 3$	1.7×10^{15}–6.7×10^{15}
	$6 \times$ Tianle LYSO	$10 \times 10 \times 3$	1.7×10^{15}–6.7×10^{15}
	$8 \times$ BGRI BaF$_2$	$10 \times 10 \times 2$	1.7×10^{15}–6.7×10^{15}
	$8 \times$ SIC BaF$_2$	$10 \times 10 \times 2$	1.7×10^{15}–6.7×10^{15}
	$3 \times$ SIC LuAG	$\Phi 14.4 \times 1$	1.7×10^{15}–6.7×10^{15}

The LANSCE 800 MeV and the CERN PS-IRRAD 24-GeV proton beam show a Gaussian shape with a full width at half-maximum (FWHM) of about 25 and 12 mm, respectively. Dosimeters with a cross-section of 10×10 and 20×20 mm^2 were used to measure the proton fluence at CERN PS-IRRAD. The proton fluence at LANSCE was calculated by measuring the proton current. The systematic uncertainties of the proton fluence at CERN and LANSCE are 7% and 10%, respectively. In the neutron irradiation experiments, half of the samples used in experiments 7332 and 7638 were irradiated with a 5 mm lead shielding, the other half without lead shielding. The error for the neutron fluence is about 10%.

To avoid optical bleaching and thermal annealing, all samples were wrapped with aluminum foil during and after irradiation and were kept at room temperature after irradiation. Transmittance was measured by using a Hitachi U3210 spectrophotometer with a precision of 0.2%. The emission-weighted longitudinal transmittance (*EWLT*) was calculated by using the equation

$$EWLT = \frac{\int T(\lambda)E_m(\lambda)d\lambda}{\int E_m(\lambda)d\lambda} \, , \tag{1}$$

where $T(\lambda)$ and $E_m(\lambda)$ are the transmittance and emission spectra, respectively. The *EWLT* is a numerical value of transmittance over the entire emission spectrum. The radiation-induced absorption coefficient (*RIAC*) was obtained from

$$RIAC = \frac{1}{l}\ln\left(\frac{T_0}{T_1}\right) \, , \tag{2}$$

where l is the crystal length and T_0 and T_1 are the transmittances before and after irradiation, respectively.

The errors for the RIAC value depend on the light path length and the initial transmittance of the sample. For samples with suitable initial transparency, the errors are about 1, 3.5, and 5 m^{-1} for the thickness of 5, 1.5, and 1 mm, respectively. The RIAC errors for some LuAG:Ce ceramic samples could be larger than 5 m^{-1} because of poor initial transparency due to rough surfaces or scattering centers in the ceramic bulk [33]. A Hamamatsu R2059 PMT was used to measure the light output (LO) before and after irradiation with a grease coupling for 0.511-MeV γ-rays from a ^{22}Na source with a coincidence trigger. The uncertainty for the light output measurements is about 1%.

3. Results and Discussion

Figure 1 shows transmittance spectra for LYSO (top), BaF$_2$ (middle) and PWO crystals (bottom) measured before (black lines) and after (red lines) (a) proton irradiation with a proton fluence of 9.7×10^{14} p/cm^2 and (b) neutron irradiation [31,32] with a one-MeV-equivalent neutron fluence of 8.3×10^{15} n$_{eq}$/cm^2, respectively. Also shown in the figure are the emission spectra (blue dash lines), the numerical values of the EWLT, and the theoretical limit of transmittance (black dots) calculated according to the refractive index, assuming multiple bounces without internal absorption [34]. The results demonstrate that the radiation hardness of LYSO and BaF$_2$ crystals is much better than PWO crystals against both protons and neutrons.

Figure 2 shows LO as a function of integration time for LYSO (top), BaF$_2$ (middle), and PWO crystals (bottom) measured before (black) and after (red) (a) a proton fluence of up to 9.7×10^{14} p/cm^2 [28] and (b) a one MeV equivalent neutron fluence of up to 8.3×10^{15} n$_{eq}$/cm^2 [31,32], respectively. Also shown in the figure are the numerical values of LO (200), LO (2500), A$_0$, A$_1$, and τ, which represent the light output integrated into the time gate of 200 ns and 2500 ns, the fast component (if applicable), the slow component, and the decay time from the exponential fit. Since the light output of PWO crystals is too low after 9.7×10^{14} p/cm^2 and 8.3×10^{15} n$_{eq}$/cm^2, the data after 1.6×10^{14} p/cm^2 and 3.7×10^{15} n$_{eq}$/cm^2 are shown for PWO samples. We note that more than 91% and 77% LO remain for LYSO and BaF$_2$ crystals after 9.7×10^{14} p/cm^2 and 8.3×10^{15} n$_{eq}$/cm^2, respectively. This indicates that LYSO and BaF$_2$ crystals survive up to these hadron fluences but not PWO.

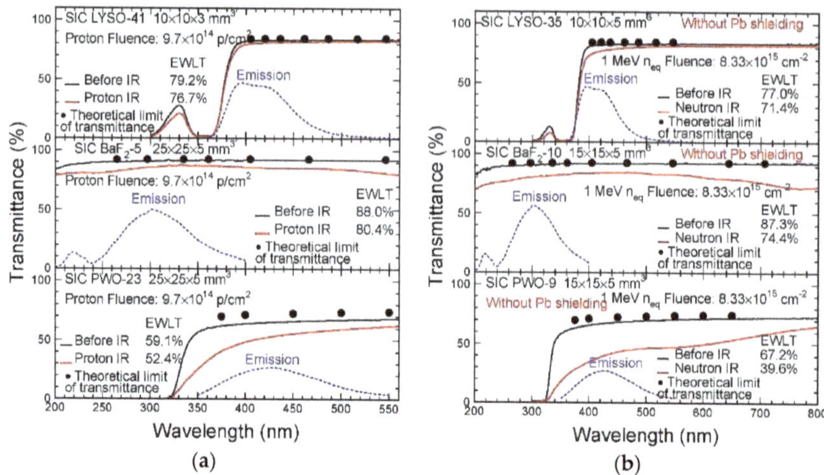

Figure 1. The transmittance spectra measured before (black lines) and after (red lines) irradiation in (**a**) the proton experiment 7324 and (**b**) the neutron experiment 7332 [31,32] are shown for LYSO (**top**), BaF$_2$ (**middle**), and PWO crystals (**bottom**).

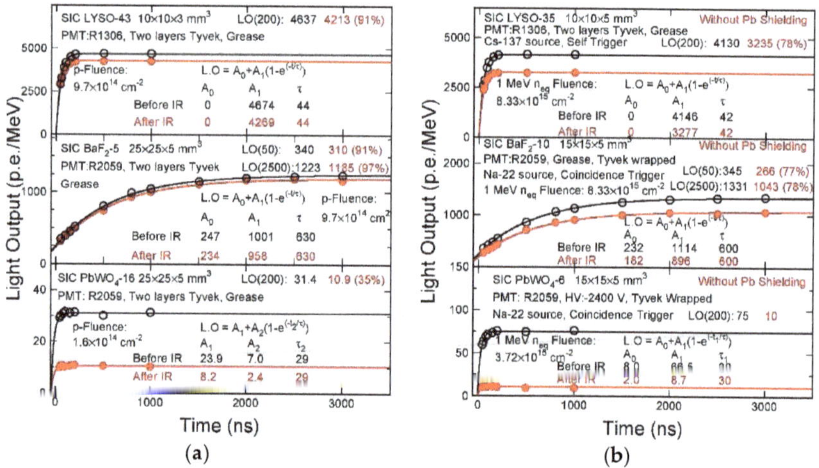

Figure 2. LO measured before (black) and after (red) irradiation in (**a**) the proton experiment 7324 and (**b**) the neutron experiment 7332 [31,32] is shown as a function of the integration time for LYSO (**top**), BaF$_2$ (**middle**), and PWO crystals (**bottom**).

Figure 3 shows the normalized LO integrated in 50 ns gate as a function of the emission-weighted RIAC (EWRIAC) of the 220-nm peak for BaF$_2$ samples used in (a) the proton experiment 7324 and the neutron experiment 7332 and (b) the neutron experiment 7332 and 7638, respectively. Correlation coefficients of 0.91, 0.95, and 0.95 were observed for 18 BaF$_2$ plates of 5 mm in Figure 3a, and 12 BaF$_2$ plates of $15 \times 15 \times 5$ mm^3 and 16 BaF$_2$ plates of $10 \times 10 \times 2$ mm^3 in Figure 3b. It is interesting to note that the relative LO loss can be ascribed to the radiation-induced absorption for both proton and neutron irradiation. It is also interesting to note that the mean light path length, which is L, shown in these figures, depends on the sample thickness.

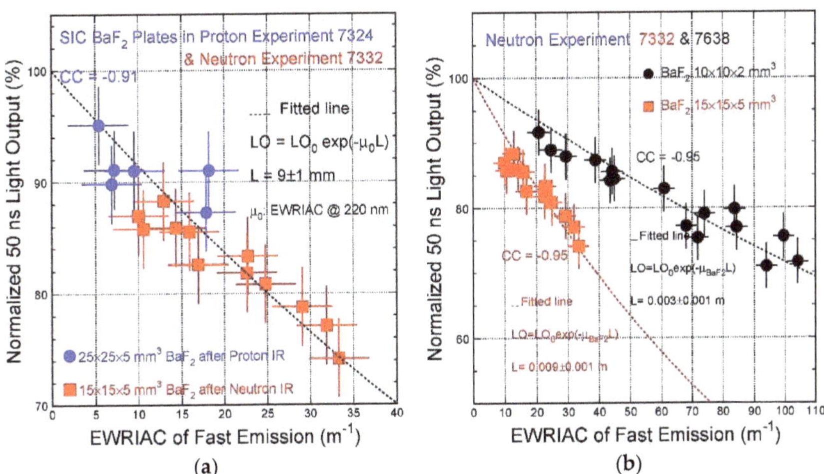

Figure 3. The normalized LO integrated in 50 ns gate is shown as a function of EWRIAC of the 220nm peak for the BaF$_2$ samples used in (**a**) the proton experiment 7324 and the neutron experiment 7332 [28], and (**b**) the neutron experiment 7332 and 7638, respectively.

Figure 4 shows the RIAC values as a function of (a) the proton fluence [25–27,35] and (b) the one MeV equivalent neutron fluence [32] for LYSO/LFS crystals from different vendors irradiated in the proton experiments 6990 and 7324 up to 3.0×10^{15} p/cm^2 and the proton experiment at CERN PS-IRRAD up to 8.2×10^{15} p/cm^2 and the neutron experiments 6991, 7332, and 7638 up to 9.2×10^{15} n$_{eq}$/cm^2.

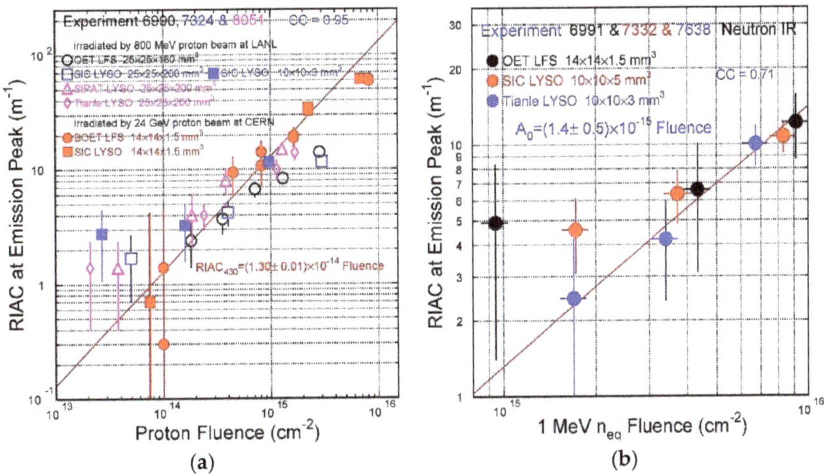

Figure 4. The RIAC values are shown as a function of (**a**) the proton fluence [25–27,35] and (**b**) the one MeV equivalent neutron fluence [32] for LYSO/LFS crystals from different vendors irradiated in the proton experiment 6990, 7324 and at the CERN PS-IRRAD experiment and the neutron experiment 6991, 7332, and 7638, respectively.

Also shown in the figure are the corresponding fits and the uncertainties of the fittings, which depend on the uncertainties of the data points. The result also shows a consistent linear relation between the RIAC values at 430 nm and the proton fluence for LYSO crystals from different vendors with a correlation coefficient (CC) of 0.95. The corresponding value is 0.71 for one MeV equivalent neutron fluence. The radiation hardness specification for the CMS BTL LYSO crystals of $3.12 \times 3.12 \times 57$ mm^3 is: RIAC should be less than 3 m^{-1} after 48 kGy, 2.5×10^{13} p/cm^2 and 3×10^{14} n$_{eq}$/cm^2. It is clear that LYSO crystals from different vendors meet the CMS BTL specification.

We also notice that the numerical RIAC values for LYSO against neutrons are a factor of ten less than that against 800 MeV protons at LANSCE and 24 GeV at CERN. This difference can be ascribed to damage by ionization energy loss from protons as compared to damage by displacement and nuclear breakup only from neutrons.

Figure 5 shows RIAC values as a function of (a) proton fluence [33] and (b) and one MeV equivalent neutron fluence [33] for LuAG:Ce ceramics, LYSO:Ce, and BaF$_2$ crystals irradiated in the proton experiment at CERN PS-IRRAD up to 8.2×10^{15} p/cm^2 and the neutron experiment 7638 up to 6.7×10^{15} n$_{eq}$/cm^2, respectively. Also shown are the corresponding fits. It is interesting to note that the RIAC values for LuAG:Ce ceramics are a factor of two smaller than that of LYSO:Ce crystals. This material thus is promising for future colliders with harsh radiation environments, such as the proposed FCC-hh. We also note that large systematic uncertainties were observed for LuAG:Ce ceramics with poor initial transparency. The surface condition of and the scattering centers inside the ceramic bulk may degrade measured transmission and thus introduce a large systematic uncertainty in the RIAC values. The EWLT values of two ceramic samples shown in Figure 5a, for example, are 67.6% and 32.1% before irradiation and 67.5% and 31.9% after 7.1×10^{13} p/cm^2 and 1.2×10^{15} p/cm^2, respectively, corresponding to RIAC values of 0.5 and 7.1 m^{-1}. Further improvement in optical quality, fast-total ratio (F/T), and radiation hardness of LuAG:Ce ceramics are important for such investigation.

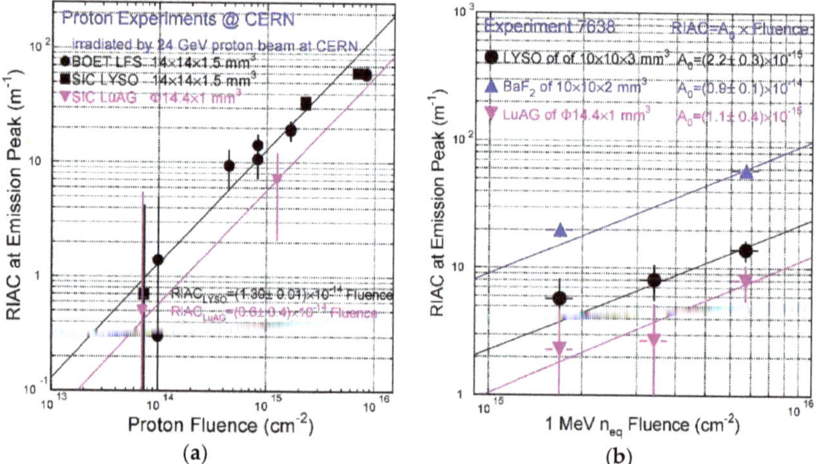

Figure 5. The RIAC values are shown as a function of (**a**) proton fluence [33] and (**b**) one MeV equivalent neutron fluence [33] for LuAG:Ce ceramics, LYSO:Ce, and BaF$_2$ crystals irradiated in the proton experiment at CERN PS-IRRAD and the neutron experiment 7638 at LANSCE, respectively.

Figure 6 shows the LO normalized to before irradiation as a function of (a) proton fluence [28] and (b) one MeV equivalent neutron fluence [32] for LYSO, BaF$_2$, and PWO plates irradiated in the proton experiment 7324 up to 9.7×10^{14} p/cm^2 and the neutron experiment 7332 up to 8.3×10^{15} n$_{eq}$/cm^2, respectively. It is interesting to note that both LYSO:Ce and BaF$_2$ plates maintain more than 85% and 75% of light output after a proton fluence of 9.7×10^{14} p/cm^2 and a one-MeV-equivalent neutron fluence of 8.3×10^{15} n$_{eq}$/cm^2. This result indicates that the radiation hardness of BaF$_2$ is similar to LYSO:Ce under high hadron fluence. BaF$_2$ crystals, however, have an issue of slow component with a decay time of 600 ns. Yttrium doping in BaF$_2$ suppresses the slow component effectively while maintaining the ultrafast light [30,36–40]. An investigation is ongoing to understand hadron-induced radiation damage in BaF$_2$:Y crystals.

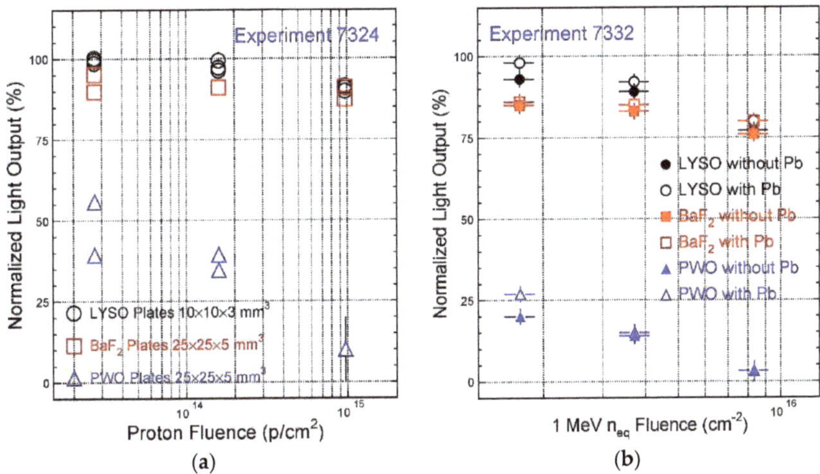

Figure 6. The LO values normalized before irradiation are shown as a function of (**a**) the proton fluence [28] and (**b**) the one MeV equivalent neutron fluence [32] for LYSO, BaF$_2$, and PWO plates irradiated in the proton experiment 7324 and the neutron experiment 7332, respectively.

4. Conclusions

Radiation damage was measured before and after proton and neutron irradiation conducted at LANSCE and CERN for LYSO, BaF$_2$, and PWO crystals and LuAG ceramics. Both LYSO and BaF$_2$ plates maintain more than 85% and 75% of light output after proton and neutron irradiation up to 3.0×10^{14} p/cm^2 and 9.2×10^{15} n$_{eq}$/cm^2, respectively. LYSO:Ce and LFS crystals from different vendors show consistent damage against protons of 800 MeV and 24 GeV. The RIAC values for LuAG:Ce ceramics are a factor of two smaller than that of LYSO:Ce crystals against both neutrons and protons. The radiation hardness of BaF$_2$ is similar to that of LYSO:Ce at high hadron fluence. We plan to improve optical quality, F/T ratio, and radiation hardness for LuAG:Ce and BaF$_2$:Y crystals. We also plan to investigate radiation damage in various fast and heavy inorganic scintillators against ionization dose, protons, and neutrons.

Author Contributions: Conceptualization, R.-Y.Z.; methodology, R.-Y.Z., L.Z. and R.N.; software, C.H., F.Y., L.Z., R.-Y.Z. and M.M.; investigation, C.H., F.Y., L.Z., R.-Y.Z., X.L. and Z.W.; resources, R.-Y.Z., J.K., R.N., M.M., S.W. and Z.W.; writing—original draft preparation, C.H.; writing—review and editing, L.Z. and R.-Y.Z. All authors have read and agreed to the published version of the manuscript.

Funding: This research was funded by the Department of Energy, Office of Science, Office of High Energy Physics, under award number DE-SC0011925 and DE-AC52-06NA25396.

Data Availability Statement: All data are available on the Web. The data presented in this paper are openly available in: http://www.hep.caltech.edu/~zhu/ accessed on 9 August 2022.

Acknowledgments: Samples used in this investigation are provided by SIC, BGRI, BOET, SIPAT, SIOM, and Tianle. The authors would like to thank the CERN PS-IRRAD proton facility for providing the 24 GeV proton beams used in this investigation.

Conflicts of Interest: The authors declare no conflict of interest.

References

1. Aleksa, M.; Allport, P.; Bosley, R.; Faltova, J.; Gentil, J.; Goncalo, R.; Helsens, C.; Henriques, A.; Karyukhin, A.; Kieseler, J.; et al. Calorimeters for the FCC-Hh. *arXiv* **2019**, arXiv:1912.09962.
2. Butler, J.N.; Tabarelli de Fatis, T. *A MIP Timing Detector for the CMS Phase-2 Upgrade*; CMS Collaboration: Geneva, Switzerland, 2019.
3. Pezzullo, G.; Budagov, J.; Carosi, R.; Cervelli, F.; Cheng, C.; Cordelli, M.; Corradi, G.; Davydov, Y.; Echenard, B.; Giovannella, S.; et al. The LYSO Crystal Calorimeter for the Mu2e Experiment. *J. Instrum.* **2014**, *9*, C03018. [CrossRef]
4. Oishi, K. An LYSO Electromagnetic Calorimeter for COMET at J-Park. In *Paper O47-4 Presented in IEEE NSS 2014*; IEEE: Seattle, WA, USA, 2014.
5. Zhang, S.N.; Adriani, O.; Albergo, S.; Ambrosi, G.; An, Q.; Bao, T.W.; Battiston, R.; Bi, X.J.; Cao, Z.; Chai, J.Y.; et al. *The High Energy Cosmic-Radiation Detection (HERD) Facility Onboard China's Space Station*; Takahashi, T., den Herder, J.-W.A., Bautz, M., Eds.; SPIE: Montreal, QC, Canada, 2014; p. 91440X.
6. Anderson, T.; Barbera, T.; Blend, D.; Chigurupati, N.; Cox, B.; Debbins, P.; Dubnowski, M.; Herrmann, M.; Hu, C.; Ford, K.; et al. RADiCAL: Precision-Timing, Ultracompact, Radiation-Hard Electromagnetic Calorimetry. *arXiv* **2022**, arXiv:2203.12806. [CrossRef]
7. Abusalma, F.; Ambrose, D.; Artikov, A.; Bernstein, R.; Blazey, G.C.; Bloise, C.; Boi, S.; Bolton, T.; Bono, J.; Bonventre, R.; et al. Expression of Interest for Evolution of the Mu2e Experiment. *arXiv* **2018**, arXiv:1802.02599.
8. Auffray, E.; Barysevich, A.; Fedorov, A.; Korjik, M.; Koschan, M.; Lucchini, M.; Mechinski, V.; Melcher, C.L.; Voitovich, A. Radiation Damage of LSO Crystals under γ- and 24GeV Protons Irradiation. *Nucl. Instrum. Methods Phys. Res. A* **2013**, *721*, 76–82. [CrossRef]
9. Derdzyan, M.V.; Ovanesyan, K.L.; Petrosyan, A.G.; Belsky, A.; Dujardin, C.; Pedrini, C.; Auffray, E.; Lecoq, P.; Lucchini, M.; Pauwels, K. Radiation Hardness of LuAG:Ce and LuAG:Pr Scintillator Crystals. *J. Cryst. Growth* **2012**, *361*, 212–216. [CrossRef]
10. Petrosyan, A.G.; Ovanesyan, K.L.; Derdzyan, M.V.; Ghambaryan, I.; Patton, G.; Moretti, F.; Auffray, E.; Lecoq, P.; Lucchini, M.; Pauwels, K.; et al. A Study of Radiation Effects on LuAG:Ce(Pr) Co-Activated with Ca. *J. Cryst. Growth* **2015**, *430*, 46–51. [CrossRef]
11. Shen, Y.; Feng, X.; Shi, Y.; Vedda, A.; Moretti, F.; Hu, C.; Liu, S.; Pan, Y.; Kou, H.; Wu, L. The Radiation Hardness of Pr:LuAG Scintillating Ceramics. *Ceram. Int.* **2014**, *40*, 3715–3719. [CrossRef]
12. Dissertori, G.; Luckey, D.; Nessi-Tedaldi, F.; Pauss, F.; Wallny, R. Performance Studies of Scintillating Ceramic Samples Exposed to Ionizing Radiation. In Proceedings of the 2012 IEEE Nuclear Science Symposium and Medical Imaging Conference Record (NSS/MIC), Anaheim, CA, USA, 29 October–3 November 2012; pp. 305–307.
13. Dormenev, V.; Korjik, M.; Kuske, T.; Mechinski, V.; Novotny, R.W. Comparison of Radiation Damage Effects in PWO Crystals Under 150 MeV and 24 GeV High Fluence Proton Irradiation. *IEEE Trans. Nucl. Sci.* **2014**, *61*, 501–506. [CrossRef]
14. The CMS Electromagnetic Calorimeter Group; Adzic, P.; Almeida, N.; Andelin, D.; Anicin, I.; Antunovic, Z.; Arcidiacono, R.; Arenton, M.W.; Auffray, E.; Argiro, S.; et al. Radiation Hardness Qualification of PbWO$_4$ Scintillation Crystals for the CMS Electromagnetic Calorimeter. *J. Instrum.* **2010**, *5*, P03010. [CrossRef]
15. Dissertori, G.; Perez, C.M.; Nessi-Tedaldi, F. A FLUKA Study towards Predicting Hadron-Specific Damage Due to High-Energy Hadrons in Inorganic Crystals for Calorimetry. *J. Instrum.* **2020**, *15*, P06006. [CrossRef]
16. Dissertori, G.; Luckey, D.; Nessi-Tedaldi, F.; Pauss, F.; Quittnat, M.; Wallny, R.; Glaser, M. Results on Damage Induced by High-Energy Protons in LYSO Calorimeter Crystals. *Nucl. Instrum. Methods Phys. Res. A* **2014**, *745*, 1–6. [CrossRef]
17. Dissertori, G.; Lecomte, P.; Luckey, D.; Nessi-Tedaldi, F.; Pauss, F.; Otto, Th.; Roesler, S.; Urscheler, Ch. A Study of High-Energy Proton Induced Damage in Cerium Fluoride in Comparison with Measurements in Lead Tungstate Calorimeter Crystals. *Nucl. Instrum. Methods Phys. Res. A* **2010**, *622*, 41–48. [CrossRef]
18. Lucchini, M.T.; Pauwels, K.; Blazek, K.; Ochesanu, S.; Auffray, E. Radiation Tolerance of LuAG:Ce and YAG:Ce Crystals Under High Levels of Gamma- and Proton-Irradiation. *IEEE Trans. Nucl. Sci.* **2016**, *63*, 586–590. [CrossRef]
19. Nessi-Tedaldi, F. Studies of the Effect of Charged Hadrons on Lead Tungstate Crystals. *J. Phys. Conf. Ser.* **2009**, *160*, 012013. [CrossRef]
20. Dissertori, G.; Luckey, D.; Nessi-Tedaldi, F.; Pauss, F.; Wallny, R.; Spikings, R.; van der Lelij, R.; Arnau Izquierdo, G. A Visualization of the Damage in Lead Tungstate Calorimeter Crystals after Exposure to High-Energy Hadrons. *Nucl. Instrum. Methods Phys. Res. A* **2012**, *684*, 57–62. [CrossRef]
21. Lecomte, P.; Luckey, D.; Nessi-Tedaldi, F.; Pauss, F. High-Energy Proton Induced Damage Study of Scintillation Light Output from Calorimeter Crystals. *Nucl. Instrum. Methods Phys. Res. A* **2006**, *564*, 164–168. [CrossRef]
22. Huhtinen, M.; Lecomte, P.; Luckey, D.; Nessi-Tedaldi, F.; Pauss, F. High-Energy Proton Induced Damage in PbWO$_4$ Calorimeter Crystals. *Nucl. Instrum. Methods Phys. Res. A* **2005**, *545*, 63–87. [CrossRef]
23. Auffray, E.; Korjik, M.; Singovski, A. Experimental Study of Lead Tungstate Scintillator Proton-Induced Damage and Recovery. *IEEE Trans. Nucl. Sci.* **2012**, *59*, 2219–2223. [CrossRef]
24. Batarin, V.A.; Brennan, T.; Butler, J.; Cheung, H.; Datsko, V.S.; Davidenko, A.M.; Derevschikov, A.A.; Dzhelyadin, R.I.; Fomin, Y.V.; Frolov, V.; et al. Study of Radiation Damage in Lead Tungstate Crystals Using Intense High-Energy Beams. *Nucl. Instrum. Methods Phys. Res. A* **2003**, *512*, 488–505. [CrossRef]
25. Yang, F.; Zhang, L.; Zhu, R.-Y.; Kapustinsky, J.; Nelson, R.; Wang, Z. Proton Induced Radiation Damage in Fast Crystal Scintillators. *Nucl. Instrum. Methods Phys. Res. A* **2016**, *824*, 726–728. [CrossRef]

26. Yang, F.; Zhang, L.; Zhu, R.-Y.; Kapustinsky, J.; Nelson, R.; Wang, Z. Proton-Induced Radiation Damage in Fast Crystal Scintillators. *IEEE Trans. Nucl. Sci.* **2017**, *64*, 665–672. [CrossRef]
27. Yang, F.; Zhang, L.; Zhu, R.-Y.; Kapustinsky, J.; Nelson, R.; Wang, Z. Proton-Induced Radiation Damage in BGO, LFS, PWO and a LFS/W/Quartz Capillary Shashlik Cell. In Proceedings of the 2016 IEEE Nuclear Science Symposium, Medical Imaging Conference and Room-Temperature Semiconductor Detector Workshop (NSS/MIC/RTSD), Strasbourg, France, 29 October–6 November 2016; pp. 1–4.
28. Hu, C.; Yang, F.; Zhang, L.; Zhu, R.-Y.; Kapustinsky, J.; Nelson, R.; Wang, Z. Proton-Induced Radiation Damage in BaF_2, LYSO, and PWO Crystal Scintillators. *IEEE Trans. Nucl. Sci.* **2018**, *65*, 1018–1024. [CrossRef]
29. Chipaux, R.; Borizevich, A.; Dujardin, C.; Lecocq, P.; Korzhik, M.V. Behaviour of PWO Scintillators after High Fluence Neutron Irradiation. In Proceedings of the 8th International Conference on Inorganic Scintillators and Their Applications, Alushta, Ukraine, 19–23 September 2005; pp. 369–371.
30. Baranov, V.; Davydov, Y.I.; Vasilyev, I.I. Light Outputs of Yttrium Doped BaF_2 Crystals Irradiated with Neutrons. *J. Instrum.* **2022**, *17*, P01036. [CrossRef]
31. Hu, C.; Yang, F.; Zhang, L.; Zhu, R.-Y.; Kapustinsky, J.; Mocko, M.; Nelson, R.; Wang, Z. Neutron-Induced Radiation Damage in BaF_2, LYSO/LFS and PWO Crystals. *J. Phys. Conf. Ser.* **2019**, *1162*, 012020. [CrossRef]
32. Hu, C.; Yang, F.; Zhang, L.; Zhu, R.-Y.; Kapustinsky, J.; Mocko, M.; Nelson, R.; Wang, Z. Neutron-Induced Radiation Damage in LYSO, BaF_2, and PWO Crystals. *IEEE Trans. Nucl. Sci.* **2020**, *67*, 1086–1092. [CrossRef]
33. Hu, C.; Zhang, L.; Zhu, R.-Y.; Li, J.; Jiang, B.; Kapustinsky, J.; Mocko, M.; Nelson, R.; Li, X.; Wang, Z. Hadron-Induced Radiation Damage in LuAG:Ce Scintillating Ceramics. *IEEE Trans. Nucl. Sci.* **2022**, *69*, 181–186. [CrossRef]
34. Ma, D.; Zhu, R. Light Attenuation Length of Barium Fluoride Crystals. *Nucl. Instrum. Methods Phys. Res. A* **1993**, *333*, 422–424. [CrossRef]
35. Yang, F.; Zhang, L.; Zhu, R.-Y. Gamma-Ray Induced Radiation Damage Up to 340 Mrad in Various Scintillation Crystals. *IEEE Trans. Nucl. Sci.* **2016**, *63*, 612–619. [CrossRef]
36. Sobolev, B.P.; Krivandina, E.A.; Derenzo, S.E.; Moses, W.W.; West, A.C. Suppression of BaF_2 Slow Component of X-RAY Luminescence in Non-Stoichiometric $Ba_{0.9}R_{0.1}F_{2.1}$ Crystals (*R* = Rare Earth Element). *MRS Proc.* **1994**, *348*, 277. [CrossRef]
37. Radzhabov, E.; Istomin, A.; Nepomnyashikh, A.; Egranov, A.; Ivashechkin, V. Exciton Interaction with Impurity in Barium Fluoride Crystals. *Nucl. Instrum. Methods Phys. Res. A* **2005**, *537*, 71–75. [CrossRef]
38. Myasnikova, A.S.; Radzhabov, E.A.; Egranov, A.v. Extrinsic Luminescence of BaF_2:R^{3+} Crystals (R^{3+} = La^{3+}, Y^{3+}, Yb^{3+}). *Phys. Solid State* **2008**, *50*, 1644–1647. [CrossRef]
39. Chen, J.; Yang, F.; Zhang, L.; Zhu, R.-Y.; Du, Y.; Wang, S.; Sun, S.; Li, X. Slow Scintillation Suppression in Yttrium Doped BaF_2 Crystals. *IEEE Trans. Nucl. Sci.* **2018**, *65*, 2147–2151. [CrossRef]
40. Gundacker, S.; Pots, R.H.; Nepomnyashchikh, A.; Radzhabov, E.; Shendrik, R.; Omelkov, S.; Kirm, M.; Acerbi, F.; Capasso, M.; Paternoster, G.; et al. Vacuum Ultraviolet Silicon Photomultipliers Applied to BaF_2 Cross-Luminescence Detection for High-Rate Ultrafast Timing Applications. *Phys. Med. Biol.* **2021**, *66*, 114002. [CrossRef] [PubMed]

 *instruments* 

Communication

Enhanced Proton Tracking with ASTRA Using Calorimetry and Deep Learning

César Jesús-Valls [1,*,†], Marc Granado-González [2,†], Thorsten Lux [1], Tony Price [2] and Federico Sánchez [3]

1   Institut de Física d'Altes Energies (IFAE)—The Barcelona Institute of Science and Technology (BIST), Campus UAB, 08193 Bellaterra, Spain
2   School of Physics and Astronomy, University of Birmingham, Edgbaston, Birmingham B15 2TT, UK
3   The Département de Physique Nucléaire et Corpusculaire (DPNC), University of Geneva, 1205 Genève, Switzerland
*   Correspondence: cesar.jesus@cern.ch
†   These authors contributed equally to this work.

Abstract: Recently, we proposed a novel range detector concept named ASTRA. ASTRA is optimized to accurately measure (better than 1%) the residual energy of protons with kinetic energies in the range from tens to a few hundred MeVs at a very high rate of $O(100 \text{ MHz})$. These combined performances are aimed at achieving fast and high-quality proton Computerized Tomography (pCT), which is crucial to correctly assessing treatment planning in proton beam therapy. Despite being a range telescope, ASTRA is also a calorimeter, opening the door to enhanced tracking possibilities based on deep learning. Here, we review the ASTRA concept, and we study an alternative tracking method that exploits calorimetry. In particular, we study the potential of ASTRA to deal with pile-up protons by means of a novel tracking method based on semantic segmentation, a deep learning network architecture that performs classification at the pixel level.

Keywords: proton CT; image reconstruction; proton tracking; deep learning

Citation: Jesús-Valls, C.; Granado-González, M.; Lux, T.; Price, T.; Sánchez, F. Enhanced Proton Tracking with ASTRA Using Calorimetry and Deep Learning. *Instruments* **2022**, 6, 58. https://doi.org/10.3390/instruments6040058

Academic Editors: Fabrizio Salvatore, Alessandro Cerri, Antonella De Santo and Iacopo Vivarelli

Received: 31 July 2022
Accepted: 29 September 2022
Published: 8 October 2022

Publisher's Note: MDPI stays neutral with regard to jurisdictional claims in published maps and institutional affiliations.

1. Introduction

Radiation therapy consists of the targeted destruction of malignant tissue by means of controlled beams of particles or photons, the latter being the most widespread solution [1]. However, photon energy deposition decays exponentially with distance, so to treat a patient with photons, a non-negligible dose of radiation is delivered to healthy tissue. The stopping power of a proton, on the other hand, increases with distance and is maximized at the stopping point, known as the Bragg peak [2]. Consequently, proton beam therapy (PBT) is an attractive treatment alternative; see Ref. [3] for a review.

To reliably plan PBT treatment, it is important to create a tomographic image of the body in terms of its relative stopping power (RSP), which is indicative of how much the protons will slow down as they travel through the patient. The most widespread solution is to generate these images using X-rays (X-ray CT). However, photon imaging to address proton treatment introduces uncertainties that limit the potential of PBT [4]. To overcome this barrier, research has been conducted for decades with the goal of achieving high-quality proton computed tomography (pCT). Various designs have been proposed over the years, see Refs. [5–10], to pave the way forward. Recently, A Super Thin RAnge (ASTRA) telescope has been proposed as a next-generation detector for pCT, its main advantages being its speed (aims at 100 MHz) and its fine segmentation ($3 \times 3 \text{ mm}^2$ bars) meant to accurately reconstruct the proton energies by range and to efficiently deal with pile-up.

Here, we review the most prominent features of ASTRA as presented in Ref. [11] and extend the capabilities presented there by proposing a new tracking method based on semantic segmentation.

1.1. Detector Concept

The concept of the detector is illustrated in Figure 1. It consists of an upstream tracker made up of four pixel sensors, two before and two after the phantom to be imaged, and ASTRA located downstream.

The main role of the front tracker is to very precisely identify the path of the protons within the phantom being imaged. A possible solution could be to use large area depleted monolithic active pixel sensors (DMAPS) [12] covering a surface of 10×10 cm^2, similar to those in Ref. [13], with 2500×2500 silicon pixels of 40×40 µm^2.

ASTRA is made up of layers of plastic scintillators positioned perpendicular to the proton beam. Each layer consists of bars of $3 \times 3 \times 96$ mm^3, and bars in consecutive layers are rotated by $90°$. To achieve a very fast response, ASTRA bars could be made up of EJ-200 plastic with 0.9 ns scintillation rise time and 2.1 ns decay time (for 1 MeV electrons [14]) and an attenuation length of 380 cm [14]. To match ASTRAs fast plastic scintillation, fast photosensors capable of providing a full waveform in a few nanoseconds would be used, e.g., MicroFJ SiPM [15]. Finally, custom electronics would be implemented, taking as a reference the performance of the CITIROC ASIC that provides a dead-time free readout at a sampling frequency of 0.4 GHz [16].

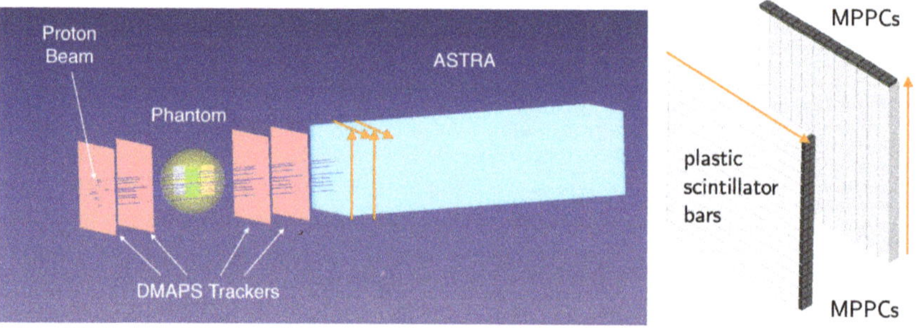

Figure 1. (**Left**): Sketch of the simulated pCT system, including a front tracker made up of four DMAPS and a proton energy tagger named ASTRA. (**Right**): Detailed view of two exploded layers of ASTRA showing the relative orientation of bars in consecutive layers and the placement of multi photon pixel counters (MPPCs).

1.2. Tracking and Energy Reconstruction

To assess the potential performances of ASTRA, we designed and tested custom reconstruction algorithms in Ref. [11]. The most relevant conclusions and characteristics of these studies are summarized below. ASTRA's fine segmentation allows multiple protons to be identified when they cannot be separated by time alone. This is crucial to reduce the inefficiencies caused by pile-up, which, for a beam tuned to provide a single proton per time frame, are approximately (assuming Poisson statistics in the distribution of protons per time frame: $(1 - P\{\mu = 1, x = 0\} - P\{\mu = 1, x = 1\})/(1 - P\{\mu = 1, x = 0\}) \approx 0.4$) 40% of all events. When working with data, the number of proton trajectories in a single time frame is expected to be known reliably from the number of isolated clusters recorded in the first front-tracker plane.

Regarding the proton energy reconstruction, in Ref. [11], a range-based method was considered, which mapped the reconstructed range for the tracked protons measured in ASTRA to a reconstructed value for the kinetic energy. This method proved to be successful as it resulted in energy resolutions of up to 0.7% for the energies of interest. However, in Ref. [11], it was discussed that such a method worked only for protons without inelastic interactions ($\sim$70% at $E \approx 180$ MeV), which in most cases significantly shortened the range with respect to the expectation for a given initial kinetic energy, forcing to consider alternatives, including calorimetric information for a better result.

As we anticipated in Ref. [11], the proposed tracking and energy reconstruction methods were primarily aimed at demonstrating the potential of ASTRA by showing a lower bound of the detector's capabilities; however, we planned from the beginning to test alternative solutions, which are now under development.

2. Towards an Enhanced Proton Tracking

Improving the performances presented in Ref. [11] requires exploiting all the information provided by ASTRA. In particular, the addition of high-quality calorimetric information is expected to improve the tracking capabilities of the detector.

For a set of proton trajectories recorded in ASTRA over the same time frame, a major problem is identifying which bar hits are associated with each trajectory. This step is crucial: wrong tracking outputs directly translate into energy smearing, both degrading the image quality and increasing inefficiencies. However, correctly labeling the hits for pile-up events is a big challenge as the narrow beam width of $\sigma = 1$ cm [17] makes overlaps at the hit level very common.

To overcome this issue, an algorithm that exploits the proton ionization continuity over consecutive hits can be used in order to classify individual hits and break tracking ambiguities. This has the additional benefit of allowing to perform stand-alone reconstruction for ASTRA, which otherwise needs additional inputs from the DMAPS tracker.

The rise of deep learning in recent years opens up a whole set of new possibilities for designing novel reconstruction methods. Semantic segmentation [18], which emerged in the field of computer vision, is a branch of deep learning that enables image classification at the pixel level. Therefore, a tracking solution could be to build images with event displays from ASTRA using one pixel per ASTRA bar and classifying the pixels into different categories, such as `track-1`, `track-2` and `overlap`, for events with two proton tracks. This has two obvious advantages. First, semantic segmentation algorithms are capable of learning non-trivial transformations and combining local and long-distance information to classify recurring image patterns with very high performance [19]. Second, deep-learning-based tracking algorithms do not require defining custom decision rules or manually modifying parameters, as the algorithm optimization is handled directly by training on labeled examples that can be obtained straightforwardly once a simulation framework is available.

To enhance the proton tracking in ASTRA, a U-shaped convolutional neural network, so-called UNet [20], is being considered. UNets are a well-spread, robust, high-performance deep-learning architecture used to realize semantic segmentation. The algorithm takes images generated with the GEANT4-based Monte Carlo (MC) simulation described in Ref. [11] as the input. The simulation uses uniformly distributed protons in the energy range of 80 to 180 MeV, secondary particles are included, and all events are considered without rejecting any event based on true MC information. Each image consists of 64×60 pixels that correspond to the positions of the bars and signals measured in each of the 60 ASTRA layers, with 32 bars per layer. To combine both planes, the images are merged in a vertical stack of 32×2 bars. The algorithm is trained using labels (`track-1`, `track-2` and `overlap`) obtained from the true information of the simulation and, so far, we worked exclusively with events with two simultaneous protons. The predicted labels are used to identify the tracks, which are split into `track-1` and `track-2` images, including all hits classified as `overlap` on both. Illustrative examples are presented in Figure 2.

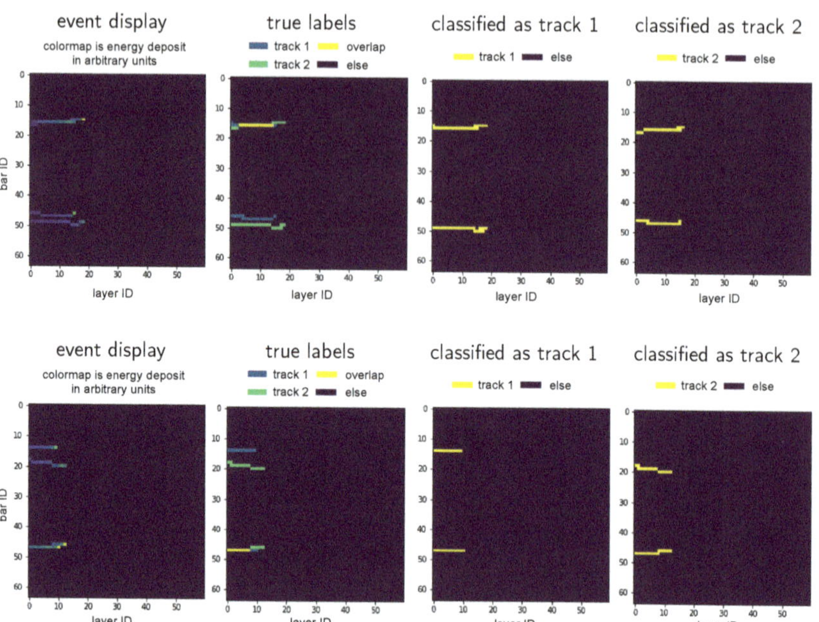

Figure 2. Examples of two events, one per row, including the input event display, the true labels and the reconstructed tracks based on the UNet output. Bar IDs 0–31 (32–63) correspond to the top (side) view of the ASTRA detector.

To evaluate the performance of the algorithm, the true and reconstructed Euclidean range from the first to the last track hit was computed from the predicted pixel labels and compared to the range calculated with perfect pixel classification. This intermediate step allows us to directly assess the potential of the algorithm not at the pixel but at the track level, which is the most relevant for our purposes. The preliminary results obtained from individually analyzing all reconstructed tracks are presented in Figure 3. As can be seen, near-perfect regression performance is achieved in the range, with about 98% of the events with an error equal to or better than 3 mm, the width of one ASTRA bar. For 75% of tracks, the range is perfectly reconstructed.

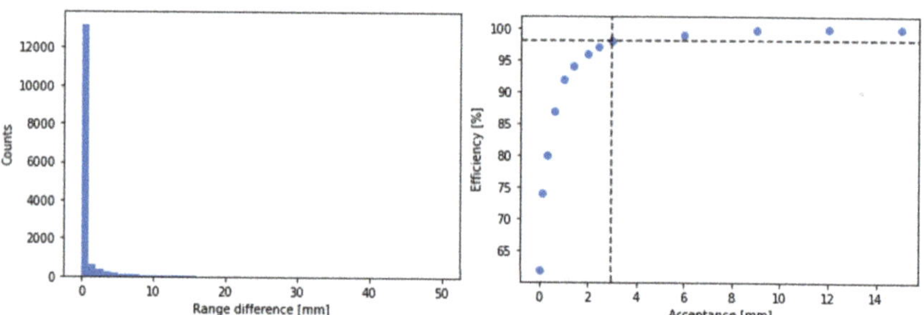

Figure 3. (**Left**): Distribution of the difference between the reconstructed range using the true pixel information compared to that calculated using predictions from the UNet-based tracking algorithm. (**Right**): Fraction of tracks with an error smaller than an acceptance cut for the distribution on the left.

To translate the range into a reconstructed energy, we follow the method we previously presented in Ref. [11], i.e., using Monte Carlo true information, we parameterize what typical energy is associated with each true range and use it to map a reconstructed range into a reconstructed energy. For a collection of protons with the same initial true energy, we study the associated reconstructed energy, and we fit the central peak with a Gaussian distribution. All proton trajectories within 2σ are selected as good for imaging. Under this criterion, the new algorithm significantly outperforms the metrics reported in Ref. [11]. In particular, it significantly increases the fraction of protons good for imaging in events with two piled-up protons. For instance, for protons with an energy similar to 150 MeV, it increases this fraction from about 55% (reported in Ref. [11]) to 68%, much closer to the 80% of protons good for imaging in events without pile-up (reported in Ref. [11]). The remaining 20% is known to be poorly reconstructed due to inelastic interactions. To overcome this limitation, tests are underway to reconstruct the energy not by range but directly from the input event display images using convolutional neural networks (CNN). By exploiting the correlated information of the proton energy deposits and their trajectories, a significant increase in performance is expected. Going in this direction, we already presented the first tests using a Boost Decision Tree (BDT) that combined range and calorimetry in Ref. [11] and achieved an important enhancement in the energy resolution from 0.7% to 0.5% for events without inelastic interactions.

3. Conclusions

The design of the ASTRA range telescope has been reviewed, and alternatives to its mainstream reconstruction strategy have been presented. A UNet-based tracking algorithm has been tested as an alternative to enhance the reconstruction of events with piled-up protons. The preliminary results are very promising, significantly outperforming those in Ref. [11]. Additional deep learning methods to reconstruct protons energy are being evaluated with the primary goal of improving reconstruction metrics associated with protons undergoing inelastic interactions.

Author Contributions: Conceptualization, C.J.-V., M.G.-G. and F.S.; methodology, C.J.-V.; Software, C.J.-V. and M.G.-G.; formal analysis, M.G.-G.; writing, C.J.-V.; Supervision, F.S., T.L. and T.P. All authors have read and agreed to the published version of the manuscript.

Funding: This research was funded by SEIDI-MINECO grants number PID2019-107564GB-I00 and SEV-2016-0588, the SNF grant number 200021_85012 and the EPSRC grant number EP/R023220/1.

Data Availability Statement: The data presented in this study are available on request from the corresponding author.

Conflicts of Interest: The authors declare no conflict of interest.

References

1. Bryant, A.K.; Banegas, M.P.; Martinez, M.E.; Mell, L.K.; Murphy, J.D. Trends in Radiation Therapy among Cancer Survivors in the United States, 2000–2030. *Cancer Epidemiol. Biomarkers Prev.* **2017**, *26*, 963–970. [CrossRef] [PubMed]
2. Bragg, W.H.; Kleeman, R. LXXIV. On the ionization curves of radium. *Lond. Edinb. Dublin Philos. Mag. J. Sci.* **1904**, *8*, 726–738. [CrossRef]
3. Johnson, R.P. Review of medical radiography and tomography with proton beams. *Rep. Prog. Phys.* **2017**, *81*, 016701. [CrossRef]
4. Bär, E.; Lalonde, A.; Zhang, R.; Jee, K.W.; Yang, K.; Sharp, G.; Liu, B.; Royle, G.; Bouchard, H.; Lu, H.M. Experimental validation of two dual-energy CT methods for proton therapy using heterogeneous tissue samples. *Med. Phys.* **2018**, *45*, 48–59. [CrossRef] [PubMed]
5. Plautz, T.; Bashkirov, V.; Feng, V.; Hurley, F.; Johnson, R.P.; Leary, C.; Macafee, S.; Plumb, A.; Rykalin, V.; Sadrozinski, H.W.; et al. 200 MeV Proton Radiography Studies with a Hand Phantom Using a Prototype Proton CT Scanner. *IEEE Trans. Med. Imaging* **2014**, *33*, 875–881. [CrossRef] [PubMed]
6. Johnson, R.P.; Bashkirov, V.; Giacometti, V.; Hurley, R.F.; Piersimoni, P.; Plautz, T.E.; Sadrozinski, H.F.; Schubert, K.; Schulte, R.; Schultze, B.; et al. A Fast Experimental Scanner for Proton CT: Technical Performance and First Experience with Phantom Scans. *IEEE Trans. Nucl. Sci.* **2016**, *63*, 52–60. [CrossRef]
7. Esposito, M.; Waltham, C.; Taylor, J.T.; Manger, S.; Phoenix, B.; Price, T.; Poludniowski, G.; Green, S.; Evans, P.M.; Allport, P.P.; et al. PRaVDA: The first solid-state system for proton computed tomography. *Phys. Medica* **2018**, *55*, 149–154. [CrossRef] [PubMed]
8. DeJongh, E.A.; DeJongh, D.F.; Polnyi, I.; Rykalin, V.; Sarosiek, C.; Coutrakon, G.; Duffin, K.L.; Karonis, N.T.; Ordoñez, C.E.; Pankuch, M.; et al. A fast and monolithic prototype clinical proton radiography system optimized for pencil beam scanning. *Phys. Medica* **2021**, *48*, 1356–1364. [CrossRef] [PubMed]

9. Baruffaldi, F.; Iuppa, R.; Ricci, E.; Snoeys, W. iMPACT: an innovative tracker and calorimeter for proton computed tomography. *IEEE Trans. Radiat. Plasma Med. Sci.* **2018**, *2*, 345–352. [CrossRef]
10. Ackernley, T.; Casse, G.; Cristoforetti, M. Proton path reconstruction for proton computed tomography using neural networks. *Phys. Med. Biol.* **2021**, *66*, 075015. [CrossRef] [PubMed]
11. Granado-González, M.; Jesús-Valls, C.; Lux, T.; Price, T.; Sánchez, F. A novel range telescope concept for proton CT. *Phys. Med. Biol.* **2022**, *67*, 035013. [CrossRef] [PubMed]
12. Pernegger, H.; Bates, R.; Buttar, C.; Dalla, M.; Van Hoorne, J.W.; Kugathasan, T.; Maneuski, D.; Musa, L.; Riedler, P.; Riegel, C.; et al. First tests of a novel radiation hard CMOS sensor process for Depleted Monolithic Active Pixel Sensors. *JINST* **2017**, *12*, P06008. [CrossRef]
13. Neubüser, C.; Corradino, T.; Dalla Betta, G.F.; De Cilladi, L.; Pancheri, L. Sensor Design Optimization of Innovative Low-Power, Large Area FD-MAPS for HEP and Applied Science. *Front. Phys.* **2021**, *9*, 625401. [CrossRef]
14. Eljen Technology—General Purpose EJ-200, EJ-204, EJ-208, EJ-212. Available online: https://eljentechnology.com/products/plastic-scintillators/ej-200-ej-204-ej-208-ej-212 (accessed on 29 July 2022).
15. J-Series SiPM Sensors. Available online: https://www.onsemi.com/pdf/datasheet/microj-series-d.pdf (accessed on 29 July 2022).
16. Blondel, A.; Bogomilov, M.; Bordoni, S.; Cadoux, F.; Douqa, D.; Dugas, K.; Ekelof, T.; Favre, Y.; Fedotov, S.; Fransson, K.; et al. The SuperFGD Prototype Charged Particle Beam Tests. *JINST* **2020**, *15*, P12003. [CrossRef]
17. Esposito, M.; Anaxagoras, T.; Evans, P.M.; Green, S.; Manolopoulos, S.; Nieto-Camero, J.; Parker, D.J.; Poludniowski, G.; Price, T.; Waltham, C.; et al. CMOS Active Pixel Sensors as energy-range detectors for proton Computed Tomography. *JINST* **2020**, *10*, C06001. [CrossRef] [PubMed]
18. Long, J.; Shelhamer, E.; Darrell, T. Fully convolutional networks for semantic segmentation. In Proceedings of the IEEE Conference on Computer Vision and Pattern Recognition, Boston, MA, USA, 7–12 June 2015; pp. 3431–3440.
19. Wang, P.; Chen, P.; Yuan, Y.; Liu, D.; Huang, Z.; Hou, X.; Cottrell, G. Understanding convolution for semantic segmentation. In Proceedings of the IEEE Winter Conference on Applications of Computer Vision (WACV), Lake Tahoe, NV, USA, 12–15 March 2018; pp. 1451–1460.
20. Ronneberger, O.; Fischer, P.; Brox, T. U-net: Convolutional networks for biomedical image segmentation. In Proceedings of the International Conference on Medical Image Computing and Computer-Assisted Intervention, Munich, Germany, 5–9 October 2015; pp. 234–241.

 *instruments*

Article

SiPMs for Dual-Readout Calorimetry

Romualdo Santoro [1,2] on behalf of the IDEA Dual-Readout Group

1 Department of Science, University of Insubria, Via Valleggio 11, 22100 Como, Italy
2 INFN Milano, Via Celoria 16, 20133 Milano, Italy

Abstract: A new fibre-sampling dual-readout calorimeter prototype has been qualified on beam at two facilities (DESY and CERN) using electrons from 1 to 100 GeV. The prototype was designed to almost fully contain electromagnetic showers and a central module (highly granular readout) was equipped with 320 Silicon Photomultipliers (SiPMs) spaced by 2 mm and individually read out. The test beams performed in 2021, allowed to qualify the readout boards used to operate the SiPMs, to define the calibration procedure and to measure the light yield for scintillating and Cherenkov signals produced by the shower development. This paper reports the first results obtained with the highly granular readout and discusses the ongoing R&D to address some open questions concerning the mechanical integration and the scalable readout scheme that will allow to build and operate the next prototype designed for hadronic showers containment.

Keywords: SiPMs; calorimetry for high energy physics; dual readout detector R&D

Citation: Santoro, R., on behalf of the IDEA Dual-Readout Group. SiPMs for Dual-Readout Calorimetry. *Instruments* **2022**, *6*, 59. https://doi.org/10.3390/instruments6040059

Academic Editors: Fabrizio Salvatore, Alessandro Cerri, Antonella De Santo and Iacopo Vivarelli

Received: 8 August 2022
Accepted: 3 October 2022
Published: 8 October 2022

Publisher's Note: MDPI stays neutral with regard to jurisdictional claims in published maps and institutional affiliations.

1. Introduction

Experiments operating at future e+e− circular colliders (such as, for example, FCC-ee and CEPC) must cope with a rich and complex high precision physics program [1,2]. The abundance of hadronic final states from collisions at centre-of-mass energies ranging from 90 to 365 GeV will require superior jet reconstruction capabilities. Hadronic showers develop both an electromagnetic and a hadronic components which are usually detected with a different response (non-compensation). As a result, the fluctuations among the two components constitute one of the most limiting factors for the hadronic energy resolution. Dual readout is a calorimetric technique able to overcome the limits due to non-compensation by simultaneously detecting scintillation and Cherenkov lights. Scintillating photons provide a signal related to the energy deposition in the calorimeter by all ionising particles, while Cherenkov photons provide a signal almost exclusively related to the shower electromagnetic component. In fact, by looking at the two independent signals, it is possible to measure, event by event, the electromagnetic shower component and to properly reconstruct the primary hadron energy. Several prototypes were constructed based on different active media and absorber materials. The 20-year-long research programme by the DREAM/RD52 collaboration on dual readout calorimetry has provided a technology that is now mature for application [3]. The performance studies based on a Monte Carlo simulation reproducing a full experiment geometry have been recently summarised [4] even though, the modules are not identical to the ones tested on beam in 2021. Accordingly to these studies, we could target a hadronic energy resolution of $\approx 30\%/\sqrt{E}$ for single hadron and $\approx 38\%/\sqrt{E}$ for jets. In addition, using SiPMs instead of Photomultipliers (PMTs), we could further improve the excellent particle-ID capability by adding a projective image of the shower [5,6].

In this paper, the first results obtained with the core of the electromagnetic-size prototype, sensed with SiPMs, will be discussed together with the R&D strategy ongoing for building a new demonstrator capable of fully contain hadronic showers and measure the energy resolution.

2. Experimental Setup

The prototype qualified on beam in 2021 is shown in Figure 1. It is 1 m long with a cross section of 10×10 cm^2. The Moliere radius (R_M) is 23.8 mm while the effective radiation length (X_0) is 22.7 mm. It consists of 9 modules, each made of 320 brass capillaries (outer diameter = 2 mm and inner diameter = 1.1 mm) equipped, alternatively, with scintillating (BC-10 from Saint Gobain) and clear (SK-40 from Mitsubishi) fibres to allow the dual sampling. The external modules are instrumented with R-8900 PMTs. The scintillating and clear fibres are separated and bundled in two groups on the back side of each module to match the PMTs' window. A yellow filter (Kodak, Wratten nr 3, with nominal transmission of $\approx$7% at 425 nm and $\approx$90% at 550 nm) is placed between the scintillating fibres and the detector to cut off the short wavelength component of the scintillating signal (standard configuration). In fact, yellow filters reduce the calorimeter response dependence on the shower starting point by filtering the component of the light more affected by attenuation in the fibres. The PMTs are read out with V792AC QDC modules produced by CAEN s.p.a.

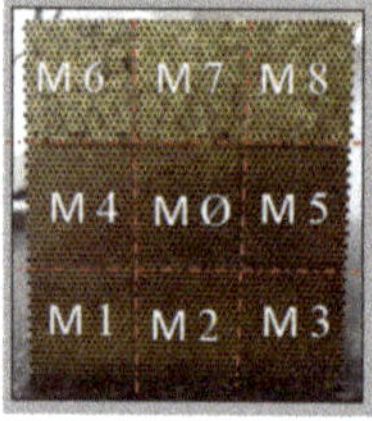

Figure 1. A front view (**left**) and a side view (**right**) of the em-size prototype before the connection to the light sensors (PMTs and SiPMs) are shown in picture. The fibres from the external modules (M1–M8) are bundled to match the PMTs' window while the longer fibres from the central module (MØ) are connected to a patch panel to be interfaced with SiPMs.

The central module, namely the highly granular module, has each fibre connected to a SiPM by Hamamatsu (S14160-1315 PS) with a sensitive area of 1.3×1.3 mm^2 read out independently. Since almost 10% of the entire energy is released within one mm from the core of the shower (1–2 fibres) [6], SiPMs with a wide dynamic range (i.e., 7284 cells, 15 μm pitch) were selected. Unfortunately, SiPMs with compact packaging were not available at the time of the construction. For this reason, we were forced to fan-out the fibres on the back sides of the calorimeter to match the front-end boards housing 64 SiPMs. The SiPMs on the front-end board are separated in two groups (32 SiPMs each) insulated with a light tight frame (Figure 2) to avoid light contamination between Cherenkov and scintillating signals. As for the external module, yellow filters are placed between the scintillating fibres and the SiPMs. In addition, a transparent paper is used between the clear fibres and the SiPMs for mechanical reasons and to avoid any air gap between the fibres and the light sensors.

The SiPM readout is based on the FERS-System produced by CAEN s.p.a. (https://www.caen.it/products/a5202/, accessed on 1 August 2022) to fully exploit the Citiroc-1A (https://www.weeroc.com/products/sipm-read-out/citiroc-1a, accessed on 1 August 2022) performance: i.e., wide dynamic range, linearity and multi-photon quality even with SiPMs with small pitch size and small gain (1–3 $\times$ 10^5 at nominal settings). Each readout board (A5202) is equipped with two Citiroc-1A to operate 64 SiPMs. The signal produced by each SiPM feeds, at the same time, two charge amplifiers with tunable gains. The range accessible by one of the two amplifiers (namely the High Gain—HG) is almost 10 times higher than the other (Low Gain—LG). This characteristic allows to store on disk two spectra per each SiPM. The first (HG) useful to analyse the multi-photon and to extract the ADC to photo-electrons (ph-e) constant, and the second (LG) needed to extend the overall dynamic range. The settings for the two charge amplifiers were chosen to guarantee: (1) good quality HG spectra, (2) an overlap region between the HG and LG spectra used for the calibration, and (3) a wide dynamic range. These settings allow to read signals from

1 to almost 4000 ph-e which is ≈55% of the SiPM occupancy considering the microcells available in the sensitive area.

The prototype was qualified on beam at DESY and at CERN with small differences in the setup configuration.

2.1. DESY Setup Configuration

The trigger scheme was based on the coincidence from two scintillator counters placed in front of the calorimeter and the signal produced by the A5202 boards, running in self-trigger mode with a majority algorithm. The trigger received by the scintillators was used by the A5202 boards to accept and store on disk (event accept) the data produced by the SiPMs. The majority algorithm was running independently on each readout board (64-SiPMs) with a coincidence of 3 SiPMs detecting a signal over the threshold set at 3.5 ph-e. The event building was performed off-line correlating the trigger ID from the different A5202 boards. Data produced by electrons with energy range from 1 to 6 GeV were acquired and used in the analysis. For this test beam, the scintillating and the clear fibres were directly connected to the SiPMs. We decided to remove the yellow filters because we had access to low energy electrons and we were also interested in measuring the small signals released in the tail of the shower to tune the Monte Carlo simulation.

2.2. CERN Setup Configuration

The trigger was provided by three scintillator counters. The first two counters were used in coincidence and the third (the scintillator had a hole with a diameter of 10 mm) was used as veto. Additional detectors were included in the data taking (i.e., two delay wire chambers, a pre-shower and a muon detector) to determine the impact point of the beam particles and to flag electrons with off-line analysis. The latter selection was important because the electron beam had muon and hadron contamination with a ratio depending on the beam line extraction and beam energy. The SiPM readout schema was the same as the one used at the DESY test beam. We were running two independent data acquisition systems sharing the same trigger. The off-line synchronisation is performed by using the trigger ID. Data produced by electrons from 6 to 100 GeV are used in the analysis. For this test beam, we used yellow filters between the scintillating fibres and the SiPMs (standard configuration).

Figure 2. The picture on the (**left**) shows the back side of the highly granular module and a front-end board with 64 SiPMs before the installation. The system, ready to take data and connected to the five A5202 boards requested to operate the SiPMs, is shown on the (**right**).

3. Test Beam Data Analysis

The two test beams allowed to qualify the readout system for the highly granular module, to define the calibration procedure and to measure the light yield for the scintillating and Cherenkov signals for this prototype.

3.1. Equalisation and Calibration

Before the installation in the experimental area, all SiPMs were qualified in the lab with an ultra-fast LED emitting at 420 nm. The SiPM response was equalised by applying the same over-voltage and by tuning the amplifier settings. The procedure allowed to operate all SiPMs with the same photon detection efficiency (PDE) and with signals equalised in amplitude. The latter requirement is needed because the system, working in self-trigger mode, has the same threshold for all 64 SiPMs served by one A5202 board. There is the possibility to adjust the threshold for each channel but the fine tuning covers 1 DAC only of the global threshold. The SiPMs were operated with a voltage set at +7 V over the breakdown. Even if it is not a typical setting for a SiPM, it guarantees a stable PDE (against small temperature variation) and a multiplication factor in the avalanche region of the order of 0.5×10^6 per each detected photon, allowing to set a trigger threshold at the level of a single ph-e together with good quality multi-photon spectra. The spurious effects (i.e., dark count rate and crosstalk) have limited impact on the measurements. Even if the readout system was running in self-trigger mode, the majority algorithm plus the event-accept technique allowed to maintain the spurious events at the sub-Hz level and a crosstalk at the percent level.

Calibration in ph-e of the LG response is required in order to sum signals from different SiPMs and to correct for non-linearity due to the limited number of cells available in each detector, if needed. Figure 3 shows the typical HG (on the left) and LG (on the right) spectra measured by one SiPM connected to a scintillating fibre in response to 6 GeV electrons at the DESY test beam. The pedestal, the multi-photon, and the ADC saturation are clearly visible in the plot on the left. The saturation is not affecting the measurement since the information is still available in the LG spectrum and the strategy used to calibrate both spectra in ph-e is the following:

- The pedestal and the multi-photon are fitted with Gaussian functions. The results are used to convert the ADC channels in ph-e by using the mean value of the pedestal and the average peak-to-peak distance obtained by fitting three consecutive peaks (Figure 4 left plot).
- The HG values, converted in ph-e, are correlated to the ADC counts of the LG channel (Figure 4 right plot). The points in the plot exceeding 125 ph-e are not considered in the fit and the slope is used to extract the ADC to ph-e conversion for the LG even if the multi-photon structure is not accessible in this regime. The typical conversion factor is $\approx$1 ph-e/ADC with a few per mille of uncertainty.

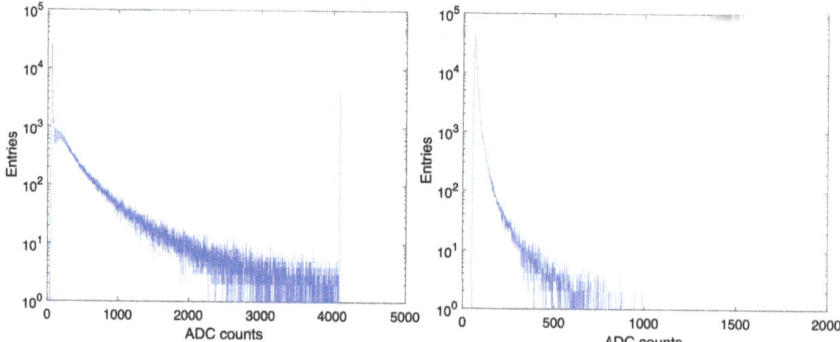

Figure 3. Two spectra obtained for the same SiPM with 6 GeV electrons at the DESY test beam. The spectrum on the left refers to the HG amplifiers while the results for the LG are shown on the right.

The procedure is performed for all SiPMs and the parameters are extracted on a run by run basis, without having the need to collect dedicated data for calibration. The analysis performed on these data allowed to verify the system stability in different runs. The Figure 5 shows the baseline (on the left) and the peak-to-peak difference (on the right) measured, for one SiPM in each readout board, using the HG data collected during a series of consecutive runs. Variations never exceeding 2% were measured for the baseline and peak-to-peak difference for all SiPMs. In fact a single calibration constant per SiPM was used for all the datasets analysed.

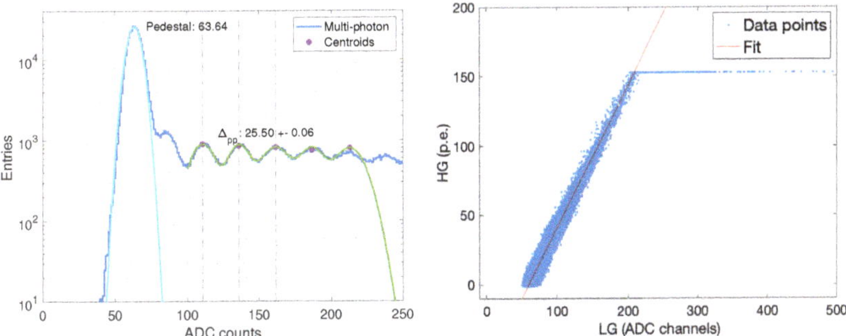

Figure 4. The spectrum on the (**left**) is an x-zoom of the spectrum in Figure 3 to appreciate the quality of the multi-photon and the fitting procedure. The HG signals, converted in ph-e, are correlated to the LG signals on the (**right**) plot.

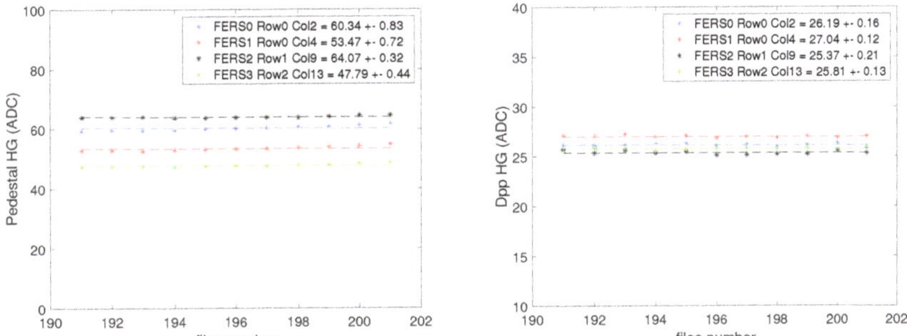

Figure 5. Baseline (on the (**left**)) and peak-to-peak difference (on the (**right**)) measured for one SiPM per each readout board during a series of consecutive runs used in the data analysis. These plots refer to the DESY test beam data.

3.2. Light Collection

Once the signals from all SiPMs are calibrated in ph-e, they are summed event by event to obtain the total distribution. Only events hitting the central part of the module equipped with SiPMs are selected. The selection is performed by using the centre of gravity $(\bar{x}, \bar{y})$ estimated with the energy deposited (E_i) in the 160 scintillating fibres (x_i, y_i) by using the following Formula:

$$\bar{x} = \frac{\sum_i x_i E_i}{\sum_i E_i}; \qquad \bar{y} = \frac{\sum_i y_i E_i}{\sum_i E_i} \qquad (1)$$

and the cut in the central region is defined by a square of 4×4 fibres ($\approx 8 \times 7$ mm^2).

Figure 6 (on the left) shows the average number of detected photons produced by the Cherenkov and scintillating process in the central module as a function of the electron beam energy, divided by the beam energy. The preliminary results obtained in the small energy range available at DESY show an average value of 42.9 ± 0.6 ph-e/GeV for the Cherenkov light and 460.5 ± 5.5 ph-e/GeV for the scintillating light. Once corrected for the shower containment ($\approx 72\%$), estimated with a detailed Monte Carlo simulation that describes the experimental setup, we obtain ≈ 60 ph-e/GeV for the Cherenkov light and ≈ 640 ph-e/GeV for the scintillating light. The value measured for the Cherenkov light is similar to the result obtained with the first measurement performed using a small module

and mounted with the spacing matching the fibre pitch. Figure 8 shows the rear part of the mini-module designed for this scope and the main components. The SiPMs are directly connected to the fibres and the capillaries have different lengths (in alternating rows) to avoid the light contamination previously mentioned. This design requires compact SiPMs, PCB boards and connectors to fit in the limited space available on the back side of the calorimeter.

- As discussed, it would be extremely useful to include the time information in the readout. This requirement would allow to add a longitudinal segmentation of few centimetres to the calorimeter with a time stamping $\leq$100 ps. This is difficult to be achieved with the actual system even though some preliminary investigation can be done. A possible solution could be based on the ASIC RADIOROC [8] from WEEROC supposed to have a compatible performance in terms of energy spectrum plus improved timing information ($\approx$35 ps FWHM on single ph-e). This possibility will be investigated in the near future.

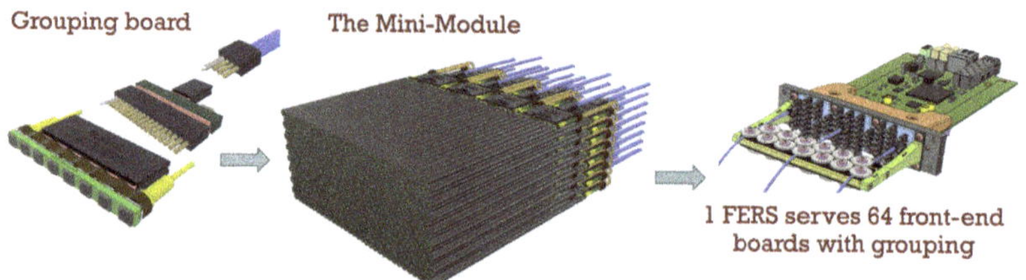

Figure 8. The main components designed for the next generation of prototype are shown. The PCB board equipped with 8 SiPM and the grouping board is shown on the (**left**). The grouping board is used to sum the analogue signals produced by 8 SiPMs with the goal of reducing the number of channels to be read out. The mini-module is shown in the (**middle**) and the new design should allow to connect the SiPMs directly to the capillaries avoiding light contamination. The new patch panel, where the 512 SiPMs from the mini-module (64 signals after the grouping) are interfaced to one A5202 board, is shown on the (**right**).

A larger prototype, capable of containing hadronic showers, will be built in the next two years. As for the electromagnetic prototype, only the central part will be equipped with SiPMs but, in this case, the collaboration wants to target a real scalable solution from both the mechanical and readout point of view. This prototype will be equipped with more than 10,000 SiPMs with the final goal to target the study of the hadronic performance of the dual readout technique.

Funding: This project has received funding from the European Union's Horizon 2020 research and innovation program under grant agreement No. 669014.

Data Availability Statement: Not applicable.

Acknowledgments: The measurements leading to these results have been performed at the Test Beam Facility at DESY Hamburg (Germany) [9] and at the Super Proton Synchrotron (SPS) at CERN.

Conflicts of Interest: The authors declare no conflict of interest.

References

1. Abada, A.; Abbrescia, M.; AbdusSalam, S.S.; Abdyukhanov, I.; Fernandez, J.A.; Abramov, A.; Aburaia, M.; Acar, A.O.; Adzic, P.R.; Agrawal, P. FCC-ee: The Lepton Collider. Future Circular Collider Conceptual Design Report Volume 2. *Eur. Phys. J. Spec. Top.* **2019**, *228*, 261–623. [CrossRef]
2. The CEPC Study Group. CEPC Conceptual Design Report: Volume 2—Physics & Detector. *arXiv* **2018**, arXiv:1811.10545.
3. Lee, S.; Livan, M.; Wig-Mans, R. Dual-readout calorimetry. *Rev. Mod. Phys.* **2018**, *90*, 025002. [CrossRef]
4. Pezzoti, L. Particle Detectors R&D: Dual-Readout Calorimetry for Future Colliders and MicroMegas Chambers for the ATLAS New Small Wheel Upgrade. Ph.D. Thesis, The University of Pavia, Pavia, Italy, 2021. Available online: https://iris.unipv.it/handle/11571/1429275?mode=full.4142 (accessed on 1 August 2021).

5.   Antonello, M.; Caccia, M.; Cascella, M.; Dunser, M.; Ferrari, R.; Franchino, S.; Gaudio, G.; Hall, K.; Hauptman, J.; Jo, H. Tests of dual-readout fiber calorimeter with SiPM light sensor. *Nucl. Instrum. Methods Phys. Res. Sect. A Accel. Spectrometers Detect. Assoc. Equip.* **2018**, *899*, 52–64. [CrossRef]
6.   Pezzotti, L.; Ferrari, R.; Antonello, M.; Caccia, M.; Santoro, R. Dual-readout fiber-sampling calorimeter with SiPM light sensors. *J. Phys. Conf. Ser.* **2019**, *1162*, 012014. [CrossRef]
7.   Karadzhinova-Ferrer, A.; Agarwala, J.; Ampilogov, N.; Chmill, V.; Ferrari, R.; Freddi, A.; Gaudio, G.; Gigli, S.G.; Samec, Ž.; Santoro, R.; et al. Novel Prototype Tower Structure for the Dual-Readout Fiber Calorimeter. *JINST* **2022**, *17*, T09007. [CrossRef]
8.   Contino, G.; Catalano, O.; Sottile, G.; Sangiorgi, P.; Capalbi, M.; Osteria, G.; Scotti, V.; Miyamoto, H.; Vigorito, C.; Casolino, M.; et al. An ASIC front-end for fluorescence and Cherenkov light detection with SiPM for space and ground applications. *Nucl. Instrum. Methods Phys. Res. Sect. A Accel. Spectrometers Detect. Assoc. Equip.* **2020**, *980*, 164510. [CrossRef]
9.   Diener, R.; Dreyling-Eschweiler, J.; Ehrlichmann, H.; Gregor, I.M.; Kötz, U.; Krämer, U.; Meyners, N.; Potylitsina-Kube, N.; Schütz, A.; Schütze, P.; et al. The DESY II test beam facility. *Nucl. Instrum. Methods Phys. Res. Sect. A Accel. Spectrometers Detect. Assoc. Equip.* **2019**, *922*, 265. [CrossRef]

 *instruments*

Article

The Mu2e Crystal Calorimeter: An Overview

Nikolay Atanov [1], Vladimir Baranov [1], Leo Borrel [2], Caterina Bloise [3], Julian Budagov [1], Sergio Ceravolo [3], Franco Cervelli [4], Francesco Colao [3,5], Marco Cordelli [3], Giovanni Corradi [3], Yuri Davydov [1], Stefano Di Falco [4], Eleonora Diociaiuti [3], Simone Donati [4,6], Bertrand Echenard [2], Carlo Ferrari [4], Antonio Gioiosa [4], Simona Giovannella [3], Valerio Giusti [4], Vladimir Glagolev [1], Francesco Grancagnolo [7], Dariush Hampai [3], Fabio Happacher [3], David Hitlin [2], Matteo Martini [3,8], Sophie Middleton [2], Stefano Miscetti [3,*], Luca Morescalchi [4], Daniele Paesani [3], Daniele Pasciuto [4,6], Elena Pedreschi [4], Frank Porter [2], Fabrizio Raffaelli [4], Alessandro Saputi [9], Ivano Sarra [3], Franco Spinella [4], Alessandra Taffara [4], Anna Maria Zanetti [10] and Ren Yuan Zhu [2]

1   Joint Institute for Nuclear Research, 141980 Dubna, Russia
2   California Institute of Technology, Pasadena, CA 91125, USA
3   Laboratori Nazionali di Frascati dell'INFN, 00044 Frascati, Italy
4   INFN—Sezione di Pisa, 56100 Pisa, Italy
5   ENEA—Frascati, 00044 Frascati, Italy
6   Department of Physics, University of Pisa, 56100 Pisa, Italy
7   INFN—Sezione di Lecce, 73100 Lecce, Italy
8   Department of Engineering Sciences, Guglielmo Marconi University, 00193 Roma, Italy
9   INFN—Sezione di Ferrara, 44100 Ferrara, Italy
10   INFN—Sezione di Trieste, 34149 Trieste, Italy
*   Correspondence: stefano.miscetti@lnf.infn.it

Citation: Atanov, N.; Baranov, V.; Borrel, L.; Bloise, C.; Budagov, J.; Ceravolo, S.; Cervelli, F.; Colao, F.; Cordelli, M.; Corradi, G.; et al. The Mu2e Crystal Calorimeter: An Overview. *Instruments* 2022, 6, 60. https://doi.org/10.3390/instruments6040060

Academic Editors: Fabrizio Salvatore, Alessandro Cerri, Antonella De Santo and Iacopo Vivarelli

Received: 5 September 2022
Accepted: 20 September 2022
Published: 9 October 2022

Abstract: The Mu2e experiment at Fermilab will search for the standard model-forbidden, charged lepton flavour-violating conversion of a negative muon into an electron in the field of an aluminium nucleus. The distinctive signal signature is represented by a mono-energetic electron with an energy near the muon's rest mass. The experiment aims to improve the current single-event sensitivity by four orders of magnitude by means of a high-intensity pulsed muon beam and a high-precision tracking system. The electromagnetic calorimeter complements the tracker by providing high rejection power in muon to electron identification and a seed for track reconstruction while working in vacuum in presence of a 1 T axial magnetic field and in a harsh radiation environment. For 100 MeV electrons, the calorimeter should achieve: (a) a time resolution better than 0.5 ns, (b) an energy resolution <10%, and (c) a position resolution of 1 cm. The calorimeter design consists of two disks, each loaded with 674 undoped CsI crystals read out by two large-area arrays of UV-extended SiPMs and custom analogue and digital electronics. We describe here the status of construction for all calorimeter components and the performance measurements conducted on the large-sized prototype with electron beams and minimum ionizing particles at a cosmic ray test stand. A discussion of the calorimeter's engineering aspects and the on-going assembly is also reported.

Keywords: scintillation; crystals; SiPM; calorimetry

1. Introduction

The Mu2e experiment [1] at Fermilab aims to improve, by four orders of magnitude, the current single-event sensitivity in searching for the yet unobserved charged lepton flavour violating (CLFV) neutrino-less conversion of a negative muon into an electron in the field of an aluminium nucleus. Such a process, forbidden in the standard model, has a clear signature provided by the identification of a mono-energetic conversion electron (CE) with an energy slightly below the muon's rest mass (104.97 MeV). Even assuming neutrino oscillations, CLFV processes in the muon system remain completely negligible, $BR(\mu \to e\,\gamma) = 10^{-52}$ [2]. Observing CLFV candidates will indicate physics beyond the standard model. If conversions are not observed, Mu2e will set a 90% upper limit on the ratio between the conversion and capture rates ($R_{\mu e}$) at $< 8 \times 10^{-17}$.

The Mu2e layout, based on an original concept by V. Lobashev and R. Djilibaev [3], is shown in Figure 1. The large solenoidal system is designed to largely increase the number of μ^- arriving at the stopping target (ST). A 8 GeV pulsed proton beam sent at the tungsten target inside the production solenoid (PS) produces low-momentum pions that are funnelled, by the graded field, inside the S-shaped transport solenoid (TS). Here, the pions decay to muons and are charge selected by a middle section collimator. At the end of the transport chain, a very intense negative muon beam ($\sim 10^{10}$ μ/s) enters the detector solenoid (DS) and is stopped at an aluminium target. In its lifetime, the experiment plans to collect 6×10^{17} muon stops to reach its sensitivity goal. Decay products are analysed by the tracker [4] and calorimeter [5] systems. Cosmic ray muons can produce fake CE candidates when interacting in the DS. To reduce their contribution, the external area of the DS and part of the TS are covered by a cosmic ray veto (CRV) [6] system.

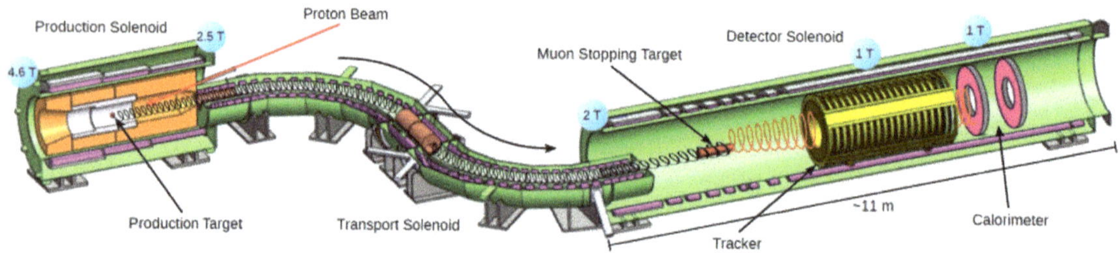

Figure 1. Layout of the Mu2e experiment: PS, DS, and TS solenoids are indicated in the picture. The cosmic ray veto, surrounding the DS and part of the TS solenoids, is not shown.

Muons stopped in the aluminium target form a muonic atom and cascade to the 1S ground state, with 39% decaying in orbit (DIO) and 61% captured by the nucleus. Low energy protons, neutrons, and photons are emitted in the nuclear capture process, thus originating both a large neutron fluence and, together with the flash of particles accompanying the beam, the bulk of the ionizing dose observed in the detectors. The tracker, composed of $\sim$20,000 low-mass straw drift tubes, measures the charged particles' momenta by reconstructing their trajectories in the magnetic field with the detected hits. Full simulation shows that a momentum resolution of O(160 keV) can be reached, thus separating the CE line from the fast-falling spectrum of the DIO electrons.

2. Materials and Methods

In this section, we describe the calorimeter system in more detail, starting from requirements and technical choices down to its engineering design.

2.1. Calorimeter Requirements

The calorimeter complements the CE identification tracker by providing a high μ/e rejection better than 200, a fast online trigger filter and a seed for track reconstruction [7]. To fulfil these tasks, simulation guided us to define the reconstruction requirements for 105 MeV electrons, which are summarized by this short list: (a) a large acceptance, (b) a time resolution better than 0.5 ns, (d) an energy resolution $<10\%$, and (d) a position resolution of 1 cm. Moreover, the calorimeter should maintain its functionality when operating inside the DS without interruption for one year in a harsh radiation environment in the presence of 1 T axial magnetic field and in a region evacuated to 10^{-4} Torr. This asked for a high reliability, redundancy, and a high level of radiation hardness on all calorimeter components.

2.2. Technical Choices for Crystals and Photo-Sensors

Our solution was to design a high-quality crystal calorimeter with silicon photomultiplier (SiPMs) readout and a geometry organized into two annular disks (Figure 2) to maximize acceptance for spiralling electrons. The crystals had to provide a high light yield of at least 20 photoelectrons (p.e.)/MeV per single SiPM readout. To handle the pileup of particles, fast signals were needed, thus asking for crystals with a decay time (τ) better than 40 ns and front end electronics (FEE) providing fast amplified signals to be sampled at 200 Msps (5 ns binning) by the digitization system. The selected crystals, SiPMs had to sustain a total ionization dose up to 1000 Gy (900 Gy) and a neutron fluence of up to 3, 1.2×10^{12} n/cm^2 respectively. All of this without deteriorating the calorimeter performance. The high redundancy and reliability required translated into having two independent SiPMs, FEE boards/crystal, and an independent digitization system for the two readout lines. A detailed simulation quantified that the typical mean time to failure (MTTF) needed to be $\sim 10^6$ h/component.

Figure 2. (Left), CAD of the two calorimeter disks. (**Top right**), few unwrapped calorimeter crystals with their own parallelepiped shape, and (**bottom right**), two SiPM arrays glued onto copper holder on the left and one readout unit formed by two SiPM arrays and two FEE boards mounted on its copper holder and on the right.

At the end of the R&D program [8–10], undoped CsI crystals were chosen as the best compromise between cost, performance, and reliability, being sufficiently radiation hard for our task and having a fast emission time and an acceptable light yield. Since the main scintillation component has a wavelength of 310 nm to well match the SiPM photon detection efficiency (PDE) as a function of wavelength, we selected Hamamatsu UV extended SiPMs, where the front window epoxy was replaced by a silicon resin to achieve >20% PDE down to 280 nm. To operate in a vacuum and minimize outgassing contributions, the crystal SiPM coupling was performed without any optical grease. This choice reduced the light collected by the SiPM, so we opted to build a very large area (12×16 mm^2) SiPM array. In Figure 2 (bottom right), a picture of two Mu2e SiPMs glued to a copper holder is shown, each one consisting of the parallel of two series of three 6×6 mm^2 monolithic Hamamatsu surface mount SiPMs, model S13360-6050PE, with 50 μm pixel size. This configuration reduced the array capacitance and quenching time while simplifying the FEE design.

2.3. Electronics Scheme

The electronics is based on analogue FEE cards directly connected to the SiPM pins and a digital readout part distributed on the crates surrounding the disk.

The FEE shapes and amplifies the signal while locally regulating the supplied voltage. The two SiPMs glued to a copper holder, assembled together with two FEE boards and a copper Faraday cage, constitute a readout unit (ROU).

The digital boards are subdivided in a mezzanine (MB) for controlling the HV and monitor the current and temperature of SiPMs and a readout and digitization board (DIRAC) that performs the zero suppression and samples the signals with 5 ns binning. Each digital board is able to handle 20 channels. Much effort was dedicated to design and produce boards that work well in a magnetic field and in the radiation-harsh Mu2e environment. A description of all the details of the electronics scheme, the radiation hardness program, and the quality tests performed can be found elsewhere [11,12].

2.4. Breakdown of Mechanical Layout

Figure 3 shows a breakdown of the calorimeter structure.

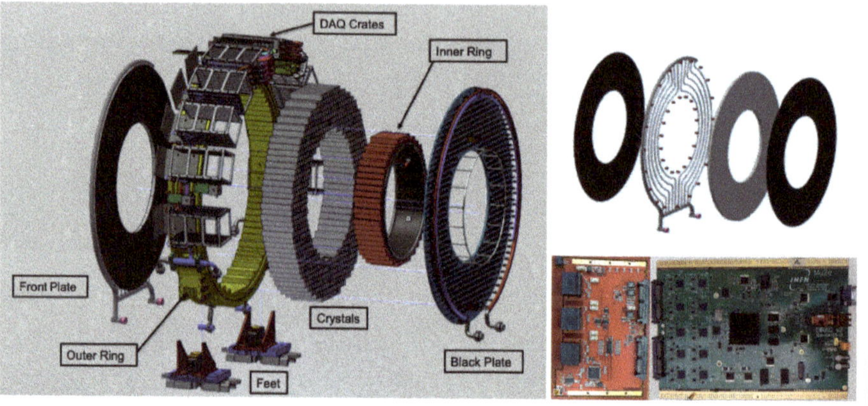

Figure 3. (**Left**) breakout of calorimeter mechanical components; (**top right**), breakdown of front panel plate embedding source tubing; and (**bottom right**), the mezzanine and DIRAC boards.

Each disk is filled with a matrix of 674 parallelepiped undoped CsI crystals ($34 \times 34 \times 200$ mm^3) for an inner/outer diameter of 650 mm/1314 mm. The crystals are wrapped in 150 μm thick Tyvek foils and separated from each other with 50 μm thick Tedlar foil to make the optical cross talk negligible. The crystal matrix is supported externally by an aluminium shaped outer ring, with an outer diameter of 1460 mm and a thickness of 146 mm to provide the required stiffness. The disk is milled to shape to form the lateral steps where crystal rows are positioned and aligned. The outer ring also provides support and the place where all other components are fastened. It hosts the custom DAQ crates in its external surface and their cooling manifold.

In order to minimize the energy loss of particles arriving to the crystals, the inner ring and the front plate, which are traversed by particles, are made in carbon fibre with embedded light aluminium honeycomb structures, thus achieving the right stiffness while avoiding vacuum virtual leaks. The inner ring occupies the inner bore surface, sustains the crystal vertical load, and grants a reference for the positioning of the crystal matrix. The front plate is the frontal protection cover of the crystal matrix and hosts 10 thin-wall aluminium tubes, symmetrically arranged on each disk, to flow the calibration source fluid (FC-770).

Finally, in the back of the crystals, the back plate constitutes the rear mechanical enclosure of the calorimeter disk. The back plate is created from a 20 mm thick PEEK plate milled to shape to host and support the 674 ROUs. PEEK has been chosen for its good outgassing characteristics and to optimize the thermal isolation of the electronics. The FEE plate embeds a network of 38 parallel vacuum brazed copper lines, where a cooling liquid (3M NOVEC HFE 1700) will be circulated at −15 °C to cool down the SiPMs in the ROUs. The latter ones will be fastened on the cooling lines to optimize the thermal conductivity. The stainless steel manifolds, placed on the outer plate border, distribute the cooling fluid to the copper lines. More details on the mechanical components can be found elsewhere [13].

2.5. Calibration Systems

The crystal-by-crystal energy equalization is obtained by means of a calibration system, formerly devised for the BaBar calorimeter [14], where a 6.13 MeV photon line is obtained from a short-lived ^{16}O transition. The decay chain comes from a Fluorinert^{TM} coolant liquid (FC-770) that is activated by fast neutrons produced by a DT generator. The activated liquid circulates in the aluminium tubes positioned in the front plate to uniformly illuminate each crystal face. The source system is accompanied by a laser monitoring system that provides a continuous monitoring of the sensor gains and of each channel timing offsets. Each crystal is illuminated by the laser light coming from the laser head via primary and secondary distribution systems, through an optical fibre whose needle is inserted in the ROU structure. The laser offers also a simple method to monitor variations in the energy and timing resolutions. Usage of cosmic ray and DIO events is foreseen for a continuous in situ energy and timing calibration during operation.

3. Results

3.1. Calorimeter Qualification with Module-0

Before starting production, a large size prototype, dubbed Module-0, was assembled to mimic the calorimeter disk and confirm technical choices, drive the readout electronics development, and test its performance. The prototype, assembled in May 2017 with 51 crystals and 102 SiPMs, was equipped with the first FEE version and tested with an electron beam at BTF (Frascati) soon after its assembly. The digital readout was based on commercial CAEN digitizers. A detailed description of the test beam can be found elsewhere [15]. The main results are summarized by an energy and timing resolution parametrized as the quadrature of stochastic $a/\sqrt{(E/GeV)}$, noise, $b/(E/GeV)$, and constant, c, terms, as in Table 1. The main conclusion for the energy was that the stochastic term was consistent with a light yield of O(20 pe/MeV/SiPM), the noise term was attributed to an electronic noise of O(400) keV/channel, and a coherent noise related to the used digitizers, while the c term, the dominant one, was due to shower leakage effects, as demonstrated by a Geant4 simulation. The runs at $0°$ had energy loss due to longitudinal leakage. The runs at $50°$ exacerbated the b and c terms due to longer clusters and higher transversal leakage. Overall, a resolution better than 5 (7.5)% was achieved at 100 MeV for runs at normal ($50°$) incidence. The timing resolution, determined by time difference between the two SiPMs/crystals, showed a constant term of 91 (118) ps and a noise term of 6.8 (8.9) ps/E/GeV at normal (at $50°$) incidence, granting a timing resolution better than 200 ps at 100 MeV. These results fully satisfied the calorimeter requirements and provided a green light for the production of components. In the last five years, we have used Module-0 also to study the behaviour in a vacuum, at low temperature, and for carrying out vertical slice tests of increasing complexity at a cosmic ray test stand. In Figure 4 (left), a picture of Module-0 inside the vacuum vessel can be seen. In Figure 4 (right), the distribution of the time difference between all pairs of crystals, with energy deposition consistent with an MIP, is shown before and after the T_0 calibration procedure. A clear gaussian peak is observed, consistent with a mean time resolution of 300 ps for a 20 MeV energy deposition. Time improvements depend on the fit procedure of the pulse shape and are presented elsewhere [12].

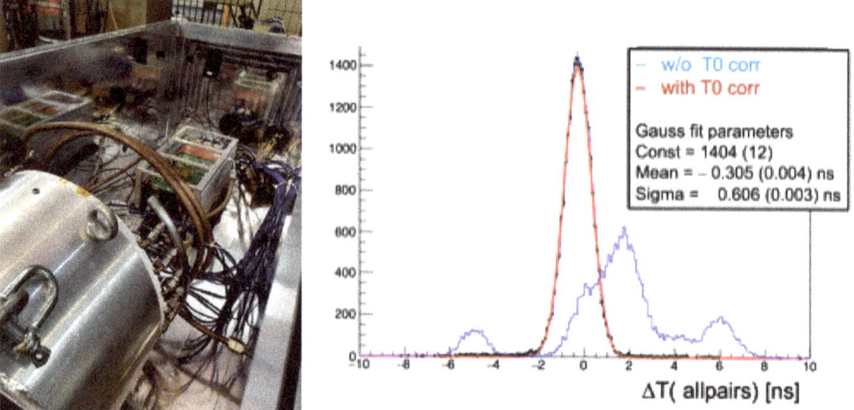

Figure 4. Operations with Module-0: (**left**), picture of Module-0 inside the vacuum vessel readout with MB and DIRAC boards; (**right**), distribution of time differences between all crystal pairs, before (blue) and after (red) T0's calibration.

Table 1. Parametrization of energy and time resolution for Module-0 electron test beam. The numbers in parentheses represent one sigma uncertainties.

Params	$\frac{\sigma_E}{E}$ (%) at 0°	$\frac{\sigma_E}{E}$ (%) at 50°	σ_t (ps) at 0°	σ_t (ps) at 50°
a	0.6	0.6	-	-
b	0.27 (0.03)	0.37 (0.04)	6.8 (0.1)	8.9 (0.2)
c	4.05 (0.27)	5.86 (0.39)	91 (4)	118 (7)

3.2. Production of Crystals, SiPMs and FEE

The production of basic components started in 2018. We successfully completed procurement and quality testing of 4000 SiPMs in 2019, 1500 crystals in 2020, and 3300 FEE boards in 2021. The electronics production was delayed by the pandemic and by the need to make new electronics releases for improving its radiation hardness, as reported elsewhere [12].

The crystals were produced by SICCAS (China) and St.Gobain (France), while the SiPMs were produced by Hamamatsu (Japan). The quality of crystal and SiPM production was excellent, as shown by the reference pictures in Figure 5 (left,center). The optical parameters of the production crystals from both producers were acceptable [16], while St.Gobain's crystals evidenced some difficulties in matching the 100 μm precision of the mechanical realization of the parts. In the end, 8% of the crystals were replaced by relying only on SICCAS for a new production of the final batches. The SiPM performed as expected with a very high quality [17,18] on gain, PDE, and dark current values and a rejection factor smaller than 2%.

The FEE was produced by ARTEL (Italy) with a negligible rejection factor, albeit all boards underwent a burn-in test at 65 °C in a climatic chamber in JINR (Dubna, Ru), followed by a calibration phase for both HV, gain, and differential linearity parameters (see Figure 5 (right)). At the moment of writing, we have reacquired 2500 FEE units from JINR after burn-in and calibration.

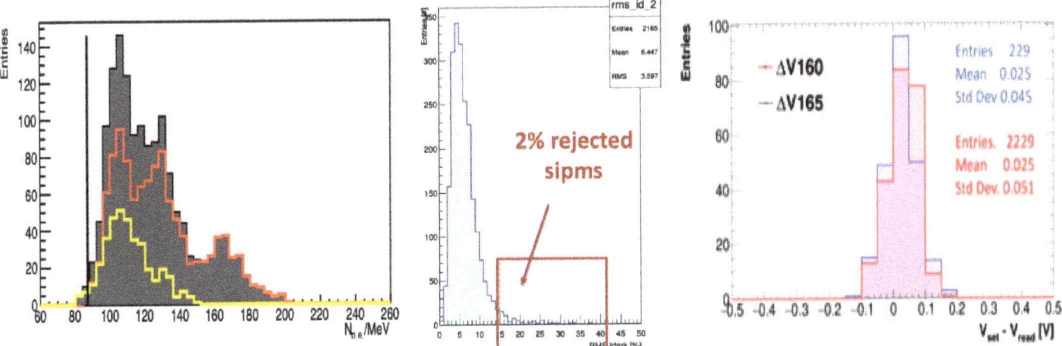

Figure 5. Crystals, SIPMs, and FEE production: (**left**), light yield of crystals (Npe/MeV) as obtained with a large-area PMT fully covering the crystal readout face, the red (yellow) distribution is for the SICCAS (St.Gobain) production crystals; (**center**), RMS of Idark (%) of each SiPM array; (**right**), precision of HV settings for the FEE boards.

3.3. Preparation and Test of the Readout Units

The preparation and test of the readout units (ROUs) is advanced. Two Mu2e SiPMs were glued to each copper holder, as shown in Figure 2 (bottom right), by relying on the EP30AN MasterBond thermal glue with a good performance in vacuum. In total, 3000 production SiPMs were glued to 1500 SiPM holders with a precise glue distribution machine developed at INFN Frascati (LNF). After this operation, we started assembling the calibrated FEE. So far, we have assembled 70% of the entire production lot.

In the last year, we have also dedicated a lot of effort to design, realize, and put into operation a semi-automated quality station to calibrate the ROUs. In a 10 min run, the gain, PDE, and charge response in a $(-4:+2)$ V region around the operational voltage, V_{op}, were evaluated. Currently, we have performed a quality test for 600 ROUs, achieving a gain reproducibility better than 2%, with a channel-by-channel gain spread at a level of 3–4% along production. A much more detailed explanation of the station functionality and its results can be found elsewhere [12,19].

3.4. Production and Assembly of the Calorimeter Mechanical Structure

In the last two years, all large mechanical parts were produced by Italian firms, as shown in Figure 6. The first pieces realized were the two outer rings, produced by Cerasa Mechanics (Assisi). They were milled by a single aluminium block providing the right stiffness and excellent precision on the final step edges. The back plate was built in PEEK by CINEL (Vigonza). Its picture, shown in Figure 6 (top right), was taken during the leak test of the cooling lines with a helium sniffer. The measured leak rate was below 10^{-10} atm×cc/s. A test of temperature uniformity of the same lines was carried out in INFN Pisa laboratories by flowing HFE at 50 °C and controlling with a thermal camera. The parts with composed materials were built by CETMA (Brindisi). On Figure 6 (bottom-left), a picture of the inner ring can be seen, soon after being completed. Once at LNF, a series of stiffness tests were performed loading this structure vertically with more than 100 kg. Deformations observed were below 400 μm as expected. In Figure 6 (bottom-right), the front plate is shown. This plate was completed integrating the source aluminium tubing in grooves on the internal aluminium honeycomb. The source tubing was realized by Caltech and Fermilab.

Figure 6. Calorimeter mechanical parts: (**top left**), outer ring; (**top right**), FEE plate; (**bottom left**), inner ring; (**bottom right**), front plate with source tubing embedded in the aluminium honeycomb.

Before shipping all mechanical parts to Fermilab, a dry fit was carried out in a clean room at LNF to check that all pieces were well fitting. In this occasion, we also assembled, on the external outer ring surface, the ten crates produced by TecnoAlarm (Rome) and their cooling manifolds (Figure 7 (left)). A careful vacuum leak test was performed on both the crate manifolds and the elbow joints between the crates and manifolds. Only two small leaks were detected in two of the lowest elbows and fixed locally by Tig welding. At the end of this operation, a maximum leak rate of 10^{-10} atm $\times$ cc/s was achieved, satisfying the experiment's requirements. Other components, such as the feet, the disk support stands, and other smaller mechanical parts, have been built by Italian firms and completed with the support of the LNF mechanical shops.

In June 2022, we started the final assembly of the downstream disk in a ISO-7-class clean room located at SIDET, Fermilab. The operation sequence started with mounting and aligning the outer ring over its stand, and progressed with alignment of the inner ring. As soon as this was completed and the survey indicated a reproducible and stable detector assembly, we started stacking crystals.

The stacking proceeded from the bottom to top for increasing rows, following an optimized crystal placement obtained by examining the whole crystal production parameters and positioning: (i) the ones with higher (lower) light yield, faster (slower) signals, and reduced (increased) radiation-induced currents in the innermost (outermost) rings where more (less) radiation dose is expected and (ii) the remaining ones in the central regions. While selecting the crystals/rows we also minimized the flatness of the row thickness. Before stacking the crystals on the outer disk, a set of two-day-long outgassing runs were performed in a dedicated vacuum vessel to reduce the single-crystal outgassing level to below 10^{-7} Torr $\times$ l/s. Between each stacked row, an additional 50 µm Tedlar layer was placed. To keep the crystal matrix solidly connected, each row was compressed using screws pushing on plastic shims at both row ends. The accuracy of crystal stacking was checked while operating using a high-precision bubble level and then confirmed offline with a laser tracker. The crystal stacking proceeded steadily, having previously completed the outgassing operation, so that in less than a month (see Figure 7 (right)), the downstream disk had all of the crystal matrix inserted.

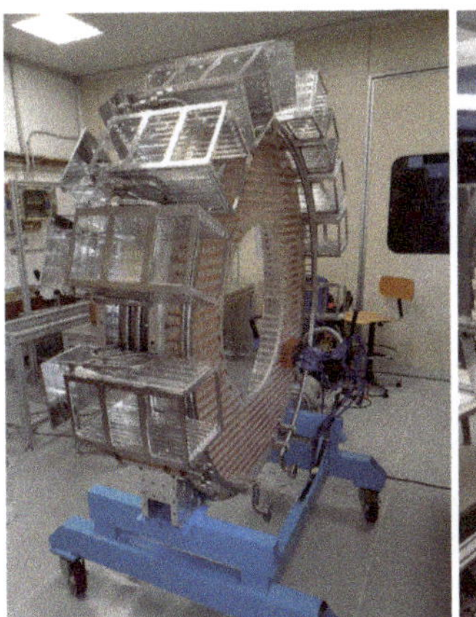

Figure 7. Current status of calorimeter assembly: (**left**), a dry fit of the upstream disk mechanical structure in the LNF clean room; (**right**), the downstream disk with all 674 crystals inserted at Fermilab SIDET clean room.

4. Discussion

The Mu2e calorimeter demonstrated that its requirements are satisfied through the tests carried out with Module-0 (see Section 3.1). However, it could face some difficulties in maintaining the required resolutions, while operating in a vacuum, and in a radiation-harsh environment. We solved these problems by combining a mix of technical choices to a high level of engineering. The crystals, SiPMs, and FEE have proven to be radiation hard up to the maximum level of TID and neutron fluence expected at the end of the experimental lifetime [12]. The most relevant effect we will observe while running will be the rapid increase in SiPM leakage current (Idark) under neutron irradiation, requiring us to cool down the SiPMs to $-10\,°$C to maintain operation. To control the correctness of our choices, we carried out two dedicated measurements:

(1) Measurement of the Idark increase as a function of the neutron fluence. For each SiPM production batch, we randomly chose five units for irradiation purposes. While the TID test indicated no relevant increase in Idark, irradiation with neutrons showed otherwise. Two neutron irradiation facilities, one at EPOS (HZDR, Dresden) and one at Enea-FNG (Frascati), were used. The two measurements indicated, for the same irradiation level, a O(1.5) difference in the increase in Idark. While data are still being analysed, we conservatively used the case with the largest Idark increase, looking at several SiPMs exposed at different fluences at FNG. A first summary result is reported in Figure 8 (left), where the average Idark is shown as a function of temperature at different overvoltages ($\Delta V = V - V_{brk}$) for two different fluences. Please note that the supplied bias is three times larger than the one of a monolithic Hamamatsu SiPMs due to the series configuration. These data confirm that decreasing the temperature by $10\,°$C corresponds to a decrease in Idark by a factor of two, and that this current is linearly proportional to the fluence. The bottom plot guides the running condition in the experiment. We expect to operate the SiPMs at $-10\,°$C to keep the leakage current inside our maximum operation limit of 2 mA/SiPM, with a minor bias adjustment.

(2) Measurement of the time resolution achieved with irradiated sensors. The second measurement tested the effect of neutron irradiated sensors on the resolution. For our benchmark, we used sensors exposed to a fluence of 5×10^{11} n/cm^2 and kept at 0 °C, that are equivalent to the sensors exposed to the fluence expected at experiment lifetime 1.2×10^{12} n/cm^2, if kept at $-10\,°$C. While for the energy, we estimated that 2 mA current corresponds to a noise level of O(1.5 MeV) to be added independently to each fired channel, it was more difficult to evaluate the effect on time resolution, since a fit to the waveform signals was needed. A systematic test was carried out with a 50 picosecond Hamamatsu

laser illuminating crystals readout with one irradiated and one non-irradiated SiPM. The resolution obtained as a function of the reconstructed charge is shown in Figure 8 (right). To provide an indication of the energy scale, 2500 pC corresponded to 1 MiP energy deposition, i.e., 20 MeV. The result of 700 ps timing resolution achieved with one SiPM at a fluence of 5×10^{11} n_{1MeVeq}/cm^2, and the nice trend with energy, demonstrated that even with only one irradiated sensor, we can reach a resolution better than 500 ps for a 100 MeV energy deposition.

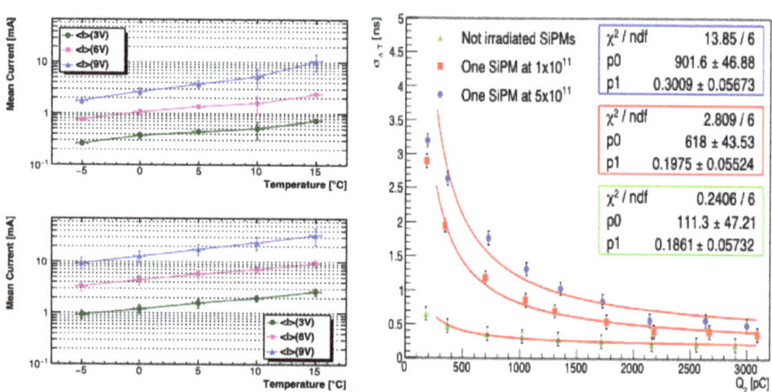

Figure 8. (**Left**) Idark dependence on SiPM temperature for different levels of neutron fluence (top, 10^{11}, bottom 5×10^{11} n_{1MeVeq}/cm^2) and overvoltages. (**Right**), time resolution dependence on charge for SiPM irradiated at different neutron fluences and running at 0 °C.

5. Conclusions

In this paper, we summarized the construction status of the Mu2e calorimeter that will complement the tracking and CRV systems in identifying conversion electrons.

The chosen undoped CsI crystals proved to have fast signals and a large light yield when readout with our custom design UV-extended SiPMs. The production of 1500 crystals and 4000 Mu2e SiPM was successfully completed before the pandemic started, satisfying all required production parameters. In the last two years, we have also completed the development and production of 3300 radiation-hard FEE boards as well as the production of all mechanical components. A lot of effort has been dedicated to the assembly and test of the ROUs, which is more than halfway completed. All the mechanical components have been successfully first assembled in a dry fit at LNF and then at Fermilab for the crystal stacking operation. The crystal stacking of the first disk has been completed at the moment of writing. We are now waiting the shipment from INFN to Fermilab of the front plate in composite material, where the source aluminium tubes have also been integrated. The front plate will conclude the assembly of the mechanical parts.

Meanwhile, the production of the digital electronics is under way. Due to the pandemic, there are a lot of delays to the FPGA delivery time, so we will start cabling the detector from the FEE to the mezzanine boards in the fall of this year and postpone the integration of the digital readout boards (DIRAC) to the spring of 2023. In parallel, we plan to start the construction of the second disk this fall. Finally, we foresee the performance of an integrated test of calorimeter readout in summer 2023, before preparing for installation on the detector rails. In parallel, we will complete the infrastructure for the calibration systems and for the cooling station, needed for operation in Mu2e, looking forward to a successful experiment in the following years.

Author Contributions: Conceptualization, writing, project administration: S.M. (Stefano Miscetti); the rest of the work was really done as a common work with all other authors. All authors have read and agreed to the published version of the manuscript.

Funding: This research was funded and supported bu all institutions listed on the acknowledgments section.

Data Availability Statement: Not applicable.

Acknowledgments: We are grateful for the vital contributions of the Fermilab staff and the technical staff of the participating institutions. This work was supported by the US Department of Energy; the Istituto Nazionale di Fisica Nucleare, Italy; the Science and Technology Facilities Council, UK; the Ministry of Education and Science, Russian Federation; the National Science Foundation, USA; the Thousand Talents Plan, China; the Helmholtz Association, Germany; and the EU Horizon 2020 Research and Innovation Program under Marie Sklodowska-Curie Grant Agreement Nos. 101003460, 101006726, 734303, 822185, and 858199. This document was prepared by members of the Mu2e Collaboration using the resources of the Fermi National Accelerator Laboratory (Fermilab), U.S. Department of Energy, Office of Science, HEP User Facility. Fermilab is managed by Fermi Research Alliance, LLC (FRA), acting under Contract No. DE-AC02-07CH11359.

Conflicts of Interest: The authors declare no conflict of interest.

References

1. Bartoszek, L.; Barnes, E.; Miller, J.P.; Mott, J.; Palladino, A.; Quirk, J.; Roberts, B.L.; Crnkovic, J.; Polychronakos, V.; Tishchenko, V.; et al. The Mu2e Collaboration. Mu2e technical design report. *arXiv* **2015**, arXiv:1501.05241.
2. Bilenky, S.M.; Petcov, S.T.; Pontecorvo, B. Lepton mixing, $\mu \rightarrow e\gamma$ decay and neutrino oscillations. *Phys. Lett. B* **1977**, *67*, 309–312. [CrossRef]
3. Lobashev, V.; Djilibaev, R. On the search for $\mu \rightarrow e$ conversion on Nuclei. *Sov. J. Nucl. Phys.* **1989**, *49*, 384–385.
4. Lee, M.J. Mu2e Collaboration. The straw tube tracker for the Mu2e experiment. *Nucl. Phys. B Proc. Suppl.* **2016**, *273–275*, 2530–2532.
5. Atanov, N.; Baranov, V.; Bloise, C.; Budagov, J.; Cervelli, F.; Ceravolo, S.; Colao, F.; Cordelli, M.; Corradi, G.; Davydov, Y.I.; et al. Design and status of the Mu2e calorimeter. *IEEE Trans. Nucl. Sci.* **2018**, *65*, 2073–2080. [CrossRef]
6. Artikov, A.; Baranov, V.; Blazey, G.C.; Chen, N.; Chokheli, D.; Davydov, Y.; Dukes, E.C.; Dychkant, A.; Ehrlich, R.; Francis, K.; et al. Photoelectron Yield od scintillating counters with embedded wavelength shifting fibers read out with silicon photomultipliers. *Nucl. Instrum. Meth. A* **2018**, *890*, 84–95. [CrossRef]
7. Atanov, N.; Baranov, V.; Budagov, J.; Ceravolo, S.; Cervelli, F.; Colao, F.; Cordelli, M.; Corradi, G.; Dane, E.; Davydov, Y.; et al. The Mu2e calorimeter final design report. *arXiv* **2018**, arXiv:1802.06341.
8. Atanov, N.; Baranov, V.; Colao, F.; Cordelli, M.; Corradi, G.; Dane, E.; Davydov, Y.I.; Flood, K.; Giovannella, S.; Glagolev, V.; et al. Measurement of time resolution of the Mu2e LYSO calorimeter prototype. *Nucl. Instrum. Meth. A* **2016**, *812*, 104–111. [CrossRef]
9. Atanov, N.; Baranov, V.; Colao, F.; Cordelli, M.; Corradi, G.; Dane, E.; Davydov, Y.I.; Flood, K.; Giovannella, S.; Glagolev, V.; et al. Energy and time resolution of a LYSO matrix prototype for the Mu2e experiment. *Nucl. Instrum. Meth. A* **2016**, *824*, 684–685. [CrossRef]
10. Atanov, N.; Baranov, V.; Budagov, J.; Carosi, R.; Cervelli, F.; Colao, F.; Cordelli, M.; Corradi, G.; Dane, E.; Davydov, Y.I.; et al. Design and status of the Mu2e electromagnetic calorimeter. *Nucl. Instrum. Meth. A* **2016**, *824*, 695–698. [CrossRef]
11. Ceravolo, S.; Colao, F.; Diociaiuti, E.; Corradi, G.; Di Falco, S.; Donati, S.; Fiore, S.; Ferrari, A.; Gioiosa, A. Design and qualification of the Mu2e electromagnetic calorimeter electronics system. In Proceedings of the 15th Pisa Meeting on Advanced Detectors, Isola d'Elba, Livorno, Italy, 22–27 May 2022.
12. Atanov, N.; Baranov, V.; Borrel, L.; Bloise, C.; Budagov, J.; Ceravolo, S.; Cervelli, F.; Colao, F.; Cordelli, M.; Corradi, G.; et al. Mu2e crystal calorimetry front-end electronics: Design and characterization. In Proceedings of the Special Issue of Instruments 2022 Selected Papers from the 19th International Conference on Calorimetry in Particle Physics (Calor 2022), Brighton, UK, 16–20 May 2022.
13. Atanov, N.; Baranov, V.; Borrel, L.; Bloise, C.; Budagov, J.; Ceravolo, S.; Cervelli, F.; Colao, F.; Cordelli, M.; Corradi, G.; et al. Development, construction and test of the Mu2e electromagnetic calorimeter mechanical structures. *JINST* **2022**, *17*, C01007. [CrossRef]
14. Aubert, B.; Bazan, A.; Boucham, A.; Boutigny, D.; De Bonis, I.; Favier, J.; Gaillard, J.-M.; Jeremie, A.; Karyotakis, Y.; Le Flour, T.; et al. The BABAR detector. *Nucl. Instrum. Meth. A* **2002**, *479*, 1–116. [CrossRef]
15. Atanov, N.; Baranov, A.; Budagov, A.; Caiulo, D.; Cervelli, F.; Colao, F.; Cordelli, M.; Davydov, Y.I.; Di Falco, S.; Diociaiuti, E.; et al. Design and test of the Mu2e un-doped CsI + SiPM crystal calorimeter. *Nucl. Instrum. Meth. A* **2019**, *936*, 94–97. [CrossRef]
16. Atanov, N.; Baranov, V.; Budagov, J.; Davydov, Y.I.; Glagolev, V.; Tereshchenko, V.; Usubov, Z.; Cervelli, F.; Di Falco, S.; Donati, S.; et al. Quality assurance on un-doped CsI crystals for the Mu2e experiment. In Proceedings of the 2017 IEEE Nuclear Science Symposium and Medical Imaging Conference (NSS/MIC), Atlanta, GA, USA, 21–28 October 2017; pp. 752–757.
17. Cordelli, M.; Cervelli, F.; Diociaiuti, E.; Donati, S.; Donghia, R.; Di Falco, S.; Ferrari, A.; Giovannella, S.; Happacher, F.; Martini, M.; et al. Pre-production and quality assurance of the Mu2e calorimeter Silicon Photomultipliers. *Nucl. Instrum. Meth. A* **2018**, *912*, 347–349. [CrossRef]

18. Atanov, N.; Baranov, V.; Budagov, J.; Caiulo, D.; Cervelli, F.; Colao, F.; Cordelli, M.; Corradi, G.; Davydov, Y.I.; Di Falco, S.; et al. The Mu2e calorimeter: Quality assurance of production crystals and SiPMs. *Nucl. Instrum. Meth. A* **2019**, *936*, 154–155. [CrossRef]
19. Sanzani, E.; Bloise, C.; Ceravolo, S.; Cervelli, F.; Colao, F.; Cordelli, M.; Corradi, G.; Falco, S.D.; Diociaiuti, E. An automated QC station for the calibration of the Mu2e calorimeter Readout Units. In Proceedings of the 1th Pisa Meeting on Advanced Detectors, Isola d'Elba, Livorno, Italy, 22–27 May 2022.

 *instruments*

Article

Crilin: A Semi-Homogeneous Calorimeter for a Future Muon Collider

Sergio Ceravolo [1], Francesco Colao [2], Camilla Curatolo [3], Elisa Di Meco [1,*], Eleonora Diociaiuti [1], Donatella Lucchesi [4], Daniele Paesani [1], Nadia Pastrone [5], Gianantonio Pezzullo [6], Alessandro Saputi [7], Ivano Sarra [1], Lorenzo Sestini [4] and Diego Tagnani [8]

1   Istituto Nazionale di Fisica Nucleare, Laboratori Nazionali di Frascati, Via Enrico Fermi 54, 00054 Frascati, Italy
2   Enea Frascati, Via Enrico Fermi 45, 00044 Frascati, Italy
3   Istituto Nazionale di Fisica Nucleare, Sezione di Milano, Via Celoria 16, 20133 Milano, Italy
4   Istituto Nazionale di Fisica Nucleare, Sezione di Padova, Via Francesco Marzolo 8, 35131 Padova, Italy
5   Istituto Nazionale di Fisica Nucleare, Sezione di Torino, Via Pietro Giuria 1, 10125 Torino, Italy
6   Department of Physics, Yale University, New Heaven, CT 06511, USA
7   Istituto Nazionale di Fisica Nucleare, Sezione di Ferrara, Via Saragat 1I, 44122 Ferrara, Italy
8   Istituto Nazionale di Fisica Nucleare, Sezione di Roma Tre, Via della Vasca Navale 84, 00146 Roma, Italy
*   Correspondence: elisa.dimeco@lnf.infn.it

Abstract: Calorimeters, as other detectors, have to face the increasing performance demands of the new energy frontier experiments. For a future Muon Collider the main challenge is given by the Beam Induced Background that may pose limitations to the physics performance. However, it is possible to reduce the BIB impact by exploiting some of its characteristics by ensuring high granularity, excellent timing, longitudinal segmentation and good energy resolution. The proposed design, the Crilin calorimeter, is an alternative semi-homogeneous ECAL barrel for the Muon Collider based on Lead Fluoride Crystals (PbF_2) with a surface-mount UV-extended Silicon Photomultipliers (SiPMs) readout with an optimized design for a future Muon Collider.

Keywords: calorimeters; PbF_2; SiPM; crystals; high granularity

Citation: Ceravolo, S.; Colao, F.; Curatolo, C.; Di Meco, E.; Diociaiuti, E.; Lucchesi, D.; Paesani, D.; Pastrone, N.; Pezzullo, G.; Saputi, A.; et al. Crilin: A Semi-Homogeneous Calorimeter for a Future Muon Collider. *Instruments* **2022**, *6*, 62. https://doi.org/10.3390/instruments6040062

Academic Editors: Fabrizio Salvatore, Alessandro Cerri, Antonella De Santo and Iacopo Vivarelli

Received: 16 September 2022
Accepted: 8 October 2022
Published: 11 October 2022

Publisher's Note: MDPI stays neutral with regard to jurisdictional claims in published maps and institutional affiliations.

1. Introduction

The idea of developing a Muon Collider facility was born as an alternative to electron and hadron accelerators with the aim to reach the Multi-TeV range and unlock new discoveries. This choice would have unique advantages, since clean events as in electron-positron colliders are possible, and high collision energy as in hadron colliders could be reached due to negligible beam radiation losses.

Unfortunately the expected environment is not truly clean because of the presence of the beam-induced background (BIB), produced by the decay of Muons and subsequent interactions with the machine elements. This might reduce the physics performance, however it is possible to reduce the effect of the BIB resolution degradation by exploiting some characteristics. The BIB arrival hit arrival time is expected to be out-of-time with respect to the bunch crossing. In addition, the longitudinal energy distribution of the BIB particles will be different from particles from the primary interaction which are expected to propagate deeper into the detector. Another one is the longitudinal energy deposit distribution of the BIB particles that is expected to be deposited in the innermost layers of the calorimeter, while particles coming from the primary interaction propagate deeper in the detector. The technology and the design of the calorimeters should be chosen to reduce the effect of the BIB, while keeping good physics performance. This can be achieved with four main requirements:

- High granularity to reduce the overlap of BIB particles in the same calorimeter cell.

- Good timing resolution can be utilised to reduce the out-of-time component of the BIB (Figure 1-left). A time resolution of about $\sigma_t = 80$ ps should be achieved.
- Longitudinal segmentation since the signal energy profile in the longitudinal direction is different from the BIB one, hence a segmentation of the calorimeter can help distinguishing the signal showers from the fake showers produced by the BIB (Figure 1-right).
- Good energy resolution, $10\%/\sqrt{E}$ in the ECAL system is expected to be enough to obtain good physics performance.

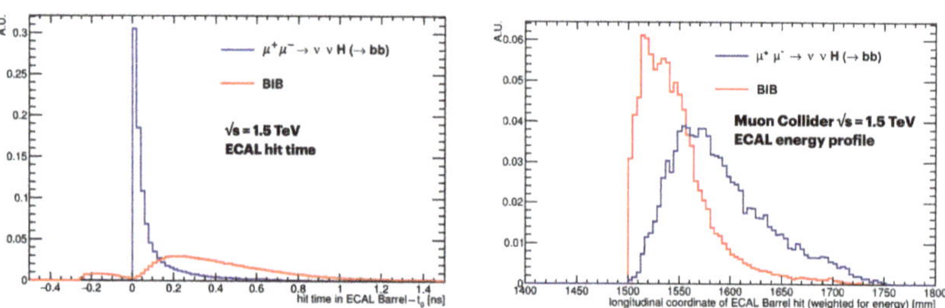

Figure 1. The (**left**) panel shows the time of the ECAL barrel hits with respect to bunch crossing time, for BIB and H$\rightarrow b\bar{b}$ signal. In the (**right**) one the energy distribution of ECAL barrel hits as a function of the distance with respect to the beam axis is presented [1].

The design described in the following pages is an alternative electromagnetic calorimeter for the Muon Collider: Crilin, a crystal calorimeter with longitudinal information. Crilin has a modular architecture made of stackable and interchangeable sub-modules composed of matrices of Lead Fluoride (PbF$_2$) crystals, where each crystal is individually readout by 2 series of 2 UV-extended surface mount Silicon Photomultipliers (SiPMs) each and represents a valid and cheaper alternative to the baseline W-Si Muon Collider ECAL barrel.

2. The Crilin Calorimeter

Crilin is a semi-homogeneous calorimeter made of sub-modules of PbF$_2$ crystals matrices. Each of these crystals is individually readout by two series of two UV-extended surface mounted Silicon Photomultipliers (SiPMs). The essential advantages of this choice consist in an excellent compromise of the following features:

- As the Cherenkov light production in PbF$_2$ is instantaneous with respect to the particle passage this leads to an inherently fast detector response and excellent timing resolution.
- Narrow signals hence an excellent ability to temporally resolve close events at high rate.
- A good light collection that enables a fine energy resolution throughout the whole dynamic range.
- Resistance to radiation.
- Fine granularity that scales with the SiPMs dimensions.

All these advantages are combined with a substantial reduction of the costs compared to other technologies, together with longitudinal segmentation and excellent time resolution. The Crilin ECAL barrel design for the Muon Collider, simulated in Figure 2, consists of five layers with 40 mm thick, and 10×10 mm^2 of cell area PbF$_2$ crystals provided by SICCAS readout with a matrix (2×2) of 4 Hamamatsu SiPMs per crystal.

In order to demonstrate the viability of the Crilin technology, the simulation framework of the International Muon Collider Collaboration has been employed. With this purpose the W-Si ECAL barrel used in this simulation framework has been substituted with the Crilin calorimeter and the performances have been compared. This was carried out on objects of primary interest for Muon Collider physics: hadronic jets.

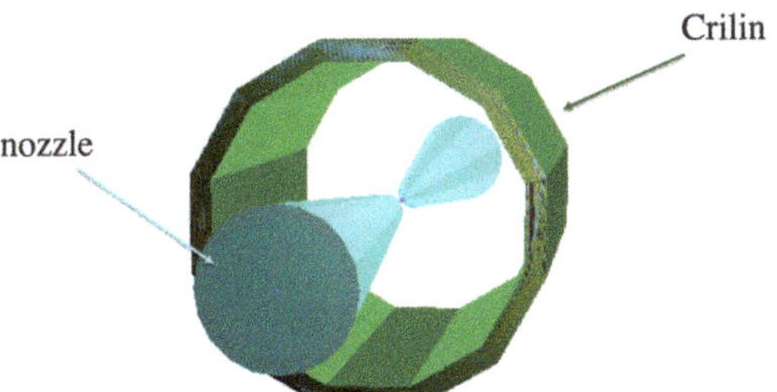

Figure 2. Geant4 realisation of Crilin using a dodecahedra geometry together with nozzles.

A Particle Flow algorithm [2], not yet fully optimized, was employed for jet reconstruction. The full simulation of *b*-jets has been used for this purpose, and the beam-induced background as been simulated as well. The jet reconstruction efficiency and jet p_T resolutions are presented in Figure 3. It can be noticed that the performance is similar in the two cases but, at the same time, the money cost of Crilin is a factor 10 less.

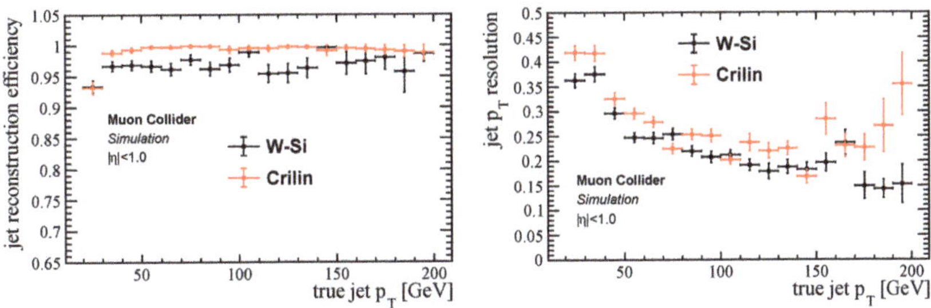

Figure 3. (**Left**): jet reconstruction efficiency as a function of the jet p_T, obtained by using the Crilin ECAL barrel and the W-Si ECAL barrel. (**Right**): jet p_T resolution as a function of the jet p_T, obtained by using the Crilin ECAL barrel and the W-Si ECAL barrel.

Radiation Hardness and Crystals Characterisation

As anticipated the beam induced background represents a real challenge for the Muon Collider detectors not only from the point of view of resolution performances but also in terms of radiation hardness. This is why the BIB at $\sqrt{s} = 1.5$ TeV has been simulated by means of FLUKA on a simplified collider geometry. The dose maps are available in Figure 4 where both of them are normalised to one year of operation (200 days/year) for a 2.5 km circumference ring with 5 Hz injection frequency. The 1-MeV-neq fluence is expected to be $\sim 10^{14}$ cm^{-2}y^{-1} in the electromagnetic calorimeter, with a steeply decreasing radial dependence beyond it. The total ionizing dose is instead $\sim 10^{-4}$ Grad/y on the electromagnetic calorimeter.

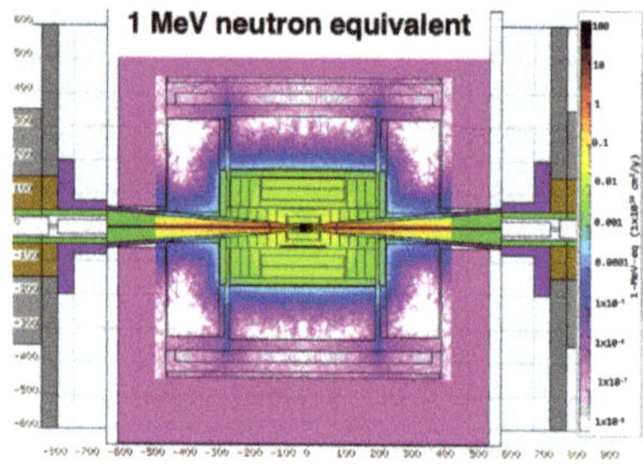

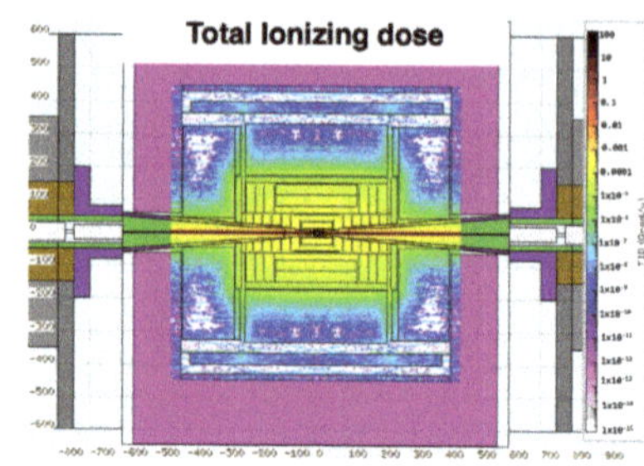

Figure 4. (**Left**): Map of the 1-MeV-neq fluence in the detector region for a Muon Collider operating at $\sqrt{s} = 1.5$ TeV with the parameters described in [3], shown as a function of the position along the beam axis and the radius. (**Right**): Map of the TID in the detector region shown as a function of the position along the beam axis and the radius.

Once the BIB effects were predicted it was necessary to evaluate the consequent deterioration of the single components of the Crilin ECAL, starting with a crystal characterization campaign. The aim was to compare the transmittance spectrum of two PbF_2 crystals. They were sized $5 \times 5 \times 40$ mm^3 and manufactured by SICCAS using a melt growth process, thus resulting in a cubic form (β-PbF_2). Measurements were performed with and without a Mylar wrapping, before and after irradiation with both photons and neutrons. The photon irradiation phase was carried out at ENEA-Calliope, a pool-type gamma irradiation facility equipped with a ^{60}Co radio-isotopic source array producing photons with E_γ = 1.25 MeV. The two crystals were exposed to different irradiation steps during three days and in Figure 5 it is possible to observe the longitudinal transmittance spectra obtained for the two crystals. After a TID of approximately 80 krad, close to the Muon Collider expected one, no significant decrease in transmittance was observed. It is worth noticing that after this dose a saturation effect associated with the damage mechanism is shown, as already observed in [4], and that the maximum degradation is at the level of 40% [5].

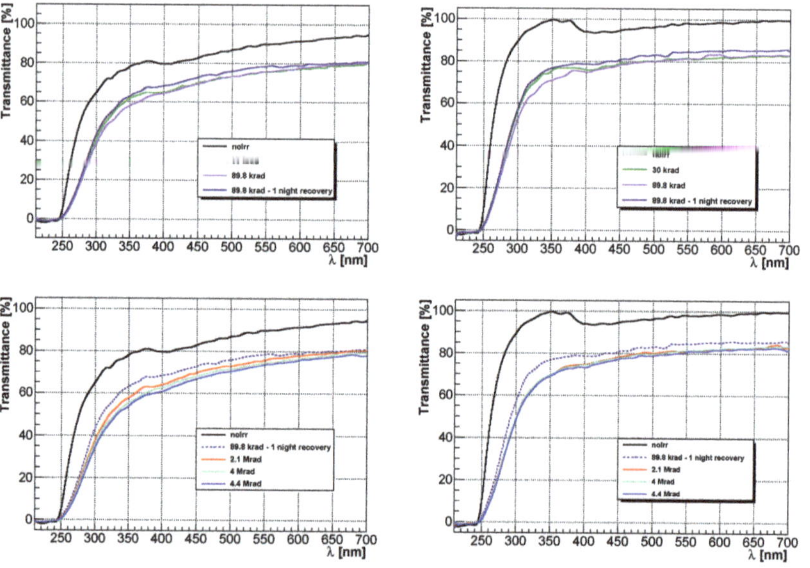

Figure 5. Transmission spectra obtained in the different irradiation steps for the naked crystal (**top** and **bottom left**) and the crystal with Mylar wrapping (**top** and **bottom right**).

The second characterization has been carried out with neutron irradiation at the the ENEA Frascati Neutron Generator facility, based on the T(d,n)α fusion reaction. The source provided 14 MeV neutrons with a total fluence of 10^{13} n/cm^2 for a total amount of time of one hour and 30 min. However, because of some technical time related to logistics and shipment of the crystals, transmittance measurements could only be performed 14 days after the irradiation and showed no alteration in the transmittance spectrum as reported in Figure 6. This result indeed highlights the natural annealing of these crystals.

A neutron irradiation campaign has been very recently carried out also for the SiPMs in order to test the radiation hardness of two different pixel dimension Hamamatsu SMD sensors choices (10 μ and 15 μ) however the analysis is still ongoing.

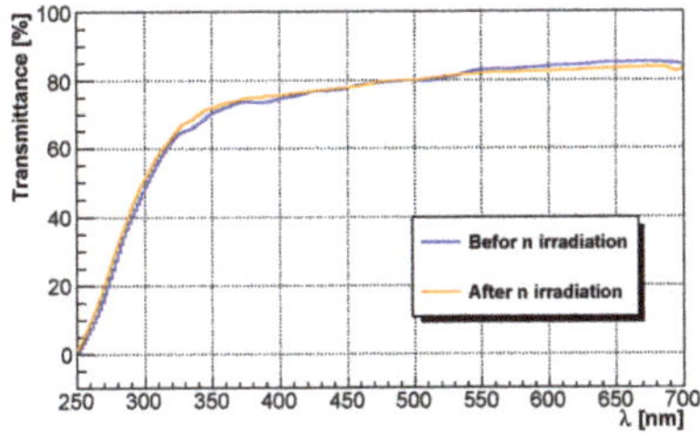
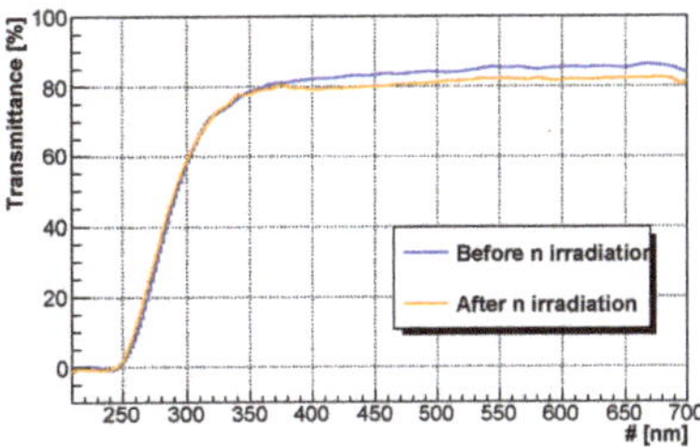

Figure 6. Transmission spectra obtained after the irradiation at FNG with 14 MeV neutrons for a total fluence of 10^{13}n/cm^2 (orange line) compared with the results after the 16 hours optical bleaching (blue line) for the naked crystal (**left**) and for the crystal with Mylar wrapping (**right**).

3. The Crilin Prototype

The full prototype design of the Crilin ECAL for the Muon Collider will consist of four layers of 5 × 5 PbF$_2$ crystals each readout by thin SMD SiPMs by Hamamatsu. And will operate at a temperature of 0/−10 °C. A one layer 2 crystals preliminary prototype (Proto-0) has already been built in 2021 and tested at Beam Test Facility of Laboratori Nazionali of Frascati with 500 MeV electrons in July 2021 and at H2 test facility of CERN with 120 GeV electrons in August 2021. The ongoing analysis is already showing promising results both in terms of time resolution (less than 100 ps for deposit energies greater than 1 GeV) and energy resolution (1 p.e./MeV of light yield) however further improvements of these results can be achieved. Is now under construction a larger and improved prototype, Proto-1, made of two sub-modules, each composed of a 3 × 3 crystals matrix and with a new choice of SMD sensors, the Hamamatsu S14160-3015PS [6] ones. A rendering of Proto-1 is represented in Figure 7 together with a close up look at the single module (on the right). The SiPMs have already been tested with the new front end electronics and results will be shown in the following pages.

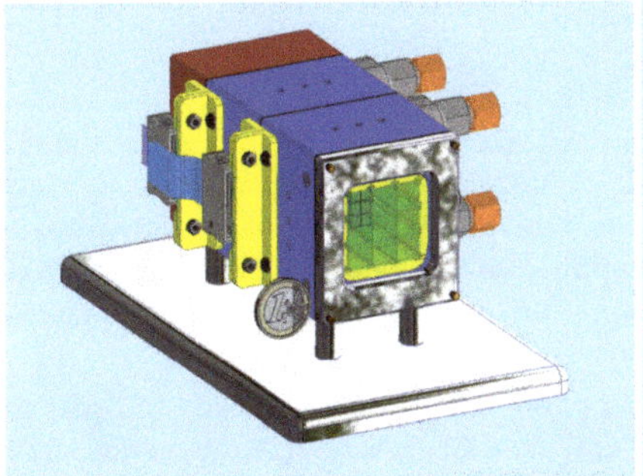
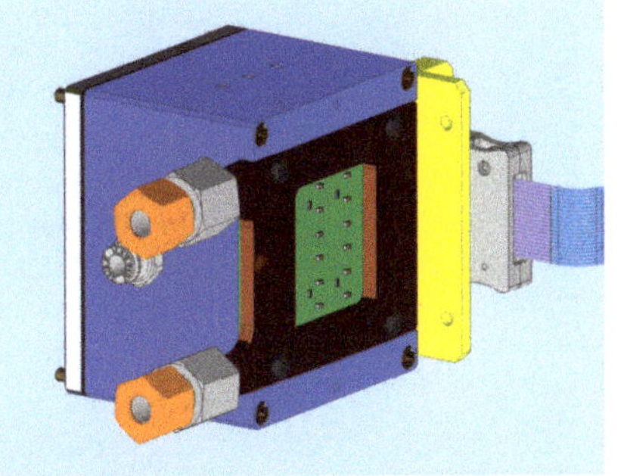

Figure 7. CAD 3D model of Crilin Prototype (Proto-1) on the (**left**). On the (**right**) a single module detail showing the cold plate heat exchanger mounted over the electronic board.

Thanks to the tests and analysis performed on Proto-0 it was possible to set some requirements in order to improve the electronic system. The new front end electronics is now made of two distinct parts: the SiPM boards and the Mezzanine boards. Each SiPM board, in Figure 8-left, is made of 36 photo-sensors so that each crystal in the matrix has two separate and independent readout channels, consisting in a series of two 15 μm pixel-size SMD S14160-3015PS SiPMs. These were chosen for their high-speed response, narrow signals and radiation hardness. Four 0603 SMD blue LEDs were also placed in between the SiPMs matrices in order to perform in-situ calibration, diagnostic and monitoring. The SiPMs are then connected via 50 Ω micro-coaxial transmission lines to a microprocessor-controlled Mezzanine Board (right panel in Figure 8).

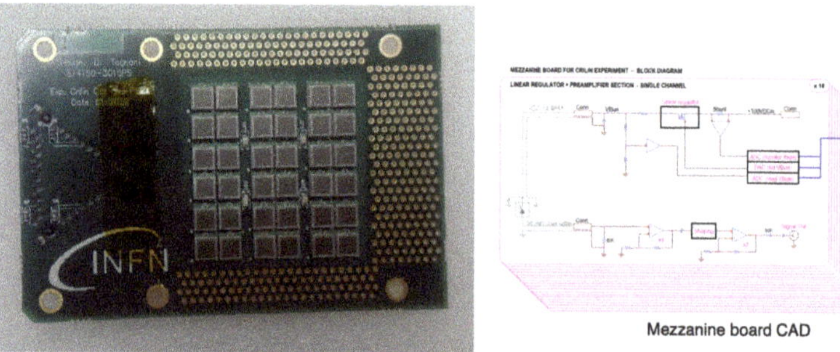

Figure 8. (**Left**): SiPMs board detail, showing the sensor matrices and the LEDs. (**Right**): Mezzanine Boards CAD rendering.

The Mezzanine Board provides signal amplification and shaping, along with all slow control functions for all the 18 readout channels. Signals are shaped and amplified by means of two non inverting amplification stages and a pole-zero cancellation network with a dynamic range of 2 V and an overall gain of 8. The SiPMs biasing is controlled by 12-bit DACs while regulated voltages, bias currents, and the temperature of the SiPM matrix are sensed via dedicated 12-bit ADC channels. The slow control routines are than handled by an onboard Cortex M4 microprocessor.

The prototype consists of two sub-modules, each composed of a 3 × 3 crystals matrix and arranged in a series and assembled by bolting, leading to a compact and small calorimeter (Figure 7-left). The operational temperature for this prototype is going to be 0 °C. Since the total heat load has been estimated to be 350 mW per crystals (per two channels) a cooling system was necessary. This system consists of a cooling plant and a cold plate heat exchanger in direct contact with the electronic board. A cold plate heat exchanger, made of copper, is mounted over the electronic board and a glycol based water solution, supplied by the cooling plant, passes through the deep drilled channels to absorb the SiPMs generated heat (Figure 7-right).

FEE and SiPMs Tests

A first prototype of the front-end electronics was tested by exposing two 15 μm SiPMs to a picosecond UV laser source while signals were digitised using a 40 GS/s oscilloscope. The aim was to evaluate the time resolution performances by reconstructing timing using a log-normal fit applied to SiPM pulse rising edge and a constant fraction technique. A digitised waveform is presented in Figure 9-left. The time resolution was evaluated in three different situations:

1. Constant laser pulse amplitude (adjusted in order to have signals with 1 V peak amplitude) and fixed 40 Gsps sample rate, while laser repetition rate was increased from 50 kHz up to 5 MHz.
2. Fixed laser amplitude (as before) and fixed 100 kHz laser repetition rate, while the oscilloscope sample rate was swept in the range 2.5 to 40 Gsps.

3. Fixed laser repetition rate and sampling frequency, while the waveform peak amplitude was swept over the FEE dynamic range.

The first measurements the waveform peak amplitude and the sampling rate were set to 1 V and 40 Gsps. Figure 9-right actually shows that the waveform profiles remained unchanged throughout the test and no effect on the timing resolution was found for laser repetition rates in the range 50 kHz–5 MHz.

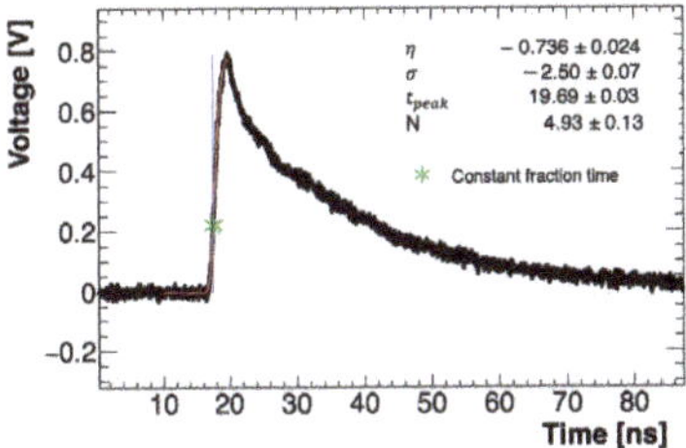
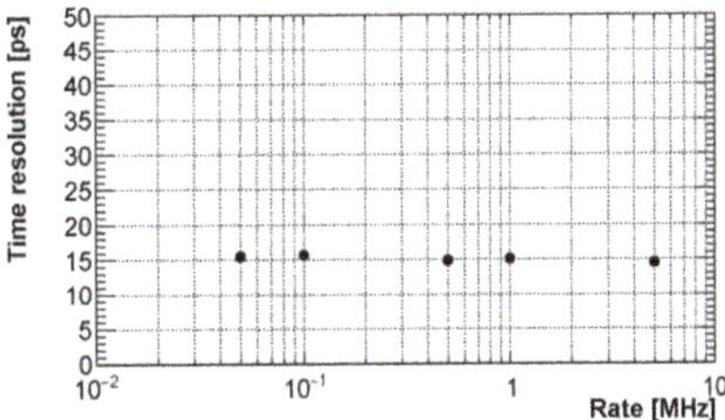

Figure 9. (**Left**): SiPM waveform in response to a laser pulse, sampled at 40 GS/s. Log-normal fit on the rising edge is overlaid. (**Right**): Effect of laser repetition rate on the time resolution.

The second dataset was carried out by changing the oscilloscope sample rate in the range 2.5 Gsps–40 Gsps while the waveform peak amplitude and the laser repetition rate were fixed at 1 V and 100 kHz. The effect of the sampling rate is summarised in Figure 10-left, which shows the strong dependence of the time resolution form the digitizer sample rate since it scales from the worst-case $\sigma_t \sim 32$ ps obtained at 2.5 GS/s to $\sigma_t \sim 15$ ps at 40 GS/s.

The charge dependence of the time resolution was evaluated with a dedicated set of measurements carried out at 40 GS/s, using fixed laser repetition rate of 100 kHz, and six different laser amplitude settings. Pulse charges were evaluated by integrating each waveform taking into account the 50 Ω oscilloscope input impedance. For each of the six runs the charge distribution was Gaussian therefore the reference value was evaluated as the mean value of a normal fit. Taking into account the SiPMs and FEE gains it was possible to convert the charge value to the corresponding number of photo-electrons $N_{p.e.}$ thus resulting in a conversion factor of $\sim$2.48 photo-electrons per pC. The time resolution dependence on charge and $N_{p.e.}$ was then evaluated in (Figure 10-right).

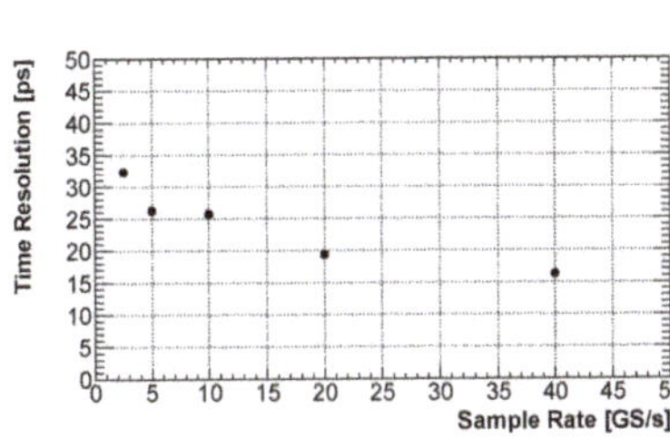
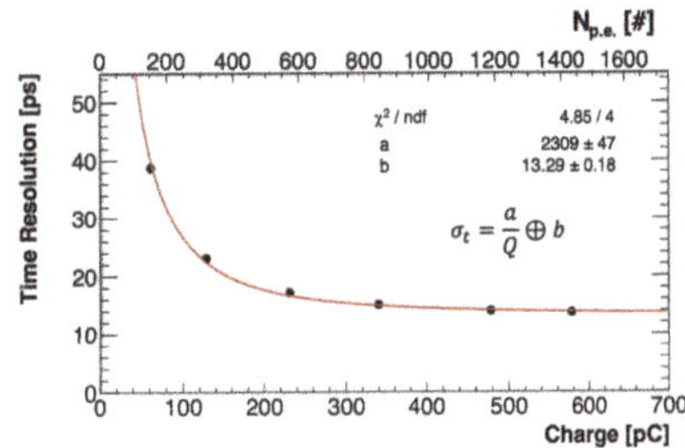

Figure 10. Effect of sampling frequency on time resolution (**left**). Time resolution as a function of charge and $N_{p.e.}$ (**right**).

This last scan points up the high performances that the calorimeter electronics can reach by showing a time resolution that is already less than 40 ps even at low charges (50 pC–124 photo-electrons) and an impressive constant term b of $\sim$13 ps.

4. Conclusions

High intensity experiments need high performance detectors capable of supporting innovative reconstruction techniques, enabling superior signal extraction from harsh and high-rate backgrounds. The described solution represents a valid and cheaper alternative to the baseline W-Si ECAL barrel of the Muon Collider ensuring, through a semi-homogeneous calorimeter, longitudinal segmentation and a great timing resolution while enabling superior event reconstruction strategies, thanks to the use of 5D discrimination.

Single components were already characterised by estimating the radiation damages on crystals and SiPMs though irradiation studies but also with a preliminary two crystal prototype test beam carried out at BTF with 500 MeV in July 2021 and at CERN in August 2021.

Crilin Proto-1, made of 2 layers of 3×3 PbF_2 crystals will be shortly assembled. The aim is to test its performances in a new test beam at BFT with 500 MeV electrons and at H2 test facility at CERN with a high energy beam (>100 GeV) before the end of 2022 and to set new requirements for future prototypes.

Author Contributions: Writing—original draft, S.C., F.C., C.C., E.D.M., E.D., D.L., D.P., N.P., G.P., A.S., I.S., L.S. and D.T. All authors have read and agreed to the published version of the manuscript.

Funding: This research received no external funding.

Data Availability Statement: The data presented in this study are available on request from the corresponding author.

Acknowledgments: This work was developed within the framework of the International Muon Collider Collaboration (https://muoncollider.web.cern.ch, accessed on 7 October 2022), where the Physics and Detector Group aims to evaluate potential detector R&D to optimize experiment design in the multi-TeV energy regime. This project has received support (funding) from the European Union's Horizon 2020 Research and Innovation program under Grant Agreement No 101004761. The authors wish to thank the LNF Division Research and SPCM departments for their technical and logistic support and are grateful to the people in Enea Casaccia and Enea FNG facilities for their extensive collaboration during the irradiation campaign of crystals and photo-sensors.

Conflicts of Interest: The authors declare no conflict of interest.

References

1. Sestini, L.; Sarra, I.; Andreetto, P.; Gianelle, A.; Lucchesi, D.; Buonincontri, L.; Zuliani, D.; Casarsa, M.; Bartosik, N.; Pastrone, N.; et al. Design a calorimeter system for the Muon Collider experiment. *PoS* **2022**, *EPS-HEP2021*, 776. [CrossRef]
2. Bartosik, N.; Krizka, K.; Griso, S.P.; Aimè, C.; Apyan, A.; Mahmoud, M.A.; Bertolin, A.; Braghieri, A.; Buonincontri, L.; Calzaferri, S.; et al. Simulated Detector Performance at the Muon Collider. *arXiv* **2022**, arXiv:2203.07964. [CrossRef]
3. Collamati, F.; Curatolo, C.; Lucchesi, D.; Mereghetti, A.; Mokhov, N.; Palmer, M.; Sala, P. Advanced assessment of beam-induced background at a muon collider. *J. Instrum.* **2021**, *16*, P11009. [CrossRef]
4. Zhu, R.; Ma, D.; Newman, H.; Woody, C.; Kierstead, J.; Stoll, S.; Levy, P. A study on the properties of lead tungstate crystals. *Nucl. Instrum. Methods Phys. Res. Sect. A Accel. Spectrometers Detect. Assoc. Equip.* **1996**, *376*, 319–334. [CrossRef]
5. Cemmi, A.; Colangeli, A.; D'orsi, B.; Sarcina, I.D.; Diociaiuti, E.; Fiore, S.; Paesani, D.; Pillon, M.; Saputi, A.; Sarra, I.; et al. Radiation study of Lead Fluoride crystals. *J. Instrum.* **2022**, *17*, T05015. [CrossRef]
6. Hamamatsu SiPMs Datasheet. Available online: https://www.hamamatsu.com/content/dam/hamamatsu-photonics/sites/documents/99_SALES_LIBRARY/ssd/s14160-1310ps_etc_kapd1070e.pdf (accessed on 7 October 2022).

instruments

Article

Mechanical Design of an Electromagnetic Calorimeter Prototype for a Future Muon Collider

Daniele Paesani [1], Alessandro Saputi [2,*] and Ivano Sarra [1]

[1]   Laboratori Nazionali di Frascati, National Institute of Nuclear Physics, 00044 Frascati, Italy
[2]   Sezione di Ferrara, National Institute of Nuclear Physics, 00044 Ferrara, Italy
*   Correspondence: alessandro.saputi@fe.infn.it

Abstract: Measurement of physics processes at new energy frontier experiments requires excellent spatial, time, and energy resolutions to resolve the structure of collimated high-energy jets. In a future Muon Collider, beam-induced backgrounds (BIB) represent the main challenge in the design of the detectors and of the event reconstruction algorithms. The technology and the design of the calorimeters should be chosen to reduce the effect of the BIB, while keeping good physics performance. Several requirements can be inferred: (i) high granularity to reduce the overlap of BIB particles in the same calorimeter cell; (ii) excellent timing (of the order of 100 ps) to reduce the out-of-time component of the BIB; (iii) longitudinal segmentation to distinguish the signal showers from the fake showers produced by the BIB. Moreover, the calorimeter should operate in a very harsh radiation environment, withstanding yearly a neutron flux of 10^{14} n1MeV/cm^2 and a dose of 100 krad. Our proposal consists of a semi-homogeneous electromagnetic calorimeter based on Lead Fluoride Crystals (PbF2) readout by surface-mount UV-extended Silicon Photomultipliers (SiPMs): the Crilin calorimeter. In this paper, we report the mechanical design for the development of a small-scale prototype, consisting of 2 layers of 3 × 3 crystals.

Keywords: mechanics; crystals; SiPM; calorimetry

Citation: Paesani, D.; Saputi, A.; Sarra, I. Mechanical Design of an Electromagnetic Calorimeter Prototype for a Future Muon Collider. *Instruments* **2022**, 6, 63. https://doi.org/10.3390/instruments6040063

Academic Editors: Fabrizio Salvatore, Alessandro Cerri, Antonella De Santo and Iacopo Vivarelli

Received: 15 September 2022
Accepted: 8 October 2022
Published: 14 October 2022

Publisher's Note: MDPI stays neutral with regard to jurisdictional claims in published maps and institutional affiliations.

1. Introduction

The Muon Collider environment [1] is not so clean as one might expect, since the presence of the beam-induced background (BIB), produced by the decay of muons and subsequent interactions, may pose limitations on the physics performance [2]. Although the BIB can be partially mitigated by a proper design of the machine-detector interface, for instance using two shielding tungsten nozzles in the detector region [3], it poses requirements on the detector development [4].

BIIB particles at a Muon Collider have a number of characteristic features: low momentum, displaced origin and asynchronous time of arrival. The BIB flux has been simulated to be in the order of 300 fl/cm^2 on the surface of the electromagnetic calorimeter (ECAL), with energy spectrum peaked around 1.8 MeV. One of the most promising options for ECAL, proposed by the CALICE collaboration, is a sandwich of tungsten and silicon sensors [5] that combines a mature technology with the possibility to implement fine segmentation. However, if this technology well sited the Muon Collider environment, its implementation in a barrel calorimeter needs of about 64 million silicon sensors. Moreover, future developments should implement a precise timing measurement in these sensors (<100 ps) in order to make them usable at a Muon Collider.

In this paper, we propose a cheaper alternative as electromagnetic barrel calorimeter for the Muon Collider: Crilin [], a semi-homogeneous crystal calorimeter with longitudinal information. It is based on Lead Fluoride (PbF2) crystals readout by surface mounted UV extended Silicon Photomultipliers (SiPMs). Crilin has a modular architecture made of stackable and interchangeable sub-modules composed of matrices of 10 × 10 × 40 mm^3 PbF2 crystals, where each crystal is individually readout by 2 series of 2 UV-extended surface mount SiPMs each. It can provide: high response speed, good pileup capability, great light collection hence good energy resolution throughout the whole dynamic range, resistance to radiation, and fine granularity which is also scalable with SiPMs pixel dimensions.

2. The Crilin Prototype

First studies on ECAL crystal dimensions have shown that a basic configuration with $10 \times 10 \times 40$ mm^3 allows a good separation of BIB from signal with O(5 GeV) energy deposit per crystal. In this regard, the choice of SiPMs with 15 μm pixels, with the measured Light Yield (LY) of 1 p.e./MeV at an earlier existing 2-crystals prototype (Proto-0 [6]), would guarantee an excellent linearity in the response.

In order to validate the design choices, the proposal is to build a larger prototype, called Proto-1. The design will be optimized with the simulation studies starting from dimensions of 0.7 R_M and 8.5 X_0 ($\sim$0.3 λ). This size comes from a compromise of an acceptable containment of 100 GeV electrons and cost constraints. Results will be extrapolated to the optimum length of the Muon Collider calorimeter of the order of 20 X_0. The proposal is to build Proto-1 with two layers of 3×3 PbF2 crystals, each readout with UV-extended SiPMs (Hamamatsu S14160-3015PS SMD sensors [7]), as already done in Proto-0. These new SiPMs were already tested with an ultra-fast blue laser (400 nm, 100 ps) and new electronics front-end (FEE) that showed a dynamic range from 0 to 2 V, a rise time of $\sim$2 ns with full signal in $\sim$70 ns and a σ_t less than 50 ps even at charge as low as 100 pC ($\sim$250 Np.e.) [8].

Proto-1 operational temperature will be 0/$-$10 °C and the performance will be validated in a dedicated test beam.

Specifically, our goals are: (1) perform a complete operational test of the prototype, including operation with cooling; (2) obtain data for a complete analysis of digitized signals from the detector for electrons and minimum-ionising particles; (3) test the cluster reconstruction capability and measure the time resolution; (4) measure longitudinal and transverse shower profile and compare with results obtained in simulation.

3. Mechanics

In the current design, the prototype consists of two sub-modules, each composed of a 3×3 crystal matrix. The modules are arranged in a series and assembled by bolting, thus obtaining a compact and small calorimeter, as shown in Figure 1.

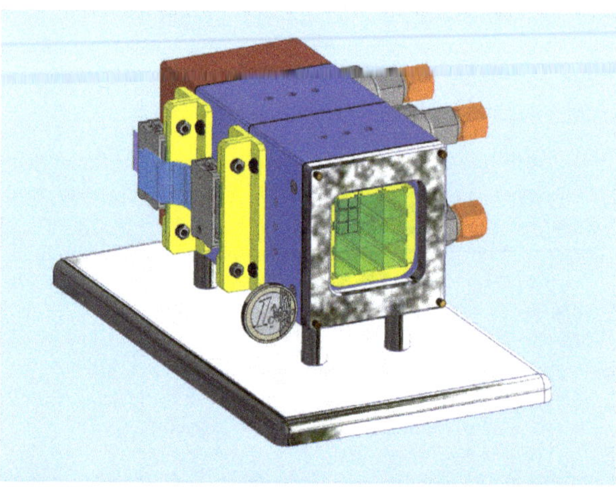

Figure 1. CAD 3D model of Crilin Prototype (Proto-1).

The mechanics have been realized from the mechanical workshop of the Laboratori Nazionali di Frascati of INFN and assembled with fake aluminium crystals, as shown in Figure 2.

Figure 2. Mechanics of the Proto-1 asembled with fake aluminium crystals.

Each crystal matrix is housed in a light-tight case which also embeds the front-end electronic boards and the heat exchange needed to cool down the SiPMs. The mechanical architecture of the prototype comprises the following key elements:

- The cases, which house each crystal matrix and embed the front-end electronic boards. They are manufactured in common acrylonitrile butadiene styrene (ABS) plastic to minimize the thermal exchange with both the external environment and between the modules.
- The locking plates, to which the positioning and blocking of crystals are entrusted, also manufactured in ABS. This solution eases the assembling, positioning, and locking of the crystal matrix.
- The hydraulic connectors, which transport dry gas into the individually sealed modules; the dry gas is circulated inside the active volume of the prototype to prevent condensation.
- In between the modules are installed seals, which make each sub-module light-tight. The modules are bolted together using special screws that allow assembling the modules in series. Tedlar® windows close the calorimeter at either end.

The on-detector electronics and SiPMs must be cooled during operation, so as to improve and stabilize the performance of SiPMs against irradiation. Our design is capable of removing the heat load due to the increased photosensor leakage current after exposure to the expected 10^{14} n1MeV $/cm^2$ fluence, along with the power dissipated by the amplification circuitry. The total heat load was estimated as 350 mW per channel. The Crilin cooling system consists of a cooling plant and a cold plate heat exchanger (see Figure 3), in direct contact with the electronic board. It will provide the optimum operating temperature for the electronics and SiPMs at $0/-10\ °C$.

Figure 3. For comparison the cooling exchangers have been realized at CERN in 3D metal printer technology (**left**) and at the mechanical workshop of Sezione of Ferrara of INFN with the Computer Numerical Control (CNC) milling machine (**right**).

To improve the thermal performance of the cold plate, a complex cross-section pin fin arrays for forced convection heat exchange has been chosen to provide high thermal performance in a compact size. Pin fin arrays are formed on both top and bottom side of the cold plate; the cold plate is made by brazing top and bottom plates. The coolant inlet pipe and the outlet pipe are also connected to the cold plate by brazing. The use of pin fins increases heat transfer through two main mechanisms: increasing the wetted surface area over which convective heat transfer can occur, and promoting turbulent flow in the inter-fin region of the array. This turbulence is generated mainly by interrupting boundary layers on the channel walls, and inducing vorticity at the base and in the wake of the fins. Four geometrically different cross sections micro pin-fins are under investigation using experimental and computation means: a conventional circular shape, a hydrofoil shape, a modified hydrofoil shape and a symmetric convex lens shape.

The cooling plant supplies the cold plate with a glycol-based water solution at the required flow, temperature, and pressure. Hydraulic connectors, transport dry gas into the individually sealed modules. The dry gas is fluxed inside the active volume of the prototype to prevent condensation, see Figure 4.

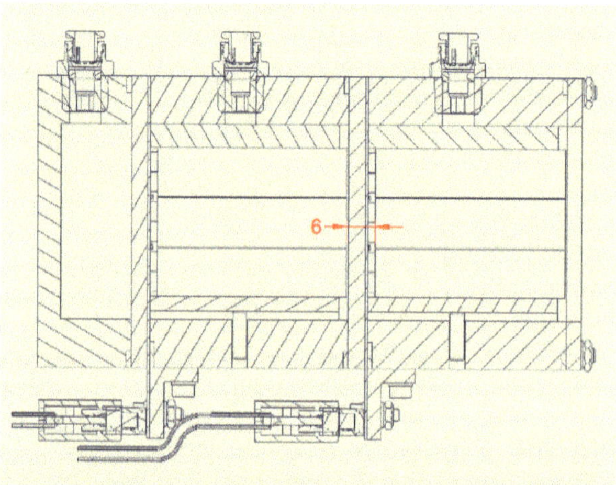

Figure 4. Top view of the calorimeter.

4. Conclusions

Crilin is a semi-homogeneous calorimeter with longitudinal segmentation and superior timing resolution (less then 100 ps for each individual readout channel), capable to work in a very hard radiation environment. A Crilin prototype, composed of two layers of nine crystals each and operating at $-10/0\ ^\circ$C, will be built during 2022. Our goal is to test its performance with 500 MeV electrons at BTF and with a high energy beam ($>$100 GeV) at CERN at end of 2022.

Author Contributions: Methodology, D.P.; Writing—original draft, A.S. and I.S. All authors have read and agreed to the published version of the manuscript.

Funding: This project has received funding from the European Union's Horizon 2020 Research and Innovation program under GA no 101004761.

Data Availability Statement: The data presented in this study are available on request from the corresponding author.

Acknowledgments: This work was developed within the framework of the International Muon Collider Collaboration (https://muoncollider.web.cern.ch, accessed on 14 September 2022), where the Physics and Detector Group aims to evaluate potential detector R&D to optimize experiment design in the multi-TeV energy regime.

Conflicts of Interest: The authors declare no conflict of interest.

References

1. Muon Collider Web Page. Available online: https://muoncollider.web.cern.ch/ (accessed on 14 September 2022).
2. Muon Accelerator Program. Available online: https://map.fnal.gov/ (accessed on 14 September 2022).
3. Mokhov, N.V.; Striganov, S.I. Detector Backgrounds at Muon Colliders. *Phys. Procedia* **2012**, *37*, 2015–2022. [CrossRef]
4. Mokhov, N.V.; James, C.C. Technical Publications: Fermilab-FN-1058-APC. 2018. Available online: https://lss.fnal.gov/lists/fermilab-reports-fn.html (accessed on 14 September 2022).
5. Linssen, L.; Miyamoto, A.; Stanitzki, M.; Weerts, H. Physics and Detectors at CLIC: CLIC Conceptual Design Report. *arXiv* **2012**, arXiv:1202.5940.
6. Moulson, M. Talk Given at AIDAinnova 1st Annual Meeting, 28–31 March 2002. Available online: https://indico.cern.ch/event/1104064/contributions/4801240/attachments/2417197/4136687/Moulson_220329WP8.pdf (accessed on 14 September 2022).
7. Hamamatsu SiPMs Datasheet. Available online: https://www.hamamatsu.com/content/dam/hamamatsu-photonics/sites/documents/99_SALES_LIBRARY/ssd/s14160-1310ps_etc_kapd1070e.pdf (accessed on 14 September 2022).
8. Ceravolo, S.; Colao, F.; Curatolo, C.; Di Meco, E.; Diociaiuti, E.; Lucchesi, D.; Paesani, D.; Pastrone, N.; Saputi, A.; Sarra, I.; et al. Crilin: A CRystal calorImeter with Longitudinal InformatioN for a future Muon Collider. *J. Instrum.* **2022**, *17*, P09033. [CrossRef]

instruments

MDPI

Article

The CMS Level-1 Calorimeter Trigger for the HL-LHC

Piyush Kumar and Bhawna Gomber * on behalf of the CMS Collaboration

The Centre for Advanced Studies in Electronics Science and Technology (CASEST), School of Physics, University of Hyderabad, Gachibowli, Hyderabad 500046, Telangana, India
* Correspondence: bhawna.gomber@cern.ch; Tel.: +91-9873-416-985

Abstract: The High-Luminosity LHC (HL-LHC) provides an opportunity for a pioneering physics program to harness an integrated luminosity of 4000 fb^{-1} of ten years of operations. This large volume of collision data will help in high precision measurements of the Standard Model (SM) and the search for new and rare physics phenomena. The harsh environment of 200 proton–proton interactions poses a substantial challenge in the collection of these large datasets. The HL-LHC CMS Level-1 (L1) trigger, including the calorimeter trigger, will receive a massive upgrade to tackle the challenge of a high-bandwidth and high pileup environment. The L1 trigger is planned to handle a very high bandwidth (∼63 Tb/s) with an output rate of 750 kHz, and the desired latency budget is 12.5 µs. The calorimeter trigger aims to process the high-granular information from the new end-cap detector called the high-granularity calorimeter (HGCAL) and the barrel calorimeter. The HL-LHC trigger prototyped boards are equipped with large modern-day FPGAs and high-speed optical links (∼28 Gb/s), which helps in the parallel and rapid computation of the calorimeter trigger algorithms. This article discusses the proposed design and expected performance of the upgraded CMS Level-1 calorimeter trigger system.

Keywords: LHC; calorimeter; trigger; FPGA; SLR

Citation: Kumar, P.; Gomber, B., on behalf of the CMS Collaboration. The CMS Level-1 Calorimeter Trigger for the HL-LHC. *Instruments* **2022**, *6*, 64. https://doi.org/10.3390/instruments6040064

Academic Editors: Fabrizio Salvatore, Alessandro Cerri, Antonella De Santo and Iacopo Vivarelli

Received: 31 July 2022
Accepted: 19 September 2022
Published: 17 October 2022

Publisher's Note: MDPI stays neutral with regard to jurisdictional claims in published maps and institutional affiliations.

1. Introduction

The High-Luminosity LHC (HL-LHC) will broaden the prospects of the current LHC in terms of new physics discoveries. The desired instantaneous luminosity of the HL-LHC is 7.5 × 10^{34} cm^{-2}s^{-1}, seven times the LHC's original parameter (1 × 10^{34} cm^{-2}s^{-1}). The availability of this increased datasets will help in the high precision measurements of the Standard Model (SM), and in the search of new territories beyond the SM (BSM). Figure 1 shows the LHC's schedule and the timeline for the HL-LHC.

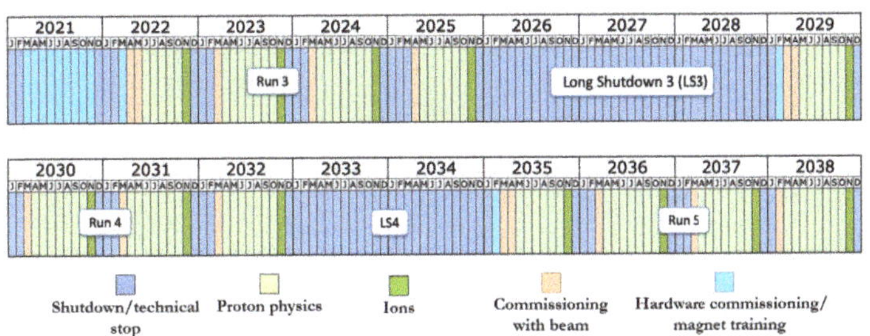

Figure 1. The LHC timeline, and its evolution to the HL-LHC (May 2029 to October 2038) after Long Shutdown 3.

The CMS detector will encounter a massive upgrade to avail the benefits and sustain the high radiation of the HL-LHC phase. This includes the replacement of the pixel and strip tracking detector, replacing the end-cap calorimeter with a more radiation tolerant high-granularity calorimeter (HGCAL), and completely new back-end and front-end electronics for the barrel calorimeter to attain finer granularity.

The Phase-2 upgrade of the CMS trigger system will maintain the two-level triggering strategy during the HL-LHC. The Level-1 trigger comprises custom design electronics, and at the second stage, a CPU farm-based High-Level Trigger (HLT) is employed [1]. During HL-LHC, the desired latency of the L1-trigger will increase from 4 μs to 12.5 μs. This increased latency permits the inclusion of the tracker data and high granular information from the HGCAL. The prominent features of the Phase-2 Level-1 trigger are [2]:

- The large-scale use of Xilinx Stacked Silicon Interconnect (SSI) technology-based FPGA. SSI-based FPGAs are fabricated by stacking several FPGA dies or super logic regions (SLRs). These large FPGAs can handle the significant challenges of the L1 trigger in terms of high speed serial communications, efficient reconstruction, latency, and resource constraints.
- The employment of the high-speed optical link (28 Gb/s) to meet the high bandwidth requirement of the HL-LHC. These high-speed links will assist in rapidly relaying the data from the detector back-end system to the L1 trigger chain.
- A flexible, modular, and scalable implementation of the calorimeter trigger algorithm. This approach is advantageous to address the HL-LHC dynamic running condition, changes in the hardware choices, and physics needs.

In this paper, we will discuss the Level-1 calorimeter trigger system and its hardware aspects, firmware implementation, latency, scalability, and the bitstream test on the prototype board.

2. Level-1 Calorimeter Trigger

The Level-1 calorimeter trigger processes the four calorimeter sub-detector system. The barrel part comprises the electromagnetic calorimeter (ECAL) and the hadronic calorimeter (HCAL), and the forward area includes the HGCAL and the forward hadronic calorimeter (HF). The key calorimeter trigger objects are: jets, photons, electrons, energy sums, and hadronically decaying taus. Figure 2 illustrates the block diagram of Phase-2 Level-1 trigger.

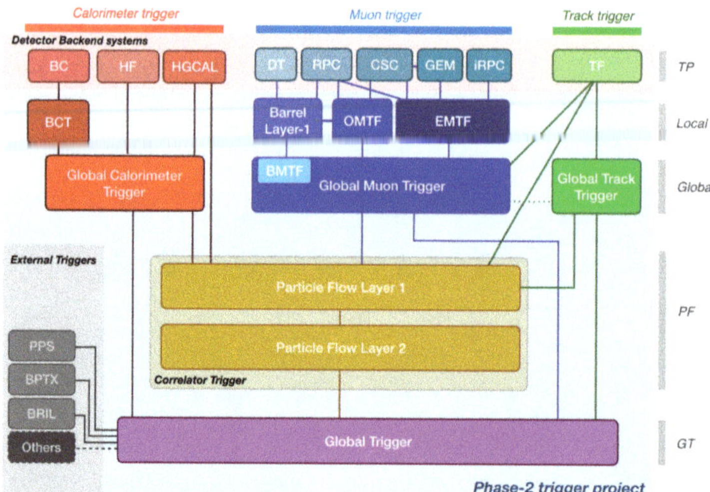

Figure 2. Block diagram of Phase-2 Level-1 trigger system [2]. It processes the information from sub-detectors, such as calorimeter, muon, and tracker. The trigger system is divided into five layers:

backend (generate the trigger input known as trigger primitives), local, global, particle-flow (PF), and global trigger (GT). The calorimeter trigger is implemented in two steps: barrel calorimeter trigger (BCT) and global calorimeter trigger (GCT). The muon trigger system takes input from muon spectrometers: the drift tube (DT), resistive plate chamber (RPC), cathode ray strip (CSC), and gas electron multiplier (GEM). The muon trigger reconstructs the muon tracks using the barrel muon track finder (BMTF), overlap muon track finder (OMTF), and endcap muon track finder (EMTF). It combines the muon's information in a global muon trigger (GMT). The track trigger comprised the backend track finder (TF) and global track trigger (GTT). The correlator trigger (CT) is implemented in two layers, viz. particle-flow layer-1 and particle-flow layer-2. The GT is planned to possibly explore triggering by taking information from external components such as precision proton spectrometer (PPS) [3], beam position and timing monitors (BPTX), and luminosity and beam monitoring detectors (BRIL) [4].

The detector back-end system collects the raw collision data from the front-end electronics and delivers the trigger primitive (input) to the downstream processing board. The central barrel calorimeter (ECAL and HCAL) is processed in two steps. The barrel calorimeter trigger (BCT), also known as the regional calorimeter trigger (RCT), creates the electron/photon clusters (accumulation of ECAL crystal energies) and towers (a group of 25 ECAL crystals) and sends them to the global calorimeter trigger (GCT). Figure 3 demonstrates the geometry of the barrel and the coverage of the RCT and the GCT FPGA cards.

A single RCT card covers a region of $17\eta \times 4\phi$ towers. A total of 36 RCT cards are required to process the complete barrel information. A single GCT card evaluates 16 RCT cards, of which 12 are unique while the other 4 are the neighboring RCT card to share the boundary condition. It requires three GCT cards to cover the complete barrel. The calorimeter trigger uses a time-multiplex scheme to transmit data between the trigger sub-systems [5]. It increases the system's flexibility and removes the constraint of boundary sharing between the FPGAs. The two primary tasks of the GCT algorithm are:

- Prepare and transmit the RCT and HF information in a time-multiplex manner to the correlator trigger (CT).
- Demultiplex the incoming HGCAL time-multiplexed data, merge it with the calorimeter-wide signals from ECAL, HCAL, and HF, and send it to the global trigger (GT).

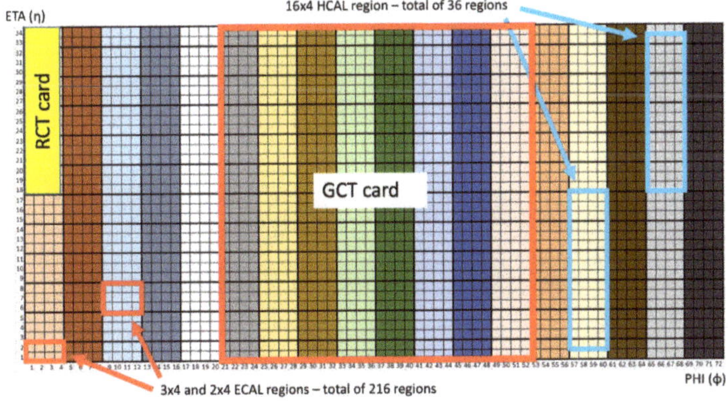

Figure 3. CMS barrel calorimeter segmentation. The *x-axis* represents the integer azimuthal angle (each integer represents 5 degrees in ϕ). The *y-axis* represents the integer η or pseudorapidity (derived from the polar angle of the LHC coordinate system) [6]. This region represents the ECAL ($34\eta \times 72\phi$) and HCAL ($32\eta \times 72\phi$) barrel geometry. Each small square represents one tower of HCAL and ECAL. For ECAL, one tower represents 25 ECAL crystal. The geometry of $17\eta \times 4\phi$ is the processing region for one RCT card.

The desired latency budget to provide input to the CT via GCT is 5 μs, and the GCT must send the output to the GT within 9 μs from the bunch-crossing. Figure 4 represents the L1 calorimeter trigger architecture and its time-multiplexing and latency scheme.

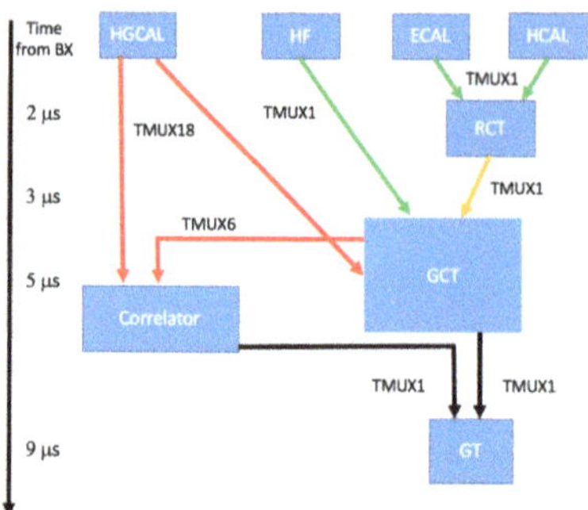

Figure 4. Calorimeter trigger architecture. It depicts the desired latency budget and time-multiplexing scheme employed for each calorimeter sub-system in Phase-2.

3. Trigger Algorithms and Hardware Implementation

The calorimeter trigger algorithms are implemented in the Xilinx XCVU9P FPGA, and their latency and resource utilization are presented. In the Phase-2 upgrade, the RCT algorithm receives the input data from each ECAL crystal, increasing the granularity 25 times compared to the Phase-1 trigger system. Similarly, the GCT algorithm for the HGCAL (endcap calorimeter) has the advantage of using high-granular input information from the HGCAL back-end system. The following section describes the calorimeter trigger development and floor-planning, RCT algorithm, and the GCT algorithm.

3.1. Calorimeter Trigger Development and Floorplanning

The calorimeter trigger algorithms are developed with the help of the high-level synthesis (HLS) tools such as Vivado-HLS [7]. This tool synthesizes the algorithms written in a higher-level sequential language such as C++ and generates the hardware description language (HDL). Vivado-HLS also provides the early estimation of the latency and utilization. One of the main advantages of Vivado-HLS is the rapid prototyping of the trigger algorithms, and ease of performing firmware evaluation with the emulator (C++) generated output. The HDL wrapper integrates the HLS-generated IP with the firmware shell in the downstream implementation. The multi-gigabit transceivers (MGTs) used for the trigger algorithms inputs and outputs are spread across the FPGA boundary. Without crossing the SLR boundary, each SLR possesses a fraction of these MGTs. For example, the SLR1 of XCVU9P FPGA can access only 40 (out of 96 available) MGTs. The division of the HLS algorithms in SLRs reflects the constraint of these distributed MGTs. Figure 5 reflects the floorplan of the calorimeter trigger algorithm.

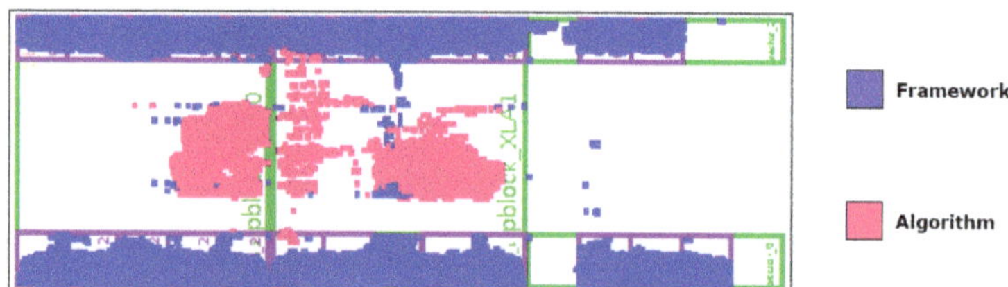

Figure 5. Calorimeter trigger algorithm floorplan for XCVU9P C2104 package FPGA. The trigger algorithms are placed in the middle of the SLR (magenta color). The firmware shell, which includes the MGTs firmware, is placed at the edges, which corresponds to the MGT ports (blue color).

3.2. Regional Calorimeter Trigger (RCT)

The RCT algorithm processes the entire geometry of the barrel ECAL and HCAL. It creates the electron/photon clusters and towers and forwards them to the global calorimeter trigger (GCT). The RCT algorithm processes the region of $17\eta \times 4\phi$ of ECAL and $16\eta \times 4\phi$ of HCAL. The critical step of this algorithm is to find the seed crystal, build a cluster of size $3\eta \times 5\phi$ (crystal level) around the seed, build the bremsstrahlung area (two $3\eta \times 5\phi$ regions around the central cluster in both left and right of ϕ), and build the shape area (two $2\eta \times 5\phi$ areas, to differentiate between electrons and hadrons). The RCT algorithm employs a stitching logic that merges the lower cluster energy into the higher one and nullifies the energy of the lower cluster based on their position in η and ϕ. Figure 6 illustrates the RCT coverage and clusters geometry which is described in [2].

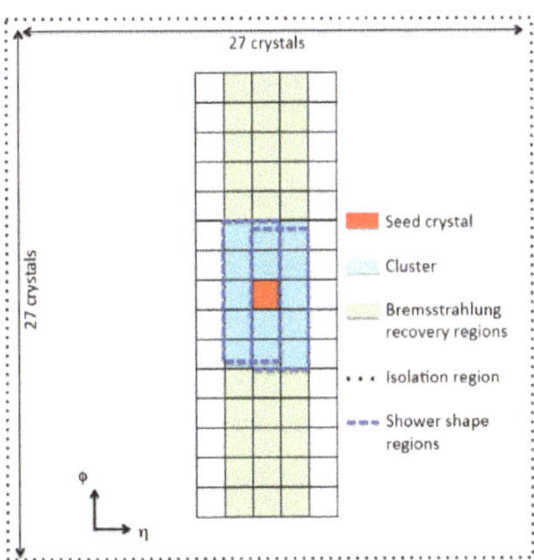

Figure 6. RCT algorithm coverage and clusters formation. Each small square represents one ECAL crystal. The seed represents the crystal with highest energy deposition by electron/photon.

The input and output fiber bandwidth is 16 Gbps which carries 384 bits/bunch-crossing (bits/BX) of data. The RCT region is divided into two SLRs (SLR1 and SLR2) to ease the MGTs requirement while saving the SLR0 for future use. The algorithm processes the region of $8\eta \times 4\phi$ (RCT8×4) in SLR1 and $9\eta \times 4\phi$ (RCT9×4) in SLR2. The RCT8×4 and RCT9×4 process 68 input ECAL links (each link carries 25 crystals information), and RCT9×4 considers two additional HCAL links. RCT8×4 output in SLR1 is routed to the SLR2 using the super long line (SLL) of the XCVU9P FPGA. In contrast, the RCT9×4 output is buffered to match the additional routing delay of the RCT8×4 output. This buffering ensures that the RCTSUM algorithm in SLR2 receives both outputs simultaneously. The RCTSUM algorithm stitches the RCT8×4 and RCT9×4 algorithm at its η boundary and computes the h/e (HCAL to ECAL energy ratio) over the entire RCT geometry, and prepares the four output links to the GCT.

3.3. Global Calorimeter Trigger (GCT)

GCT is a collection of several calorimeter trigger algorithms to process various physics objects. It includes the jet, taus, missing transverse energy (MET), η, and ϕ stitch algorithm. The ϕ stitch (GCT) and RCT algorithms are implemented in SLR0, SLR1, and SLR2, respectively. This implementation demonstrates the working of RCT and GCT together in a single FPGA card. Therefore, the RCT output is replicated five times at the SLR0 to mimic the testing of a GCT algorithm processing five RCT cards simultaneously. The four output links from RCTSUM are buffered while routing it from SLR2 to SLR0. This buffering helps meet the timing constraint of the 240 MHz algorithm clock and prevents any setup time violation in SLR0. Figure 7 demonstrates the single card test implementation of the RCT and GCT algorithm.

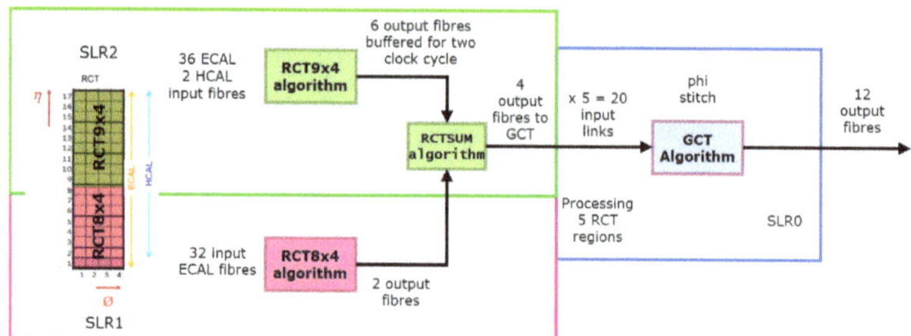

Figure 7. Integrated RCT and GCT algorithm block diagram. The RCT9×4, RCTSUM, RCT8×4, and the GCT phi stitch algorithms are implemented in SLR2, SLR1, and SLR0, respectively.

4. Results and Tests

The RCT and GCT algorithms are implemented in Vivado-HLS with a clock frequency of 240 MHz and a pipeline interval of 6 clock cycles. The bitstream for the XCVU9P C2104 package FPGA is generated by integrating the HLS IP with the firmware shell using the HDL wrapper. The latency of the RCT and GCT algorithms is 230 and 120 clock-cycles (combined latency of 1.458 μs), respectively. The device utilization of the integrated RCT and GCT algorithms along with the firmware shell is listed in Table 1.

Table 1. Device utilization of the integrated RCT and GCT algorithms for the XCVU9P C2104 package FPGA.

Look-Up Tables (LUT) %	Flip-Flop %	Block-RAM %
42	26	24

The RCT algorithm is implemented in XCVU9P FPGA utilizing only two SLRs (SLR2 and SLR1) while leaving the SLR0 for the GCT phi stitch algorithm. The bitstream is generated and successfully tested on the first advanced processor demonstrator (APd1) board, which hosts the XCVU9P FPGA. Figure 8 shows the APd1 board and the placement of the integrated RCT and GCT algorithm on XCVU9P FPGA.

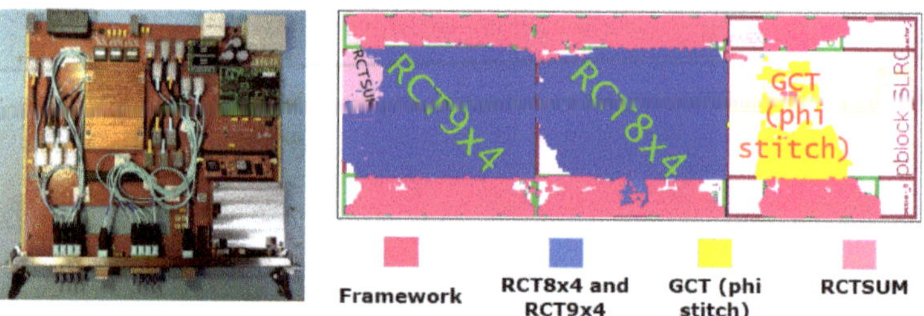

Figure 8. (Left): prototype APd1 board hosting the Xilinx XCVU9P C2104 package FPGA. **(Right):** Floor-planning and placement of the RCT and GCT algorithm on XCVU9P FPGA.

5. Conclusions

The HL-LHC calorimeter trigger algorithm for the barrel calorimeter (RCT and GCT) is developed using the Vivado-HLS tool. The algorithms are implemented and tested on the APd1 board. The combined latency is 1.458 μs (RCT and GCT phi stitch), which is within the desired budget of 2 μs. The resource utilization is 42% of the LUT, 26% of the flip-flop, and 24% of the BRAM. Several algorithms at the GCT, such as jet, taus, and missing transverse momentum (MET), are being developed.

Author Contributions: Conceptualization and Methodology—B.G., and P.K.; Validation: B.G., and P.K.; Software: B.G.; Formal Analysis: B.G., and P.K.; Data Curation—P.K.; Project Administration—B.G.; Funding Acquisition—B.G. All authors have read and agreed to the published version of the manuscript.

Funding: This research was funded by IOE, University of Hyderabad, grant number UOH-IOE-RC2-21-006.

Institutional Review Board Statement: Not applicable.

Informed Consent Statement: Not applicable.

Data Availability Statement: Not applicable.

Conflicts of Interest: The authors declare no conflict of interest.

References

1. Khachatryan, V.; Sirunyan, A.M.; Tumasyan, A.; Adam, W.; Asilar, E.; Bergauer, T.; Brandstetter, J.; Brondolin, E.; Dragicevic, M.; Erö, J.; et al. The CMS trigger system. *J. Instrum.* **2017**, *12*, P01020. [CrossRef]
2. *The Phase-2 Upgrade of the CMS Level-1 Trigger*; Technical Report CERN-LHCC-2020-004. CMS-TDR-021; CERN: Geneva, Switzerland, April 2020. Available online: http://cds.cern.ch/record/2714892 (accessed on 30 July 2022).
3. The CMS Collaboration. *The CMS Precision Proton Spectrometer at the HL-LHC Expression of Interest*; Technical Report; CERN: Geneva, Switzerland, November 2020. Available online: https://cds.cern.ch/record/2750358 (accessed on 30 July 2022).
4. Karacheban, O. Performance of the BRIL luminometers at CMS in Run 2. In Proceedings of the European Physical Society Conference on High Energy Physics (EPS-HEP) 2019, Ghent, Belgium, 10–17 July 2019; Volume 194. [CrossRef]
5. RicFrazier, R.; Fayer, S.; Hall, G.; Hunt, C.; Iles, G.; Newbold, D.; Rose, A. A demonstration of a time multiplexed trigger for the cms experiment. *J. Instrum.* **2012**, *7*, C01060. [CrossRef]
6. Wong, C.Y. *Introduction to High-Energy Heavy-Ion Collisions*; World Scientific: Singapore, 1994.
7. O'Loughlin, D.; Coffey, A.; Callaly, F.; Lyons, D.; Morgan, F. Xilinx vivado high level synthesis: Case studies. In Proceedings of the 25th IET Irish Signals Systems Conference 2014 and 2014 China-Ireland International Conference on Information and Communications Technologies (ISSC 2014/CIICT 2014), Limerick, Ireland, 26–27 June 2014; pp. 352–356. [CrossRef]

Article

Towards a Large Calorimeter Based on Lyso Crystals for Future High Energy Physics

Patrick Schwendimann [1,*], Andrea Gurgone [2,3] and Angela Papa [4,5,6]

1 Department of Physics, University of Washington, Seattle, WA 98195-1560, USA
2 Dipartimento di Fisica, Università di Pavia, Via Agostino Bassi 6, 27100 Pavia, Italy
3 INFN, Sezione di Pavia, Via Agostino Bassi 6, 27100 Pavia, Italy
4 Paul Scherrer Institute, 5232 Villigen, Switzerland
5 INFN, Sezione di Pisa, Largo B. Pontecorvo 3, 56127 Pisa, Italy
6 Dipartimento di Fisica, Università di Pisa, Largo B. Pontecorvo 3, 56127 Pisa, Italy
* Correspondence: schwenpa@uw.edu

Abstract: The state-of-the-art research at the intensity frontier of particle physics aims to find evidence for new physics beyond the Standard Model by searching for faint signals in a vast amount of background. To this end, detectors with excellent resolution in all kinematic variables are required. For future calorimeters, a very promising material is LYSO, due to its short radiation length, fast decay time and good light yield. In this article, the simulation of a calorimeter assembled from multiple large LYSO crystals is presented. Although there is still a long way to go before crystals of that size can be produced, the results suggest an energy resolution of 1%, a position resolution around 5 mm and a time resolution of about 30 ps for photons and positrons with an energy of 55 MeV. These results would put such a calorimeter at the technology forefront in precision particle physics.

Keywords: calorimeter; LYSO; simulation; intensity frontier

---

1. Introduction

Research at the intensity frontier in particle physics aims to find evidence for physics beyond the Standard Model (SM) by comparing its high-precision predictions with equally precise experimental measurements. This requires a detector system capable of measuring low signals in a large amount of background with unprecedented accuracy.

One of the main investigations in this field is the search for charged Lepton Flavour Violation (cLFV) in processes that are forbidden or highly suppressed in the SM [1]. In this regard, muons are a very sensitive probe, as they are fairly simple to produce and can be transported at low energies from the production target to the experiment, without receiving significant contamination from other particles.

Some of these processes, such as $\mu \rightarrow e\gamma$, contain one or more photons in the final state with an energy scale of 0–100 MeV. The state-of-the-art method to detect such particles are calorimeters based on a high-density scintillating material coupled to photosensors of various kind, such as Silicon Photomultipliers (SiPMs). The same method can be applied to electrons and positrons, eventually in combination with a more accurate tracking system.

The photon detection in the MEG II experiment relies on a calorimeter based on a single large volume of liquid Xenon, which has been estimated to provide an energy resolution of 1.7%, a time resolution of about 40 ps and a position resolution of about 2.5 mm for signal photons at 53 MeV [2].

Other experiments at the precision frontier rely on calorimeters built from large crystals. As an example, the PIENU experiment uses a single NaI crystal and reports a detector resolution of 2.2% FWHM for 70 MeV/c positrons [3].

Citation: Schwendimann, P.; Gurgone, A.; Papa, A. Towards a Large Calorimeter Based on Lyso Crystals for Future High Energy Physics. *Instruments* **2022**, *6*, 65. https://doi.org/10.3390/instruments6040065

Academic Editors: Fabrizio Salvatore, Alessandro Cerri, Antonella De Santo and Iacopo Vivarelli

Received: 22 August 2022
Accepted: 10 October 2022
Published: 18 October 2022

Publisher's Note: MDPI stays neutral with regard to jurisdictional claims in published maps and institutional affiliations.

The importance of the calorimeter in this kind of experiments can be easily understood by looking at its requirements. In MEG II one of the limiting factor is the accidental background rate, which is estimated to be

$$R_{\text{acc}} \propto R_\mu^2 \cdot \Delta E_\gamma^2 \cdot \Delta p_e \cdot \Delta\Theta_{e\gamma}^2 \cdot \Delta t_{e\gamma} \, , \tag{1}$$

where only the muon beam rate R_μ and the positron momentum resolution Δp_e do not depend on the calorimeter performance. The photon energy resolution ΔE_γ enters directly, while position and time resolutions enter through the relative angle resolution $\Delta\Theta_{e\gamma}$ the relative timing resolution $\Delta t_{e\gamma}$, respectively.

For experiments such as PIENU, the situation is different. Here, the importance of the energy resolution is crucial to reliably separate the different decay channels and suppress the low energy tail of the $\pi \to e\nu$ decay channel, while in this situation the time resolution does not directly affect the result, a fast detector allows to take data at higher rate and thus acquire the required statistics in shorter time.

The development towards future calorimeters for the next generation of precision experiments is currently ongoing. Thanks to its high density, good light yield and fast decay time, LYSO is promising material. A comparison with other commonly used scintillators is given in Table 1.

Table 1. Properties of commonly used scintillators.

	Density ρ (g/cm^3)	Light Yield *LY* (ph/keV)	Decay Time τ (ns)	Radiation Length X_0 (cm)
LaBr$_3$(Ce)	5.08 [1]	63 [1]	16 [1]	2.1 [2]
LYSO	7.1 [3]	29 [3]	41 [3]	1.21 [4]
LXe	2.95 [5]	40 [6]	45 [6]	2.9 [5]
NaI(Tl)	3.67	38	245	2.59
BGO	7.13	9	300	1.12

The information was taken from Review of Particle Physics ([4]) unless specified otherwise: [1] Manufacturer's Datasheet [5]. [2] Private communication [6]. [3] Manufacturer's datasheet [7]. [4] Geant4-based estimate. [5] PDG Online [8]. [6] MEG II values for the LXe Calorimeter [2].

LYSO was first used in medical applications but soon attracted attention from the HEP community. In the past decade, multiple tests using thin crystals with a front area up to 2.5 cm × 2.5 cm and a length between 10 and 20 cm have been made [9–12]. Limited by the crystal size, these experimental tests found resolutions on the order of 4% for particles below 100 MeV. In particular, the size of the used crystals was comparable to the Moliere radius of LYSO, which is approximately 2 cm.

In addition, LaBr$_3$(Ce) is a very promising material as well. However, earlier studies [13] found that the higher light yield and faster decay time cannot compensate for the worse energy containment due to the longer radiation length of LaBr$_3$(Ce) compared to LYSO.

Both materials are currently limited by the single crystal size. However, the constant progress in the crystal growing process suggests that larger crystals will be available in the future. This perspective makes LYSO a viable choice of material for the precision frontier in particle physics. In this regard, a prototype made of a single LYSO crystal of 10 cm length and 7.5 cm diameter is currently under construction.

Focusing on future developments, larger crystals are assumed in this study. Improving the crystal size is crucial to counteract the energy leakage and improve the containment of the shower. Given the larger surface of these crystals, it is crucial to provide a bulk SiPM readout to have an appreciable light collection. The high granularity also provides a handle for position reconstruction, improving the spatial resolution.

2. Materials and Methods

Motivated by the construction of the prototype, a large scale calorimeter built of several LYSO crystals is simulated using Geant4 [14] and analysed with ROOT [15]. Each crystal has a size of 25 cm × 25 cm × 15 cm.

A representation of the simulated detector geometry is shown in Figure 1. Eight crystals are arranged in an octagonal structure with the 25 cm × 25 cm surfaces oriented perpendicular to the radial direction. Three of these octagonal structures are stacked next to each other to obtain a cylindrical structure of 24 crystals in total. This geometry results in a calorimeter with approximately $12X_0$ depth and an inner volume with about 60 cm diameter for further instrumentation, such as stopping targets or trackers.

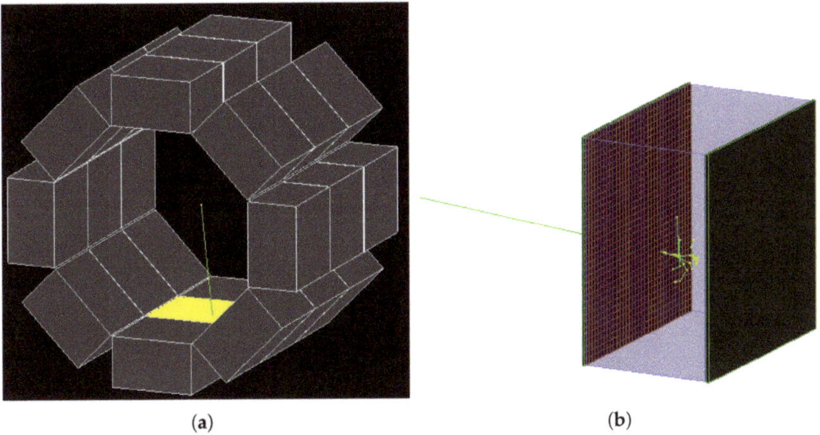

(a)   (b)

Figure 1. Representation of the simulated calorimeter. A set of 24 large crystals are arranged around the vertex. Inner and outer surface of each crystal are covered with SiPMs. (**a**) Total View. (**b**) Single Crystal.

Every crystal features its independent readout on the inner (facing the centre of the octagon) and outer surface. The lateral surfaces are sealed with a thin sub-millimetre aluminum layer to protect the crystal itself and avoid optical cross-talk.

The readout consists of a matrix of SiPMs with an active area of 6 mm × 6 mm and a slightly larger support structure, mounted on PCBs. A carbon fibre layer is also used to close the system on both sides. Given the small thickness of the readout system, its effect on incoming photons is negligible. Hence, the inner readout does not significantly affect the energy deposited in the crystal.

The LYSO crystals are simulated almost fully transparent with an order of magnitude estimate of the bulk absorption length of 1 m. The refractive index of LYSO is assumed to be 1.81. The aluminum coated surfaces are considered to be polished with a reflectivity of 95%. The SiPM entrance window is simulated with a refractive index of 1.55 and a dielectric boundary towards the LYSO crystal. The active area of the SiPM is simulated with a high refractive index (4) and a ultra-short absorption length.

The small size of the SiPMs results in a huge number of channels, which is on the order of a thousand per crystal, while this produces a massive amount of data, it also provides a high granularity that can be used for the geometrical reconstruction of the event.

For this study, photon or positron events at the centre of the geometry with an energy of 55 MeV and an isotropic angular distribution in the detector acceptance are simulated. The energy has been chosen to resemble a $\mu \rightarrow e\gamma$ signal, as in the previous studies [13]. It also corresponds to the energy at which the prototype will be tested.

The standard Geant4 algorithms are used to simulate the shower inside the crystals and propagate optical photons from production through scintillation to absorption in the SiPM active area. For a photon absorbed inside the active area of a SiPM, the (x, y) position is used to determine the SiPM pixel if any, thus accounting for the fill factor. If a photon is absorbed within a pixel, a hit is generated with a probability corresponding to the quantum efficiency of the SiPM.

The simulation is also complemented with measurements performed to characterise the SiPMs to be used for the prototype. Namely, the digitised waveforms obtained for the Hamamatsu SiPM S13360-6025PE are used to simulate the detector and electronics response to a single photon hit in the post-processing. This is done by iterating over all hits in each SiPM, sorted by pixel and time, and accumulating the single photon responses. A second hit in the same pixel is only considered if the time separation is larger than the dead time of 40 ns of the SiPM, thus accounting for saturation effects.

After distorting the obtained waveform with gaussian noise for each channel, charge integration and constant fraction timing is performed for each obtained waveform. This information is then used along with the position of each SiPM to reconstruct the events in terms of energy, time and three-dimensional position.

In addition to the analysis of each individual channel, the waveforms for all channels on the inner readout of each crystal are summed up and the constant fraction method is applied to obtain an averaged time t_i for the inner readout of the crystal. The same procedures is applied to the channels on the outer readout to obtain the outer time t_o.

These parameters are used to estimate the entrance time t of the original particle as

$$t = \frac{(n-1)t_i + (n+1)t_o - L/c(n^2 + n)}{2n} ,\tag{2}$$

where $L = 15\,\text{cm}$ stands for the thickness of the crystal, $n = 1.81$ for its refractive index and c for the speed of light. This formula is obtained by assuming in a one dimensional geometry that the incident particle propagates at the speed of light c and the optical photons propagate at a reduced velocity c/n. In order to extract the intrinsic time of the event, the estimated entrance time has to be corrected with the time of flight between the vertex and the crystal entrance point.

The depth (z-coordinate) of the first interaction between the incoming particle and the crystal is reconstructed as

$$z = \frac{1}{2}\left(\frac{c}{n} \cdot (t_i - t_o) + L\right)\tag{3}$$

where again the solution of the one dimensional problem is used. The position in the other two dimensions (x, y) is estimated by analysing the distribution of the charge collected by each SiPM and using the high granularity to fit it with a Gaussian. The coordinates (x, y) are then used to estimate the distance between the centre of the setup and entrance location on the crystal.

In first approximation, the total energy is assumed to be proportional to the charge collected by all SiPMs of one crystal, i.e.,

$$E \propto Q_{\text{tot}} = \sum Q_i .\tag{4}$$

To account for the position dependence, an approximate correction was introduced to modify the charge depending on the hit position

$$Q_{\text{tot}}^{(2)} = \frac{Q_{\text{tot}}}{1 - a(x^2 + y^2)} ,\tag{5}$$

where a is a geometry-dependent coefficient to be determined and x, y are obtained from the reconstructed position of the event. Note that this correction corresponds to the simplest term that respects the symmetry of the geometry, while a more sophisticated correction could be considered, this form was found to be sufficient for the current application.

3. Results

For each simulated event, the waveform of every channel is analysed to extract its charge and time. An example of charge distribution obtained at the inner readout for one crystal is shown in Figure 2. This is the starting point for all further reconstructions.

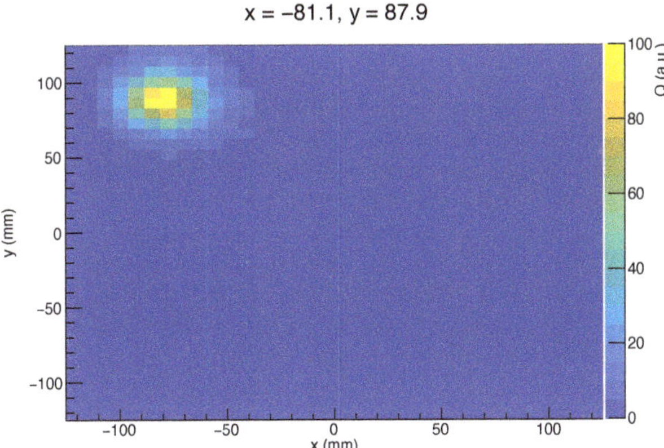

Figure 2. Charge distribution on the inner readout for one event. The initial photon entered at $x = -81.1$ mm, $y = 87.9$ mm (Monte Carlo truth value). The charge is arbitrarily scaled.

3.1. Time Reconstruction

Using the time reconstruction described in Section 2, the distribution shown in Figure 3 is obtained for photons and positrons, respectively. The result includes the time of flight correction due to the distance between the event vertex and the calorimeter hit.

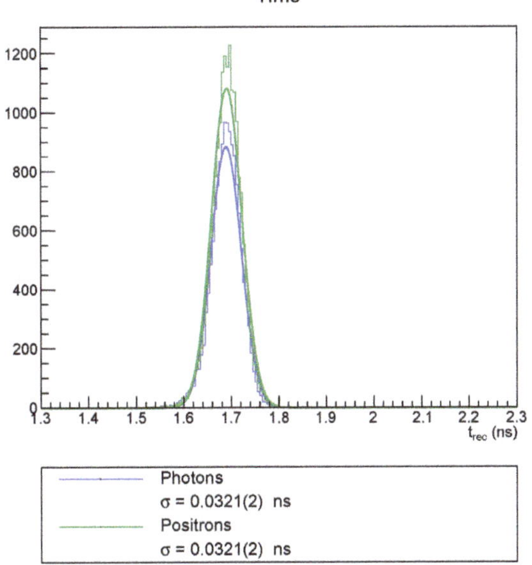

Figure 3. Time reconstruction for photons and positrons of 55 MeV. The correction accounting for the particle time of flight is applied. The resolution σ is obtained from a Gaussian fit.

In addition, a Gaussian fit is shown for each distribution and a standard deviation of about 30 ps is obtained. This value can be compared to the reported time resolution of 40 ps of the MEG II calorimeter [2].

Both particles behave fairly similar with respect to time reconstruction. The time offset is due to the fact that the reconstruction does not account for the time delay between the energy deposit and the light emission, as well as the time required to build up the readout waveforms. However, since this is a constant offset, it can be neglected for this simulation and canceled by adequate calibrations in an actual experiment.

3.2. Position Reconstruction

The position of first interaction in the directions x and y are reconstructed from a Gaussian fit of the charge distribution as observed by the inner readout plane. The distribution of the deviation between the reconstructed position x_{rec} and the Monte Carlo truth value x_{sim} is shown in Figure 4.

These distributions are fitted with the sum of two Gaussians, one for the core distribution and one to account for the tail. The standard deviations for the core fit are reported in the legend of Figure 4. One can observe that the reconstruction for photons is more precise than for positrons. An excellent resolution below 5 mm is observed for photons and a resolution around 5 mm for positrons. For the chosen geometry a spatial resolution of 5 mm corresponds to an angular resolution below 20 mrad.

These numbers compare unfavourably to the resolutions obtained by the MEG II calorimeter of about 2.5 mm for the vertex position and below 10 mrad for the angle between photon and positron [2]. In the context of the angular resolution, it has to be considered that the MEG II calorimeter is located further away from the vertex. Moreover, it consists of only one singular volume, thus reducing the effects of the boundaries.

Due to the symmetric nature of the crystals, a very similar behaviour along the y-direction is observed. Applying the same methods for the z-direction yields to depth resolutions for the point of first interaction on the same scale.

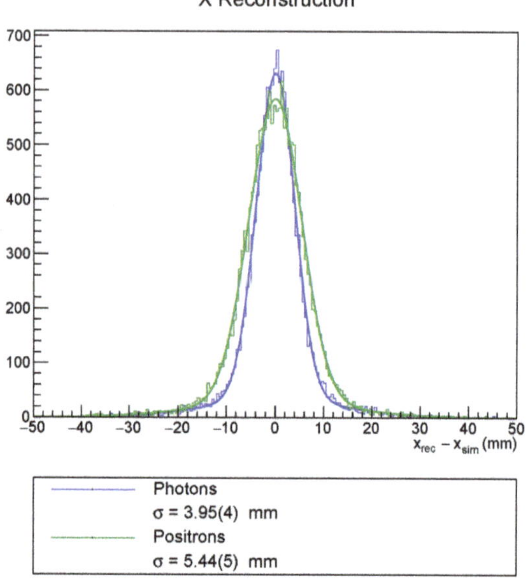

Figure 4. Position reconstruction in the x-direction for photons and positrons of 55 MeV. The resolution σ is obtained from a Gaussian fit.

3.3. Energy Reconstruction

The effect of correcting the charge collected based on the hit position reconstructed is shown in Figure 5. Note that for this visualisation, the geometrical position (Pos) on the x-axis is a relative value that describes how centred the event was. It is computed as

$$\text{Pos} = \max\left(\frac{|x_{\text{rec}}|}{x_{\text{max}}}, \frac{|y_{\text{rec}}|}{y_{\text{max}}}\right) \tag{6}$$

with x_{max}, y_{max} being the position of the crystal edges, i.e., half the crystal length in the corresponding direction. Thus, a value of 0 represents a perfectly central hit and a value of 1 is a hit at the very edge of the crystal.

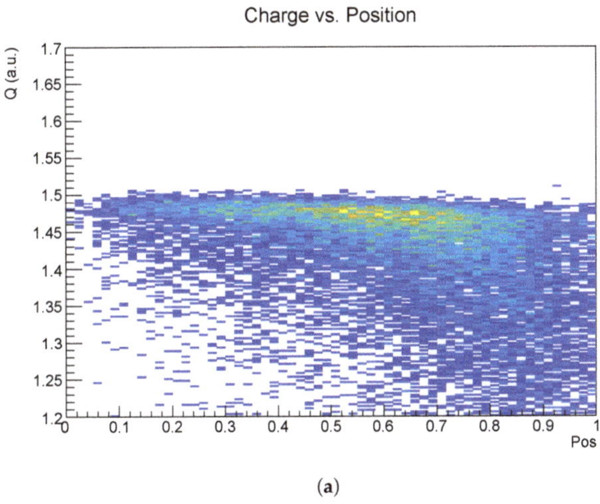

(a)

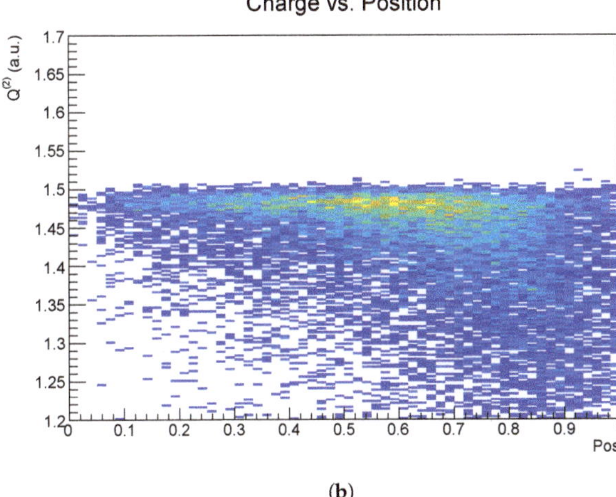

(b)

Figure 5. Charge versus reconstructed position of the hit. The position is chosen such that 0 refers to the centre of the crystal and 1 to the edge. See (6) for details. (**a**) Uncorrected. (**b**) Corrected.

While for corrected and uncorrected charge it is clearly visible that the distribution widens up for more lateral hits, the uncorrected charge distribution is slightly tilted. This effect is mostly removed by applying the correction to the charge.

The distribution of the corrected charge $Q^{(2)}$ without any geometrical selection is shown in Figure 6 for both photons and positrons. There is a main peak at around $Q^{(2)} = 1.5$, corresponding to the 55 MeV particle energy. The secondary peak at $Q^{(2)} = 0$ is due to energy leakage from neighbouring crystals.

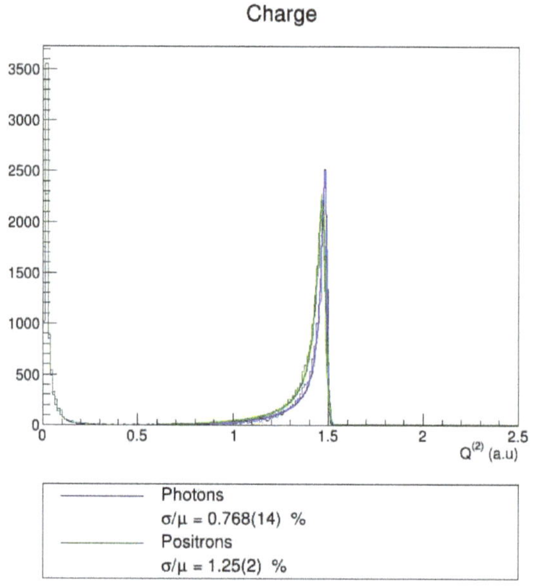

Figure 6. Corrected charge reconstruction for photons and positrons of 55 MeV. The relative resolution σ/μ is obtained from a fit with a tailed Gaussian function, defined in (7).

The main peak at 55 MeV is fitted with a tailed Gaussian function

$$f(x|N,\mu,\sigma_1,\sigma_2,\sigma_3) = \begin{cases} N \exp\left(-\frac{(x-\mu)^2}{2(\sigma_1)^2}\right) & \text{if } x > \mu \\ N \exp\left(-\frac{(x-\mu)^2}{2(\sigma_1+\sigma_2(x-\mu)+\sigma_3(x-\mu)^2)^2}\right) & \text{if } x < \mu \end{cases} \tag{7}$$

where μ stands for the peak position, σ_1 for the Gaussian standard deviation and σ_2, σ_3 are parameters to model the tail. The relative resolution is estimated as σ_1/μ.

Doing so for 55 MeV photons yields an energy resolution below 1%. For 55 MeV positrons, the resolution estimated is just above 1%. One can further notice that the positron peak is slightly shifted to lower charges compared to the photon peak. This is due to energy losses in the readout layer for the positron, whereas the photon passes straight through to the scintillating crystal.

This resolution can be compared to the MC simulation of the MEG II calorimeter, which suggests a resolution of about 1.1% for 52.8 MeV photons, or the recently measured value of 1.7% [2]. This suggests that a LYSO calorimeter is likely to provide a better energy resolution compared to this kind of currently running calorimeters.

As can be seen from Figure 5, events further out contribute more prominently to the tail of the distribution and thus affect the overall resolution adversely. Hence, it can be considered to remove the worst of those events by applying a geometrical cut based on the reconstructed position.

This possibility is studied systematically and the key findings are shown in Figure 7. The more stringent the cut, the more events get rejected and the overall efficiency gets reduced. One can see that the resolution tends to improve at the expense of reconstruction efficiency. This effect is enhanced for the uncorrected charge, while it is shown as well that the correction applied does not cancel all geometrical effects. However, it is clear that the correction is more effective for the most lateral events, that are cut first.

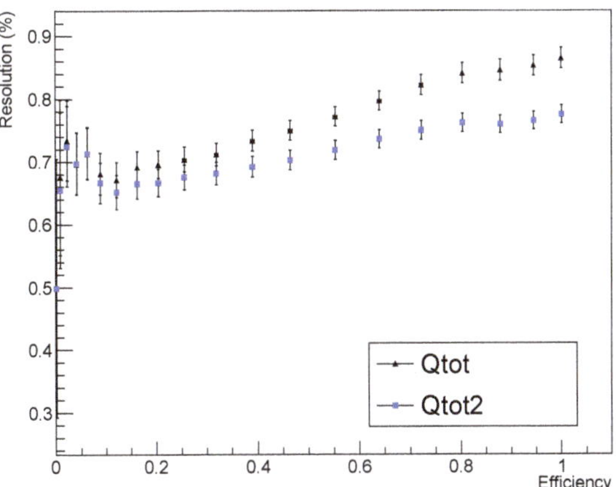

Figure 7. Application of a geometrical cut for photons to reject events close to the edge, based on the position reconstruction.

4. Discussion

In this study, simulation results for a large calorimeter based on multiple LYSO crystals are presented. Unfortunately, such crystals cannot yet be produced with the required uniformity, but the constant progress in the crystal growing process suggests that this might be possible in the future. Until then, LYSO remains an interesting candidate for high-precision measurements by assembling a calorimeter from a multitude of smaller tapered crystals, as proposed for the PIONEER experiment [16].

The readout is based on a matrix of closely packed SiPM, covering both the inner and outer surface of each crystal. This allows a high-precision reconstruction of energy, time and position for each individual particle hit. For 55 MeV photons, an energy resolution below 1%, a time resolution around 30 ps and a position resolution of 5 mm was obtained. For positrons of the same energy, similar values for time and position were found, while the energy resolution resulted slightly worse.

Comparing these numbers to the performance of the MEG II calorimeter, which operates at very similar energies, it shows that this calorimeter offers slightly better timing resolution (32 ps this study, 40 ps for MEG II) and clearly improved energy resolution up to a factor two (clearly below 1% for 55 MeV photons compared to 1.7% for MEG II) at the cost of position resolution (4 mm compared to 2.5 mm for MEG II). However, the position reconstruction used in this study is still to be optimised and thus offers opportunities for improvements.

In addition, the geometrical reconstruction was found to be good enough to allow an event selection that rejects lateral events, characterised by a broader charge distribution which worsens the overall energy resolution. This cut has to be tuned on the experimental context in which the calorimeter is used, in order to balance the improvement in energy resolution with the consequent efficiency loss.

In conclusion, the energy resolution obtained for the studied crystal size is clearly improved for 55 MeV photons compared to state-of-the-art detectors and it appears to be slightly better when it comes to timing. With the current position reconstruction algorithms, one may not yet be able to match the position resolutions of the MEG II calorimeter which has been used as comparison here. However, it has to be considered that the geometry of the detector can be adapted to the specific situation. For example, placing the crystals further away, would improve the angular resolutions at the cost of detector acceptance. Moreover, one must not forget about the other advantages such as these crystals do not require a cryogenic environment opposed to liquid Xenon and the fact that the events will distribute over all crystals instead of only one Xenon volume, thus reducing pileup. These results would put such a calorimeter at the technology forefront in precision particle physics.

Author Contributions: Conceptualization, P.S. and A.P.; methodology, P.S., A.G. and A.P.; software, P.S.; validation, P.S. and A.G.; formal analysis, P.S.; investigation, P.S., A.G. and A.P.; resources, A.P.; writing—original draft preparation, P.S.; writing—review and editing, P.S., A.G. and A.P.; supervision, A.P.; project administration, A.P.; funding acquisition, A.P. All authors have read and agreed to the published version of the manuscript.

Funding: This research was funded by SNF n. 200020_172706 and MIUR Montalcini D.M. 2014 n. 975.

Data Availability Statement: Not applicable.

Acknowledgments: The authors thank the Paul Scherrer Institute in Villigen for hosting the research facilities.

Conflicts of Interest: The authors declare no conflict of interest. The funders had no role in the design of the study; in the collection, analyses, or interpretation of data; in the writing of the manuscript; or in the decision to publish the results.

Abbreviations

The following abbreviations are used in this article:

cLFV	charged Lepton Flavour Violation
HEP	High Energy Physics
LYSO	Lutetium Yttrium Oxyorthosilicate
PCB	Printed Circuit Board
PDG	Particle Data Group
SiPM	Silicon Photomultiplier
SM	Standard Model

References

1. Calibbi, L.; Signorelli, G. Charged Lepton Flavour Violation: An Experimental and Theoretical Introduction. *La Riv. Nuovo Cim.* **2018**, *41*, 71–174.
2. Baldini, A.M.; Baranov, V.; Biasotti, M.; Boca, G.; Cattaneo, P.W.; Cavoto, G.; Cei, F.; Chiappini, M.; Chiarello, G.; Corvaglia, A.; et al. The Search for $\mu^+ \to e^+\gamma$ with 10^{-14} Sensitivity: The Upgrade of the MEG Experiment. *Symmetry* **2021**, *13*, 1591. [CrossRef]
3. Aguilar-Arevalo, A.A.; Aoki, M.; Blecher, M.; Vom Bruch, D.; Bryman, D.; Comfort, J.; Cuen-Rochin, S.; Doria, L.; Gumplinger, P.; Hussein, A.; et al. Detector for measuring the $\pi^+ \to e^+\nu$ branching fraction. *Nucl. Instrum. Methods Phys. Res. Sect. A* **2015**, *791*, 38–46. [CrossRef]
4. Particle Data Group. Review of Particle Physics. *Chin. Phys. C* **2016**, *40*, 100001. [CrossRef]
5. Saint-Gobain Ceramics & Plastics Inc. *Lanthanum Bromide and Enhanced Lanthanum Bromide Scintillation Materials*; Saint-Gobain Ceramics & Plastics Inc.: Malvern, PA, USA, 2017.
6. Papa, A. (Paul Scherrer Institute, 5232 Villigen, Switzerland); Saint-Gobain (02410 Paris, France). Private communication, 2018.
7. EPIC-Crystal. LYSO(Ce) Scintillator. Available online: https://www.epic-crystal.com/oxide-scintillators/lyso-ce-scintillator.html (accessed on 8 June 2021).
8. Groom, D. Atomic and Nuclear Properties of Materials. Available online: https://pdg.lbl.gov/2021/AtomicNuclearProperties/index.html (accessed on 8 June 2021).
9. Eigen, G.; Zhou, Z.; Chao, D.; Cheng, C.H.; Echenard, B.; Flood, K.T.; Hitlin, D.G.; Porter, F.C.; Zhu, R.Y.; De Nardo, G.; et al. A LYSO calorimeter for the SuperB factory. *Nucl. Instrum. Methods Phys. Res. Sect. A* **2013**, *718*, 107–109. [CrossRef]
10. Cordelli, M.; Happacher, F.; Martini, M.; Miscetti, S.; Sarra, I.; Schioppa, M.; Stucci, S. CCALT: A crystal calorimeter for the KLOE-2 experiment. *J. Phys. Conf. Ser.* **2011**, *293*, 012010. [CrossRef]
11. Pezzullo, G.; Budagov, J.; Carosi, R.; Cervelli, F.; Cheng, C.; Cordelli, M.; Corradi, G.; Davydov, Y.; Echenard, B.; Giovannella, S.; et al. The LYSO crystal calorimeter for the Mu2e experiment. *J. Instrum.* **2014**, *9*, C03018. [CrossRef]
12. Oishi, K. Development of Electromagnetic Calorimeter Using LYSO Crystals for the COMET Experiment at J-PARC. *Proc. Sci.* **2018**, *314*, 800.
13. Schwendimann, P.; Papa, A. Study of 3D calorimetry based on LYSO or LaBr$_3$:Ce crystals for future high energy precision physics. *J. Instrum.* **2020**, *15*, C06018. [CrossRef]

14. Agostinelli, S.; Allison, J.; Amako, K.; Apostolakis, J.; Araujo, H.; Arce, P.; Asai, M.; Axen, D.; Banerjee, S.; Barrand, G.; et al. GEANT4—A simulation toolkit. *Nucl. Instrum. Methods Phys. Res. Sect. A* **2003**, *506*, 250–303. [CrossRef]
15. Brun, R.; Rademakers, F. ROOT: An object oriented data analysis framework. *Nucl. Instrum.Methods Phys. Res. Sect. A* **1997**, *389*, 81–86. [CrossRef]
16. Altmannshofer, W.; Binney, H.; Blucher, E.; Bryman, D.; Caminada, L.; Chen, S.; Cirigliano, V.; Corrodi, S.; Crivellin, A.; Cuen-Rochin, S.; et al. PSI Ring Cyclotron Proposal R-22-01.1. *arXiv* **2022**, arXiv:2203.01981.

MDPI
St. Alban-Anlage 66
4052 Basel
Switzerland
Tel. +41 61 683 77 34
Fax +41 61 302 89 18
www.mdpi.com

Instruments Editorial Office
E-mail: instruments@mdpi.com
www.mdpi.com/journal/instruments